City Development: Issues and Best Practices

Editors-in-Chief

Huhua Cao, Geography, Environment and Geomatics, University of Ottawa, Ottawa, ON, Canada

John Zacharias, College of Architecture and Landscape, Peking University, Beijing, China

Claude Ngomsi, United Nations Human Settlements Programme (UN-Habitat), Nairobi, Kenya

The current rate of urbanization is unprecedented and poses enormous challenges for governing bodies. New approaches to development and urban management are needed in the context of globalization and the need for local sustainability. While the developing world itself offers an abundance of lessons, case studies, and best practices, these have rarely been positioned as cutting-edge contributions to reformed practices in city development. It is well recognized that the experience of the developed world is an incomplete guide to the new challenges posed by urbanization in the contemporary world.

The "City Development: Issues and Best Practices" book series includes academic research, comparative and applied research, and case studies at the scale of the neighborhood, city, region, nation, and supranational levels. This series will offer an opportunity to present the latest academic research and best practices in urban development with the goal of promoting sustainable and inclusive development, learning from the diverse and complementary experiences of rapidly urbanizing areas of the world. Although this book series focuses primarily on the developing world, we intend to include the latest academic research and evolving best practices from developed countries.

The series is intended for geographers, planners, engineers, urban designers, architects, political scientists, sociologists, and economists, as well as policy makers and representatives from government, civil society, industry, etc. who are interested in the developing world. The series will also interest students seeking a foundation in the comparative analysis of key issues facing rapidly urbanizing areas of the world. It will include monographs, edited volumes, and textbooks. Book proposals and final manuscripts will be peer-reviewed.

The areas to be covered in the series include, but are not limited to, the following:

- Urban Equity and Inclusivity
- Reforming Informal Settlements
- Climate Change and Adaptation of the Built Environment
- Health Crises and Urban Risk Management
- Privacy, Surveillance, Security and Collective Wellbeing
- Governance and Urban Resilience
- Participatory Policy and the Right to the City
- Indigenous and Ethnic Minority Urbanization
- Mobility and Urban Transformation
- Land Use and Urban Landscapes
- Urban Infrastructure and Transport Systems
- Urban Intelligence and Technologies
- Participatory Budgeting and Community Building
- Industrial Parks and Agro-Processing Zones
- Housing and Land Tenure Issues

Seyed Navid Mashhadi Moghaddam · Huhua Cao

Artificial Intelligence-Driven Geographies

Revolutionizing Urban Studies

Springer

Seyed Navid Mashhadi Moghaddam
Department of Geography
Environment and Geomatics
University of Ottawa
Ottawa, ON, Canada

Huhua Cao
Department of Geography
Environment and Geomatics
University of Ottawa
Ottawa, ON, Canada

ISSN 2731-7773 ISSN 2731-7781 (electronic)
City Development: Issues and Best Practices
ISBN 978-981-97-5115-0 ISBN 978-981-97-5116-7 (eBook)
https://doi.org/10.1007/978-981-97-5116-7

This Springer imprint is published by the registered company Springer Nature Singapore Pte Ltd.
The registered company address is: 152 Beach Road, #21-01/04 Gateway East, Singapore 189721, Singapore

Contents

Chapter 1
Introduction

1.1 Brief Overview of AI and Its Growing Role in Various Fields

Artificial Intelligence (AI) has been a subject of interest since the inception of computing. The term 'Artificial Intelligence' was first coined by John McCarthy in 1956, during the Dartmouth Conference [30]. Over the years, AI has progressed from being a theoretical concept to becoming a driving force behind numerous technological advancements. The rapid development of AI in recent years can be attributed to factors such as increased computational power, access to vast amounts of data, and advancements in machine learning algorithms [7].

AI can be broadly defined as the development of computer systems that can perform tasks that would typically require human intelligence, such as visual perception, speech recognition, decision-making, and natural language understanding [32]. Machine learning (ML), a subfield of AI, focuses on the development of algorithms that enable computers to learn from data and improve their performance over time [21]. Deep learning, a subset of ML, is particularly noteworthy for its ability to process high-dimensional and complex data, such as images and speech, through the use of artificial neural networks [26].

The growing role of AI in various fields can be seen in numerous applications, such as (Table 1.1):

1. Healthcare: AI has been used to improve diagnostics, enhance treatment options, and optimize hospital workflows. For instance, deep learning algorithms have been developed to analyze medical images and detect diseases like cancer with accuracy comparable to, or sometimes even surpassing, human radiologists [17]. AI is also employed in drug discovery to identify potential drug candidates and optimize the drug development process [37].
2. Finance: AI has revolutionized the financial sector through applications such as fraud detection, risk assessment, and algorithmic trading. Machine learning

S. N. M. Moghaddam and H. Cao, *Artificial Intelligence-Driven Geographies*, City Development: Issues and Best Practices, https://doi.org/10.1007/978-981-97-5116-7_1

Table 1.1 Brief overview of AI and its growing role in various fields

Field	Examples of AI applications
Healthcare	– Improved diagnostics
	– Enhanced treatment options
	– Optimization of hospital workflows
	– Medical image analysis for disease detection
	– Drug discovery
Finance	– Fraud detection
	– Risk assessment
	– Algorithmic trading
	– Credit risk analysis
Transportation	– Autonomous vehicles navigation
	– Traffic signal optimization
	– Traffic pattern prediction
	– Public transportation services improvement
Manufacturing	– Increased productivity
	– Cost reduction
	– Production process optimization
	– Defect identification
	– Quality control
Retail and E-commerce	– Personalized marketing
	– Supply chain management
	– Customer service improvement
	– Product recommendation
Entertainment	– Realistic visual effects
	– CGI creation

models have been developed to analyze patterns in large datasets, enabling banks and other financial institutions to detect fraudulent transactions or assess credit risk more efficiently [4].

3. Transportation: Autonomous vehicles, which rely on AI to navigate and make decisions, have the potential to reduce traffic accidents and increase the efficiency of transportation systems [34]. Additionally, AI algorithms are being used to optimize traffic signal timings, predict traffic patterns, and improve public transportation services [38].
4. Manufacturing: AI-powered robots and automation systems have increased productivity and reduced costs in manufacturing. These systems can learn from data to optimize production processes, identify defects, and perform quality control [31].
5. Retail and E-commerce: AI has transformed the retail industry by enabling personalized marketing, improving supply chain management, and enhancing

customer service. Machine learning algorithms can analyze customer data to generate personalized product recommendations and promotions, while AI-powered chatbots and virtual assistants help provide customer support [22].
6. Entertainment: AI has been utilized in the creation of realistic visual effects, computer-generated imagery (CGI), and even the generation of original content, such as music, art, and stories. Machine learning models have been employed to generate realistic human faces, synthesize speech, and even create entirely new musical compositions [6].

The growing role of AI in various fields has been accompanied by concerns regarding its ethical implications, potential biases, and the future of employment in these industries [10]. For example, as AI algorithms are often trained on historical data, they may inadvertently perpetuate existing biases and discrimination if not carefully designed and monitored [3]. Additionally, the widespread adoption of AI may lead to job displacement in some sectors, raising concerns about the future of work and the need for re-skilling and up-skilling programs to help workers adapt to these changes [1].

Despite these challenges, AI continues to have a transformative impact on various fields, including human geography and urban planning, which are the focus of this book. By leveraging AI techniques, researchers and practitioners in these disciplines can gain new insights, improve decision-making processes, and ultimately create more sustainable, resilient, and equitable urban environments.

1.2 Importance of Human Geography and Urban Planning

Human geography and urban planning are interdisciplinary fields that focus on understanding and shaping the spatial organization of human activities, the built environment, and their interactions with the natural world. Both disciplines play crucial roles in addressing contemporary challenges such as rapid urbanization, climate change, socio-economic inequalities, and resource management [25]. This section provides an overview of the importance of human geography and urban planning, highlighting their key concepts, goals, and contributions to sustainable development.

Human geography is a subfield of geography that examines the distribution, patterns, and processes of human populations, activities, and settlements on Earth's surface [25]. It is concerned with the study of diverse aspects of human society, such as culture, economy, politics, and the environment. Human geography provides essential insights into the spatial organization of human life, offering valuable perspectives on issues such as migration, urbanization, resource allocation, and environmental sustainability [15].

The importance of human geography can be seen in its contributions to several critical areas, including:

1. Population studies: Human geographers analyze population distribution, density, and growth patterns to inform policies and interventions related to housing, transportation, healthcare, and other public services [29].
2. Economic geography: This subfield investigates the spatial organization of economic activities, including the distribution of resources, industries, and trade. Human geographers contribute to understanding regional development, economic inequality, and the role of globalization in shaping local economies [13].
3. Political geography: Human geographers study the spatial aspects of political systems, such as the organization of states, territories, and boundaries. This knowledge is essential for understanding geopolitical conflicts, electoral processes, and the impact of political decisions on regional and global scales [19].
4. Cultural geography: This subfield explores the spatial dimensions of culture, including language, religion, and other cultural practices. Human geographers contribute to our understanding of cultural identity, diversity, and the complex relationships between people and places [14].
5. Environmental geography: Human geographers examine the interactions between humans and the environment, focusing on issues such as resource consumption, pollution, land-use change, and climate change adaptation [9].

Urban planning, on the other hand, is a professional discipline that focuses on designing, managing, and shaping urban spaces to promote sustainable and equitable development [18]. Urban planners work with various stakeholders, including governments, businesses, and communities, to create comprehensive plans that guide land use, transportation, housing, infrastructure, and public services. The importance of urban planning lies in its ability to address the complex challenges associated with urbanization and contribute to the creation of resilient, inclusive, and sustainable cities [39].

Key areas where urban planning plays a significant role include:

1. Land-use planning: Urban planners develop land-use policies and zoning regulations to guide the spatial organization of cities, balancing competing demands for residential, commercial, industrial, and recreational spaces [27].
2. Transportation planning: This aspect of urban planning focuses on the design and management of transportation systems, including roads, public transit, and active transportation options such as walking and cycling. Effective transportation planning is crucial for reducing congestion, enhancing mobility, and minimizing environmental impacts [11].
3. Housing and community development: Urban planners are responsible for ensuring the availability of affordable, adequate, and diverse housing options for all residents. They also work towards fostering cohesive and inclusive communities by promoting social equity and addressing issues such as gentrification, segregation, and homelessness [35].
4. Environmental planning: Urban planners integrate environmental considerations into urban development strategies to minimize negative impacts on natural

resources, ecosystems, and public health. They play a critical role in promoting sustainable urban design, green infrastructure, and climate change adaptation [8].
5. Economic development: Urban planners contribute to the creation of vibrant and prosperous urban economies by encouraging business growth, supporting workforce development, and fostering innovation. They help create strategies for attracting investments, stimulating job creation, and fostering equitable economic growth [5].
6. Public space and urban design: Urban planners focus on designing attractive, functional, and accessible public spaces that contribute to the quality of urban life. They are involved in the planning and design of parks, plazas, streetscapes, and other public spaces that facilitate social interaction, promote active lifestyles, and enhance the urban environment [20].

As the world continues to urbanize, with over two-thirds of the global population projected to live in urban areas by 2050 [36], the importance of human geography and urban planning becomes increasingly evident. Both disciplines provide critical insights and tools for addressing the complex challenges associated with rapid urbanization, such as providing adequate housing, infrastructure, and services for growing populations, mitigating the environmental impacts of urban growth, and promoting social equity and cohesion.

The integration of AI technologies into human geography and urban planning offers exciting new possibilities for advancing these fields and addressing contemporary urban challenges more effectively. AI can help researchers and practitioners make sense of complex, large-scale datasets, uncover previously unrecognized patterns and relationships, and support more informed and data-driven decision-making processes. As this book will demonstrate, AI has the potential to revolutionize human geography and urban planning by enhancing the capacity of these disciplines to contribute to the creation of more sustainable, resilient, and equitable urban environments.

1.3 The Potential of AI in Revolutionizing These Disciplines

Artificial intelligence has the potential to revolutionize human geography and urban planning by providing new analytical tools, insights, and techniques that can enhance the understanding and management of complex urban systems. AI can help tackle some of the most pressing challenges faced by these disciplines, such as making sense of vast amounts of data, optimizing decision-making processes, and developing more sustainable and equitable urban environments. This section will explore the potential of AI in revolutionizing human geography and urban planning, focusing on several key areas where AI can have a significant impact.

1. Data analysis and visualization: Human geography and urban planning are inherently data-driven fields that require the analysis and interpretation of large and diverse datasets, such as census data, land-use information, and transportation

networks. AI techniques, particularly machine learning and deep learning, can help researchers and practitioners analyze and visualize complex spatial data more effectively and efficiently, uncovering previously unrecognized patterns, trends, and relationships [24].

2. Predictive modeling: AI has the potential to enhance predictive modeling in human geography and urban planning, enabling more accurate forecasts of future urban growth, land-use change, and infrastructure needs. For example, machine learning algorithms can be employed to predict population growth, housing demand, and traffic patterns, providing valuable insights for planning and policy-making [40].
3. Optimization and decision-making: AI can support urban planners in optimizing decision-making processes by evaluating multiple scenarios and identifying the most effective strategies for achieving specific goals, such as reducing congestion, increasing affordable housing, or improving air quality. AI-based optimization techniques, such as genetic algorithms and multi-objective optimization, can help planners navigate complex trade-offs and make more informed, data-driven decisions [33].
4. Real-time monitoring and adaptive management: The integration of AI technologies with sensors, Internet of Things (IoT) devices, and other data sources can enable real-time monitoring and adaptive management of urban systems, such as transportation networks, energy grids, and water infrastructure. AI can help analyze real-time data, identify emerging issues, and provide timely feedback to support adaptive management and improve urban resilience [23].
5. Citizen engagement and participation: AI has the potential to enhance citizen engagement and participation in urban planning processes by providing more accessible, interactive, and personalized tools for communication, collaboration, and decision-making. For example, AI-powered chatbots and virtual assistants can help facilitate public consultations, gather feedback, and answer questions about planning proposals, making it easier for citizens to get involved and have their voices heard [16].
6. Equity and social justice: AI can contribute to promoting equity and social justice in human geography and urban planning by providing tools for identifying and addressing spatial inequalities, such as disparities in access to housing, transportation, and public services. Machine learning algorithms can be employed to analyze spatial data and identify patterns of segregation, gentrification, and environmental injustice, informing targeted interventions and policies aimed at promoting more equitable urban development [12].
7. Environmental sustainability: AI can play a significant role in promoting environmental sustainability in urban planning by supporting the development of more energy-efficient, low-carbon, and resilient urban systems. AI techniques can be used to optimize the design and operation of green infrastructure, renewable energy systems, and waste management facilities, as well as to enhance climate change adaptation and mitigation strategies [28].

However, the integration of AI into human geography and urban planning also raises several challenges and concerns, such as issues related to data privacy, security, and ethical considerations. Moreover, there is the potential for biases in AI algorithms, which can exacerbate existing inequalities and lead to unintended negative consequences for certain populations [2]. It is essential for researchers and practitioners to be aware of these challenges and develop strategies to address them while harnessing the transformative potential of AI in these disciplines.

In conclusion, AI has the potential to revolutionize human geography and urban planning by providing new tools, insights, and techniques that can enhance the understanding and management of complex urban systems. The integration of AI technologies in these disciplines offers exciting opportunities for improving data analysis and visualization, predictive modeling, optimization and decision-making, real-time monitoring and adaptive management, citizen engagement and participation, equity and social justice, and environmental sustainability. However, it is crucial to be mindful of the challenges and concerns associated with the use of AI, such as data privacy, security, ethical considerations, and potential biases. By addressing these challenges, human geography and urban planning can fully harness the transformative potential of AI to contribute to the creation of more sustainable, resilient, and equitable urban environments.

1.4 Scope and Structure of the Book

The purpose of this book is to provide a comprehensive examination of the role of artificial intelligence (AI) in human geography and urban planning. It aims to explore the potential of AI in revolutionizing these disciplines and address the challenges, opportunities, and ethical considerations associated with its implementation. The book is designed for researchers, practitioners, and decision-makers in the fields of human geography, urban planning, and related disciplines who are interested in understanding the transformative potential of AI and its applications.

The book is organized into fifteen chapters, each focusing on different aspects of AI-driven geographies and urban planning. The following is an overview of the scope and structure of the book:

This chapter: Introduction

This section introduces the reader to the growing role of AI in various fields and its potential to revolutionize human geography and urban planning. The importance of these disciplines in understanding and addressing complex urban challenges is discussed. Furthermore, the scope and structure of the book are outlined to guide the reader through the subsequent chapters.

Chapter 2: Artificial Intelligence

This chapter provides a comprehensive overview of AI, including its history and fundamental concepts. It covers machine learning, deep learning, and recurrent learning techniques that form the backbone of modern AI applications.

Chapter 3: *Data Sources and Processing*

This chapter discusses traditional and emerging data sources in human geography and urban planning, as well as the processing techniques required for AI-driven analyses. The importance of data cleaning, preprocessing, and integration is emphasized, and the significance of geospatial data is explored.

Part 1: *AI Applications in Human Geography*

Chapters 4 to 8 focus on the various applications of AI in human geography, including population distribution and migration patterns, land use and land cover change detection, environmental risk assessment and climate change impacts, socioeconomic inequality and spatial analysis, and health and disease mapping.

Part 2: *AI Applications in Urban Planning*

Chapters 9 to 13 delve into the applications of AI in urban planning, covering topics such as smart cities and IoT integration, transportation and traffic management, urban growth and sprawl prediction, housing affordability and real estate market analysis, and sustainable development and resource management.

Chapter 14: *Ethical Considerations and Challenges*

This chapter addresses the ethical challenges related to AI in human geography and urban planning, discussing data privacy and security, algorithmic bias and fairness, the digital divide, public participation and engagement in AI-driven planning, and the future of employment in these fields.

Chapter 15: *Conclusion and Future Prospects*

The concluding chapter provides a summary of AI's impact on human geography and urban planning and discusses the potential for further integration and advancement. It also highlights future research directions and challenges that need to be addressed to fully harness the transformative power of AI in these disciplines.

Overall, this book presents a comprehensive and up-to-date overview of AI's role in human geography and urban planning. By examining the various applications, challenges, and ethical considerations associated with AI, it provides a solid foundation for researchers, practitioners, and decision-makers to understand and harness the potential of AI in these fields.

As the field of AI-driven geographies and urban planning continues to grow and evolve, new challenges and opportunities will undoubtedly emerge. This book serves as a starting point for further research and collaboration, encouraging readers to engage in interdisciplinary work, share best practices, and develop innovative solutions to address the complex urban challenges that our societies face.

By fostering a better understanding of AI's transformative power and potential, this book seeks to contribute to the development of more sustainable, equitable, and resilient urban environments, ultimately improving the quality of life for people around the world.

References

1. Arntz, M., Gregory, T., & Zierahn, U. (2016). The risk of automation for jobs in OECD countries: A comparative analysis. In *OECD Social, Employment, and Migration Working Papers*, No. 189.
2. Barocas, S., & Selbst, A. D. (2016). Big data's disparate impact. *California Law Review, 104*, 671–732.
3. Barocas, S., Hardt, M., & Narayanan, A. (2019). Fairness and machine learning. Fairmlbook.org.
4. Bholowalia, P., & Kumar, A. (2014). EBK-means: A clustering technique based on elbow method and k-means in WSN. *International Journal of Computer Applications, 105*(9), 17–24.
5. Bingham, R. D., & Mier, R. (1993). *Theories of local economic development: Perspectives from across the disciplines*. Sage Publications.
6. Briot, J. P., Hadjeres, G., & Pachet, F. (2019). Deep learning techniques for music generation-a survey. arXiv:1709.01620
7. Brynjolfsson, E., & McAfee, A. (2014). *The second machine age: Work, progress, and prosperity in a time of brilliant technologies*. WW Norton & Company.
8. Campbell, S. (2016). The planner's triangle revisited: Sustainability and the evolution of a planning ideal that can't stand still. *Journal of the American Planning Association, 82*(2), 161–169.
9. Castree, N., Hulme, M., & Proctor, J. D. (2018). *Companion to environmental studies*. Routledge.
10. Cath, C., Wachter, S., Mittelstadt, B., Taddeo, M., & Floridi, L. (2018). Artificial intelligence and the 'good society': The US, EU, and UK approach. *Science and Engineering Ethics, 24*(2), 505–528.
11. Cervero, R. (2013). Linking urban transport and land use in developing countries. *Journal of Transport and Land Use, 6*(1), 7–24.
12. Chetty, R., Hendren, N., & Katz, L. F. (2016). The effects of exposure to better neighborhoods on children: New evidence from the moving to opportunity experiment. *American Economic Review, 106*(4), 855–902.
13. Coe, N. M., Kelly, P. F., & Yeung, H. W. C. (2013). *Economic geography: A contemporary introduction*. Wiley.
14. Crang, M., & Thrift, N. (2000). *Thinking space*. Routledge.
15. Daniels, P., Bradshaw, M., Shaw, D., & Sidaway, J. (2019). *An introduction to human geography: Issues for the 21st century*. Pearson.
16. Desouza, K. C., & Flanery, T. H. (2013). Designing, planning, and managing resilient cities: A conceptual framework. *Cities, 35*, 89–99.
17. Esteva, A., Kuprel, B., Novoa, R. A., Ko, J., Swetter, S. M., Blau, H. M., & Thrun, S. (2017). Dermatologist-level classification of skin cancer with deep neural networks. *Nature, 542*(7639), 115–118.
18. Fainstein, S. S., & DeFilippis, J. (2015). *Readings in planning theory*. Wiley.
19. Flint, C. (2016). *Introduction to geopolitics*. Routledge.
20. Gehl, J. (2011). *Life between buildings: Using public space*. Island Press.
21. Goodfellow, I., Bengio, Y., & Courville, A. (2016). Deep learning. MIT press.
22. Gupta, S., Garg, V., & Mittal, A. (2020). Applications of artificial intelligence & associated technologies in retail marketing: A review. *Procedia Computer Science, 167*, 2385–2394.
23. Hashem, I. A. T., Chang, V., Anuar, N. B., Adewole, K., Yaqoob, I., Gani, A., Chiroma, H., et al. (2016). The role of big data in smart city. *International Journal of Information Management, 36*(5), 748–758.

24. Kitchin, R. (2014). *The data revolution: Big data, open data, data infrastructures and their consequences*. Sage.
25. Knox, P. L., & Marston, S. A. (2018). *Human geography: Places and regions in global context*. Pearson.
26. LeCun, Y., Bengio, Y., & Hinton, G. (2015). Deep learning. *Nature, 521*(7553), 436–444.
27. Levy, J. M. (2017). *Contemporary urban planning*. Routledge.
28. Makarova, I., Aksenov, A., Sboychakov, K., & Kremlev, A. (2020). Machine learning applications for sustainable development of smart cities: A review. *Sustainability, 12*(16), 6714.
29. Moseley, W. G. (2014). *An introduction to human-environment geography: Local dynamics and global processes*. Wiley.
30. Nilsson, N. J. (2010). *The quest for artificial intelligence*. Cambridge University Press.
31. Rajkomar, A., Dean, J., & Kohane, I. (2019). Machine learning in medicine. *New England Journal of Medicine, 380*(14), 1347–1358.
32. Russell, S., & Norvig, P. (2016). *Artificial intelligence: A modern approach*. Pearson Education Limited.
33. Shen, Z., Yang, L., & Chen, Y. (2020). Urban planning decision-making: Multi-objective optimization and a case study. *Land Use Policy, 90*, 104303.
34. Shladover, S. E. (2018). Connected and automated vehicle systems: Introduction and overview. *Journal of Intelligent Transportation Systems, 22*(3), 190–200.
35. Talen, E., & Ellis, C. (2002). Beyond relativism: Reasserting the role of the local planner. *Journal of the American Planning Association, 68*(4), 381–393.
36. United Nations. (2018). *World urbanization prospects: The 2018 revision*. United Nations Department of Economic and Social Affairs, Population Division.
37. Vamathevan, J., Clark, D., Czodrowski, P., Dunham, I., Ferran, E., Lee, G., Packer, J., et al. (2019). Applications of machine learning in drug discovery and development. *Nature Reviews Drug Discovery, 18*(6), 463–477.
38. Vlahogianni, E. I., Karlaftis, M. G., & Golias, J. C. (2014). Short-term traffic forecasting: Where we are and where we're going. *Transportation Research Part C: Emerging Technologies, 43*, 3–19.
39. Watson, V. (2009). Seeing from the South: Refocusing urban planning on the globe's central urban issues. *Urban Studies, 46*(11), 2259–2275.
40. Zhang, K., & Batterman, S. A. (2013). Air pollution and health risks due to vehicle traffic. *Science of the Total Environment, 450*, 307–316.

Chapter 2
Artificial Intelligence

2.1 History of Artificial Intelligence

The history of artificial intelligence (AI) spans over several decades and encompasses a variety of ideas, approaches, and milestones. This section traces the development of AI from its inception to its present-day advancements, examining the contributions of key figures, breakthroughs, and challenges that have shaped the field.

2.1.1 Early Foundations and Pioneers

The roots of AI can be traced back to ancient civilizations, where philosophers and mathematicians pondered the nature of human intelligence and the possibility of creating machines that could think and reason [125]. However, the foundations of modern AI were laid during the first half of the 20th century with the advent of formal logic, digital computers, and theories of computation (see Fig. 2.1).

In the 1930s, British mathematician Alan Turing developed the concept of the Turing machine, a hypothetical device that could perform any computation that could be represented by a set of rules [187]. Turing's work laid the groundwork for the modern theory of computation and ultimately led to the development of the first electronic computers. In 1950, Turing published his influential paper, "Computing Machinery and Intelligence," in which he proposed the Turing Test as a criterion for determining whether a machine can exhibit intelligent behavior [188].

During this period, other researchers also made significant contributions to the development of AI. For example, in 1943, Warren McCulloch and Walter Pitts introduced the concept of artificial neurons, which formed the basis for later work on artificial neural networks [127]. In 1949, Donald Hebb proposed the Hebbian learning rule, a fundamental concept in neural network learning [74].

S. N. M. Moghaddam and H. Cao, *Artificial Intelligence-Driven Geographies*, City Development: Issues and Best Practices, https://doi.org/10.1007/978-981-97-5116-7_2

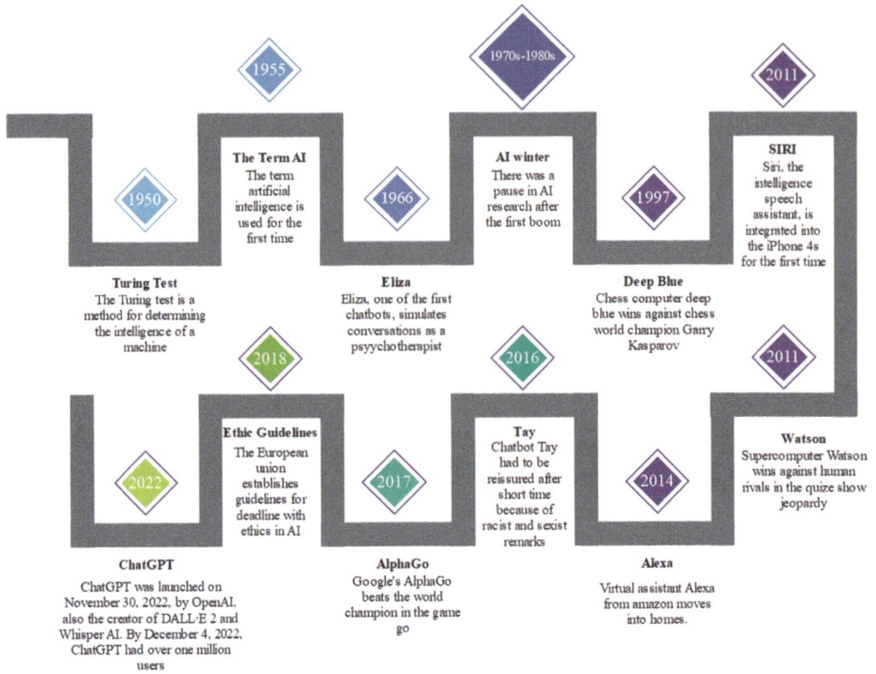

Fig. 2.1 History of AI development

2.1.2 *Birth of AI and Early Approaches*

The term "artificial intelligence" was created in 1956 by John McCarthy, Marvin Minsky, Nathaniel Rochester, and Claude Shannon during the Dartmouth Conference, which marked the birth of AI as a distinct field of study [125]. Early AI research focused on symbolic approaches and rule-based systems, such as the General Problem Solver developed by Allen Newell and Herbert A. Simon in 1959 [139]. This approach, known as "good old-fashioned AI" (GOFAI), aimed to model human reasoning through formal logic and explicit rules.

During the 1960s and 1970s, AI research expanded to include other areas, such as natural language processing, robotics, and computer vision. Early successes in these domains included SHRDLU, a natural language understanding system developed by Terry Winograd [198], and Shakey the Robot, a mobile robot that could perform simple tasks and navigate its environment autonomously [142].

2.1.3 The Rise of Machine Learning and Expert Systems

In the 1980s, AI research began to shift towards more data-driven and probabilistic approaches, giving rise to the field of machine learning (ML). ML techniques, such as decision trees, neural networks, and genetic algorithms, enabled computers to learn from data and make predictions or decisions without being explicitly programmed [163, 166].

During this period, expert systems also gained prominence as a practical application of AI. Expert systems are computer programs that mimic the decision-making process of a human expert by using a knowledge base and a set of inference rules. One of the most well-known expert systems is MYCIN, developed at Stanford University, which was designed to diagnose and recommend treatments for bacterial infections [171]. The success of MYCIN and other expert systems led to increased interest in AI and its potential applications across various industries.

2.1.4 The AI Winter and Revival

Despite the early progress in AI, the field experienced a period of reduced funding and interest during the 1980s and 1990s, often referred to as the "AI winter." This decline was partly due to the limitations of early AI techniques, which struggled to scale up to real-world problems and failed to meet overly optimistic expectations [155].

However, the AI winter eventually gave way to a resurgence of interest in the field, driven by several factors. Firstly, the development of more powerful and affordable computing hardware allowed researchers to tackle more complex problems and process larger datasets. Secondly, new machine learning techniques, such as support vector machines [189] and ensemble methods [43], emerged and demonstrated improved performance on a range of tasks.

2.1.5 The Deep Learning Revolution

The 21st century has seen a revolution in AI with the advent of deep learning, a subfield of machine learning that focuses on deep neural networks with multiple layers of interconnected nodes [106]. Deep learning has been particularly successful in tasks such as image and speech recognition, natural language processing, and game playing, outperforming traditional machine learning algorithms and, in some cases, even human performance.

One of the earliest milestones in deep learning was the development of the convolutional neural network (CNN) by Yann LeCun and colleagues in the 1990s, which proved highly effective for image recognition tasks [107]. In 2012, a deep

CNN called AlexNet, developed by Alex Krizhevsky, Ilya Sutskever, and Geoffrey Hinton, achieved a breakthrough performance on the ImageNet Large Scale Visual Recognition Challenge, significantly outperforming all other competing methods and sparking a renewed interest in deep learning [102].

Since then, deep learning has continued to advance at an impressive pace, with notable achievements such as Google DeepMind's AlphaGo, which defeated the world champion of the board game Go in 2016 [172], and OpenAI's GPT series of language models, which have demonstrated a remarkable ability to generate human-like text and perform a wide range of natural language processing tasks [154].

2.1.6 AI Today and Beyond

Today, AI is a rapidly growing field with applications across numerous domains, from healthcare and finance to transportation and entertainment. The continued development of AI techniques, including advancements in machine learning, deep learning, and recurrent learning, has fueled its increasing role in various fields, including human geography and urban planning.

As AI continues to evolve, researchers are exploring new frontiers, such as transfer learning, which aims to enable AI models to learn from one task and apply that knowledge to other related tasks [146], and explainable AI, which seeks to make AI models more interpretable and transparent to human users [62].

In summary, the history of AI has been characterized by a series of milestones, challenges, and breakthroughs that have shaped the field's development and set the stage for its growing impact on various disciplines. From its early beginnings with formal logic and Turing machines to the modern era of deep learning and beyond, AI has come a long way and holds great promise for the future.

2.2 Machine Learning

2.2.1 Introduction to Machine Learning

Machine learning is a subfield of artificial intelligence that focuses on the development of algorithms that can learn from and make predictions or decisions based on data [130]. The goal of machine learning is to create models that can generalize from a given set of training data to make accurate predictions on previously unseen data. This ability to learn from data and adapt to new situations without explicit programming makes machine learning an essential component of modern AI systems.

2.2.2 *Supervised, Unsupervised, and Reinforcement Learning*

Machine learning techniques can be broadly categorized into three main types: supervised learning, unsupervised learning, and reinforcement learning (Table 2.2). Each type of learning has unique characteristics and applications relevant to urban studies and human geography.

Supervised Learning

Supervised learning involves training a model using labeled data, where each input is associated with a corresponding output or target variable. The goal of supervised learning is to learn a mapping from inputs to outputs that can be used to make accurate predictions on new, unseen data. Common supervised learning tasks include classification, where the target variable is a discrete category, and regression, where the target variable is continuous [17].

Supervised learning techniques are widely used in urban studies and human geography for predictive modeling and analysis. Some examples of supervised learning applications in these fields include:

1. Land use and land cover classification: Supervised learning algorithms, such as support vector machines (SVMs) and random forests, can be used to classify

Table 2.2 Learning types for machine learning

Learning type	Description	Goals	Tasks
Supervised learning	Uses labeled data to train a model that can make predictions or classifications on new data	Learn a mapping from inputs to outputs for accurate predictions	– Land use and land cover classification – Housing price prediction – Crime prediction and hotspot detection – Population density estimation
Unsupervised learning	Deals with unlabeled data to uncover hidden patterns or structures without prior knowledge of target variables	Uncover hidden patterns or structures in data	– Urban structure identification – Socioeconomic segregation analysis – Environmental pattern recognition – Traffic pattern analysis
Reinforcement learning	An agent learns to make decisions by interacting with an environment, aiming to maximize cumulative rewards	Maximize cumulative rewards over time by learning through trial and error	– Traffic signal control – Public transportation routing – Urban growth modeling – Disaster response and management optimization

satellite images or aerial photographs into different land use or land cover categories (e.g., urban, agricultural, forest, water) based on training data with known labels [50, 133].
2. Housing price prediction: Regression techniques like linear regression, decision trees, or neural networks can be employed to predict housing prices based on various input features, such as location, property size, and neighborhood characteristics [19].
3. Crime prediction and hotspot detection: Supervised learning can help identify areas with higher crime rates or specific types of crimes based on historical crime data and other relevant factors, such as socioeconomic status, population density, and proximity to public transportation.
4. Population density estimation: Supervised learning methods can be used to estimate population density in a given area based on features like building density, road networks, and land use patterns derived from remote sensing data or GIS [58].

Unsupervised Learning

Unsupervised learning deals with unlabeled data, where the goal is to uncover hidden patterns or structures in the data without any prior knowledge of the target variables. This can include tasks such as clustering, where the goal is to group similar data points together, and dimensionality reduction, which aims to project high-dimensional data onto a lower-dimensional space while preserving its essential structure [17].

Unsupervised learning techniques are used in urban studies and human geography to reveal patterns or relationships that may not be evident with traditional analysis methods. Some examples of unsupervised learning applications in these fields include:

1. Urban structure identification: Clustering algorithms like k-means or hierarchical clustering can be employed to identify distinct urban structures, such as residential, commercial, or industrial areas, based on land use, population density, or other relevant features [186].
2. Socioeconomic segregation analysis: Unsupervised learning can be used to identify patterns of socioeconomic segregation within cities by clustering neighborhoods based on demographic and socioeconomic variables [157].
3. Environmental pattern recognition: Unsupervised learning techniques, such as principal component analysis (PCA) or self-organizing maps (SOMs), can be applied to analyze environmental data, such as air pollution or climate variables, to identify spatial patterns and trends [34].
4. Traffic pattern analysis: Unsupervised learning methods can be used to analyze traffic data, such as vehicle counts, speeds, and travel times, to identify distinct traffic patterns, congestion zones, or potential bottlenecks in transportation networks.

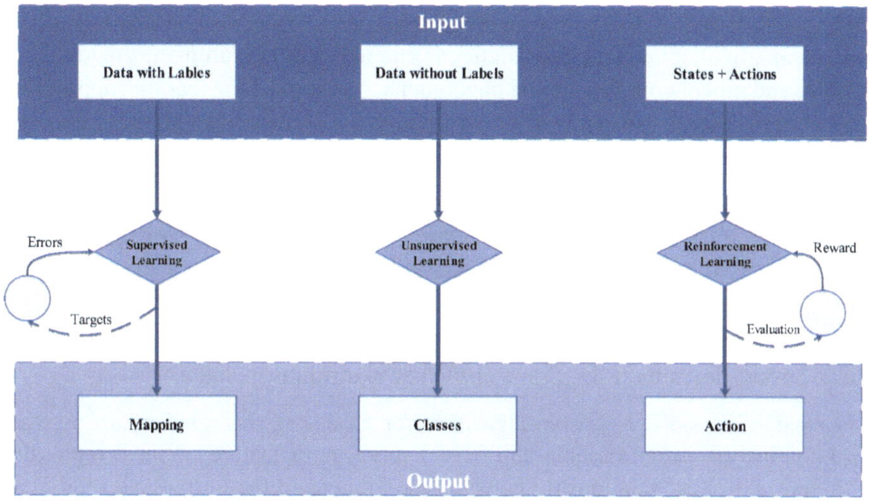

Fig. 2.2 The relationships between supervised, unsupervised, and reinforcement learning

Reinforcement Learning

Reinforcement learning (RL) is a type of machine learning where an agent learns to make decisions by interacting with an environment. The agent receives feedback in the form of rewards or penalties and aims to maximize the cumulative reward over time. Unlike supervised learning, which relies on labeled data, reinforcement learning focuses on learning through trial and error, making it particularly suitable for dynamic and uncertain environments [177] (Fig. 2.2).

Reinforcement learning has the potential to revolutionize various aspects of urban studies and human geography by enabling the development of adaptive and optimized decision-making systems. Some examples of RL applications in these fields include:

1. Traffic signal control: Reinforcement learning algorithms can be used to optimize traffic signal timings in real-time based on current traffic conditions, leading to reduced congestion, lower travel times, and improved fuel efficiency [123].
2. Public transportation routing: RL can help design adaptive public transportation routes and schedules that dynamically adjust to changing demand patterns, improving service quality and efficiency [118].
3. Urban growth modeling: Reinforcement learning can be applied to simulate and predict urban growth patterns by modeling the decisions of various stakeholders (e.g., developers, policymakers) and their interactions with the environment [185].
4. Disaster response and management: RL can be used to develop decision-support systems for disaster response and management, optimizing the allocation of resources and emergency response strategies in complex and uncertain environments [49].

In conclusion, supervised, unsupervised, and reinforcement learning techniques have diverse applications in urban studies and human geography, offering new ways to analyze and model complex spatial phenomena. These machine learning approaches have the potential to enhance our understanding of urban and geographic processes and contribute to more effective and sustainable planning and decision-making.

2.2.3 Key Machine Learning Algorithms

There is a wide array of machine learning algorithms that have been developed to tackle various learning tasks. Some of the key algorithms include:

- Linear regression: A simple algorithm for modeling the relationship between a continuous target variable and one or more input features. Linear regression assumes a linear relationship between the inputs and the output and minimizes the mean squared error between the predicted and true output values [53].
- Logistic regression: A widely used algorithm for binary classification tasks, which models the probability of a binary target variable given the input features. Logistic regression uses the logistic function to map input features to the probability of the target variable taking a specific value [53].
- Support vector machines (SVMs): A powerful and flexible algorithm for classification and regression tasks, SVMs aim to find the optimal separating hyperplane between different classes or to model the relationship between input features and a continuous target variable. SVMs can be extended to handle non-linear relationships using kernel functions [189].
- Decision trees: A popular method for both classification and regression tasks, decision trees recursively split the input space based on feature values to create a tree-like structure that can be used for making predictions. Decision trees can be prone to overfitting, but this issue can be mitigated by techniques such as pruning or by using ensemble methods like random forests and boosting [21, 152].
- Neural networks: A class of algorithms inspired by the structure and function of biological neural networks, artificial neural networks consist of interconnected layers of nodes or neurons. Neural networks can learn complex, non-linear relationships between inputs and outputs through the process of backpropagation and gradient descent [163]. The recent resurgence of neural networks, particularly deep neural networks with many hidden layers, has led to significant advancements in a wide range of AI applications, including computer vision, natural language processing, and speech recognition [106].
- k-Nearest Neighbors (k-NN): A simple yet effective algorithm for classification and regression tasks that makes predictions based on the k nearest training examples in the input space. The k-NN algorithm calculates the distance between input features and assigns a predicted value or class label based on the majority vote or weighted average of the nearest neighbors [40].

- Principal Component Analysis (PCA): A widely used unsupervised learning technique for dimensionality reduction, PCA aims to project high-dimensional data onto a lower-dimensional space while preserving the maximum variance in the data. PCA can be used for visualization, data compression, and as a pre-processing step for other machine learning algorithms [93].
- k-Means: A popular clustering algorithm that partitions data points into k clusters based on their similarity in the input space. The k-means algorithm iteratively updates the cluster centroids to minimize the within-cluster sum of squared distances [119].

Linear Regression

Linear regression is a foundational machine learning algorithm that models the relationship between a dependent variable and one or more independent variables. The algorithm assumes that the relationship between the variables is linear, and it aims to find the best-fitting line (in the case of one independent variable) or hyperplane (in the case of multiple independent variables) that minimizes the sum of the squared residuals or prediction errors. The resulting linear model can be used for prediction, explanation, or control [89].

Linear regression is relatively simple, interpretable, and computationally efficient, making it a popular choice for many applications in urban studies and human geography. The following sections will provide an overview of the linear regression algorithm and discuss its relevance and use in these fields.

The goal of linear regression is to find a linear model that can accurately predict the dependent variable (also known as the response or target variable) based on the independent variables (also known as predictors or features Fig. 2.3). The linear model can be expressed as:

$$y = \beta 0 + \beta 1x1 + \beta 2x2 + \cdots + \beta pxp + \varepsilon$$

Here, y is the dependent variable, x1, x2, ..., xp are the independent variables, β0, β1, ..., βp are the coefficients to be estimated, and ε represents the error term. The coefficients are estimated using a method called ordinary least squares (OLS), which minimizes the sum of the squared residuals or differences between the observed and predicted values [89].

Linear regression has been widely applied in urban studies and human geography, offering a valuable tool for understanding relationships between variables and making predictions. Some examples of linear regression applications in these fields include:

1. Housing price prediction: Linear regression can be used to model the relationship between housing prices and various features such as location, property size, and neighborhood characteristics. The resulting model can be used to predict the price of a house given its features or to identify the most important factors that influence housing prices [19].
2. Transportation demand modeling: Linear regression can help estimate the demand for transportation services (e.g., public transit ridership, vehicle miles

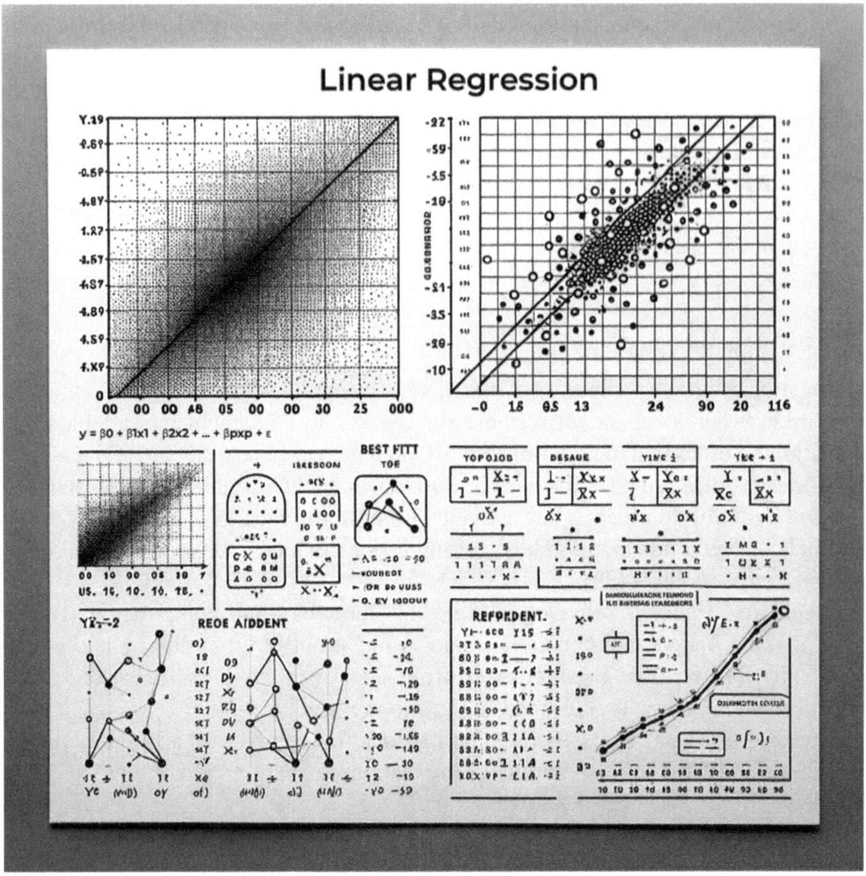

Fig. 2.3 An overview of the linear regression process, including a scatter plot with data points, a best-fitting line, and the linear regression equation

traveled) based on factors like population density, land use, and transportation infrastructure. These models can inform transportation planning and investment decisions [18].

3. Economic growth analysis: Linear regression can be employed to investigate the determinants of economic growth in cities or regions, such as education levels, infrastructure investments, and industrial composition. The results can help policymakers identify potential drivers of growth and design targeted policies to promote economic development [13].
4. Environmental impact assessment: Linear regression can be used to model the relationship between human activities (e.g., urban development, industrial production) and environmental outcomes (e.g., air quality, water quality). This can help identify the most significant factors contributing to environmental degradation and inform the design of effective mitigation measures [33].

5. Social inequality analysis: Linear regression can be applied to explore the relationship between socioeconomic variables (e.g., income, education, race) and various outcomes of interest (e.g., health status, access to public services) in urban and regional contexts. This can help identify patterns of social inequality and inform policies aimed at addressing these disparities [156].

While linear regression is a powerful and versatile tool for urban studies and human geography, it has several limitations. One of the main limitations is the assumption of linearity, which may not always hold in real-world applications. Additionally, linear regression is sensitive to multicollinearity (i.e., high correlation between independent variables) and may produce unstable or unreliable coefficient estimates in such cases. Furthermore, the algorithm assumes that the error terms are normally distributed and have constant variance, which may not always be true [89].

Despite these limitations, several extensions of linear regression have been developed to address its shortcomings and broaden its applicability. Some of these extensions include:

1. Polynomial regression: Polynomial regression extends the linear model by including higher-order terms of the independent variables, allowing for more complex relationships between variables [71].
2. Ridge regression and Lasso regression: Ridge and Lasso regression are regularization techniques that introduce a penalty term to the OLS objective function, which helps reduce overfitting and improves the stability of coefficient estimates in the presence of multicollinearity [77, 182].
3. Generalized linear models (GLMs): GLMs extend linear regression to accommodate non-normal error distributions and non-linear relationships between variables by applying a link function to the dependent variable [126].
4. Spatial regression: Spatial regression models account for spatial autocorrelation or the tendency of observations close in space to be more similar than those further apart. These models can help improve the accuracy and validity of regression analyses in geographic contexts [5].

In conclusion, linear regression is a fundamental machine learning algorithm with widespread applications in urban studies and human geography. While it has certain limitations, various extensions have been developed to address these issues and enhance the algorithm's applicability. Linear regression and its extensions continue to be valuable tools for understanding and predicting complex spatial phenomena in these fields.

2.2.4 Logistic Regression

Logistic regression is a widely used machine learning algorithm for classification tasks, particularly when the response variable is binary (i.e., it has two possible outcomes). Logistic regression models the probability of the response variable belonging to a certain class, based on the values of the independent variables. Like linear regression, logistic regression is easy to implement, interpret, and computationally efficient. It has been applied in numerous urban studies and human geography applications. This section provides an overview of logistic regression and explains its relevance and use in these fields.

Logistic regression is an extension of linear regression that models the probability of a binary response variable, Y, taking the value of 1 given the values of the independent variables, X:

$$P(Y = 1|X) = 1/(1 + \exp(-(\beta 0 + \beta 1X1 + \beta 2X2 + \cdots + \beta pXp)))$$

The logistic function, also known as the sigmoid function, transforms the linear combination of the independent variables into a probability value between 0 and 1. The coefficients, $\beta 0, \beta 1, \ldots, \beta p$, are estimated using the maximum likelihood method, which maximizes the likelihood of observing the given data [78].

Logistic regression is an essential tool for urban studies and human geography due to its ability to model binary outcomes, which are common in these fields. Some examples of logistic regression applications in urban studies and human geography include:

1. Land-use change prediction: Logistic regression can be used to predict land-use changes based on various factors such as population density, proximity to infrastructure, and terrain characteristics (Fig. 2.4). For example, researchers might use logistic regression to model the likelihood of agricultural land converting to urban land over time [149].
2. Transportation mode choice modeling: Logistic regression can help estimate the probability of individuals choosing different transportation modes (e.g., car, public transit, walking, cycling) based on factors like trip distance, travel time, and socioeconomic characteristics. These models can inform transportation planning and policy development [14].
3. Crime hotspot identification: Logistic regression can be employed to predict the likelihood of crime occurring in specific locations based on factors such as land use, socioeconomic variables, and the presence of crime attractors (e.g., bars, shopping centers). These models can support targeted crime prevention and law enforcement strategies [26].
4. Environmental risk assessment: Logistic regression can be used to model the probability of environmental hazards (e.g., floods, landslides) occurring in particular areas based on factors like topography, soil type, and land cover. This can help identify high-risk zones and inform disaster mitigation and management efforts [9].
5. Health and disease mapping: Logistic regression can be applied to model the relationship between health outcomes (e.g., disease prevalence, mortality rates) and various risk factors (e.g., socioeconomic status, environmental exposures). This can help identify patterns of health disparities and inform public health interventions [192].

Fig. 2.4 A sample of output of binary classification map of an urban area, highlighting areas predicted for land-use change and those predicted to remain unchanged

Despite its usefulness in urban studies and human geography, logistic regression has some limitations. One key limitation is its assumption that the relationship between the logit of the response variable and the independent variables is linear. This assumption might not always hold true in real-world applications. Additionally, logistic regression is sensitive to multicollinearity, which can produce unstable or unreliable coefficient estimates when independent variables are highly correlated. Moreover, logistic regression assumes that observations are independent, which may not be the case in spatial data, where nearby observations often exhibit spatial autocorrelation [78].

Despite these limitations, several extensions of logistic regression have been developed to address its shortcomings and broaden its applicability. Some of these extensions include:

1. Multinomial logistic regression: Multinomial logistic regression extends logistic regression to handle response variables with more than two categories. This can be particularly useful in urban studies and human geography when modeling

outcomes with multiple classes, such as land-use types or transportation mode choices [120].

2. Ordered logistic regression: Ordered logistic regression is designed for ordinal response variables, where the categories have a natural order (e.g., low, medium, high). This method can be applied in cases where the dependent variable represents ordered categories, such as levels of urbanization or socioeconomic status [3].
3. Penalized logistic regression: Penalized logistic regression methods, such as Lasso and Ridge logistic regression, introduce regularization terms to the maximum likelihood estimation, which helps address issues related to multicollinearity and overfitting [54].
4. Spatial logistic regression: Spatial logistic regression models account for spatial autocorrelation in the data, which can lead to more accurate and reliable coefficient estimates for spatially dependent observations. This is particularly relevant for geographic analyses, where spatial autocorrelation is common [108].

In conclusion, logistic regression is a powerful and versatile machine learning algorithm that has found extensive applications in urban studies and human geography. Its ability to model binary and categorical outcomes makes it an essential tool for understanding and predicting complex spatial phenomena. Although logistic regression has certain limitations, various extensions have been developed to address these issues and enhance its applicability in these fields.

Support Vector Machines (SVMs)

Support Vector Machines (SVMs) are powerful and versatile supervised machine learning algorithms used for classification and regression tasks. SVMs have gained popularity in various fields, including urban studies and human geography, due to their ability to handle high-dimensional data and produce accurate predictions. This section provides an overview of SVMs, describes their relevance to urban studies and human geography, and presents examples of their use in these fields.

Support Vector Machines were first introduced by Vapnik [189] and have since become a popular choice for machine learning practitioners. In the context of classification, SVMs aim to find the optimal hyperplane that separates data points from different classes with the maximum margin. The margin is defined as the distance between the hyperplane and the closest data points, called support vectors. The optimal hyperplane is the one that maximizes this margin, ensuring the best possible separation between the classes [39].

In cases where the data is not linearly separable, SVMs employ kernel functions to map the original data into a higher-dimensional space where a linear separation is possible. Some common kernel functions used in SVMs include linear, polynomial, radial basis function (RBF), and sigmoid kernels [168].

For regression tasks, SVMs seek to find a function that fits the data with an error margin below a specified threshold while maximizing the flatness of the function. This is known as Support Vector Regression (SVR) and shares several similarities with the classification variant [44].

Support Vector Machines have gained traction in urban studies and human geography due to their ability to handle high-dimensional data, tolerate noise, and produce accurate predictions. Some examples of SVM applications in urban studies and human geography include:

1. Land use and land cover classification: SVMs have been widely used to classify land use and land cover types based on remote sensing data, such as satellite imagery and aerial photographs. SVMs have shown excellent performance in these tasks, often outperforming other classification algorithms [80, 133].
2. Urban growth modeling and prediction: SVMs can be employed to model and predict urban growth patterns based on factors such as population density, proximity to transportation infrastructure, and land suitability. Researchers have used SVMs to develop urban growth models and forecast future urbanization patterns, often with high accuracy [180, 217].
3. Transportation mode choice modeling: SVMs have been applied to model and predict individual transportation mode choices based on factors such as travel distance, travel time, and personal attributes. SVMs have demonstrated superior predictive performance compared to traditional statistical methods in these tasks [11, 207].
4. Environmental risk assessment: SVMs can be used to model the likelihood of environmental hazards, such as landslides or floods, based on factors like topography, soil type, and land cover. SVMs have shown high accuracy in these tasks, often outperforming other machine learning algorithms [151, 204].
5. Socioeconomic analysis and mapping: SVMs have been employed in various socioeconomic analyses, such as predicting income levels, educational attainment, and crime rates based on geospatial data and other relevant variables. Their ability to handle high-dimensional data and produce accurate predictions makes them well-suited for these applications [26, 116].
6. Health and disease mapping: SVMs have been utilized to predict and map the spatial distribution of diseases, such as malaria or dengue, based on environmental factors, demographic data, and other relevant variables. SVMs have proven effective in these tasks, offering valuable insights for public health planning and intervention [70, 161].

Despite their strengths, SVMs have certain limitations that need to be considered when applying them to urban studies and human geography:

1. Interpretability: SVMs are considered "black-box" models, as the decision-making process and the relationships between input variables and the output are not easily interpretable. This lack of interpretability can pose challenges when communicating results to non-experts or policymakers [68].
2. Scalability: SVMs can have difficulties scaling to large datasets due to the quadratic or cubic complexity of the training process. This can be particularly problematic when dealing with large geospatial datasets, often requiring the use of efficient training algorithms or dimensionality reduction techniques [79].

3. Parameter tuning: SVMs rely on several parameters, such as the regularization parameter and kernel function parameters, which need to be tuned to obtain optimal performance. This process can be time-consuming and may require the use of grid search or other optimization methods [79].
4. Spatial autocorrelation: SVMs assume that observations are independent, which may not hold for spatial data, where nearby observations often exhibit spatial autocorrelation. Extensions of SVMs that account for spatial autocorrelation may be necessary to improve the model's performance in these cases [82].

In conclusion, Support Vector Machines are powerful and versatile machine learning algorithms that have found widespread use in urban studies and human geography. Their ability to handle high-dimensional data, tolerate noise, and produce accurate predictions makes them well-suited for various applications in these fields. Despite their limitations, SVMs continue to serve as valuable tools for researchers and practitioners working in urban studies and human geography, contributing to a better understanding of complex spatial relationships and driving evidence-based decision-making.

Decision Trees

Decision trees are a popular machine learning algorithm that can be used for both classification and regression tasks. The primary advantage of decision trees is their interpretability, as they can represent complex decision-making processes in a hierarchical, tree-like structure that can be easily visualized and understood by humans [152]. Decision trees recursively split the input space into subsets based on the values of the input features, eventually leading to a prediction at the terminal nodes or leaves (Fig. 2.5) of the tree [22]. This section discusses the fundamentals of decision trees, their applications in urban studies and human geography, and the associated challenges and limitations.

A decision tree is constructed by recursively partitioning the input space into non-overlapping regions based on the values of one or more input features. The decision tree algorithm selects the best feature and split point at each node of the tree to minimize a predefined criterion, such as the Gini impurity for classification tasks or the mean squared error for regression tasks [22]. The splitting process continues until a predefined stopping criterion is met, such as reaching a maximum tree depth or a minimum number of samples per leaf.

The resulting decision tree can be visualized as a flowchart, with each internal node representing a decision based on an input feature's value, and each leaf node representing the final prediction. To make a prediction for a new instance, the instance is passed through the tree from the root node to a leaf node, following the appropriate decision path based on the instance's feature values.

There are several algorithms for constructing decision trees, such as ID3, C4.5, and CART [22, 152]. The main difference between these algorithms lies in the way they select the best feature and split point at each node and how they handle continuous features, missing values, and categorical features with multiple levels.

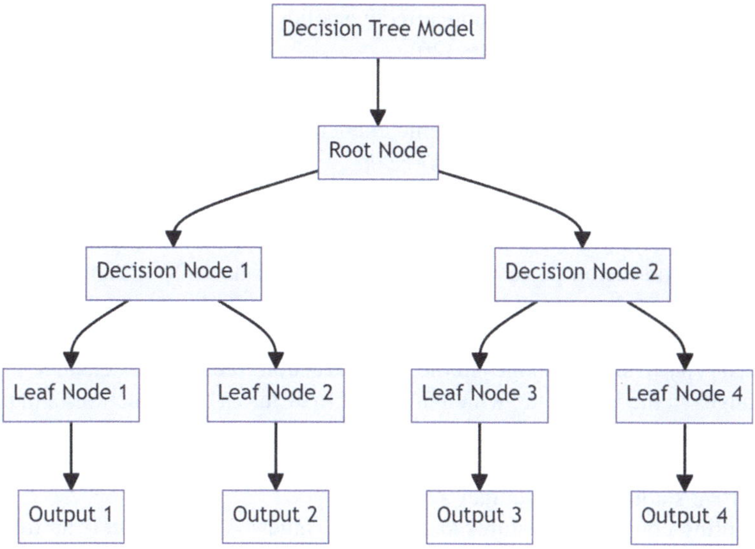

Fig. 2.5 An illustration of a clean decision tree diagram showcasing its hierarchical structure

Decision trees have been widely applied in various urban studies and human geography applications due to their interpretability, ease of use, and ability to handle a mix of continuous and categorical input features. Some notable examples include:

1. Land use and land cover classification: Decision trees are commonly used for land use and land cover classification tasks based on remote sensing data, such as satellite imagery or aerial photographs. They can efficiently handle high-dimensional data, handle missing values, and produce accurate and interpretable classification rules [52, 81].
2. Urban growth and sprawl prediction: Decision trees can be used to model the complex relationships between urban growth, socioeconomic factors, and environmental constraints, providing valuable insights for urban planning and policy-making [90, 179].
3. Transportation and traffic management: Decision trees have been used to model transportation demand, predict traffic congestion, and identify factors affecting transportation mode choice, contributing to more effective transportation planning and management [1, 11].
4. Environmental risk assessment: Decision trees can be employed to predict the spatial distribution of environmental risks, such as landslides, floods, or soil erosion, based on a combination of geospatial, environmental, and socioeconomic variables [151, 211].
5. Socioeconomic analysis and mapping: Decision trees have been applied to various socioeconomic analysis tasks, such as predicting poverty levels, identifying determinants of housing affordability, and analyzing spatial patterns of crime or health disparities. By providing interpretable and actionable insights,

decision trees contribute to more effective policy-making and resource allocation in urban and regional contexts [26].

While decision trees offer several advantages in urban studies and human geography applications, they also have some challenges and limitations that researchers and practitioners should be aware of:

1. Overfitting: Decision trees are prone to overfitting, especially when the tree is deep or when there are a large number of input features. Overfitting can lead to poor generalization performance on new, unseen data. To mitigate overfitting, techniques such as pruning, early stopping, or ensemble methods like random forests can be employed [21].
2. Sensitivity to small changes in the data: Decision trees can be sensitive to small changes in the input data, which may lead to different tree structures and predictions. This sensitivity can be addressed by using ensemble methods, such as bagging or boosting, which combine the predictions from multiple trees to improve stability and accuracy [20, 51].
3. Handling continuous features and missing values: Although decision tree algorithms, such as CART and C4.5, can handle continuous features and missing values, the process can be computationally expensive and may require additional preprocessing steps, such as discretization or imputation [22, 153].
4. Bias towards features with more levels or categories: Decision tree algorithms may be biased towards selecting features with more levels or categories, potentially leading to less accurate or interpretable models. This issue can be addressed by using techniques such as feature selection, feature weighting, or feature scaling to balance the contribution of different features [101].

Neural Networks

Artificial neural networks (ANNs) are a class of machine learning algorithms inspired by the structure and function of the human brain. ANNs are composed of interconnected nodes or neurons, which are organized into layers. The network typically consists of an input layer, one or more hidden layers, and an output layer [72]. Each neuron in a layer is connected to the neurons in the adjacent layers through weighted connections, and the weights are adjusted during the training process to minimize the error between the network's predictions and the actual output [163] (Fig. 2.6)

ANNs are capable of learning complex, nonlinear relationships in large and high-dimensional datasets, making them suitable for various applications in urban studies and human geography. Some of the key advantages of ANNs include their ability to handle noisy or incomplete data, adapt to changing environments, and generalize from learned examples to new, unseen data [72].

There are several types of neural networks that have been developed for different tasks and applications. Some of the most common types include:

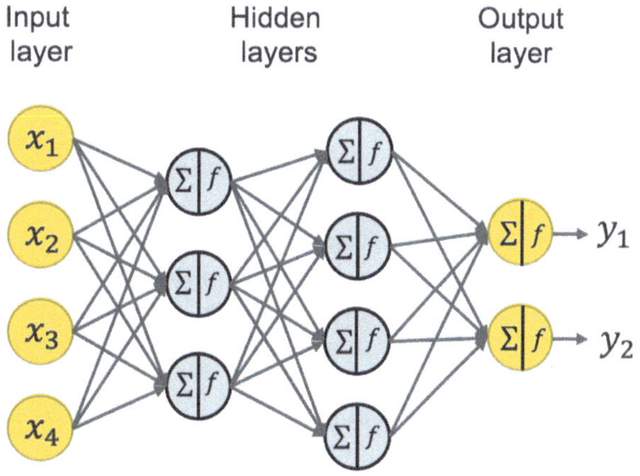

Fig. 2.6 A diagram illustrating the architecture of a deep neural network, including input, hidden, and output layers [129]

1. Feedforward Neural Networks (FNNs): In FNNs, the connections between the neurons are unidirectional, meaning that information flows only in one direction, from the input layer to the output layer. FNNs are widely used for pattern recognition and classification tasks [16].
2. Recurrent Neural Networks (RNNs): Unlike FNNs, RNNs have feedback connections, allowing information to flow in both directions. This feature enables RNNs to learn and model temporal dependencies in sequential data, making them suitable for applications involving time series data, such as forecasting in urban studies and human geography [46].
3. Convolutional Neural Networks (CNNs): CNNs are a specialized type of neural network designed for processing grid-like data structures, such as images. CNNs utilize convolutional layers and pooling layers to learn spatial hierarchies of features, making them well-suited for tasks involving spatial data, such as land use and land cover classification from satellite imagery [102].
4. Autoencoders: Autoencoders are a type of unsupervised neural network used for dimensionality reduction and feature learning. They consist of an encoder that maps the input data to a lower-dimensional latent space and a decoder that reconstructs the input from the latent representation. Autoencoders can be used for data compression, denoising, and feature extraction in urban studies and human geography applications [75].

ANNs have been widely adopted in urban studies and human geography due to their ability to model complex relationships and patterns in spatial data. Some notable applications include:

1. Land use and land cover classification: ANNs, particularly CNNs, have been successfully applied to classify land use and land cover types from remotely sensed data, such as satellite imagery and aerial photographs. By learning hierarchical representations of spatial features, CNNs can achieve high classification accuracy and generalize well to new, unseen data [213].

2. Urban growth and sprawl prediction: ANNs have been used to predict urban growth and sprawl patterns by modeling the complex relationships between various socioeconomic, demographic, and environmental factors. By capturing the nonlinearity and interactions between these factors, ANNs can provide more accurate predictions than traditional linear regression models [203].
3. Transportation and traffic management: ANNs have been employed to predict traffic congestion, estimate travel demand, and optimize traffic signal timings, contributing to more efficient transportation systems in urban areas. For instance, RNNs have been used to model temporal dependencies in traffic flow data, enabling accurate short-term traffic predictions and facilitating real-time traffic management [41].
4. Population distribution and migration patterns: ANNs can be used to estimate population distribution at a fine spatial resolution by combining various geospatial data sources, such as satellite imagery, census data, and points of interest. By modeling the relationships between these data sources, ANNs can generate high-resolution population maps that can inform urban planning, disaster response, and other applications [57].
5. Environmental risk assessment and climate change impacts: ANNs have been employed to model the complex relationships between environmental factors and their impacts on urban areas, such as flood risk, air pollution, and heat island effects. By capturing the nonlinearity and interactions between these factors, ANNs can improve the accuracy and reliability of environmental risk assessments and inform climate change adaptation strategies [174].

k-Nearest Neighbors (k-NN)

The k-Nearest Neighbors (k-NN) algorithm is a simple, yet powerful, non-parametric machine learning method used for classification and regression tasks. k-NN is based on the principle that similar data points tend to be close to one another in the feature space [40]. Given an input data point, the k-NN algorithm finds the k nearest training data points in the feature space and assigns the input data point to the majority class among these neighbors for classification tasks, or calculates the average of the neighbors' target values for regression tasks [4].

The k-NN algorithm is highly adaptable and can handle non-linear relationships between features and target variables. Its simplicity and ease of implementation make it an attractive choice for various applications in urban studies and human geography, where the relationships between variables are often complex and non-linear [136].

The performance of the k-NN algorithm heavily depends on the choice of distance metric used to determine the nearest neighbors and the number of neighbors (k). Commonly used distance metrics include Euclidean distance, Manhattan distance, and Minkowski distance [25]. Choosing the appropriate distance metric depends on the nature of the data and the problem domain.

Selecting the optimal value for k is crucial for achieving good performance in k-NN. A small value of k may result in overfitting, while a large value of k may lead to underfitting. Cross-validation can be used to find the optimal k value by comparing the model's performance for different k values on a validation dataset [89].

The k-NN algorithm has been applied to various problems in urban studies and human geography due to its ability to model complex relationships in spatial data. Some notable applications include:

1. Land use and land cover classification: k-NN has been used to classify land use and land cover types from remotely sensed data, such as satellite imagery and aerial photographs. By considering the spatial context and incorporating the information from neighboring pixels, k-NN can achieve high classification accuracy [133].
2. Urban growth and sprawl prediction: k-NN has been applied to predict urban growth and sprawl patterns by modeling the relationships between various socioeconomic, demographic, and environmental factors. The non-parametric nature of k-NN allows it to capture the nonlinearity and interactions between these factors, resulting in more accurate predictions than traditional linear regression models [61].
3. Transportation and traffic management: k-NN has been employed for predicting traffic congestion, estimating travel demand, and optimizing traffic signal timings, contributing to more efficient transportation systems in urban areas. The algorithm's ability to handle non-linear relationships between variables enables accurate short-term traffic predictions and real-time traffic management [220].
4. Population distribution and migration patterns: k-NN can be used to estimate population distribution at a fine spatial resolution by combining various geospatial data sources, such as satellite imagery, census data, and points of interest. By modeling the relationships between these data sources, k-NN can generate high-resolution population maps that can inform urban planning, disaster response, and other applications [178].
5. Environmental risk assessment and climate change impacts: k-NN has been employed to model the complex relationships between environmental factors and their impacts on urban areas, such as flood risk, air pollution, and heat island effects. By capturing the nonlinearity and interactions between these factors, k-NN can improve the accuracy and reliability of environmental risk assessments and inform climate change adaptation strategies [115].

While k-NN has proven to be useful in various applications in urban studies and human geography, it also has some limitations and challenges:

1. Scalability: k-NN can be computationally expensive for large datasets, as it requires calculating the distance between each data point and its neighbors. This can be especially problematic in applications involving high-resolution geospatial data, such as satellite imagery and LiDAR data. Several approximate nearest neighbor search techniques, such as KD-trees and ball trees, have been proposed to address this issue [135].
2. Feature selection and normalization: The performance of k-NN is sensitive to the choice of features and their scales. Irrelevant or redundant features may negatively

affect the algorithm's performance, while features with different scales can lead to biased distance calculations. Feature selection and normalization techniques can help mitigate these issues.
3. Sensitivity to noise and outliers: k-NN is sensitive to noise and outliers in the data, as they can influence the assignment of class labels or target values. Robust k-NN algorithms, which incorporate outlier detection and noise removal techniques, have been proposed to address this issue [162].

Principal Component Analysis (PCA)

Principal Component Analysis (PCA) is a widely used statistical method for dimensionality reduction, which aims to simplify complex datasets by extracting the most significant features while minimizing the loss of information. PCA is particularly useful in urban studies and human geography, where researchers often work with high-dimensional data derived from multiple sources, such as demographic, socioeconomic, environmental, and remote sensing data [94].

PCA works by transforming the original data into a new set of orthogonal (uncorrelated) variables called principal components (PCs), which are linear combinations of the original features. The first principal component (PC1) accounts for the largest amount of variance in the data, while each subsequent component explains a smaller proportion of the variance, subject to the constraint that it is orthogonal to the preceding components. The resulting PCs can be used as input features for further analysis or visualization, reducing the complexity of the dataset while retaining most of its original information [199].

PCA has been applied in a variety of contexts within urban studies and human geography, such as:

1. Land use and land cover classification: PCA can be used to reduce the dimensionality of high-resolution remote sensing data, such as multispectral and hyperspectral images, by extracting the most important spectral components. This facilitates more efficient and accurate land use and land cover classification, as it minimizes the effects of noise and multicollinearity [170].
2. Socioeconomic analysis: Researchers often use PCA to derive composite indices that capture the underlying structure of multiple socioeconomic variables, such as income, education, and employment. These indices can be used to examine patterns of social inequality and segregation in urban areas and inform the development of targeted policies and interventions [144].
3. Environmental risk assessment: PCA can help identify the main sources of environmental pollution and assess their impacts on human health and ecosystems. By reducing the dimensionality of large environmental datasets, PCA facilitates the identification of spatial patterns and correlations between different pollutants, enabling a better understanding of their sources and effects [29].
4. Transportation and traffic management: PCA can be used to analyze and visualize complex transportation datasets, such as traffic flow, travel demand, and

congestion patterns. By simplifying the data, PCA allows researchers and policymakers to identify key factors affecting transportation systems and develop more effective strategies for managing urban mobility [219].

Here are some examples of how PCA has been used in urban studies and human geography:

1. In a study by Seto and Kaufmann [170], PCA was employed to analyze Landsat Thematic Mapper (TM) satellite images for land use and land cover classification in the Pearl River Delta, China. The authors found that PCA significantly improved the accuracy of land use classification compared to the use of individual spectral bands.
2. Noble et al. [144] utilized PCA to create a composite index of multiple deprivation for small areas in England, incorporating information on income, employment, education, health, housing, and access to services. The index was used to identify areas with high levels of deprivation and inform the allocation of resources for targeted interventions.
3. Chen et al. [29] applied PCA to analyze air quality data from 367 cities in China, focusing on six major pollutants: sulfur dioxide, nitrogen dioxide, carbon monoxide, ozone, particulate matter (PM10), and fine particulate matter (PM2.5). The authors identified three principal components associated with different pollution sources, including industrial emissions, vehicle emissions, and natural processes. This information helped to inform policy recommendations for improving air quality and reducing the health impacts of pollution.
4. Zheng et al. [219] used PCA to analyze and visualize traffic flow data from a large urban road network in Beijing, China. By reducing the dimensionality of the data, the authors were able to identify distinct patterns of congestion and travel demand during different times of the day and week. This information can be used to inform the design of more efficient and sustainable transportation systems.

Despite its numerous advantages and applications, PCA also has some limitations and challenges in the context of urban studies and human geography:

1. Linearity assumption: PCA assumes that the underlying structure of the data can be represented by linear combinations of the original features. However, this may not always be the case, especially when dealing with complex spatial datasets that exhibit non-linear patterns and relationships [94].
2. Interpretability: While PCA simplifies high-dimensional datasets by extracting the most important components, the resulting principal components may not have a clear or meaningful interpretation in terms of the original variables. This can make it difficult to communicate the results of PCA to non-experts and inform policy decisions [2].
3. Sensitivity to scale and outliers: PCA is sensitive to the scale of the input variables, and it may be influenced by extreme values or outliers in the data. Researchers need to carefully preprocess and normalize the data to ensure that the PCA results accurately reflect the underlying structure of the dataset [94].

4. Choice of the number of components: Selecting the appropriate number of principal components to retain in the analysis can be challenging, as it involves balancing the trade-off between reducing dimensionality and retaining information. Various methods have been proposed to determine the optimal number of components, but there is no universally accepted approach [23].

In conclusion, PCA is a powerful and versatile method for dimensionality reduction that has been widely applied in urban studies and human geography. By simplifying complex datasets and extracting the most important features, PCA can facilitate more efficient and accurate analysis, visualization, and modeling of spatial data. However, researchers need to be aware of the limitations and challenges of PCA and carefully preprocess and interpret the data to ensure meaningful and robust results.

k-Means

k-Means is a widely used clustering algorithm in machine learning, which aims to partition a dataset into k distinct clusters based on their similarity [119]. The algorithm iteratively assigns each data point to the cluster whose centroid (mean) is nearest to the point, updating the centroids until convergence is reached or a maximum number of iterations is performed. k-Means is simple, fast, and scalable, making it suitable for large datasets and a wide range of applications, including urban studies and human geography.

The k-Means algorithm consists of the following steps [87]:

1. Initialization: Select k initial centroids, either randomly or using a heuristic method (e.g., k-Means++).
2. Assignment: Assign each data point to the nearest centroid.
3. Update: Recalculate the centroids by computing the mean of all data points in each cluster.
4. Convergence: Repeat steps 2 and 3 until the centroids do not change significantly or a predefined stopping criterion is reached (e.g., maximum number of iterations).

The quality of the final clustering depends on the initial choice of centroids, and different initializations can lead to different clusterings. Several techniques have been proposed to improve the initialization process and the overall performance of the k-Means algorithm, such as k-Means++ [7] and the use of parallel and distributed computing [218].

k-Means has been applied to various problems in urban studies and human geography, including the following examples:

1. Urban land use classification: Zhang et al. [214] used k-Means to classify urban land use types based on remote sensing data, including spectral indices, texture features, and morphological attributes. The authors found that k-Means was effective in detecting different land use patterns and provided valuable insights for urban planning and management.
2. Socioeconomic clustering: Guo and Wang [69] employed k-Means to cluster Chinese cities based on socioeconomic indicators, such as GDP, population, and

infrastructure. The results revealed distinct groups of cities with different levels of economic development and urbanization, which can inform policy decisions and regional development strategies.
3. Traffic analysis zones: Li et al. [114] applied k-Means to partition urban areas into traffic analysis zones (TAZs) using travel demand data, road network attributes, and land use information. The TAZs generated by k-Means were more homogeneous and representative than those produced by traditional methods, improving the accuracy of traffic demand models and predictions.
4. Environmental monitoring: Kumar et al. [104] used k-Means to identify distinct air pollution patterns in Delhi, India, based on air quality monitoring data. The k-Means clusters helped to identify spatial and temporal variations in air pollution levels and potential sources of emissions, guiding targeted interventions to improve air quality.

k-Means is a powerful and flexible clustering algorithm, but it also has some limitations and challenges in the context of urban studies and human geography:

1. Choice of k: Selecting the appropriate number of clusters (k) is a critical and challenging aspect of the k-Means algorithm, as it directly affects the quality of the clustering results. Various techniques have been proposed to determine the optimal k, such as the elbow method, silhouette scores, and gap statistics [183]. However, the choice of k may still be subjective and depend on the specific problem and dataset.
2. Sensitivity to initialization: As mentioned earlier, the k-Means algorithm is sensitive to the initial centroids' selection, which can lead to different clusterings and local optima. Improved initialization techniques like k-Means++ can mitigate this issue, but it remains a challenge in some cases.
3. Spherical clusters assumption: k-Means assumes that clusters are spherical and have similar sizes and densities, which may not always hold in real-world datasets, especially in urban studies and human geography where spatial patterns can be complex and irregular. Alternative clustering algorithms, such as DBSCAN [47] or Gaussian Mixture Models [128], can better handle non-spherical clusters.
4. Handling categorical data: k-Means is designed for continuous numerical data and relies on the Euclidean distance metric. When dealing with categorical data or mixed data types, other distance measures (e.g., Gower distance) or clustering algorithms (e.g., k-Modes, [83]) should be considered.
5. Scalability: Although k-Means is generally fast and scalable, it can become computationally expensive for very large datasets or high-dimensional data, which are common in urban studies and human geography. Dimensionality reduction techniques (e.g., PCA) and parallel or distributed implementations of k-Means can help address this issue.

2.2.5 Model Evaluation and Selection

An essential aspect of machine learning is evaluating the performance of models and selecting the best model for a given task. Commonly used evaluation metrics include accuracy, precision, recall, F1 score, and the area under the receiver operating characteristic (ROC) curve for classification tasks, and mean squared error, mean absolute error, and R-squared for regression tasks [48, 84].

Model selection typically involves splitting the available data into training, validation, and test sets. The training set is used to fit the model, while the validation set is used to fine-tune the model's hyperparameters and select the best model. Finally, the test set is used to assess the model's performance on unseen data [71].

Cross-validation is another widely used technique for model evaluation and selection, in which the data is partitioned into k folds, and the model is trained and evaluated k times, each time using a different fold as the validation set. The average performance across the k iterations is used as an estimate of the model's performance on unseen data [99].

2.2.6 Challenges and Future Directions

Despite the significant advancements in machine learning, several challenges remain, including dealing with imbalanced data, handling missing or noisy data, and addressing issues of overfitting and underfitting. Additionally, the interpretability of machine learning models, particularly complex models like deep neural networks, is an area of ongoing research and development [132].

As machine learning continues to evolve, new techniques and algorithms will be developed to address these challenges and improve the performance of models in various applications. Furthermore, the integration of machine learning with other AI subfields, such as knowledge representation and reasoning, and the development of hybrid models that combine the strengths of different algorithms, are promising avenues for future research [43].

2.3 Deep Learning

Deep learning is a subfield of machine learning that focuses on neural networks with many layers, known as deep neural networks (DNNs) [106]. These networks have the ability to learn complex and hierarchical representations of input data, enabling them to achieve state-of-the-art performance in a wide range of tasks, including image and speech recognition, natural language processing, and game playing (Schmidhuber, 2015). In recent years, deep learning has been increasingly applied to urban studies and human geography, where it has demonstrated significant potential for solving complex spatial problems and analyzing large-scale geospatial data [212].

2.3.1 The Architecture of Deep Neural Networks

Deep neural networks (DNNs) are composed of multiple layers of interconnected neurons, which are organized hierarchically [65]. Each layer receives input from the previous layer, performs a series of computations, and passes the results to the next layer. This section will provide an overview of the architecture of deep neural networks, including their key components and organization.

Layers

A DNN consists of several layers, including the input layer, hidden layers, and output layer. The input layer is responsible for receiving the raw data, while the output layer produces the final predictions or classifications. The hidden layers, which are sandwiched between the input and output layers, are responsible for transforming the input data into high-level representations that can be used to make accurate predictions.

- Input Layer: The input layer is the first layer of a DNN and is responsible for receiving the raw input data, such as images, text, or other forms of data. The number of neurons in the input layer typically corresponds to the dimensionality of the input data.
- Hidden Layers: Hidden layers are the intermediate layers of a DNN and are responsible for transforming the input data into higher-level representations. Each hidden layer consists of a number of neurons, which are connected to the neurons in the previous and next layers. The depth of a DNN, which is the number of hidden layers, is a key factor in determining the capacity of the network to learn complex and hierarchical representations.
- Output Layer: The output layer is the final layer of a DNN and is responsible for producing the predictions or classifications based on the high-level representations learned by the hidden layers. The number of neurons in the output layer typically corresponds to the number of classes or targets in the prediction task.

Neurons

A neuron, or node, is the fundamental building block of a deep neural network. Each neuron receives input from the neurons in the previous layer, computes a weighted sum of the inputs, adds a bias term, and applies a non-linear activation function to produce an output. The output of the neuron is then passed to the neurons in the next layer.

- Weights and Biases: The weights and biases are the parameters of a DNN that need to be learned during the training process. The weights represent the strength of the connections between neurons, while the biases control the threshold at which a neuron becomes active. The combination of weights and biases allows a DNN to learn complex and non-linear relationships between the input data and the target predictions.

- Activation Functions: Activation functions are non-linear functions that are applied to the output of a neuron to introduce non-linearity into the network. Non-linear activation functions enable DNNs to learn complex and non-linear relationships between inputs and outputs, which is a key factor in their ability to model complex data. Common activation functions include the rectified linear unit (ReLU) [137], sigmoid, and hyperbolic tangent (tanh) functions [64].

Connections

In a DNN, neurons are connected to each other through weighted connections. These connections determine how the output of one neuron influences the input of another neuron. The organization of the connections in a DNN can vary depending on the architecture and design of the network.

- Fully Connected Layers: In fully connected layers, each neuron in a layer is connected to all neurons in the previous and next layers. This dense connectivity allows the network to learn complex relationships between input and output data. However, it can also result in a large number of parameters and increased computational complexity.
- Convolution al Layers: In convolutional layers, each neuron is connected only to a local receptive field in the previous layer, rather than to all neurons. This sparse connectivity reduces the number of parameters and computational complexity compared to fully connected layers, making it more efficient for processing grid-like data, such as images or spatial data [107]. Convolutional layers learn spatial hierarchies of features by scanning the input data using filters with shared weights, which enables them to exploit the local structure and invariance properties of the data.
- Recurrent Layers: In recurrent layers, connections between neurons have a temporal aspect, meaning that the output of a neuron at a given time step depends not only on its current input but also on its previous outputs [76]. This architecture allows recurrent neural networks (RNNs) to model sequences and time series data, making them particularly suitable for natural language processing, speech recognition, and other tasks that involve sequential information.

Network Topologies

The organization and connectivity of layers and neurons within a DNN can vary, leading to different network topologies. These topologies can impact the learning capacity and performance of the network, as well as its computational complexity.

- Feedforward Networks: Feedforward networks are the most common type of DNN, characterized by a one-directional flow of information from the input layer to the output layer, with no cycles or loops. This architecture is suitable for tasks

that do not require the modeling of temporal or sequential information, such as image recognition or classification tasks.
- Recurrent Networks: Recurrent networks, as mentioned earlier, have connections that include a temporal aspect, allowing them to model sequences and time series data. The most common type of recurrent network is the RNN, which can be challenging to train due to the vanishing gradient problem [15]. To address this issue, more advanced recurrent architectures, such as long short-term memory (LSTM) [76] and gated recurrent units (GRUs) [37], have been developed.
- Modular Networks: Modular networks are composed of multiple, smaller networks that are trained separately and then combined to form a larger network. This architecture can improve the learning capacity and generalization of the network while reducing computational complexity [158].
- Skip Connections: Skip connections are connections that bypass one or more layers in a DNN, allowing the output of a layer to be directly used as input for a later layer. This architecture can help to alleviate the vanishing gradient problem and improve the flow of information through the network, as demonstrated by the success of residual networks (ResNets) [73].

Training and Optimization

Training a DNN involves minimizing a loss function that measures the difference between the network's predictions and the ground truth labels [17]. The most common optimization algorithm used for training DNNs is stochastic gradient descent (SGD) [160], which updates the weights and biases based on the gradient of the loss function. More advanced optimization algorithms, such as AdaGrad [45], RMSProp [184], and Adam [97], have been developed to improve the convergence and stability of the training process.

To avoid overfitting and enhance generalization, various regularization techniques can be applied during training, including weight decay, dropout [176], and batch normalization [85]. Weight decay penalizes large weights, encouraging the network to rely on multiple features rather than just a few. Dropout involves randomly dropping out neurons during training, which forces the network to learn redundant representations and prevents overfitting. Batch normalization normalizes the input to each layer, helping to maintain a stable distribution of activation values and improving the training process.

Additionally, techniques like data augmentation, transfer learning [205], and early stopping can further improve the performance of deep learning models. Data augmentation involves generating new training examples by applying transformations to the existing data, effectively increasing the size of the training set and reducing overfitting. Transfer learning is a technique where a pre-trained model is fine-tuned on a related task, leveraging the learned features to improve performance on the new task. Early stopping is a technique that halts training when the model's performance on a validation set begins to degrade, preventing overfitting by avoiding unnecessary training epochs.

2.3.2 Learning in Deep Neural Networks

The learning process in deep neural networks involves finding the optimal weights and biases that minimize a given loss function. This process involves several key components, including the choice of the loss function, the activation function, the optimization algorithm, and regularization techniques. In this section, we will discuss these components and their relevance to deep learning.

Loss Functions

The loss function, also known as the objective or cost function, quantifies the difference between the predicted output and the ground truth. The goal of the learning process is to minimize the loss function. There are several loss functions used in deep learning, and the choice depends on the specific problem being addressed. Common loss functions include mean squared error (MSE) for regression tasks, cross-entropy loss for classification tasks, and hinge loss for support vector machines [17].

Activation Functions

Activation functions are used to introduce non-linearity into the neural network model. Non-linear activation functions allow deep neural networks to learn complex, non-linear relationships between inputs and outputs. Common activation functions include the sigmoid function, the hyperbolic tangent (tanh) function, the rectified linear unit (ReLU), and the leaky rectified linear unit (Leaky ReLU) [64]. The choice of activation function depends on the specific problem and the desired properties of the network, such as the ability to handle vanishing or exploding gradients.

Optimization Algorithms

Optimization algorithms are used to update the weights and biases of the network to minimize the loss function. Gradient-based optimization methods, such as stochastic gradient descent (SGD) and its variants, are commonly used in deep learning. The basic idea of SGD is to update the weights and biases in the direction of the negative gradient of the loss function with respect to the network parameters [160].

Several variants of SGD have been proposed to improve the convergence and stability of the learning process, such as momentum [148], Nesterov accelerated gradient [138], AdaGrad [45], RMSprop [184], and Adam [97]. These variants adapt the learning rate during training and can handle sparse gradients, making them suitable for deep learning applications.

Regularization Techniques

Regularization techniques are used to prevent overfitting in deep neural networks by adding a penalty term to the loss function. This penalty term discourages the model from fitting the noise in the training data, thus improving its generalization performance. Common regularization techniques include L1 and L2 regularization, which add the absolute value or the square of the weights, respectively, to the loss function [182].

Another popular regularization technique is dropout [176], which randomly sets a fraction of the neuron activations to zero during training. This forces the network to learn redundant representations and prevents overfitting. Batch normalization [85] is another technique that helps improve the training of deep networks by normalizing the input to each layer during training, thus reducing the internal covariate shift and improving convergence.

Backpropagation

The backpropagation algorithm [163] is the main workhorse for training deep neural networks. It is a supervised learning algorithm that computes the gradients of the loss function with respect to the network parameters using the chain rule of calculus. The gradients are then used to update the weights and biases through an optimization algorithm, such as SGD or its variants.

The backpropagation algorithm consists of two main steps: the forward pass and the backward pass. In the forward pass, the input is propagated through the network to compute the output and the loss. In the backward pass, the gradients of the loss with respect to the network parameters are computed using the chain rule, starting from the output layer and moving backward through the network.

Convolutional Neural Networks (CNNs)

Convolutional Neural Networks (CNNs) are a specific type of deep learning architecture designed to handle grid-like data, such as images, videos, or speech signals [107]. CNNs consist of convolutional layers, pooling layers, and fully connected layers. Convolutional layers apply a set of filters to the input, resulting in feature maps that capture local patterns in the data. Pooling layers reduce the spatial dimensions of the feature maps by applying a downsampling operation, such as max-pooling or average pooling [167]. Fully connected layers are used to produce the final output, such as class probabilities in a classification task.

CNNs have been widely used in various applications in urban studies and human geography. For instance, CNNs have been used to classify satellite images for land use and land cover mapping [24], detect urban change [27], and estimate population density [196].

Recurrent Neural Networks (RNNs)

Recurrent Neural Networks (RNNs) are a type of deep learning architecture designed to handle sequential data, such as time series, natural language text, or video frames [46]. RNNs consist of a hidden state that is updated at each time step, allowing the network to maintain a memory of the past inputs. This memory enables RNNs to learn and capture temporal dependencies in the data.

However, vanilla RNNs can suffer from vanishing or exploding gradients during training, making it difficult to learn long-term dependencies [15]. To address this issue, more advanced RNN architectures have been proposed, such as Long Short-Term Memory (LSTM) [76] and Gated Recurrent Unit (GRU) [37]. These architectures introduce gating mechanisms that allow the network to better control the flow of information, making it easier to learn long-term dependencies.

RNNs, especially LSTMs and GRUs, have been used in various applications in urban studies and human geography, such as traffic flow prediction [55], spatiotemporal modeling of air pollution [110], and urban event detection using social media data [35].

Autoencoders

Autoencoders are a type of unsupervised deep learning architecture that learns to encode and decode data in a lower-dimensional representation [75]. Autoencoders consist of an encoder network that maps the input data to a lower-dimensional representation, called the bottleneck or latent space, and a decoder network that reconstructs the input data from the latent space. The learning process aims to minimize the reconstruction error, typically measured using the mean squared error or cross-entropy loss.

Autoencoders have been used in various applications in urban studies and human geography, such as dimensionality reduction for visualization [122], feature learning for clustering and classification tasks, and anomaly detection in spatiotemporal data [31].

Generative Adversarial Networks (GANs)

Generative Adversarial Networks (GANs) are a type of deep learning architecture that learns to generate realistic data by training two networks, a generator and a discriminator, in a competitive setting [66]. The generator learns to produce realistic data samples, while the discriminator learns to distinguish between real and generated samples. The training process involves updating the generator to produce more realistic samples and updating the discriminator to better distinguish between real and generated samples.

GANs have been used in various applications in urban studies and human geography, such as generating realistic satellite images [210], simulating urban growth patterns [190], and generating realistic 3D building models [194].

In conclusion, the learning process in deep neural networks involves several key components, including the choice of the loss function, activation function, optimization algorithm, and regularization techniques. Deep learning has been widely used in various applications in urban studies and human geography, such as image classification, time series prediction, dimensionality reduction, and data generation. The advancements in deep learning techniques, combined with the availability of large-scale geospatial data and computational resources, offer new opportunities for researchers and practitioners in urban studies and human geography to tackle complex problems and gain insights into the underlying processes and dynamics.

2.3.3 Convolutional Neural Networks (CNNs)

Convolutional Neural Networks (CNNs) are a class of deep learning algorithms designed specifically for processing grid-like data structures, such as images, where local spatial relationships between data points are important. This type of deep learning architecture has been remarkably successful in various computer vision tasks, including image classification, object detection, and semantic segmentation [106]. The application of CNNs in urban studies and human geography has gained momentum in recent years due to the increasing availability of high-resolution satellite and aerial imagery and the need for efficient processing and analysis of these large datasets [216].

The core concept behind CNNs is the convolution operation, which involves sliding a small filter or kernel over the input data to compute a new feature map. This process enables the network to learn and detect local features, such as edges, corners, and textures, which can be combined and organized hierarchically to recognize more complex patterns and structures in the input data [102].

The architecture of a typical CNN consists of several layers, including convolutional layers, pooling layers, and fully connected layers. Convolutional layers are responsible for applying convolution operations with learned filters, while pooling layers help reduce the spatial dimensions of the feature maps and improve the computational efficiency of the network. Fully connected layers are used to combine the features learned from the previous layers and produce the final output, such as class probabilities in image classification tasks [173].

One of the key advantages of CNNs over traditional machine learning methods is their ability to learn hierarchical feature representations directly from raw data, without the need for manual feature extraction or engineering. This property makes CNNs particularly suitable for processing and analyzing large-scale geospatial data, where the identification of relevant features can be challenging and time-consuming [27] (Fig. 2.7).

There are several examples of the successful application of CNNs in urban studies and human geography. In land use and land cover classification, CNNs have been shown to achieve high accuracy in identifying different land cover types, such as

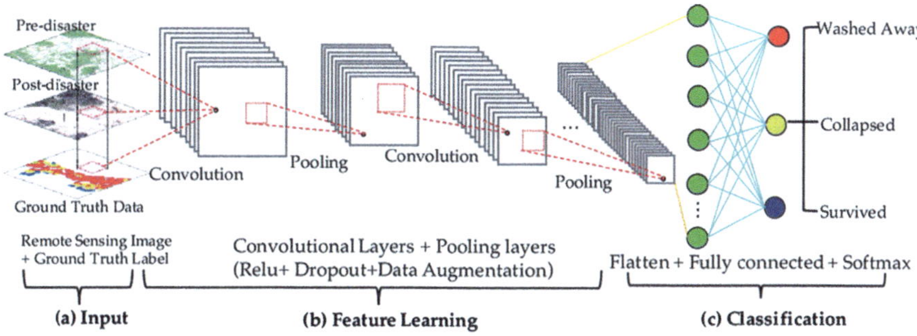

Fig. 2.7 The structure of a convolutional neural network (CNN), highlighting the convolutional, pooling, and fully connected layers [100]

urban, agricultural, and natural areas, from satellite and aerial imagery [24]. In transportation and traffic management, CNNs have been used to detect and count vehicles in aerial images, providing valuable information for traffic flow analysis and congestion mitigation [8].

In environmental monitoring and risk assessment, CNNs have been applied to analyze remote sensing data and identify areas affected by natural disasters, such as floods, landslides, and wildfires, which can help inform disaster management and mitigation efforts [30]. In socioeconomic analysis and urban planning, CNNs have been employed to estimate population density, income distribution, and other demographic variables from satellite imagery, providing a cost-effective and timely alternative to traditional survey methods [91].

Overall, CNNs have proven to be a powerful tool for analyzing geospatial data in urban studies and human geography, offering a more efficient and accurate approach to various tasks compared to traditional methods. Their ability to automatically learn and hierarchically represent relevant features from raw data has made them especially useful in processing large-scale datasets, such as satellite and aerial imagery. As the availability of high-resolution geospatial data continues to grow, and computational resources become more accessible, it is expected that the application of CNNs in urban studies and human geography will continue to expand and contribute to our understanding of complex urban processes and patterns.

2.3.4 Recurrent Neural Networks (RNNs)

Recurrent Neural Networks (RNNs) represent another important class of deep learning models that have had a significant impact on various domains, including natural language processing, time series forecasting, and speech recognition. This section will provide an overview of RNNs, discuss their architecture and learning mechanisms, and present examples of their use in urban studies and human geography.

A key characteristic of RNNs is their ability to process and model sequential data, which makes them particularly well-suited for tasks involving temporal dependencies. Unlike feedforward neural networks, such as CNNs, RNNs have recurrent connections that allow them to maintain a hidden state across time steps, thereby capturing information from previous inputs in the sequence. This architecture enables RNNs to learn and generate complex temporal patterns and dependencies, which is especially relevant in the context of urban studies and human geography, where many processes evolve over time [46, 76].

However, RNNs have certain limitations. For instance, they tend to struggle with capturing long-term dependencies due to the problem of vanishing or exploding gradients during training [15, 147]. To address this issue, more advanced RNN architectures have been proposed, such as Long Short-Term Memory (LSTM) networks [76] and Gated Recurrent Units (GRUs) [37]. Both LSTM and GRU networks employ specialized gating mechanisms to regulate the flow of information through time, enabling them to learn and model long-range dependencies more effectively.

RNNs have been applied to various tasks in urban studies and human geography that involve temporal data, such as traffic flow prediction, population dynamics, and land-use change modeling. For example, Yu et al. [208] used LSTM networks to predict urban traffic flow, leveraging the model's ability to capture complex temporal patterns and dependencies in the data. Their study showed that the LSTM-based approach outperformed traditional time series forecasting methods, demonstrating the potential of RNNs in addressing urban transportation challenges.

Another example comes from Zhang et al. [215], who employed RNNs to model the spatiotemporal dynamics of population distribution in urban areas. They combined mobile phone data with other geospatial data sources, such as points of interest and road networks, to train an RNN model that could predict population distribution at different time scales. The results of their study indicated that the RNN-based model was able to generate accurate predictions and capture the complex interactions between urban structure and human mobility patterns.

In the context of land-use change modeling, Niu et al. [143] applied RNNs to predict land-use transitions in rapidly urbanizing areas. They integrated remote sensing data with socioeconomic variables to train an LSTM network, which was able to generate accurate land-use change predictions over multiple time steps. The proposed approach demonstrated the effectiveness of RNNs in modeling the complex, dynamic processes underlying urban land-use change, offering a valuable tool for urban planners and decision-makers.

Despite their potential in urban studies and human geography, RNNs are not without challenges. For instance, the training of RNNs can be computationally intensive, especially for large-scale datasets and long sequences. Moreover, the interpretability of RNN models remains an open research question, as the internal mechanisms of these networks can be difficult to understand and explain [32]. Nevertheless, RNNs have proven to be a powerful tool for modeling temporal patterns and dependencies in various applications related to urban studies and human geography. Ongoing research and advancements in RNN architectures, training techniques, and interpretability methods are likely to further enhance their applicability and effectiveness in these fields (Fig. 2.8).

In summary, Recurrent Neural Networks (RNNs) offer a unique approach to modeling sequential data, which is especially relevant in urban studies and human geography, where many processes evolve over time (Table 2.3). RNNs have been successfully applied to various tasks, such as traffic flow prediction, population dynamics, and land-use change modeling. Despite their computational complexity and interpretability challenges, RNNs have the potential to significantly contribute to our understanding of complex, dynamic processes in urban studies and human geography.

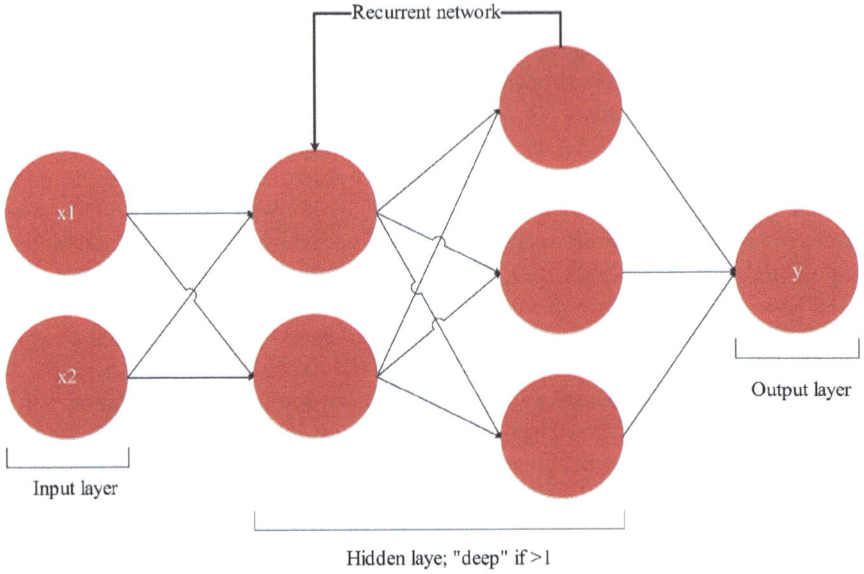

Fig. 2.8 The structure of a basic recurrent neural network

2.3.5 Generative Models

Generative models are a class of deep learning algorithms that aim to learn the underlying data distribution and generate new samples from it. They have received significant attention in recent years due to their ability to create realistic and high-quality data across various domains. Generative models can be particularly useful in urban studies and human geography for tasks such as data augmentation, simulation, and scenario analysis. This section will discuss two popular types of generative models: Variational Autoencoders (VAEs) and Generative Adversarial Networks (GANs), and their applications in urban studies and human geography.

Variational Autoencoders (VAEs)

Variational Autoencoders (VAEs) [98, 159] are a type of generative model that combine the principles of deep learning and probabilistic graphical models. VAEs consist of two main components: an encoder and a decoder. The encoder maps input data to a latent space, which is a lower-dimensional representation of the data. The decoder, on the other hand, reconstructs the original data from the latent space representation. The objective of VAEs is to minimize the reconstruction error and maximize the likelihood of the data given the latent space representation.

VAEs have been used in various applications in urban studies and human geography. For example, VAEs have been used to generate realistic land cover maps [124] and simulate urban growth patterns [150]. By learning the latent space representation of the data, VAEs can generate novel samples that maintain the spatial structure and

Table 2.3 The difference between CNNs and RNNs

Aspect	Convolutional neural networks (CNNs)	Recurrent neural networks (RNNs)
Application focus	Processing grid-like data structures, such as images, for tasks like image classification, object detection, and semantic segmentation	Modeling sequential data with temporal dependencies, relevant for tasks like traffic flow prediction, population dynamics, and land-use change modeling
Core concept	Convolution operation applied to input data to detect local features and hierarchically learn complex patterns	Recurrent connections maintain hidden states across time steps, capturing temporal dependencies in sequential data
Architecture	Consists of convolutional layers, pooling layers, and fully connected layers, enabling hierarchical feature learning from raw data	Contains recurrent connections allowing information flow across time steps, capturing temporal dependencies in sequential data
Key advantages	Automatically learns hierarchical feature representations from raw data, eliminating the need for manual feature engineering	Captures temporal patterns and dependencies in sequential data, enabling accurate modeling of dynamic processes
Examples of applications	Land use and land cover classification, vehicle detection in aerial imagery, environmental monitoring, and risk assessment	Traffic flow prediction, population dynamics modeling, land-use change prediction, and spatiotemporal data analysis
Challenges and considerations	Requires large-scale datasets for effective training, computationally intensive, interpretability of learned features may be challenging	Training can be computationally intensive, struggles with long-term dependencies, interpretability of learned patterns may be difficult

characteristics of the original data. This ability is particularly useful in data augmentation tasks, where additional samples are required for training and validation of machine learning models.

Generative Adversarial Networks (GANs)

Generative Adversarial Networks (GANs) [66] are another type of generative model that has gained significant attention in the deep learning community. GANs consist of two neural networks: a generator and a discriminator. The generator's goal is to create realistic samples from random noise, while the discriminator's goal is to distinguish between real samples and those generated by the generator. The generator and discriminator are trained simultaneously in an adversarial setting, where the generator tries to generate samples that can fool the discriminator, and the discriminator tries to improve its ability to differentiate between real and generated samples.

GANs have demonstrated impressive performance in generating high-quality, realistic images and have been applied to various tasks in urban studies and human geography. For instance, GANs have been used to generate high-resolution land cover maps [117], simulate urban growth patterns [200], and create synthetic building footprints [191]. GANs can also be used for data augmentation, generating new samples for training and validation of machine learning models, and improving the quality of satellite images [195].

In addition to their generative capabilities, GANs have been used for tasks such as domain adaptation and data fusion. For example, GANs have been used to translate satellite images from one domain (e.g., optical imagery) to another (e.g., synthetic aperture radar imagery) [206]. This capability can be particularly useful in urban studies and human geography when data from different sources and modalities need to be integrated for analysis. GANs have also been employed for fusing multi-resolution remote sensing data [109], which can improve the spatial resolution of the generated images and provide more accurate information for urban and geographical studies.

One of the challenges in applying GANs to urban studies and human geography is the need for large amounts of labeled data for training the discriminator. However, recent advances in semi-supervised and unsupervised learning techniques for GANs [145, 165] have mitigated this issue to some extent, making GANs more accessible for these disciplines.

Another challenge in using GANs is the mode collapse problem, where the generator learns to produce only a limited set of samples instead of covering the entire data distribution. This issue can be addressed through various techniques, such as using different architectures like Wasserstein GANs [6] or employing regularization methods like spectral normalization [131].

Applications in Urban Studies and Human Geography

Generative models, particularly VAEs and GANs, have found various applications in urban studies and human geography. Some of these applications include:

1. Data augmentation: Generative models can be used to create additional samples for training and validation of machine learning models, improving their performance and generalization capabilities [195].
2. Simulation and scenario analysis: Generative models can be employed to generate plausible future scenarios, such as urban growth patterns, land use changes, and environmental impacts, helping policymakers and urban planners make informed decisions [124, 200].
3. Data fusion and domain adaptation: GANs can be used to integrate data from different sources and modalities, such as remote sensing images with different resolutions or types, improving the quality and information content of the resulting images [109, 206].
4. Synthesis of geospatial data: Generative models can generate realistic geospatial data, such as building footprints, road networks, and land cover maps, which can be used for various applications in urban studies and human geography [116, 191].

In summary, generative models have shown great potential in revolutionizing urban studies and human geography by providing powerful tools for generating, augmenting, and fusing geospatial data. However, challenges remain, such as the need for large amounts of labeled data, mode collapse, and the interpretability of the generated samples. Further research is needed to address these challenges and improve the performance and applicability of generative models in urban studies and human geography.

2.4 Recurrent Learning

Recurrent learning is a concept within machine learning and artificial intelligence that refers to the use of recurrent neural networks (RNNs) for learning and prediction tasks. RNNs are a class of neural networks with loops in the network, which allows them to maintain a hidden state and effectively model sequential data [76]. This ability to handle sequential data makes RNNs suitable for various applications in human geography and urban planning, such as time series analysis, natural language processing, and spatiotemporal data modeling.

2.4.1 Recurrent Neural Networks: An Overview

Recurrent Neural Networks (RNNs) are a class of artificial neural networks designed specifically to handle sequential data. RNNs have been widely used to model temporal dependencies and patterns in various fields such as natural language processing, speech recognition, and time series analysis. In this section, we provide a comprehensive overview of RNNs, their architecture, and the essential concepts required to understand how they function.

The primary motivation behind the development of RNNs is the inherent limitation of traditional feedforward neural networks when dealing with sequential data. Feedforward networks are unable to effectively capture the temporal dependencies present in sequences because they assume that inputs are independent of each other. RNNs, on the other hand, have an internal memory that allows them to maintain information about previous inputs, making it possible to model the temporal dynamics of sequential data.

Architecture of Recurrent Neural Networks

RNNs are characterized by their unique architecture that consists of a series of hidden layers connected through time. This architecture allows RNNs to process input sequences of variable length and maintain information about previous inputs. An RNN typically consists of the following components:

1. Input Layer: This layer receives the input sequence and passes it on to the hidden layers.
2. Hidden Layers: These layers maintain the memory of the RNN and perform the necessary computations to capture the temporal dependencies in the input sequence. Each hidden layer has a set of neurons with recurrent connections, enabling them to maintain a state across time steps.
3. Output Layer: This layer produces the output of the RNN, often as a probability distribution over possible output sequences or classes.

The primary difference between RNNs and feedforward neural networks lies in the hidden layers. In an RNN, the hidden layers have recurrent connections, allowing them to maintain a state across time steps. This state, often referred to as the hidden state, is updated at each time step based on the current input and the previous hidden state. This updating process enables RNNs to learn the temporal dependencies in the input sequence.

Training Recurrent Neural Networks

Training RNNs involves learning the weights of the recurrent connections in the hidden layers. This is typically achieved using a variant of the backpropagation algorithm called Backpropagation Through Time (BPTT) [197]. BPTT works by unrolling the RNN through time, converting it into a feedforward network with multiple layers, one for each time step. The weights are then updated using the standard backpropagation algorithm, taking into account the error gradients at each time step.

Despite the effectiveness of BPTT in training RNNs, it suffers from two significant challenges: the vanishing and exploding gradient problems [15]. These issues arise when the gradients of the loss function with respect to the weights become either too small (vanishing) or too large (exploding), making it difficult for the RNN to learn long-range dependencies. To address these challenges, researchers have developed several advanced RNN architectures, such as Long Short-Term Memory (LSTM) [76] and Gated Recurrent Units (GRUs) [37], which have demonstrated improved performance in capturing long-range dependencies.

Applications of Recurrent Neural Networks in Human Geography and Urban Planning

RNNs have been increasingly applied in human geography and urban planning to model and analyze spatiotemporal data. Some of the notable applications include:

1. Traffic Flow Prediction: RNNs, particularly LSTMs and GRUs, have been employed to predict traffic flow patterns in urban areas [121, 208]. These models are capable of capturing the complex temporal dependencies in traffic data, allowing for accurate predictions that can help urban planners design better transportation systems and manage traffic congestion.
2. Human Mobility Prediction: RNNs have been used to model and predict human mobility patterns [56, 175]. By analyzing the sequence of locations visited by individuals, RNNs can learn the underlying patterns and generate predictions

for future movements. This information can be valuable for urban planners in understanding the dynamics of urban spaces and designing infrastructure that accommodates the mobility needs of the population.

3. Land Use and Land Cover Change Detection: RNNs can be employed to analyze time series of satellite images to detect and predict land use and land cover changes [86, 112]. By modeling the temporal dependencies in the image sequences, RNNs can identify changes in land use patterns, contributing to more effective land use planning and management.
4. Social Media Sentiment Analysis: RNNs have been applied to analyze the temporal patterns in social media data to study public opinion and sentiment towards various urban issues [92, 193]. These models can help urban planners and policymakers gauge public sentiment and adapt their strategies accordingly.
5. Disaster Impact Assessment: RNNs have been utilized to assess the impact of natural disasters, such as floods and earthquakes, on urban areas by analyzing spatiotemporal data, including social media posts and satellite imagery [103, 113]. This information can inform disaster management efforts and help in the development of more resilient urban environments.

Recurrent Neural Networks offer a powerful tool for modeling and analyzing spatiotemporal data in human geography and urban planning. Their unique architecture and ability to capture temporal dependencies make them well-suited for various applications in these disciplines, contributing to more informed decision-making and better urban planning outcomes.

2.4.2 Challenges with RNNs: Vanishing and Exploding Gradients

While Recurrent Neural Networks (RNNs) have shown great promise in modeling and predicting sequential data, they also come with their own set of challenges, particularly when it comes to training these models. One of the most notable issues with RNNs is the problem of vanishing and exploding gradients, which can make training deep RNNs particularly difficult.

In this section, we will discuss the nature of the vanishing and exploding gradient problems, their implications for training RNNs, and some of the techniques that have been developed to address these issues. We will also provide references to relevant literature and research that has contributed to our understanding of these challenges.

Understanding Vanishing and Exploding Gradients

The vanishing and exploding gradient problems are closely related to the process of training RNNs using backpropagation through time (BPTT), a technique that essentially unfolds the network in time to compute the gradients of the loss function with respect to the model parameters [197]. When training RNNs with long sequences, the gradients can become either very small (vanish) or very large (explode) as they are propagated back through time. This can lead to poor model performance, as the gradients may not provide useful information for updating the model parameters during training.

The vanishing gradient problem occurs when the gradients of the loss function with respect to the model parameters become very small as they are propagated back through time, leading to a lack of information for updating the model parameters. This can result in slow convergence during training and poor generalization performance [15].

The exploding gradient problem, on the other hand, occurs when the gradients become very large as they are propagated back through time, causing the model parameters to be updated with large, erratic steps during training. This can lead to instability in the training process and poor model performance [147].

Addressing the Vanishing and Exploding Gradient Problems

A number of techniques have been proposed to address the challenges posed by the vanishing and exploding gradient problems in RNNs. Some of these include:

1. Long Short-Term Memory (LSTM): LSTM is a popular variant of the RNN architecture that was specifically designed to address the vanishing gradient problem [76]. LSTM introduces a memory cell and a set of gating mechanisms that allow the network to store and access information over long time scales, effectively mitigating the vanishing gradient problem. In many applications, LSTM has been shown to outperform traditional RNNs, particularly when dealing with long sequences of data [59].
2. Gated Recurrent Unit (GRU): The GRU is another RNN variant that was developed as a simpler alternative to LSTM [37]. Like LSTM, the GRU incorporates gating mechanisms to control the flow of information through the network, helping to address the vanishing gradient problem. While GRU has fewer parameters than LSTM, it has been shown to achieve comparable performance in many applications [38].
3. Gradient clipping: Gradient clipping is a simple technique that can be applied during training to address the exploding gradient problem [147]. By limiting the maximum value of the gradients during backpropagation, gradient clipping can prevent the model parameters from being updated with large, erratic steps, leading to more stable training and improved model performance.
4. Regularization: Regularization techniques, such as L1 and L2 regularization, can also be used to address the exploding gradient problem by adding a penalty term to the loss function, effectively constraining the magnitude of the model parameters during training [140]. This can help to prevent large gradients from causing instability in the training process and improve model generalization.
5. Skip connections: Skip connections, also known as residual connections, are another technique that can be used to mitigate the vanishing gradient problem in deep RNNs [96]. By introducing direct connections between non-adjacent layers in the network, skip connections can help to maintain the flow of information

during backpropagation, preventing the gradients from vanishing as they are propagated back through time.
6. Layer normalization: Layer normalization is a technique that can help to address both the vanishing and exploding gradient problems by normalizing the activations of each layer in the network during training [10]. By ensuring that the activations have a consistent scale and distribution, layer normalization can improve the stability of the training process and the overall performance of the model.
7. Weight initialization: Proper weight initialization can also play a critical role in mitigating the vanishing and exploding gradient problems [63]. By initializing the model parameters with appropriate values, it is possible to ensure that the gradients remain well-behaved during training, reducing the likelihood of encountering vanishing or exploding gradients.

In the context of urban studies and human geography, addressing the vanishing and exploding gradient problems in RNNs is essential for effectively modeling and predicting complex spatiotemporal patterns and dynamics. For example, LSTMs have been used to model and predict taxi demand in urban areas [202], while GRUs have been applied to the analysis of social media data for disaster response [141]. By overcoming the challenges associated with vanishing and exploding gradients, researchers can harness the full potential of RNNs for a wide range of applications in urban studies and human geography.

2.4.3 *Long Short-Term Memory Networks*

Long Short-Term Memory (LSTM) networks were introduced by Hochreiter and Schmidhuber [76] as a solution to the vanishing and exploding gradient problems associated with traditional Recurrent Neural Networks (RNNs). LSTMs have since become one of the most widely used RNN architectures, particularly in applications that involve learning long-range dependencies in sequential data. In this section, we will provide an overview of the LSTM architecture, its key components, and its applications in urban studies and human geography.

LSTM Architecture

The primary innovation of the LSTM network is its unique cell structure, which replaces the standard RNN cell with a more complex architecture designed to better capture long-range dependencies. The LSTM cell consists of four main components: an input gate, a forget gate, an output gate, and a cell state. These components work together to regulate the flow of information through the network, allowing the LSTM to selectively remember and forget information over long sequences.

The input gate determines the extent to which new information from the current input is incorporated into the cell state. The forget gate controls the extent to which previous information in the cell state is retained or discarded. The output gate controls the extent to which the cell state contributes to the output at the current time step. The cell state acts as a memory buffer that retains important information over time.

These gating mechanisms enable the LSTM to learn complex temporal relationships in the input data by selectively remembering and forgetting information as needed. This makes LSTMs particularly well-suited for applications that involve learning from long sequences or time series data (Fig. 2.9).

LSTM Variants

Over the years, several variants of the original LSTM architecture have been proposed to improve its performance and adapt it to specific tasks. Some of the most notable LSTM variants include:

- Gated Recurrent Units (GRUs): A simplified version of the LSTM architecture that combines the input and forget gates into a single update gate, resulting in a more computationally efficient model [37].
- Bidirectional LSTMs (BiLSTMs): A model that processes the input sequence in both forward and backward directions, allowing it to capture both past and future context [169].

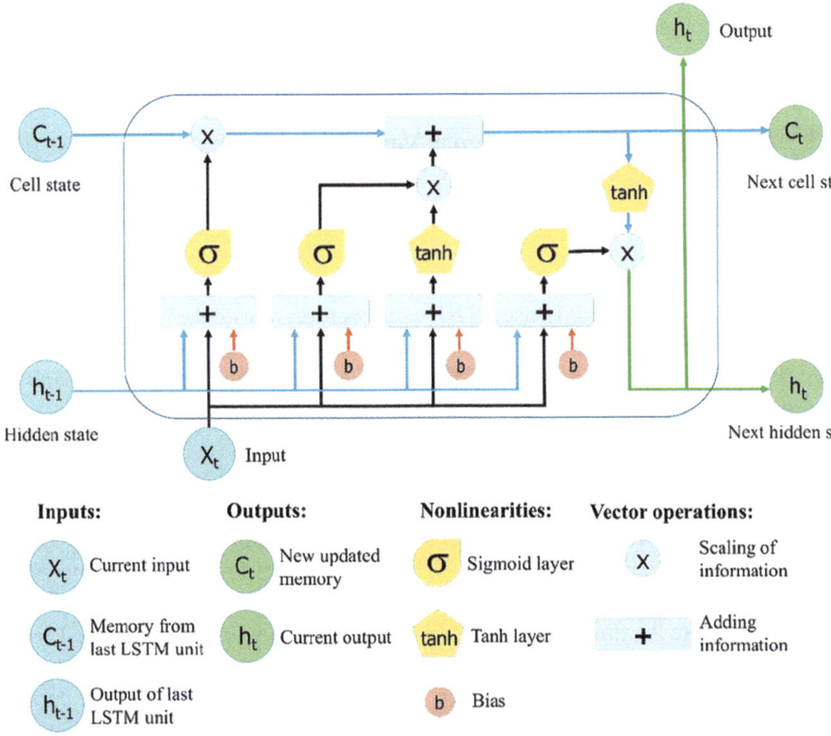

Fig. 2.9 A schematic representation of the long short-term memory (LSTM) network architecture, highlighting its memory cells and gating mechanisms [105]

- Peephole LSTMs: A modification of the original LSTM architecture that allows the gates to access the cell state directly, improving the model's ability to learn precise timing dependencies [60].
- Attention-based LSTMs: A model that incorporates an attention mechanism, enabling it to selectively focus on specific parts of the input sequence, which is particularly useful for tasks such as machine translation and text summarization [12].

Applications in Urban Studies and Human Geography

LSTMs have been applied to various urban studies and human geography problems due to their ability to model complex temporal dependencies in time series and sequential data. Some notable applications include:

- Traffic flow prediction: LSTMs have been used to predict traffic flow in urban road networks, accounting for temporal dependencies and spatial relationships between road segments [121].
- Air quality prediction: LSTMs have been employed to predict air pollution levels in cities, leveraging temporal patterns and meteorological data to improve prediction accuracy [111].
- Land use and land cover change detection: LSTMs have been utilized to model land use and land cover changes over time, incorporating both spatial and temporal dependencies in the data [134].
- Urban growth modeling: LSTMs have been applied to predict urban growth patterns by learning complex dependencies between various factors such as population density, land use, and infrastructure development [181].

Long Short-Term Memory networks have emerged as a powerful tool for modeling complex temporal dependencies in sequential data. Their unique cell structure, combined with various architectural variants and enhancements, have made them particularly well-suited for a wide range of applications in urban studies and human geography. By capturing the intricate relationships between various factors in time series and sequential data, LSTMs have the potential to significantly improve our understanding of urban and geographic processes and inform more effective and sustainable planning and decision-making.

2.4.4 Gated Recurrent Units

Gated Recurrent Units (GRUs) are a type of recurrent neural network (RNN) architecture proposed by Cho et al. [37] as a simpler alternative to Long Short-Term Memory (LSTM) networks. The primary motivation behind GRUs was to overcome the vanishing and exploding gradient problems associated with traditional RNNs while maintaining a more compact and computationally efficient structure compared to LSTMs.

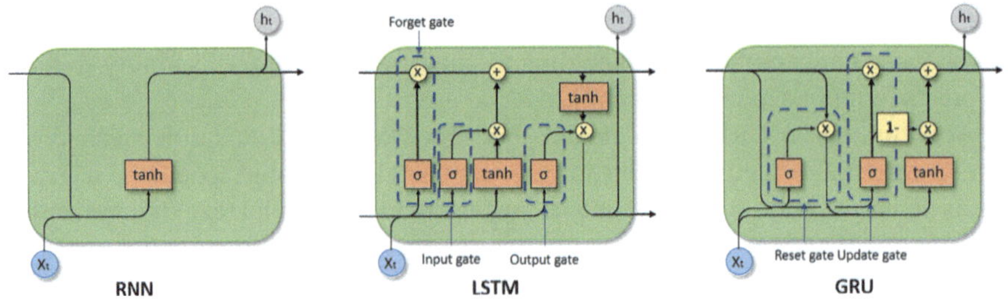

Fig. 2.10 A diagram comparing the structure of gated recurrent units (GRUs) to LSTMs [42]

GRU Architecture

GRUs, like LSTMs, are designed to capture long-term dependencies in sequential data effectively. However, they achieve this with fewer parameters and a simpler architecture. The core of the GRU architecture consists of two gates: the update gate and the reset gate. These gates are responsible for determining how much of the previous hidden state should be retained or discarded and how much of the new input should be incorporated into the current hidden state (Fig. 2.10).

Update Gate

The update gate (z) in a GRU is responsible for determining the extent to which the previous hidden state (h_{t-1}) should be carried over into the current hidden state (h_t). The update gate is computed using a sigmoid activation function, which outputs values between 0 and 1, indicating the proportion of the previous hidden state to retain:

$$z_t = \sigma(W_z * x_t + U_z * h_{t-1} + b_z)$$

Here, W_z and U_z are the weight matrices for the input x_t and previous hidden state h_{t-1}, respectively, b_z is the bias term, and σ is the sigmoid function.

Reset Gate

The reset gate (r) is responsible for determining how much of the previous hidden state should be used to compute the candidate hidden state. Like the update gate, the reset gate is computed using a sigmoid activation function:

$$r_t = \sigma(W_r * x_t + U_r * h_{t-1} + b_r)$$

Here, W_r and U_r are the weight matrices for the input x_t and previous hidden state h_{t-1}, respectively, and b_r is the bias term.

Candidate Hidden State

The candidate hidden state ($\tilde{h}_t$) is computed using the reset gate, input, and previous hidden state. The reset gate is element-wise multiplied with the previous hidden state, and this result is combined with the input to compute the candidate hidden state using a hyperbolic tangent (tanh) activation function:

$$\tilde{h}_t = \tanh(W * x_t + U * (r_t \odot h_{t-1}) + b)$$

Here, W and U are the weight matrices for the input x_t and the element-wise product of the reset gate and previous hidden state, respectively, and b is the bias term.

Current Hidden State

The current hidden state (h_t) is computed by combining the candidate hidden state (h_t) and the previous hidden state (h_{t-1}), with the update gate (z_t) determining the proportion of each:

$$h_t = (1 - z_t) \odot h_{t-1} + z_t \odot \tilde{h}_t$$

The update gate allows the GRU to retain information from previous time steps when needed, enabling the model to capture long-range dependencies effectively.

Applications of GRUs in Urban Studies and Human Geography

GRUs have been widely applied in various domains due to their ability to model complex temporal relationships in sequential data. In urban studies and human geography, GRUs can be employed to model and predict various spatiotemporal phenomena. Some examples of GRU applications in these fields include:

Traffic Flow Prediction GRUs can be used to model and predict traffic flow in urban areas. Yu et al. [208] proposed a deep learning model based on GRUs and convolutional neural networks (CNNs) to predict traffic flow using spatiotemporal data. Their model outperformed traditional methods in terms of prediction accuracy, demonstrating the potential of GRUs for traffic flow prediction.

Reference: Yu et al. [209]. Spatio-temporal graph convolutional networks: A deep learning framework for traffic forecasting. arXiv preprint arXiv:1709.04875.

Land Use and Land Cover Change Detection GRUs can be employed to model and analyze land use and land cover changes over time. For instance, Rußwurm and Körner [164] used GRUs in combination with CNNs to detect land cover changes in multitemporal remote sensing data. Their approach demonstrated high accuracy in identifying land cover changes and showcased the potential of GRUs for analyzing spatiotemporal patterns in land use and land cover.

Urban Growth Prediction GRUs can be applied to model and predict urban growth patterns by analyzing spatiotemporal data. In a study by Chandra et al. (2018), GRUs were used to predict urban growth using historical land use data, providing accurate predictions of future urban expansion. This application of GRUs can help urban planners make more informed decisions about land use management and sustainable urban development.

Reference: Chandra et al. [28]. Spatio-temporal urban growth modelling using deep GRU-LSTM network. In Proceedings of the 1st ACM SIGSPATIAL Workshop on Prediction of Human Mobility (pp. 1–4).

Gated Recurrent Units (GRUs) are a powerful RNN architecture capable of modeling complex temporal relationships in sequential data. Due to their simpler structure compared to LSTMs, they offer a more computationally efficient alternative for capturing long-range dependencies. In urban studies and human geography, GRUs have demonstrated their potential in various applications, including traffic flow prediction, land use and land cover change detection, and urban growth prediction. The use of GRUs in these fields can help researchers and practitioners gain deeper insights into spatiotemporal patterns and make more informed decisions about urban planning and sustainable development.

2.4.5 Future Directions and Challenges in Recurrent Learning

Recurrent learning, particularly through the use of recurrent neural networks (RNNs), has gained significant attention in recent years due to its ability to model and analyze sequential data. RNNs, including their variants like Long Short-Term Memory (LSTM) networks and Gated Recurrent Units (GRUs), have been successfully applied in various fields, including human geography and urban studies. However, there are still several challenges and future directions in recurrent learning that warrant further investigation.

Addressing the Limitations of RNNs

While RNNs have demonstrated their potential in modeling sequential data, they also have some limitations. One major limitation is the difficulty in training RNNs on long sequences due to the vanishing and exploding gradient problems. Although LSTMs and GRUs have been developed to alleviate these issues, more research is needed to develop new architectures and techniques that can further improve the training of RNNs for long sequences.

Incorporating Spatial Information in Recurrent Learning

Recurrent learning has been primarily focused on modeling temporal dependencies in sequential data. However, many real-world applications in urban studies and human geography involve both spatial and temporal information. Developing methods that can seamlessly integrate spatial information into recurrent learning is an important research direction. One potential approach is to combine RNNs with other neural network architectures, such as convolutional neural networks (CNNs), which have been proven to be effective in capturing spatial patterns.

Improving the Interpretability of Recurrent Learning Models

Despite their success in modeling and predicting sequential data, RNNs and their variants often suffer from a lack of interpretability. The complex nature of these models makes it difficult to understand how they arrive at their predictions. This lack of transparency can hinder the adoption of recurrent learning models in decision-making processes, particularly in fields like urban planning and human geography, where explainability is crucial. Developing methods that can improve the interpretability of recurrent learning models is an important research direction.

Leveraging Multi-Modal and Multi-Source Data

In many applications in urban studies and human geography, the data available for analysis come from various sources and in different formats. For example, data can include satellite images, social media data, and census data, each of which provides unique insights into the studied phenomena. Developing methods to effectively integrate and leverage multi-modal and multi-source data in recurrent learning models is a promising research direction that can lead to more accurate and comprehensive analyses.

Addressing Ethical and Privacy Concerns

The use of recurrent learning models in urban studies and human geography raises ethical and privacy concerns, particularly when dealing with sensitive data, such as information about individuals or specific locations. Ensuring that recurrent learning models adhere to privacy regulations and ethical guidelines is a critical research direction. Techniques such as differential privacy, federated learning, and secure multi-party computation can be employed to protect sensitive information while still allowing for effective model training and analysis.

Scalability and Efficiency in Recurrent Learning

Recurrent learning models can be computationally intensive, particularly when dealing with large-scale datasets or complex model architectures. Developing techniques to improve the scalability and efficiency of recurrent learning models is an important research direction. This could involve parallel and distributed computing approaches, model pruning and compression techniques, and hardware acceleration.

Incorporating Domain Knowledge in Recurrent Learning

Incorporating domain knowledge from urban studies and human geography can help guide the learning process of recurrent models and improve their generalizability and accuracy. Developing methods to effectively integrate domain knowledge into recurrent learning models, such as using expert knowledge to design model architectures or regularizing the model with domain-specific constraints, is a promising research direction.

Evaluating and Benchmarking Recurrent Learning Models

In order to advance the field of recurrent learning, it is essential to develop standardized benchmarks and evaluation metrics for comparing the performance of different recurrent learning models. This can help identify the strengths and weaknesses of different models and guide future research efforts. Creating comprehensive datasets and evaluation frameworks that capture the complexities of urban studies and human geography applications is an important research direction.

In conclusion, recurrent learning has shown great potential in addressing complex problems in urban studies and human geography. However, there are still several challenges and future directions that warrant further investigation. By addressing these challenges and exploring new research directions, recurrent learning can continue to revolutionize the fields of urban studies and human geography, leading to more accurate, efficient, and interpretable models.

References

1. Abdel-Aty, M., Radwan, A. E., & Lee, J. (1997). Artificial neural networks and logit models for traffic accident modeling and prediction. *Transportation Research Record, 1591*(1), 25–34.
2. Abdi, H., & Williams, L. J. (2010). Principal component analysis. *Wiley Interdisciplinary Reviews: Computational Statistics, 2*(4), 433–459.
3. Agresti, A. (2002). *Categorical data analysis*. Wiley-Interscience.
4. Altman, N. S. (1992). An introduction to kernel and nearest-neighbor nonparametric regression. *The American Statistician, 46*(3), 175–185.
5. Anselin, L. (1988). *Spatial econometrics: Methods and models*. Kluwer Academic Publishers.
6. Arjovsky, M., Chintala, S., & Bottou, L. (2017). Wasserstein GAN. arXiv:1701.07875
7. Arthur, D., & Vassilvitskii, S. (2007). k-means++: The advantages of careful seeding. In *Proceedings of the Eighteenth Annual ACM-SIAM Symposium on Discrete Algorithms* (pp. 1027–1035).
8. Audebert, N., Le Saux, B., & Lefèvre, S. (2018). Deep learning for classification of hyperspectral data: A comparative review. *IEEE Geoscience and Remote Sensing Magazine, 7*(2), 159–173.
9. Ayalew, L., & Yamagishi, H. (2005). The application of GIS-based logistic regression for landslide susceptibility mapping in the Kakuda-Yahiko Mountains. *Central Japan. Geomorphology, 65*(1–2), 15–31.
10. Ba, J. L., Kiros, J. R., & Hinton, G. E. (2016). Layer normalization. arXiv:1607.06450
11. Badoe, D. A., & Miller, E. J. (2000). Transportation-land-use interaction: Empirical findings in North America, and their implications for modeling. *Transportation Research Part D: Transport and Environment, 5*(4), 235–263.
12. Bahdanau, D., Cho, K., & Bengio, Y. (2014). Neural machine translation by jointly learning to align and translate. arXiv:1409.0473
13. Barro, R. J., & Sala-i-Martin, X. (1995). *Economic growth*. McGraw-Hill.
14. Ben-Akiva, M., & Lerman, S. R. (1985). *Discrete choice analysis: Theory and application to travel demand*. MIT Press.
15. Bengio, Y., Simard, P., & Frasconi, A. (1994). Learning long-term dependencies with gradient descent is difficult. *IEEE Transactions on Neural Networks, 5*(2), 157–166.
16. Bishop, C. M. (1995). *Neural networks for pattern recognition*. Oxford University Press.
17. Bishop, C. M. (2006). *Pattern recognition and machine learning*. Springer.
18. Boarnet, M. G., & Crane, R. (2001). The influence of land use on travel behavior: Specification and estimation strategies. *Transportation Research Part A: Policy and Practice, 35*(9), 823–845.
19. Bourassa, S. C., Hoesli, M., & Peng, V. S. (2003). Do housing submarkets really matter? *Journal of Housing Economics, 12*(1), 12–28.
20. Breiman, L. (1996). Bagging predictors. *Machine Learning, 24*(2), 123–140.
21. Breiman, L. (2001). Random forests. *Machine Learning, 45*(1), 5–32.

22. Breiman, L., Friedman, J., Olshen, R. A., & Stone, C. J. (1984). *Classification and regression trees*. CRC Press.
23. Cangelosi, R., & Goriely, A. (2007). Component retention in principal component analysis with application to cDNA microarray data. *Biology Direct, 2*(1), 2.
24. Castelluccio, M., Poggi, G., Sansone, C., & Verdoliva, L. (2015). Land use classification in remote sensing images by convolutional neural networks. arXiv:1508.00092
25. Cha, S. H. (2007). Comprehensive survey on distance/similarity measures between probability density functions. *International Journal of Mathematical Models and Methods in Applied Sciences, 1*(4), 300–307.
26. Chainey, S., Tompson, L., & Uhlig, S. (2008). The utility of hotspot mapping for predicting spatial patterns of crime. *Security Journal, 21*(1–2), 4–28.
27. Chandra, A., Bhattacharya, A., & Ghosh, S. K. (2018). Spatio-temporal urban growth modelling using deep GRU-LSTM network. In *Proceedings of the 1st ACM SIGSPATIAL Workshop on Prediction of Human Mobility* (pp. 1–4)
28. Chen, H., Teng, Y., Lu, S., Wang, Y., & Wang, J. (2010). Contamination characteristics and possible sources of PM10 and PM2.5 in different functional areas of Shanghai, China. *Atmospheric Environment, 44*(12), 1539–1547.
29. Chen, J., Chen, J., Liao, A., Cao, X., Chen, L., Chen, X., Lu, N., et al. (2018). Deep learning-based classification of hyperspectral data. *IEEE Journal of Selected Topics in Applied Earth Observations and Remote Sensing, 7*(6), 2094–2107.
30. Chen, J., Gomes, H. M., & Li, B. (2017). Outlier detection with autoencoder ensembles. *Journal of Data Science, 15*(4), 601–623.
31. Chen, J., Li, L., Guestrin, C., & Gu, T. (2018). On the interpretable of deep learning models for time series prediction. arXiv:1809.04556
32. Chen, K., Li, W., & Li, Z. (2009). A review of applications of linear regression model in air quality assessment. *Research of Environmental Sciences, 22*(2), 189–196.
33. Chen, X. (2018). A systematic comparison of spatiotemporal clustering methods: A case study of residential burglaries. *Computers, Environment and Urban Systems, 72*, 73–84.
34. Chen, X., & Cheng, L. (2018). Understanding spatiotemporal patterns of human activities in urban areas using geolocated tweets and deep learning. *Computers, Environment and Urban Systems, 72*, 7–18.
35. Chen, Y., Cheng, Y., & Li, Z. (2017). Mining the most influential k-location set from massive trajectories by a GPU-accelerated algorithm. *International Journal of Geographical Information Science, 31*(2), 417–438.
36. Cho, K., van Merrienboer, B., Gulcehre, C., Bahdanau, D., Bougares, F., Schwenk, H., & Bengio, Y. (2014). Learning phrase representations using RNN encoder-decoder for statistical machine translation. arXiv:1406.1078
37. Chung, J., Gulcehre, C., Cho, K., & Bengio, Y. (2014). Empirical evaluation of gated recurrent neural networks on sequence modeling. arXiv:1412.3555
38. Cortes, C., & Vapnik, V. (1995). Support-vector networks. *Machine Learning, 20*(3), 273–297.
39. Cover, T., & Hart, P. (1967). Nearest neighbor pattern classification. *IEEE Transactions on Information Theory, 13*(1), 21–27.
40. Dai, J., Wang, C., & Zhang, H. M. (2020). Real-time traffic prediction based on spatiotemporal convolutional neural network. *Transportation Research Part C: Emerging Technologies, 110*, 1–16.
41. Dancker, J. (2022). A brief introduction to recurrent neural networks. https://towardsdatascience.com/a-brief-introduction-to-recurrent-neural-networks-638f64a61ff4
42. Dietterich, T. G. (2000). Ensemble methods in machine learning. In J. Kittler & F. Roli (Eds.), *Multiple classifier systems: First international workshop* (pp. 1–15). Springer.
43. Drucker, H., Burges, C. J., Kaufman, L., Smola, A., & Vapnik, V. (1997). Support vector regression machines. *Advances in Neural Information Processing Systems, 9*, 155–161.
44. Duchi, J., Hazan, E., & Singer, Y. (2011). Adaptive subgradient methods for online learning and stochastic optimization. *Journal of Machine Learning Research, 12*, 2121–2159.
45. Elman, J. L. (1990). Finding structure in time. *Cognitive Science, 14*(2), 179–211.

46. Ester, M., Kriegel, H. P., Sander, J., & Xu, X. (1996). A density-based algorithm for discovering clusters in large spatial databases with noise. In *KDD* (Vol. 96, pp. 226–231).
47. Fawcett, T. (2006). An introduction to ROC analysis. *Pattern Recognition Letters, 27*(8), 861–874.
48. Ferreira, L., Chawla, N. V., & Karagiannis, G. (2019). Reinforcement learning for effective disaster management. *Proceedings of the AAAI Conference on Artificial Intelligence, 33*, 9722–9729.
49. Foody, G. M. (2002). Status of land cover classification accuracy assessment. *Remote Sensing of Environment, 80*(1), 185–201.
50. Freund, Y., & Schapire, R. E. (1997). A decision-theoretic generalization of on-line learning and an application to boosting. *Journal of Computer and System Sciences, 55*(1), 119–139.
51. Friedl, M. A., & Brodley, C. E. (1997). Decision tree classification of land cover from remotely sensed data. *Remote Sensing of Environment, 61*(3), 399–409.
52. Friedman, J., Hastie, T., & Tibshirani, R. (2001). *The elements of statistical learning*. Springer.
53. Friedman, J., Hastie, T., & Tibshirani, R. (2010). Regularization paths for generalized linear models via coordinate descent. *Journal of Statistical Software, 33*(1), 1–22.
54. Fu, R., Zhang, Z., & Li, L. (2016). Using LSTM and GRU neural network methods for traffic flow prediction. In *2016 31st Youth Academic Annual Conference of Chinese Association of Automation (YAC)* (pp. 324–328). IEEE.
55. Gao, S., Rao, J., Kang, Y., Liang, Y., & Kruse, J. (2018). Mapping fine-scale population distributions at the building level by integrating multisource geospatial big data. *International Journal of Geographical Information Science, 32*(9), 1848–1869.
56. Gaughan, A. E., Stevens, F. R., Linard, C., Jia, P., & Tatem, A. J. (2016). High-resolution population distribution maps for Southeast Asia in 2010 and 2015. *PLoS ONE, 11*(2), e0148618.
57. Gaughan, A. E., Stevens, F. R., Linard, C., Patel, N. N., & Tatem, A. J. (2015). Exploring nationally and regionally defined models for large area population mapping. *International Journal of Digital Earth, 8*(10), 989–1006.
58. Gers, F. A., Schmidhuber, J., & Cummins, F. (2000). Learning to forget: Continual prediction with LSTM. *Neural Computation, 12*(10), 2451–2471.
59. Gers, F. A., Schmidhuber, J., & Cummins, F. (2002). Learning to forget: Continual prediction with LSTM. *Neural computation, 12*(10), 2451–2471.
60. Ghosh, A., & Chakraborty, S. (2017). A survey on land use and land cover mapping using k-nearest neighbor classification technique. In *2017 IEEE Calcutta Conference (CALCON)* (pp. 120–125). IEEE.
61. Gilpin, L. H., Bau, D., Yuan, B. Z., Bajwa, A., Specter, M., & Kagal, L. (2018). Explaining explanations: An approach to evaluating interpretability of machine learning. arXiv:1806.00069
62. Glorot, X., & Bengio, Y. (2010). Understanding the difficulty of training deep feedforward neural networks. In *Proceedings of the Thirteenth International Conference on Artificial Intelligence and Statistics* (pp. 249–256).
63. Glorot, X., Bordes, A., & Bengio, Y. (2011). Deep sparse rectifier neural networks. In *Proceedings of the 14th International Conference on Artificial Intelligence and Statistics (AISTATS)*.
64. Goodfellow, I., Bengio, Y., & Courville, A. (2016). *Deep learning*. MIT Press.
65. Goodfellow, I., Pouget-Abadie, J., Mirza, M., Xu, B., Warde-Farley, D., Ozair, S., Courville, A., & Bengio, Y. (2014). Generative adversarial nets. In *Advances in neural information processing systems* (pp. 2672–2680).
66. Graif, C., Gladfelter, A. S., & Matthews, S. A. (2014). Urban poverty and neighborhood effects on crime: Incorporating spatial and network perspectives. *Sociology Compass, 8*(9), 1140–1155.
67. Guidotti, R., Monreale, A., Ruggieri, S., Turini, F., & Giannotti, F. (2018). A survey of methods for explaining black box models. *ACM Computing Surveys, 51*(5), 93.

68. Guo, D., & Wang, H. (2011). The Chinese Hukou system at 50. *Eurasian Geography and Economics, 52*(2), 250–276.
69. Guo, P., Wang, Y., Chen, J., & Liu, Y. (2011). SVM-based model for predicting Hualian outbreaks. *Computers, Environment and Urban Systems, 35*(5), 376–384.
70. Hastie, T., Tibshirani, R., & Friedman, J. (2009). *The elements of statistical learning: Data mining, inference, and prediction*. Springer.
71. Haykin, S. (2009). *Neural networks and learning machines*. Pearson Education.
72. He, K., Zhang, X., Ren, S., & Sun, J. (2016). Deep residual learning for image recognition. In *Proceedings of the IEEE Conference on Computer Vision and Pattern Recognition (CVPR)*.
73. Hebb, D. O. (1949). *The organization of behavior: A neuropsychological theory*. Wiley.
74. Hinton, G. E., & Salakhutdinov, R. R. (2006). Reducing the dimensionality of data with neural networks. *Science, 313*(5786), 504–507.
75. Hochreiter, S., & Schmidhuber, J. (1997). Long short-term memory. *Neural Computation, 9*(8), 1735–1780.
76. Hoerl, A. E., & Kennard, R. W. (1970). Ridge regression: Biased estimation for nonorthogonal problems. *Technometrics, 12*(1), 55–67.
77. Hosmer, D. W., Lemeshow, S., & Sturdivant, R. X. (2013). *Applied logistic regression*. Wiley.
78. Hsu, C. W., Chang, C. C., & Lin, C. J. (2003). *A practical guide to support vector classification*. Technical Report, Department of Computer Science and Information Engineering, National Taiwan University.
79. Huang, X., Davis, L. S., & Townshend, J. R. (2002). An assessment of support vector machines for land cover classification. *International Journal of Remote Sensing, 23*(4), 725–749.
80. Huang, X., Lu, L., & Zhang, L. (2002). A multilevel recursive partitioning algorithm for the classification of mixed land-use and land-cover data. *International Journal of Remote Sensing, 23*(17), 3429–3442.
81. Huang, Y., & Lees, B. G. (2004). Combining support vector machines with a GIS grade-of-membership analysis for mapping hardwood mortality in areas affected by 'sudden oak death.' *Photogrammetric Engineering & Remote Sensing, 70*(11), 1299–1305.
82. Huang, Z. (1998). Extensions to the k-means algorithm for clustering large data sets with categorical values. *Data Mining and Knowledge Discovery, 2*(3), 283–304.
83. Hyndman, R. J., & Koehler, A. B. (2006). Another look at measures of forecast accuracy. *International Journal of Forecasting, 22*(4), 679–688.
84. Ioffe, S., & Szegedy, C. (2015). Batch normalization: Accelerating deep network training by reducing internal covariate shift. In *Proceedings of the 32nd International Conference on Machine Learning (ICML)*.
85. Isikdogan, F., Bovik, A. C., & Passalacqua, P. (2019). R3-Net: A deep network for multiscale and hierarchical representation of vegetation. *Remote Sensing, 11*(21), 2488.
86. Jain, A. K. (2010). Data clustering: 50 years beyond K-means. *Pattern Recognition Letters, 31*(8), 651–666.
87. Jain, A. K., Duin, R. P. W., & Mao, J. (2000). Statistical pattern recognition: A review. *IEEE Transactions on Pattern Analysis and Machine Intelligence, 22*(1), 4–37.
88. James, G., Witten, D., Hastie, T., & Tibshirani, R. (2013). *An introduction to statistical learning: With Applications in R*. Springer.
89. Jantz, C. A., Goetz, S. J., & Vowinkel, E. F. (2003). Landscape structure and the spread of the exotic shrub Lonicera maackii (Amur honeysuckle) in South Central Pennsylvania. *Landscape Ecology, 18*(4), 377–391.
90. Jean, N., Burke, M., Xie, M., Davis, W. M., Lobell, D. B., & Ermon, S. (2016). Combining satellite imagery and machine learning to predict poverty. *Science, 353*(6301), 790–794.
91. Jiang, H., Qian, C., Ma, Y., Yang, J., Ma, W. Y., & Wang, J. (2019). Urban-BERT: augmenting the BERT model for urban perception. In *Proceedings of the 27th ACM International Conference on Multimedia* (pp. 1417–1425).
92. Jolliffe, I. T. (2002). *Principal component analysis*. Springer.
93. Jolliffe, I. T., & Cadima, J. (2016). Principal component analysis: A review and recent developments. *Philosophical Transactions of the Royal Society A: Mathematical, Physical and Engineering Sciences, 374*(2065), 20150202.

94. Kang, H., Park, D., & Kim, Y. (2018). A study on the prediction of crime occurrence based on the spatio-temporal pattern analysis of social media. *International Journal of Applied Engineering Research, 13*(7), 4839–4844.
95. Kim, Y., Denton, C., Hoang, L., & Rush, A. M. (2016). Structured attention networks. arXiv: 1702.00887
96. Kingma, D. P., & Ba, J. (2014). Adam: A method for stochastic optimization. arXiv:1412. 6980
97. Kingma, D. P., & Welling, M. (2013). Auto-encoding variational Bayes. arXiv:1312.6114
98. Kohavi, R. (1995). A study of cross-validation and bootstrap for accuracy estimation and model selection. In *Proceedings of the 14th International Joint Conference on Artificial Intelligence* (pp. 1137–1143). Morgan Kaufmann.
99. Koshimura, S., Moya, L., Mas, E., & Bai, Y. (2020). Tsunami damage detection with remote sensing: A review. *Geosciences, 10*(5), 177. https://www.mdpi.com/2076-3263/10/5/177
100. Kotsiantis, S. B., Zaharakis, I. D., & Pintelas, P. E. (2006). Machine learning: A review of classification and combining techniques. *Artificial Intelligence Review, 26*(3), 159–190.
101. Krizhevsky, A., Sutskever, I., & Hinton, G. E. (2012). ImageNet classification with deep convolutional neural networks. In F. Pereira, C. J. C. Burges, L. Bottou, & K. Q. Weinberger (Eds.), *Advances in neural information processing systems 25* (pp. 1097–1105). Curran Associates Inc.
102. Kryvasheyeu, Y., Chen, H., Moro, E., Van Hentenryck, P., & Cebrian, M. (2016). Performance of social network sensors during Hurricane Sandy. *PLoS ONE, 11*(2), e0147172.
103. Kumar, A., Singh, A. K., & Saksena, S. (2011). Clustering of aerosol optical depth and aerosol index in Delhi. *Environmental Monitoring and Assessment, 180*(1–4), 79–89.
104. Le, X.-H., Ho, H. V., Lee, G., & Jung, S. (2019). Application of long short-term memory (LSTM) neural network for flood forecasting. *Water, 11*(7), 1387. https://www.mdpi.com/2073-4441/11/7/1387
105. LeCun, Y., Bengio, Y., & Hinton, G. (2015). Deep learning. *Nature, 521*(7553), 436–444.
106. LeCun, Y., Bottou, L., Bengio, Y., & Haffner, P. (1998). Gradient-based learning applied to document recognition. *Proceedings of the IEEE, 86*(11), 2278–2324.
107. LeSage, J., & Pace, R. K. (2009). *Introduction to spatial econometrics.* CRC Press/Taylor & Francis Group.
108. Li, H., Lan, H., Wu, Y., Liu, Z., & Luo, X. (2019). FusionGAN: A generative adversarial network for infrared and visible image fusion. *Information Fusion, 48*, 11–26.
109. Li, X., Peng, L., Yao, X., Cui, S., Hu, Y., & You, C. (2018). Long short-term memory neural network for air pollutant concentration predictions: Method development and evaluation. *Environmental Pollution, 231*, 997–1004.
110. Li, X., Peng, L., Yao, X., Cui, S., Hu, Y., You, C., & Chi, T. (2017). Long short-term memory neural network for air pollutant concentration predictions: Method development and evaluation. *Environmental Pollution, 231*, 997–1004.
111. Li, X., Zhang, Y., Ye, X., Zhang, Q., Gao, J., & Liu, D. (2018). A review of remote sensing for urban growth and sustainable development. *Journal of Remote Sensing, 9*(5), 430.
112. Li, Y., Rao, J., Peethamparan, S., Kong, X., & Hong, Y. (2020). Deep learning-based rapid flood inundation mapping using multi-source remote sensing data. *Remote Sensing of Environment, 245*, 111848.
113. Li, Y., Wang, F., Liu, Y., & Zhang, X. (2018). A k-means-based approach to delineating traffic analysis zones using mobile phone data. *Transportation Research Part C: Emerging Technologies, 92*, 33–49.
114. Lin, Y. P., Hong, N. M., Wu, P. J., Wu, C. F., & Verburg, P. H. (2011). Impacts of land use change scenarios on hydrology and land use patterns in the Wu-Tu watershed in Northern Taiwan. *Hydrology and Earth System Sciences, 15*(5), 1597–1609.
115. Liu, Y., Huang, X., & Wang, L. (2018). Predicting urban land use change using a machine learning algorithm combined with a land use change model. *Landscape and Urban Planning, 178*, 8–18.

116. Liu, Y., Qin, H., Zhang, Z., & Zhang, Z. (2018). Multi-scale GANs for memory-efficient generation of high-resolution remote sensing images. *Remote Sensing, 10*(12), 1947.
117. Liu, Y., Tang, J., & Zhang, Y. (2019). Reinforcement learning in adaptive dynamic programming for urban rail transit scheduling problem. *Transportation Research Part C: Emerging Technologies, 98*, 169–189.
118. Lloyd, S. P. (1982). Least squares quantization in PCM. *IEEE Transactions on Information Theory, 28*(2), 129–137.
119. Long, J. S. (1997). *Regression models for categorical and limited dependent variables*. Sage Publications.
120. Lv, Y., Duan, Y., Kang, W., Li, Z., & Wang, F. Y. (2015). Traffic flow prediction with big data: A deep learning approach. *IEEE Transactions on Intelligent Transportation Systems, 16*(2), 865–873.
121. van der Maaten, L., & Hinton, G. (2008). Visualizing data using t-SNE. *Journal of Machine Learning Research, 9*, 2579–2605.
122. Mannion, P., Duggan, J., & Howley, E. (2016). An experimental review of reinforcement learning algorithms for adaptive traffic signal control. In *Autonomic road transport support systems* (pp. 47–66). Springer.
123. Márquez-Neila, P., Baumgartner, M., & Tuia, D. (2018). A deep learning-based approach for the automatic generation of land cover maps from satellite imagery. *International Journal of Applied Earth Observation and Geoinformation, 72*, 1–10.
124. McCorduck, P. (2004). *Machines who think: A personal inquiry into the history and prospects of artificial intelligence*. A K Peters/CRC Press.
125. McCullagh, P., & Nelder, J. A. (1989). *Generalized linear models*. Chapman and Hall/CRC.
126. McCulloch, W. S., & Pitts, W. (1943). A logical calculus of the ideas immanent in nervous activity. *The Bulletin of Mathematical Biophysics, 5*(4), 115–133.
127. McLachlan, G., & Peel, D. (2000). *Finite mixture models*. Wiley.
128. Melcher, K. (2021). A friendly introduction to [deep] neural networks. https://www.knime.com/blog/a-friendly-introduction-to-deep-neural-networks
129. Mitchell, T. M. (1997). *Machine learning*. McGraw-Hill.
130. Miyato, T., Kataoka, T., Koyama, M., & Yoshida, Y. (2018). Spectral normalization for generative adversarial networks. arXiv:1802.05957
131. Molnar, C. (2020). *Interpretable machine learning*. Lulu.com.
132. Mountrakis, G., Im, J., & Ogole, C. (2011). Support vector machines in remote sensing: A review. *ISPRS Journal of Photogrammetry and Remote Sensing, 66*(3), 247–259.
133. Muja, M., & Lowe, D. G. (2009). Fast approximate nearest neighbors with automatic algorithm configuration. In *International Conference on Computer Vision Theory and Applications* (pp. 331–340). Springer.
134. Mullainathan, S., & Spiess, J. (2017). Machine learning: an applied econometric approach. *Journal of Economic Perspectives, 31*(2), 87–106.
135. Nair, V., & Hinton, G. E. (2010). Rectified linear units improve restricted Boltzmann machines. In *Proceedings of the 27th International Conference on Machine Learning (ICML)*.
136. Nesterov, Y. (1983). A method for unconstrained convex minimization problem with the rate of convergence O (1/k^2). *Doklady AN USSR, 269*(3), 543–547.
137. Newell, A., & Simon, H. A. (1963). GPS, a program that simulates human thought. In E. A. Feigenbaum & J. Feldman (Eds.), *Computers and thought* (pp. 279–293). McGraw-Hill.
138. Ng, A. Y. (2004). Feature selection, L1 vs. L2 regularization, and rotational invariance. In *Proceedings of the Twenty-First International Conference on Machine Learning* (p. 78).
139. Nguyen, D. T., Alam, F., Ofli, F., & Imran, M. (2017). Automatic image filtering on social networks using deep learning and perceptual hashing during crises. In *Proceedings of the 14th International Conference on Information Systems for Crisis Response and Management* (pp. 181–189).
140. Nilsson, N. J. (1984). *Shakey the robot*. Technical Note 323, SRI International.
141. Niu, R., Zhou, Y., Zhu, X., Wang, S., & Gao, S. (2019). Spatiotemporal prediction of continuous daily PM2.5 concentrations across China using a spatially explicit machine learning algorithm. *Atmospheric Environment, 199*, 412–422.

142. Noble, M., Wright, G., Smith, G., & Dibben, C. (2006). Measuring multiple deprivation at the small-area level. *Environment and Planning A, 38*(1), 169–185.
143. Odena, A. (2016). Semi-supervised learning with generative adversarial networks. arXiv: 1606.01583
144. Pan, S. J., & Yang, Q. (2010). A survey on transfer learning. *IEEE Transactions on Knowledge and Data Engineering, 22*(10), 1345–1359.
145. Pascanu, R., Mikolov, T., & Bengio, Y. (2013). On the difficulty of training recurrent neural networks. In *International Conference on Machine Learning* (pp. 1310–1318).
146. Polyak, B. T. (1964). Some methods of speeding up the convergence of iteration methods. *USSR Computational Mathematics and Mathematical Physics, 4*(5), 1–17.
147. Pontius, R. G., & Schneider, L. C. (2001). Land-use change model validation by an ROC method for the Ipswich watershed, Massachusetts, USA. *Agriculture, Ecosystems & Environment, 85*(1–3), 239–248.
148. Poursaeed, O., Matera, T., & Belongie, S. (2018). Vision-based real estate price estimation. In *Proceedings of the IEEE Conference on Computer Vision and Pattern Recognition* (pp. 10550–10558).
149. Pradhan, B. (2013). A comparative study on the predictive ability of the decision tree, support vector machine and neuro-fuzzy models in landslide susceptibility mapping using GIS. *Computers & Geosciences, 51*, 350–365.
150. Quinlan, J. R. (1986). Induction of decision trees. *Machine Learning, 1*(1), 81–106.
151. Quinlan, J. R. (1993). *C4.5: Programs for machine learning.* Morgan Kaufmann Publishers.
152. Radford, A., Narasimhan, K., Salimans, T., & Sutskever, I. (2019). *Improving language understanding by generative pre-training.*
153. Raphael, B. (1985). *The thinking computer: Mind inside matter.* W.H. Freeman & Co Ltd.
154. Reardon, S. F., & Bischoff, K. (2011). Income inequality and income segregation. *American Journal of Sociology, 116*(4), 1092–1153.
155. Reardon, S. F., & O'Sullivan, D. (2004). Measures of spatial segregation. *Sociological Methodology, 34*(1), 121–162.
156. Reed, R., & Marks, R. J. (1999). *Neural Smithing: Supervised learning in feedforward artificial neural networks.* MIT Press.
157. Rezende, D. J., Mohamed, S., & Wierstra, D. (2014). Stochastic backpropagation and approximate inference in deep generative models. arXiv:1401.4082
158. Robbins, H., & Monro, S. (1951). A stochastic approximation method. *The Annals of Mathematical Statistics, 22*(3), 400–407.
159. Rodriguez-Galiano, V. F., Sanchez-Castillo, M., Chica-Olmo, M., & Chica-Rivas, M. (2015). Machine learning predictive models for mineral prospectivity: An evaluation of neural networks, random forest, regression trees and support vector machines. *Ore Geology Reviews, 71*, 804–818.
160. Rousseeuw, P. J., & Hubert, M. (2018). Robust statistics for outlier detection. *Wiley Interdisciplinary Reviews: Data Mining and Knowledge Discovery, 8*(3), e1236.
161. Rumelhart, D. E., Hinton, G. E., & Williams, R. J. (1986). Learning representations by back-propagating errors. *Nature, 323*(6088), 533–536.
162. Rußwurm, M., & Körner, M. (2017). Temporal vegetation modelling using long short-term memory networks for crop identification from medium-resolution multi-spectral satellite images. In *IGARSS 2017–2017 IEEE International Geoscience and Remote Sensing Symposium* (pp. 3991–3994). IEEE.
163. Salimans, T., Goodfellow, I., Zaremba, W., Cheung, V., Radford, A., & Chen, X. (2016). Improved techniques for training GANs. In *Advances in neural information processing systems* (pp. 2234–2242).
164. Samuel, A. L. (1959). Some studies in machine learning using the game of checkers. *IBM Journal of Research and Development, 3*(3), 210–229.
165. Scherer, D., Müller, A., & Behnke, S. (2010). Evaluation of pooling operations in convolutional architectures for object recognition. In *Proceedings of the International Conference on Artificial Neural Networks* (pp. 92–101). Springer.

166. Schölkopf, B., & Smola, A. J. (2001). *Learning with kernels: Support vector machines, regularization, optimization, and beyond*. MIT Press.
167. Schuster, M., & Paliwal, K. K. (1997). Bidirectional recurrent neural networks. *IEEE Transactions on Signal Processing, 45*(11), 2673–2681.
168. Seto, K. C., & Kaufmann, R. K. (2003). Modeling the drivers of urban land use change in the Pearl River Delta, China: Integrating remote sensing with socioeconomic data. *Land Economics, 79*(1), 106–121.
169. Shortliffe, E. H., & Buchanan, B. G. (1975). A model of inexact reasoning in medicine. *Mathematical Biosciences, 23*(3–4), 351–379.
170. Silver, D., Huang, A., Maddison, C. J., Guez, A., Sifre, L., van den Driessche, G., Hassabis, D., et al. (2016). Mastering the game of Go with deep neural networks and tree search. *Nature, 529*(7587), 484–489.
171. Simonyan, K., & Zisserman, A. (2014). Very deep convolutional networks for large-scale image recognition. arXiv:1409.1556
172. Solaimani, K., Yin, J., & Wang, D. (2019). A deep learning-based approach for predicting ambient ozone concentration. *Environmental Modelling & Software, 119*, 215–224.
173. Song, X., Zhang, Q., Sekimoto, Y., & Shibasaki, R. (2016). Prediction of human emergency behavior and their mobility following large-scale disaster. *Proceedings of the National Academy of Sciences, 113*(42), 11788–11793.
174. Srivastava, N., Hinton, G., Krizhevsky, A., Sutskever, I., & Salakhutdinov, R. (2014). Dropout: A simple way to prevent neural networks from overfitting. *Journal of Machine Learning Research, 15*(1), 1929–1958.
175. Sutton, R. S., & Barto, A. G. (2018). *Reinforcement learning: An introduction*. MIT Press.
176. Tatem, A. J., Noor, A. M., Hay, S. I., & Snow, R. W. (2007). Assessing the accuracy of satellite derived global and national urban maps in Kenya. *Remote Sensing of Environment, 108*(1), 133–141.
177. Tayyebi, A., Jenerette, G. D., & Buyantuyev, A. (2014). Urbanization, environmental justice, and sustainable land use: Assessing the transition in Phoenix using a decision tree model. *Applied Geography, 47*, 138–147.
178. Tayyebi, A., Pijanowski, B. C., & Tayyebi, A. H. (2014). An urban growth boundary model using neural networks, GIS, and radial parameterization: An application to Tehran, Iran. *Landscape and Urban Planning, 100*(1–2), 35–44.
179. Tian, J., Xie, H., Yang, Z., & Zhang, W. (2020). A deep learning approach for spatial–temporal modeling of urban growth: A case study of Dallas-Fort Worth, Texas. *Computers, Environment and Urban Systems, 80*, 101441.
180. Tibshirani, R. (1996). Regression shrinkage and selection via the Lasso. *Journal of the Royal Statistical Society: Series B (Methodological), 58*(1), 267–288.
181. Tibshirani, R., Walther, G., & Hastie, T. (2001). Estimating the number of clusters in a data set via the gap statistic. *Journal of the Royal Statistical Society: Series B (Statistical Methodology), 63*(2), 411–423.
182. Tieleman, T., & Hinton, G. (2012). Lecture 6.5—RmsProp: Divide the gradient by a running average of its recent magnitude. *COURSERA: Neural Networks for Machine Learning, 4*(2), 26–31.
183. Torrens, P. M. (2010). Simulating sprawl. *Annals of the Association of American Geographers, 100*(2), 371–385.
184. Tsai, Y. H. (2005). Quantifying urban form: Compactness versus sprawl. *Urban Studies, 42*(1), 141–161.
185. Turing, A. M. (1936). On computable numbers, with an application to the Entscheidungsproblem. *Proceedings of the London Mathematical Society, 2*(1), 230–265.
186. Turing, A. M. (1950). Computing machinery and intelligence. *Mind, 59*(236), 433–460.
187. Vapnik, V. N. (1995). *The nature of statistical learning theory*. Springer-Verlag.
188. Volpi, M., & Camps-Valls, G. (2018). Spectral-spatial generative adversarial networks for hyperspectral urban land cover classification. In *IGARSS 2018–2018 IEEE International Geoscience and Remote Sensing Symposium* (pp. 8142–8145). IEEE.

189. Volpi, M., Tuia, D., & Camps-Valls, G. (2020). Deep generative models for generating and augmenting geospatial data. In *Deep learning for remote sensing data* (pp. 61–76). Springer.
190. Wakefield, J., & Elliott, P. (2003). Issues in the statistical analysis of small area health data. *Statistics in Medicine, 22*(17), 2791–2815.
191. Wang, D., Liang, F., Xu, Y., Li, Y., & Zhang, Q. (2016). Geo-temporal sentiment visualization of social media data based on dynamic spatial panel data model. In *2016 IEEE Second International Conference on Multimedia Big Data (BigMM)* (pp. 33–40). IEEE.
192. Wang, N., Zhang, N., Liang, J., & Hauptmann, A. G. (2018). Conditional generative adversarial networks for face generation. In *2018 IEEE International Conference on Multimedia and Expo (ICME)* (pp. 1–6). IEEE.
193. Wang, Q., Qiu, X., & Wu, X. (2019). Generative adversarial networks for satellite image data augmentation in urban land use and land cover classification. *Computers, Environment and Urban Systems, 75*, 122–133.
194. Wang, Y., Yao, Q., Kwok, J. T., & Ni, L. M. (2016). Generalizing from a few examples: A survey on few-shot learning. arXiv:1904.05046
195. Werbos, P. J. (1990). Backpropagation through time: What it does and how to do it. *Proceedings of the IEEE, 78*(10), 1550–1560.
196. Winograd, T. (1972). Understanding natural language. *Cognitive Psychology, 3*(1), 1–191.
197. Wold, S., Esbensen, K., & Geladi, P. (1987). Principal component analysis. *Chemometrics and Intelligent Laboratory Systems, 2*(1–3), 37–52.
198. Xie, J., Xu, L., & Chen, E. (2016). Image denoising and inpainting with deep neural networks. In *Advances in neural information processing systems* (pp. 341–349).
199. Xu, T., Zhang, J., Huang, T., Zhang, Y., & He, X. (2018). BeGAN: Boundary equilibrium generative adversarial networks. In *Proceedings of the IEEE Conference on Computer Vision and Pattern Recognition* (pp. 3700–3708).
200. Yao, H., Wu, F., Ke, J., Tang, X., Jia, Y., Lu, S., Ye, J., et al. (2018). Deep multi-view spatial-temporal network for taxi demand prediction. In *Thirty-Second AAAI Conference on Artificial Intelligence.*
201. Yeh, A. G. O., & Li, X. (2001). A constrained CA model for the simulation and planning of sustainable urban forms by using GIS. *Environment and Planning B: Planning and Design, 28*(5), 733–753.
202. Yilmaz, I. (2009). Landslide susceptibility mapping using frequency ratio, logistic regression, artificial neural networks and their comparison: A case study from Kat landslides (Tokat—Turkey). *Computers & Geosciences, 35*(6), 1125–1138.
203. Yosinski, J., Clune, J., Bengio, Y., & Lipson, H. (2014). How transferable are features in deep neural networks? In *Advances in neural information processing systems (NIPS).*
204. Yu, B., Sun, X., & Zhu, Z. (2015). A comparison study of five different methods for landslide susceptibility mapping in a mountainous catchment. *Environmental Earth Sciences, 74*(7), 5801–5816.
205. Yu, B., Yin, H., & Zhu, Z. (2017). Spatiotemporal recurrent convolutional networks for traffic prediction in transportation networks. *Sensors, 17*(7), 1501.
206. Yu, B., Yin, H., & Zhu, Z. (2017). Spatio-temporal graph convolutional networks: A deep learning framework for traffic forecasting. arXiv preprint arXiv:1709.04875
207. Yu, J., Gao, Z., Li, M., Yu, H., Wang, C., & Zhang, K. (2018). Simulating urban growth using the SLEUTH model in a coastal peri-urban district in China. *Sustainability, 10*(4), 990.
208. Yu, L., Wang, N., Lai, Y., & Shi, Z. (2015). Review of decision tree optimization. *Mathematical Problems in Engineering, 2015*, 1–11.
209. Zhang, A., Lipton, Z. C., Li, M., & Smola, A. J. (2018). *Dive into deep learning.* Cambridge University Press.
210. Zhang, H., Zhang, K., Sun, G., & Zhu, X. X. (2016). Object-based building extraction from high-resolution aerial imagery and LiDAR data using the CNN. *IEEE Journal of Selected Topics in Applied Earth Observations and Remote Sensing, 9*(12), 5518–5530.
211. Zhang, H., Zhang, Y., & Lin, H. (2017). A novel approach to extract urban land use information from high-resolution remote sensing imagery. *ISPRS Journal of Photogrammetry and Remote Sensing, 126*, 170–185.

212. Zhang, J., Zheng, Y., & Qi, D. (2017). Deep spatio-temporal residual networks for citywide crowd flows prediction. In *Proceedings of the Thirty-First AAAI Conference on Artificial Intelligence* (pp. 1655–1661).
213. Zhang, Y., Du, B., Zhang, L., Xu, S., & Zhang, L. (2016). A framework for urban land use classification in China based on remote sensing and GIS. *International Journal of Remote Sensing, 37*(11), 2513–2532.
214. Zhang, Z., Li, X., & Weng, Q. (2018). A support vector machine-based method to extract urban land using Landsat 8 and HJ-1A/B data. *Remote Sensing, 10*(2), 202.
215. Zhao, W., Ma, H., & He, Q. (2009). Parallel k-means clustering based on MapReduce. In *Cloud computing* (pp. 674–679). Springer.
216. Zheng, S., Wang, J., Sun, C., Zhang, X., & Kahn, M. E. (2014). Air quality in Lhasa, Tibet: Are we all in this together? *Urban Studies, 51*(6), 1306–1319.
217. Zheng, Y., Liu, Q., Chen, E., Ge, Y., & Zhao, J. L. (2015). Time series approaches for forecasting the number of Chinese outbound tourists. *Asia Pacific Journal of Tourism Research, 20*(4), 377–395.
218. Zhong, L., Hu, T., Gao, S., Li, H., & Gao, Y. (2018). LSTM-CNNs-based deep learning for spatiotemporal sequence data: A case study on long-term land use/land cover change prediction. *IEEE Access, 6*, 61976–61985.
219. Zhou, Y., Wang, L., & Kung, H. T. (2018). Urban change detection using deep learning: A survey. In *2018 IEEE International Conference on Multimedia & Expo Workshops (ICMEW)* (pp. 1–6). IEEE.
220. Zhu, J. Y., Park, T., Isola, P., & Efros, A. A. (2017). Unpaired image-to-image translation using cycle-consistent adversarial networks. In *Proceedings of the IEEE International Conference on Computer Vision* (pp. 2223–2232).

Chapter 3
Data Sources and Processing

3.1 Traditional Data Sources in Human Geography and Urban Planning

Human geography and urban planning have long relied on a variety of data sources to understand spatial patterns, analyze demographic trends, and inform policy decisions (Table 3.1). Traditional data sources have typically included censuses, surveys, land use maps, aerial photographs, and other types of geospatial data. In this section, we will discuss the role of these traditional data sources in human geography and urban planning and how they have evolved over time.

3.1.1 Census Data

Census data has been one of the most critical sources of information for human geography and urban planning for centuries. The census is a systematic, comprehensive, and periodic enumeration of the population, typically conducted by governments worldwide. Census data provides essential demographic, social, and economic information about the population, which is crucial for understanding and addressing various urban and regional issues. This section will explore the history and significance of census data, its various types and applications, and the challenges and opportunities it presents for human geography and urban planning.

The first known census dates back to ancient civilizations, such as the Babylonians and Egyptians, who used census data for taxation, military conscription, and other administrative purposes [137]. The modern census, as we know it today, began in the late 18th and early 19th centuries when countries like the United States and the United Kingdom started conducting regular population censuses [98]. Since then, the scope and scale of census data have grown, with many countries now collecting

S. N. M. Moghaddam and H. Cao, *Artificial Intelligence-Driven Geographies*, City Development: Issues and Best Practices, https://doi.org/10.1007/978-981-97-5116-7_3

Table 3.1 Traditional data sources in human geography and urban planning

Geospatial data type	Description	Methodologies	Applications	Advantages	Disadvantages	Future directions	Relevance
Census data	Provides demographic, social, and economic information about the population	Systematic enumeration of the population	Urban and regional issues such as demographic trends, social disparities, and economic conditions	Comprehensive data source	Infrequent and resource-intensive data collection	Integration with emerging data sources for enhanced understanding and addressing challenges	Valuable for understanding demographic trends and informing policy decisions
Surveys	Offer a direct method for gathering information from individuals or groups	Various types: cross-sectional, longitudinal, mail, telephone, face-to-face, online	Identifying community needs, evaluating programs and policies, monitoring trends and changes	Direct means of data collection	Nonresponse bias, social desirability bias, questionnaire design challenges	Advances in methodologies to enhance efficiency and accuracy	Crucial for understanding relationships between people, places, and environments

(continued)

Table 3.1 (continued)

Geospatial data type	Description	Methodologies	Applications	Advantages	Disadvantages	Future directions	Relevance
Land use maps	Depict the spatial distribution of land uses and land cover types	Manual interpretation of aerial photographs, field surveys, remote sensing combined with GIS, machine learning	Urban planning, environmental management, policy-making by identifying land use patterns, assessing impacts, informing development strategies	Detailed spatial information	Labor-intensive, requires expertise in remote sensing and GIS	Leveraging emerging technologies to improve accuracy and utility	Essential for informing urban planning and environmental management
Aerial photographs	Images captured from aircraft or airborne platforms, providing unique perspectives of the Earth's surface	Captured using cameras mounted on aircraft or drones	Land use mapping, urban growth monitoring, transportation planning, environmental assessment, disaster management, risk assessment	Unique perspectives, high-resolution imagery	Limitations in image quality, temporal coverage, data processing requirements	Developments in technologies such as drones and high-resolution sensors to overcome challenges and expand applications	Offer valuable insights into land use, urbanization, and environmental features

detailed information on various aspects of the population, such as age, sex, ethnicity, education, income, and employment status.

Census data has been a vital resource for human geography and urban planning, serving several purposes. First, it provides a detailed snapshot of the population at a given point in time, which is essential for understanding demographic trends and spatial patterns [119]. This information is critical for urban and regional planning, as it helps planners and policymakers allocate resources, design public services, and develop targeted interventions to address specific community needs [26].

Second, census data serves as the basis for creating spatially disaggregated data, such as population density and other socio-economic indicators, which are vital for understanding the spatial organization of urban areas and informing land use and transportation planning [70]. These data also enable the analysis of urban and regional processes, such as residential segregation, gentrification, and spatial inequality [111].

Third, census data is often used in combination with other data sources, such as remote sensing, GIS, and big data analytics, to develop sophisticated models and simulations for understanding urban and regional dynamics [57]. This integration of data sources has opened up new possibilities for analyzing complex urban and regional issues, such as the impacts of climate change, housing affordability, and socio-economic disparities [11].

Despite its many advantages, census data also presents several challenges for human geography and urban planning. One significant challenge is the infrequent nature of the census, which typically occurs every five or ten years [98]. This means that the data may not accurately reflect the rapid changes and dynamics occurring in urban areas, limiting its utility for real-time decision-making and planning. Additionally, the census can be costly and resource-intensive to conduct, particularly for low-income countries with limited capacity and infrastructure [137].

Another challenge is the potential for data quality and accuracy issues in census data, such as undercounting, non-response, and measurement errors [140]. These issues can have significant implications for the validity and reliability of census data, particularly when used for planning and policy-making purposes. Moreover, there are concerns related to the confidentiality and privacy of census data, as the increasing granularity of the data raises questions about the potential for identifying individuals and revealing sensitive information [42].

Census data has been a vital resource for human geography and urban planning, providing essential demographic, social, and economic information about the population and enabling a deeper understanding of urban and regional processes. Its applications span various domains, including land use and transportation planning, housing policy, social service provision, and environmental management. However, the infrequent nature of the census, data quality issues, and privacy concerns present significant challenges for researchers and practitioners alike.

Despite these challenges, census data remains a valuable resource for human geography and urban planning. To address some of these limitations, researchers and practitioners are increasingly exploring alternative and supplementary data sources, such as administrative records, geospatial data, and big data from social media and other digital platforms [130]. These new data sources, when combined with traditional census data, can provide a more comprehensive and timely understanding of urban and regional dynamics, allowing for more effective planning and decision-making.

Furthermore, advances in data processing techniques, such as machine learning and spatial analysis methods, have provided new opportunities to extract valuable insights from census data and enhance its utility for human geography and urban planning [11]. These advances have also facilitated the development of innovative approaches to addressing data quality and privacy concerns, such as statistical disclosure control techniques and privacy-preserving data sharing mechanisms [42].

In the future, the role of census data in human geography and urban planning is likely to evolve as new data sources and analytical techniques become increasingly available and accessible. However, the fundamental importance of census data as a comprehensive, systematic, and periodic source of information about the population is unlikely to diminish. Instead, it will continue to serve as a critical foundation for understanding and addressing the complex challenges facing urban areas and regions in the 21st century.

3.1.2 Surveys

Surveys have long been a cornerstone of data collection in human geography and urban planning. They provide researchers and practitioners with a direct method for gathering information from individuals or groups, enabling them to gain insights into various aspects of social, economic, and environmental conditions. In this section, we will explore the role of surveys as a traditional data source in these disciplines, discussing various types of surveys, their advantages and disadvantages, and the evolution of survey methodologies over time.

Types of Surveys

Surveys can be classified into several different types, each with its own unique set of characteristics and applications in human geography and urban planning [4]:

1. Cross-sectional surveys: These surveys capture data at a single point in time, providing a snapshot of the characteristics or opinions of a population. Cross-sectional surveys are often used to assess the current state of a community or region, allowing researchers to identify patterns and relationships between various factors (e.g., income, education, housing).
2. Longitudinal surveys: Unlike cross-sectional surveys, longitudinal surveys collect data from the same respondents over an extended period. These surveys enable researchers to track changes in individual or community characteristics over time and to identify trends and causal relationships [112].
3. Mail surveys: Respondents receive a questionnaire via mail and return their completed responses to the researcher. Mail surveys are typically cost-effective and have a broader reach than other survey methods, but they can suffer from low response rates.

4. Telephone surveys: In telephone surveys, interviewers call respondents and ask them a series of questions. These surveys offer quick data collection, but response rates have declined in recent years due to the increasing prevalence of mobile phones and caller ID systems [64].
5. Face-to-face surveys: Interviewers visit respondents in person to conduct the survey. While this method can yield high-quality data and allow for complex questionnaires, it can be time-consuming and expensive.
6. Online surveys: Respondents complete the survey electronically via a website or mobile application. Online surveys have gained popularity in recent years due to their cost-effectiveness and convenience for both researchers and respondents. However, they can suffer from selection bias if certain population groups lack internet access or are less likely to participate [27].

Advantages and Disadvantages of Surveys

Surveys offer several advantages as a data source in human geography and urban planning:

1. Versatility: Surveys can be adapted to collect a wide range of data, from basic demographic information to complex opinions and attitudes. This versatility allows researchers to explore various aspects of human geography and urban planning, including housing, transportation, land use, and social issues.
2. Comparability: Standardized survey instruments can be used across different populations and geographic areas, allowing for the comparison of data across studies and over time [38].
3. Cost-effectiveness: Surveys, particularly online and mail surveys, can be relatively inexpensive compared to other data collection methods, such as in-depth interviews or participant observation.

However, surveys also have some limitations:

1. Nonresponse bias: Nonresponse bias occurs when individuals who choose not to participate in a survey differ systematically from those who do participate. This bias can lead to skewed results and reduced generalizability [64].
2. Social desirability bias: Respondents may provide answers they believe will be viewed positively by others or the researcher, rather than their true opinions or behaviors. This can result in inaccurate data [96].
3. Questionnaire design challenges: Designing clear, concise, and unbiased survey questions can be difficult, and poorly constructed questions can lead to unreliable or invalid data [38].

The Evolution of Survey Methodologies

Over time, the methodologies used in survey research have evolved to address some of the limitations and challenges associated with traditional survey methods. Some notable advancements include:

1. Mixed-mode surveys: These surveys combine multiple data collection methods (e.g., mail, telephone, and online) to reach a broader range of respondents and improve response rates [39].
2. Mobile surveys: The increasing ubiquity of smartphones has allowed researchers to develop mobile applications for survey data collection. These apps enable respondents to complete surveys on their devices at their convenience, potentially increasing response rates and reducing costs [27].
3. Computer-assisted personal interviewing (CAPI): CAPI involves interviewers using a computer or tablet to administer surveys, replacing traditional paper questionnaires. This approach can improve data quality by reducing errors associated with manual data entry and allowing for complex skip patterns and adaptive questioning [38].
4. Responsive and adaptive survey designs: These designs involve the use of paradata (i.e., data about the data collection process) to monitor survey progress and make real-time adjustments to improve response rates and data quality. For example, researchers may modify the mode of data collection, survey length, or contact strategies based on the observed response patterns [63].

Relevance of Surveys in Human Geography and Urban Planning

Surveys continue to play a crucial role in human geography and urban planning by providing essential data for understanding the complex relationships between people, places, and environments. Some specific applications of surveys in these fields include:

1. Identifying needs and priorities: Surveys can help planners and policymakers gauge public opinion and identify community needs, allowing them to make informed decisions about resource allocation and development priorities [93].
2. Evaluating programs and policies: Surveys can be used to assess the effectiveness of policies, programs, or interventions in addressing specific issues, such as housing affordability, transportation accessibility, or environmental quality [84].
3. Monitoring trends and changes: Longitudinal surveys can track changes in demographics, land use patterns, or social and economic conditions over time, helping researchers and planners identify emerging issues and adapt their strategies accordingly [112].

In conclusion, surveys remain a valuable data source in human geography and urban planning, offering a direct means of gathering information from individuals and groups. The evolution of survey methodologies has addressed some of the challenges associated with traditional survey methods, and the ongoing development of innovative techniques and technologies will continue to enhance the utility of surveys in these fields.

The ongoing development of innovative survey techniques and technologies promises to enhance the utility of surveys in human geography and urban planning. As researchers continue to refine their methods and incorporate new data collection tools, they will be better equipped to address the complex challenges and questions that define these disciplines. This progress will ultimately contribute to a deeper understanding of the relationships between people, places, and environments, helping to inform policy and planning decisions that promote sustainable, equitable, and thriving communities.

In this ever-changing landscape, it is essential for researchers and practitioners in human geography and urban planning to stay informed about new developments in survey methodologies and to be open to adopting new approaches as needed. By embracing these advances and incorporating them into their work, they can ensure that they continue to generate reliable, valid, and actionable data to inform their research and practice.

3.1.3 Land Use Maps

Land use maps have long been a crucial data source in human geography and urban planning, providing valuable information on the spatial distribution of different land uses and land cover types. These maps help researchers and practitioners understand how human activities shape the landscape and inform policy decisions regarding land use management, conservation, and urban development. This section will provide an overview of land use maps, their historical development, methodologies for creating and updating these maps, and their relevance in human geography and urban planning.

Historical Development of Land Use Maps

Land use mapping has its roots in the early 20th century, when cartographers began to systematically categorize and map different types of land uses to support planning efforts [114]. Over time, these maps have evolved from hand-drawn illustrations to detailed digital representations, facilitated by advances in remote sensing, geographic information systems (GIS), and computer-assisted mapping techniques.

Methodologies for Creating Land Use Maps

There are several methods for creating land use maps, each with its strengths and limitations. Traditional methods involve manual interpretation of aerial photographs, field surveys, and expert knowledge to classify land use types. More recent techniques rely on remote sensing data, such as satellite imagery and LiDAR, combined with GIS and machine learning algorithms to automate land use classification [149].

Updating Land Use Maps

Land use maps must be periodically updated to account for changes in land use patterns resulting from urban growth, infrastructure development, and other human activities. Updating these maps typically involves comparing new remote sensing data with previous land use maps, field surveys, and expert knowledge to identify changes in land use patterns and update the map accordingly [133].

Relevance in Human Geography and Urban Planning

Land use maps serve a variety of purposes in human geography and urban planning, including:

- Identifying patterns of urban growth and land use change [128]
- Assessing the environmental impacts of land use decisions [49]
- Evaluating the effectiveness of land use policies and regulations [10]
- Informing the design of sustainable urban development strategies [1].

Challenges and Future Directions in Land Use Mapping

Despite the advances in land use mapping, several challenges remain. These include difficulties in obtaining accurate and up-to-date remote sensing data, particularly for rapidly changing urban areas, and limitations in the spatial resolution and classification accuracy of land use maps [23]. Additionally, there is a need for better integration of land use maps with other spatial data, such as socioeconomic data, to enhance our understanding of the complex relationships between human activities and land use patterns [45].

Future research in land use mapping will likely focus on addressing these challenges, as well as leveraging emerging technologies such as high-resolution satellite imagery, crowd-sourced data, and advanced machine learning algorithms to improve the accuracy, timeliness, and utility of land use maps for human geography and urban planning applications [104].

In conclusion, land use maps have been an essential data source in human geography and urban planning for over a century. They provide valuable information on the spatial distribution of different land uses and land cover types, helping researchers and practitioners understand how human activities shape the landscape and inform policy decisions. Advances in remote sensing, GIS, and machine learning algorithms have significantly improved the accuracy and efficiency of land use mapping, but several challenges remain, such as obtaining up-to-date data and integrating land use maps with other spatial data. Future research will likely focus on addressing these challenges and leveraging emerging technologies to enhance the utility of land use maps for urban planning and human geography applications.

3.1.4 Aerial Photographs

Aerial photographs are valuable data sources in human geography and urban planning, providing insights into the spatial distribution of human activities and the environment. Aerial photographs, taken from aircraft or other airborne platforms, offer unique perspectives of the Earth's surface and facilitate the study of various geographic phenomena [61]. This section will discuss the history and development of aerial photography, its applications in human geography and urban planning, and the challenges and future directions in using aerial photographs as data sources.

History and Development of Aerial Photography

Aerial photography has its roots in the 19th century, with the first aerial photograph taken in 1858 by French photographer and balloonist Gaspard-Félix Tournachon, known as Nadar [116]. The development of aerial photography accelerated during World War I, as it was used extensively for reconnaissance and military purposes [22]. Post-war advancements in photographic and aviation technologies contributed to the increased use of aerial photographs in various fields, including agriculture, forestry, and urban planning [22].

In the second half of the 20th century, aerial photographs played a crucial role in the development of remote sensing, which is the science of obtaining information about the Earth's surface through the analysis of data acquired from airborne or spaceborne sensors [105]. With the launch of Earth observation satellites, such as Landsat in 1972, aerial photography was complemented and sometimes replaced by satellite imagery, providing more extensive and frequent coverage of the Earth's surface [105]. However, aerial photographs remain valuable data sources for local and regional studies, as they offer higher spatial resolution and more flexibility in data acquisition compared to satellite imagery [61] (Fig. 3.1).

Aerial photographs are used in various applications in human geography and urban planning, offering detailed information on the distribution and dynamics of human settlements, land use, and the environment. Some of the key applications include:

1. Land use and land cover mapping: Aerial photographs are widely used for creating and updating land use and land cover maps, which are essential for urban planning and environmental management [105]. By visually interpreting or digitally classifying aerial photographs, researchers and practitioners can identify and map different land use and land cover types, such as residential areas, commercial districts, agricultural lands, and natural habitats [105].

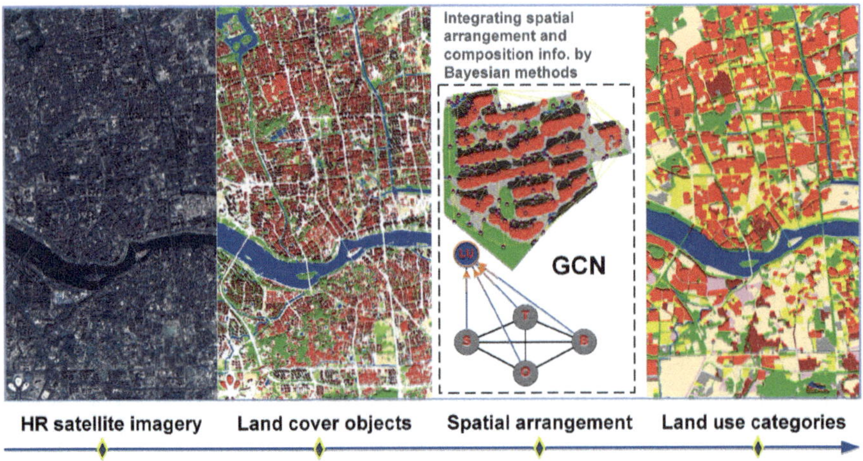

Fig. 3.1 Examples of land use maps and satellite images [103]

2. Urban growth monitoring and analysis: Aerial photographs can be used to track the expansion and transformation of urban areas over time, providing insights into urbanization patterns and processes [105]. By comparing historical and contemporary aerial photographs, researchers can analyze changes in urban morphology, density, and land use, informing urban planning policies and strategies [61].
3. Transportation planning: Aerial photographs are useful for assessing the existing transportation infrastructure, such as roads, highways, railways, and airports, and identifying potential bottlenecks, congestion, and accessibility issues [105]. Furthermore, aerial photographs can facilitate the planning and design of new transportation facilities, providing detailed information on topography, land use, and environmental constraints [105].
4. Environmental assessment and management: Aerial photographs enable the identification and monitoring of environmental features and issues, such as water bodies, vegetation, soil erosion, and pollution [105]. These data can be used for environmental impact assessments, conservation planning, and natural resource management, helping to balance urban development with ecological sustainability [105].
5. Disaster management and risk assessment: Aerial photographs are valuable tools for assessing the impacts of natural and human-induced disasters, such as floods, earthquakes, landslides, and industrial accidents [105]. By analyzing pre- and post-disaster aerial photographs, emergency managers can identify affected areas, evaluate damages, and plan recovery efforts [61]. Furthermore, aerial photographs can contribute to hazard and vulnerability assessments, supporting the development of risk reduction strategies and early warning systems [105].

Despite their numerous applications, aerial photographs present some challenges as data sources in human geography and urban planning:

1. Image quality and resolution: The quality and resolution of aerial photographs can vary depending on factors such as the camera system, altitude, and weather conditions during data acquisition [61]. Low-quality or low-resolution images may limit the accuracy and reliability of the derived information, especially in highly detailed or heterogeneous environments [105].
2. Temporal availability and consistency: Aerial photographs are typically acquired on an ad hoc basis or through periodic surveys, resulting in limited and uneven temporal coverage [61]. This may hinder the analysis of temporal trends and the comparison of different time periods, particularly when the acquisition dates, image characteristics, or environmental conditions are not consistent [105].
3. Data processing and analysis: The processing and analysis of aerial photographs can be time-consuming and labor-intensive, especially when manual visual interpretation is employed [61]. Automated image processing and classification techniques, such as object-based image analysis (OBIA) and deep learning, have the potential to improve the efficiency and accuracy of aerial photograph analysis, but they also require specialized expertise and computational resources [105].
4. Privacy and ethical concerns: Aerial photographs, particularly those with high spatial resolution, may raise privacy and ethical concerns, as they can reveal

sensitive information about individuals and their properties [61]. Researchers and practitioners should consider the implications of using aerial photographs in their work and follow relevant data protection and ethical guidelines [105].

Future developments in aerial photography and related technologies, such as unmanned aerial vehicles (UAVs) or drones, high-resolution multispectral and hyperspectral sensors, and advanced image processing algorithms, may help to overcome some of these challenges and expand the applications of aerial photographs in human geography and urban planning [61, 105].

Aerial photographs have played a significant role in the history of human geography and urban planning, offering valuable insights into the spatial distribution and dynamics of human activities and the environment. Despite the emergence of alternative data sources, such as satellite imagery and geospatial big data, aerial photographs remain essential tools for local and regional studies, providing high-resolution and flexible data acquisition. By addressing the challenges and embracing the opportunities associated with aerial photography, researchers and practitioners can continue to advance the understanding and management of urban and geographical phenomena.

3.1.5 Other Geospatial Data

Geospatial data, or spatial data, refers to information about the geographic location and characteristics of natural or constructed features and boundaries on or near the Earth's surface. In addition to the data sources mentioned earlier, several other types of geospatial data are commonly used in human geography and urban planning. These include Geographic Information System (GIS) data, administrative boundaries, transportation networks, and points of interest, among others. This section will provide an overview of these other geospatial data sources and their applications in human geography and urban planning.

Geographic Information System (GIS) Data

Geographic Information Systems (GIS) are computer-based tools that allow users to create, store, analyze, and visualize spatial data. GIS data are often available in vector or raster formats, with vector data representing discrete features such as points, lines, or polygons, and raster data representing continuous phenomena, such as elevation or land cover, in grid cells [59]. GIS data can be acquired from various sources, including government agencies, non-governmental organizations, and commercial providers.

GIS data have a wide range of applications in human geography and urban planning, such as land use and land cover mapping, population density estimation, transportation planning, and environmental management. GIS data can also be combined with other geospatial data, such as aerial photographs and remote sensing imagery, to provide a more comprehensive understanding of spatial patterns and relationships [109].

Administrative Boundaries

Administrative boundaries are used to delineate political or administrative jurisdictions, such as countries, states, provinces, or municipalities. These boundaries are essential for organizing and analyzing spatial data in human geography and urban planning, as they provide a framework for aggregating and disaggregating data at various spatial scales [117]. Administrative boundaries can be obtained from national mapping agencies, international organizations, or other sources.

Applications of administrative boundaries in human geography and urban planning include the analysis of population distribution, economic activities, social and environmental indicators, and the design and evaluation of public policies at different geographic levels [78, 117].

Transportation Networks

Transportation networks comprise data on roadways, railways, waterways, and other transportation infrastructure. These networks are crucial for understanding the connectivity and accessibility of different locations within an urban or regional context. Transportation network data can be obtained from various sources, including government agencies, commercial mapping providers, and crowdsourced platforms such as OpenStreetMap [72].

In human geography and urban planning, transportation network data are used for various purposes, such as transportation planning, traffic management, accessibility analysis, and the evaluation of the impacts of transportation infrastructure on land use, economic activities, and the environment [123].

Points of Interest

Points of interest (POIs) are specific locations or features that are of interest to various users, such as tourists, residents, or businesses. POIs can include natural landmarks, cultural and historical sites, public facilities, commercial establishments, and other essential services. POI data can be obtained from multiple sources, including government databases, commercial providers, and crowdsourced platforms like OpenStreetMap or Google Places.

In human geography and urban planning, POI data can be used to analyze the spatial distribution of services and amenities, understand urban land use patterns, assess the accessibility of essential services, and support tourism planning and management [150].

Remote Sensing Data

Remote sensing data are collected using sensors mounted on satellites, aircraft, or drones that capture information about the Earth's surface in the form of digital images or other data formats. Remote sensing data can provide valuable information about land use, land cover, vegetation, and other environmental variables at various spatial and temporal resolutions [87].

Applications of remote sensing data in human geography and urban planning include land use and land cover classification, urban growth monitoring, disaster management, environmental monitoring, and climate change impact assessment [141, 142].

In conclusion, various traditional data sources, including census data, surveys, land use maps, aerial photographs, and other geospatial data, play a crucial role in human geography and urban planning. These data sources enable researchers and practitioners to analyze and understand spatial patterns and relationships, inform policy decisions, and develop strategies for sustainable urban development. With the advancement of technology and data availability, the integration of these traditional data sources with emerging data sources, such as big data and social media, offers promising opportunities for enhancing the understanding of complex urban systems and addressing critical challenges in human geography and urban planning.

3.2 Big Data and Open Data: New Opportunities for AI-Driven Analyses

The advent of big data and open data has transformed the landscape of data sources available for human geography and urban planning research. These new data sources offer unprecedented opportunities for AI-driven analyses, enabling researchers and practitioners to explore complex urban phenomena and develop innovative solutions to urban challenges [94]. In this section, we discuss the potential of big data and open data for AI-driven analyses in the context of human geography and urban planning, as well as the challenges and limitations associated with these data sources.

3.2.1 Big Data: Sources, Characteristics, and Applications

Big data has become a crucial aspect of AI-driven analyses in various fields, including human geography and urban planning. In this section, we will discuss the sources, characteristics, and applications of big data, focusing on its potential to revolutionize human geography and urban planning.

Sources of Big Data

In this section, we delve into the various sources of big data relevant to human geography and urban planning. These sources provide the foundation for AI-driven analyses and offer unprecedented opportunities to study complex urban phenomena, develop innovative solutions to urban challenges, and support evidence-based decision-making.

Social Media Data

Social media platforms such as Twitter, Facebook, Instagram, and LinkedIn generate a wealth of information on human behavior, preferences, and social interactions.

This data can be used to study urban phenomena, such as public sentiment, mobility patterns, and the spread of information [90, 132].

a. Twitter data: Tweets, geolocation information, and other metadata associated with Twitter posts can provide insights into public opinion, events, and social dynamics within urban settings [53]. Researchers have used Twitter data to analyze urban mobility patterns, detect events and incidents, and monitor public sentiment during crises [126].
b. Facebook data: Facebook data, including posts, likes, comments, and check-ins, can offer valuable information on social networks, community engagement, and urban dynamics [145]. For example, researchers have used Facebook data to study the relationship between social media activity and urban land use patterns [62].
c. Instagram data: Instagram's image-based content, along with geolocation data and user-generated hashtags, can be used to analyze urban aesthetics, cultural trends, and spatial distribution of activities [80]. Studies have leveraged Instagram data to explore urban landscapes, tourist behaviors, and social events [131].
d. LinkedIn data: LinkedIn's professional networking data, including user profiles, connections, and job postings, can provide insights into urban labor markets, industry clusters, and regional economic development [47]. Researchers have used LinkedIn data to study spatial patterns of employment and skills distribution in cities [3].

Mobile Phone Data

Mobile phones generate a vast amount of data, including location data, call records, and text message data, providing valuable information on human mobility patterns and social networks [16]. This data can be used to study urban mobility, transportation planning, and social dynamics in urban settings [55].

a. Call Detail Records (CDRs): CDRs contain information about calls and text messages exchanged between mobile phone users, including the timestamp, duration, and location of the communication [16]. CDR data has been used to study human mobility patterns, model urban transportation systems, and analyze social interactions [2, 55].
b. Location-based services (LBS) data: Mobile applications that provide location-based services, such as Foursquare and Google Maps, generate data on user locations, points of interest, and user-generated content (e.g., reviews, ratings). This data can be used to study spatial distribution of activities, urban attractiveness, and user preferences [29, 152].

Remote Sensing Data

Satellite imagery and other remote sensing data, such as LiDAR and aerial photographs, provide detailed spatial and temporal information on land use, land cover, and urban growth [141, 142]. This data can be analyzed using AI-driven techniques to understand urban change, environmental impacts, and resource management.

a. Satellite imagery: High-resolution satellite images from sources like Landsat, Sentinel, and WorldView can be used to monitor urban growth, assess land use changes, and evaluate environmental impacts [65]. Machine learning and deep learning techniques have been applied to satellite imagery to classify land use and land cover types, detect urban sprawl, and monitor deforestation [79].
b. LiDAR data: Light Detection and Ranging (LiDAR) data provides high-resolution, three-dimensional information about Earth's surface, vegetation, and infrastructure. LiDAR data has been used to study urban morphology, estimate building heights, and monitor vegetation growth [115, 125].
c. Aerial photographs: Aerial photographs captured by drones, airplanes, or balloons can provide detailed, high-resolution images of urban landscapes. These photographs can be used to assess land use patterns, monitor construction activities, and evaluate the impacts of natural disasters [31].

Internet of Things (IoT) Data

IoT devices, such as sensors embedded in urban infrastructure, vehicles, and buildings, generate vast amounts of real-time data on various aspects of urban life, including traffic, air quality, and energy consumption [156]. This data can be used to develop smart city applications and improve urban planning and management [14].

a. Traffic data: Sensors deployed in transportation systems, such as traffic cameras, loop detectors, and GPS devices, produce data on vehicle counts, speeds, and travel times. This data can be used to monitor traffic congestion, optimize traffic signal timings, and inform the design of transportation infrastructure [159].
b. Environmental data: Air quality sensors, noise monitors, and weather stations generate data on various environmental parameters, such as air pollution levels, temperature, and precipitation. This data can be used to monitor environmental conditions, assess the impacts of urbanization on the environment, and develop sustainable urban policies [99].
c. Energy consumption data: Smart meters and other IoT devices can provide detailed information on energy consumption patterns in residential, commercial, and industrial buildings. This data can be used to optimize energy management systems, identify energy-saving opportunities, and support the development of energy-efficient buildings [48].

Big data sources offer significant potential for AI-driven analyses in human geography and urban planning. By leveraging the various sources of big data, researchers and practitioners can gain valuable insights into complex urban phenomena and develop innovative solutions to address urban challenges.

Characteristics of Big Data

Big data refers to datasets that are too large, complex, or dynamic for conventional data processing systems to handle. The characteristics of big data can be summarized using the 5V model: Volume, Velocity, Variety, Veracity, and Value [50, 100]. These characteristics make big data challenging to process, analyze, and extract meaningful insights from but also provide unprecedented opportunities for AI-driven analyses in human geography and urban planning. In this section, we will discuss each characteristic in detail and provide examples of their implications in the context of urban studies.

1. Volume: Volume refers to the sheer size of big data, which can range from terabytes to petabytes or even larger [20]. The massive scale of big data poses challenges for data storage, processing, and analysis, requiring novel techniques and tools to manage efficiently. For instance, distributed computing frameworks like Hadoop and Spark have emerged to handle large-scale data processing [154]. In urban studies, volume enables researchers to analyze patterns and trends at finer spatial and temporal resolutions, contributing to a more comprehensive understanding of urban processes [11].
2. Velocity: Velocity refers to the speed at which big data is generated, collected, and processed. High-velocity data streams, such as social media feeds or real-time sensor data, require real-time or near-real-time analysis to extract valuable insights [91]. This characteristic enables urban planners and policymakers to make timely decisions and respond to rapidly changing urban environments. For example, real-time traffic data can be used to optimize traffic flow and alleviate congestion [153].
3. Variety: Variety refers to the diverse range of data types and formats found in big data. This diversity includes structured, semi-structured, and unstructured data, as well as text, images, videos, and audio [85]. The heterogeneity of big data necessitates advanced data integration, preprocessing, and analysis techniques. For instance, geospatial data fusion can combine diverse data sources, such as remote sensing imagery and social media data, to provide a more comprehensive understanding of urban landscapes [132].
4. Veracity: Veracity refers to the quality, accuracy, and reliability of big data. Due to the diverse sources and formats of big data, issues such as noise, inconsistency, and incompleteness can impact data quality [139]. Addressing these challenges requires data cleansing, validation, and imputation techniques to ensure that the data used for analysis is reliable and accurate. In urban studies, veracity is critical for drawing robust conclusions and making informed decisions, such as predicting urban growth or assessing the impact of urban policies [14].
5. Value: Value refers to the potential insights and benefits that can be derived from big data. Although big data holds immense potential, extracting meaningful insights requires sophisticated analytics and domain expertise [50]. Machine learning and AI techniques can help uncover hidden patterns, correlations, and trends in big data, enabling researchers and practitioners to address complex urban challenges and make data-driven decisions [11] (Fig. 3.2).

In conclusion, big data's characteristics—volume, velocity, variety, veracity, and value—present both challenges and opportunities for urban studies and human geography. By harnessing the power of AI-driven analyses, researchers can leverage these

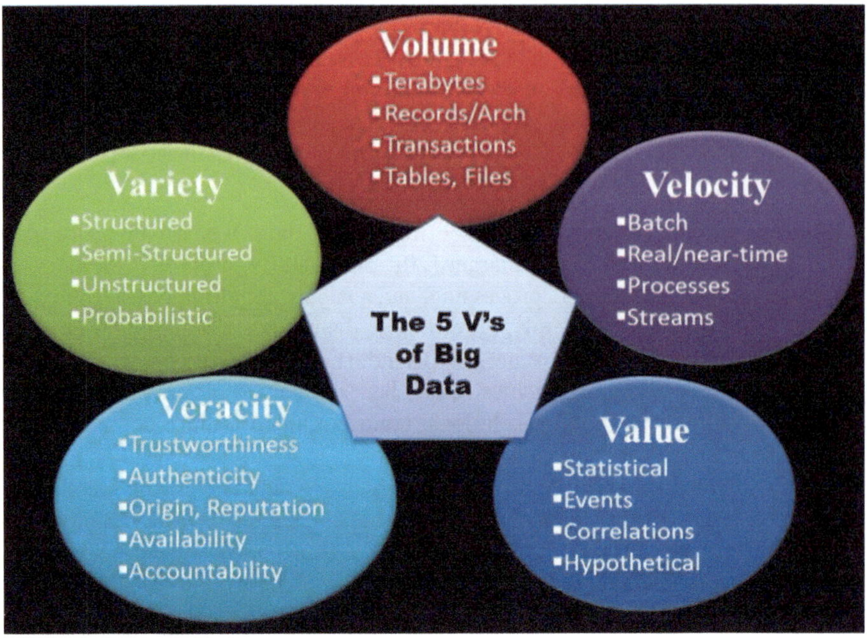

Fig. 3.2 A diagram illustrating the main characteristics of big data (volume, velocity, variety, veracity, and value) [34]

characteristics to address complex urban issues and improve urban planning and policymaking. The integration of diverse data sources and the development of innovative data processing and analysis techniques will continue to advance our understanding of urban dynamics and contribute to more sustainable, resilient, and livable cities.

Applications of Big Data in Human Geography and Urban Planning

The advent of big data has revolutionized human geography and urban planning, providing unprecedented opportunities for researchers and practitioners to study and manage complex urban processes. In this section, we will discuss some of the key applications of big data in human geography and urban planning, along with relevant literature references.

1. Urban mobility and transportation planning

Big data has significantly transformed the way we study and manage urban mobility and transportation systems. Data from sources such as GPS devices, mobile phones, and social media platforms have provided valuable insights into human mobility patterns, enabling planners to make more informed decisions. Applications of big data in transportation planning include traffic congestion analysis, public transportation planning, and understanding the impacts of new mobility services such as ride-sharing platforms [55, 110, 157].

2. Urban land use and environmental planning

Big data has also been employed to study urban land use patterns and environmental issues, enabling planners to monitor and predict changes in land use and assess the impacts of various policies on the urban environment. Satellite imagery, remote sensing data, and geolocated social media data have been used to study urban expansion, land cover changes, and green space distribution, among other applications [12, 51, 151].

3. Social and demographic analysis

Big data has proven invaluable in the analysis of social and demographic phenomena in urban environments. Researchers have used data from sources such as social media, mobile phones, and internet search queries to study social interactions, segregation, gentrification, and public health issues, among other applications [17, 129, 143].

4. Urban economics and housing market analysis

Big data has also been applied in the study of urban economics and housing markets. Researchers have used data from sources such as real estate websites, rental listings, and social media to study housing prices, affordability, and the dynamics of housing markets, providing valuable insights for policymakers and urban planners [147, 158].

5. Public safety and disaster management

The application of big data in public safety and disaster management has provided novel insights into the dynamics of urban risks and vulnerabilities. Data from sources such as social media, mobile phones, and satellite imagery have been used to monitor and predict the occurrence of natural disasters, track the spread of infectious diseases, and analyze crime patterns, among other applications [32, 97, 144].

In summary, big data has transformed the field of human geography and urban planning by providing researchers and practitioners with unprecedented access to rich, high-resolution datasets that have enabled them to study and manage complex urban processes. The applications of big data in human geography and urban planning span across various domains, such as urban mobility, land use planning, social and demographic analysis, urban economics, and public safety. As the availability and variety of big data sources continue to grow, the potential for harnessing these data to inform urban planning and policy decisions is expected to expand further.

3.2.2 Open Data: Sources, Characteristics, and Applications

Open data refers to data that is freely accessible, usable, and shareable by anyone, without restrictions on copyright, patents, or other forms of control. Open data has become increasingly important in the field of human geography and urban planning, as it promotes transparency, collaboration, and innovation in both research and practice. In this section, we discuss the sources, characteristics, and applications of open data in human geography and urban planning.

Sources of Open Data

Open data is data that is freely available, accessible, and reusable by anyone without any restrictions. It is an essential resource for research and planning in human geography and urban planning, offering a wealth of information to better understand and address urban challenges. This section will discuss the various sources of open data and their relevance to human geography and urban planning research.

1. Governmental Open Data Portals

Governments around the world have recognized the potential of open data and have created open data portals to make a wide variety of datasets available to the public. These portals often provide access to information on demographics, transportation, land use, and the environment, among other topics. Some well-known governmental open data portals include:

- data.gov (United States)
- data.gov.uk (United Kingdom)
- data.europa.eu (European Union)
- data.gov.au (Australia)
- data.gc.ca (Canada).

Researchers in human geography and urban planning can use these portals to access relevant datasets for their studies and analyses [86].

2. Non-Governmental Organizations (NGOs) and International Organizations

NGOs and international organizations also provide open data resources, often focusing on specific topics or geographic regions. Examples of these organizations include:

- World Bank Open Data (global development data)
- United Nations (UN) Data (global statistics)
- Global Biodiversity Information Facility (GBIF) (biodiversity data)
- Humanitarian Data Exchange (HDX) (humanitarian and crisis data)
- European Environment Agency (EEA) (European environmental data).

These organizations offer valuable datasets for researchers in human geography and urban planning to explore various aspects of urban and regional development [56].

3. Citizen Science and Volunteered Geographic Information (VGI)

Citizen science initiatives and Volunteered Geographic Information (VGI) platforms enable individuals and communities to contribute to the creation of open data. These platforms often involve the use of smartphones and other GPS-enabled devices to collect and share geospatial information. Examples of citizen science and VGI platforms include:

- OpenStreetMap (OSM) (community-generated map data)
- eBird (bird observation data)
- iNaturalist (biodiversity observations)
- Ushahidi (crisis mapping and crowd-sourced data).

Researchers in human geography and urban planning can leverage these platforms to access hyperlocal and up-to-date data, which can be particularly valuable in rapidly changing urban environments [71].

4. Remote Sensing Data

Remote sensing data, obtained through satellite imagery and aerial photography, can provide valuable geospatial information for urban planning and human geography research. Some organizations and platforms that offer open remote sensing data include:

- United States Geological Survey (USGS) EarthExplorer (satellite imagery)
- Copernicus Open Access Hub (European satellite data)
- Google Earth Engine (geospatial data platform).

These datasets can be used for various applications, such as land use and land cover change analysis, urban growth monitoring, and environmental assessments [148].

5. Open Access Academic Data Repositories

Academic institutions and researchers often share their datasets through open access data repositories, which can be valuable resources for human geography and urban planning research. Examples of open access data repositories include:

- Harvard Dataverse (multidisciplinary data repository)
- ICPSR (Inter-university Consortium for Political and Social Research) (social science data)
- re3data.org (global registry of research data repositories).

By accessing these repositories, researchers can build upon previous work and explore new avenues for investigation in their fields [18].

In conclusion, open data offers numerous opportunities for researchers and practitioners in human geography and urban planning. The sources of open data discussed in this section are diverse, ranging from governmental portals and international organizations to citizen science initiatives and remote sensing data. By leveraging these resources, researchers can gain insights into various aspects of urban development, land use, demographics, and environmental issues.

However, it is essential to consider the quality and reliability of open data sources when using them for research purposes. Researchers should be aware of potential biases, inaccuracies, and inconsistencies in the data and consider validating the data through other sources or methods when necessary [25].

Moreover, researchers should be mindful of ethical considerations when using open data. They should respect the privacy of individuals and communities, ensure the responsible use of sensitive information, and acknowledge the contributions of data providers and creators [160].

Overall, open data has the potential to significantly advance research and practice in human geography and urban planning by providing a wealth of information to support evidence-based decision-making and foster innovation.

Characteristics of Open Data

Open data is characterized by several distinctive features that make it a valuable resource for researchers, practitioners, and decision-makers in human geography and urban planning. In this section, we will discuss the key characteristics of open data, including accessibility, reusability, timeliness, and interoperability.

1. Accessibility

One of the main characteristics of open data is its accessibility. Open data is made available to the public without any restrictions on access, such as registration, login, or fees [36]. This allows anyone with an internet connection to access, download, and use the data for various purposes, including research, innovation, and policy-making [86].

Accessibility is crucial for promoting transparency, accountability, and trust between data providers and users. By making data openly available, governments, organizations, and individuals can better understand and address the needs and challenges of their communities [66].

2. Reusability

Open data is also characterized by its reusability, which means that it can be used, reused, and redistributed freely by anyone [36]. This characteristic is supported by open licenses, such as the Creative Commons licenses, that grant users the right to share, adapt, and build upon the data without any restrictions [30].

Reusability encourages collaboration, innovation, and the development of new applications, tools, and services based on the data. By reusing and combining different datasets, researchers can gain new insights and create novel solutions to complex problems in human geography and urban planning [86].

3. Timeliness

Timeliness is another essential characteristic of open data. It refers to the frequency and speed at which data is updated and released [136]. Timely data is crucial for decision-making and planning, as it allows users to base their decisions on the most up-to-date information available [86].

Timeliness can vary depending on the data source, type, and provider. Some datasets, such as real-time traffic data or social media data, may be updated continuously, while others, such as census data or land use maps, may be updated less frequently [66].

4. Interoperability

Interoperability is a critical characteristic of open data, referring to the ability of different datasets, systems, and tools to work together seamlessly [136]. This is achieved through the use of common data formats, standards, and protocols that enable data exchange, integration, and analysis [86].

Interoperability is essential for maximizing the value of open data, as it allows users to combine and analyze data from different sources to gain new insights and develop innovative solutions. In the context of human geography and urban planning, interoperability can facilitate the integration of various types of data, such as demographic, environmental, and spatial data, to support evidence-based decision-making and planning [66].

Despite the many advantages of open data, there are also challenges and limitations to consider. These include issues related to data quality, privacy, and security, as well as the digital divide, which may limit access to and use of open data by certain groups or communities [86]. Furthermore, the successful implementation of open data initiatives requires strong institutional support, clear governance structures, and adequate resources for data management, maintenance, and dissemination [136].

In conclusion, open data has the potential to significantly impact human geography and urban planning by providing accessible, reusable, timely, and interoperable data that can inform decision-making, planning, and research. The characteristics of open data enable a wide range of stakeholders, including researchers, policymakers, and citizens, to access and utilize data for various purposes, such as identifying trends, addressing challenges, and developing innovative solutions to complex problems.

In the context of human geography and urban planning, open data can support evidence-based decision-making, facilitate collaboration between different sectors and disciplines, and help to address key challenges such as urbanization, climate change, and inequality. By harnessing the power of open data, researchers and practitioners can develop more effective and sustainable strategies for the management and development of urban areas and contribute to the creation of more inclusive, resilient, and sustainable cities and communities.

Applications of Open Data in Human Geography and Urban Planning

The increasing availability of open data has opened up new possibilities for research, planning, and decision-making in human geography and urban planning. By providing access to a wealth of diverse data sources, open data enables researchers, practitioners, and policymakers to better understand the complex dynamics of urban environments, design more effective and sustainable solutions, and engage citizens in the planning and decision-making processes. In this section, we will discuss some of the key applications of open data in human geography and urban planning, focusing on the following areas: (i) spatial analysis and mapping; (ii) monitoring and evaluation; (iii) urban planning and design; (iv) transportation and mobility; (v) environmental management and sustainability; and (vi) public participation and engagement.

(i) Spatial analysis and mapping

Open data has transformed the field of spatial analysis and mapping, allowing researchers and practitioners to access a wide range of datasets, such as land use, topography, administrative boundaries, and socio-economic data, that can be combined, analyzed, and visualized using geographic information systems (GIS) and other spatial analysis tools [46]. These spatial data can be used to identify patterns, trends, and relationships, as well as to inform the development of spatial models and simulations, which can help to predict and assess the potential impacts of different policies, plans, and interventions [88].

(ii) Monitoring and evaluation

Open data is also increasingly being used for monitoring and evaluation purposes in human geography and urban planning. By providing access to timely and accurate data on various indicators, such as population, housing, employment, and the environment, open data enables researchers and practitioners to monitor the performance of urban areas and evaluate the effectiveness of different policies, plans, and programs [86]. This can help to identify areas of improvement, as well as to inform the allocation of resources and the design of more effective and targeted interventions.

(iii) Urban planning and design

Open data has significant potential to support the planning and design of urban areas, by providing planners, architects, and designers with a wealth of information on the existing physical, social, and economic conditions, as well as the potential impacts of different planning scenarios [67]. For example, open data can be used to assess the capacity and suitability of different sites for development, to estimate the demand for different types of land uses and services, and to evaluate the potential impacts of different design options on the urban environment, such as solar access, wind patterns, and noise levels [5].

(iv) Transportation and mobility

Open data has been widely adopted in the field of transportation and mobility, allowing researchers, practitioners, and policymakers to access a range of datasets, such as traffic counts, travel times, and public transport schedules, that can be used to analyze and model the complex dynamics of urban transportation systems [33]. These data can be used to inform the planning and design of transportation infrastructure and services, as well as to monitor and evaluate the performance of transportation systems and the impacts of different policies and interventions, such as congestion pricing, public transport subsidies, and cycling infrastructure [7].

(v) Environmental management and sustainability

Open data has also been widely adopted in the field of environmental management and sustainability, providing researchers, practitioners, and policymakers with access to a range of datasets on various environmental indicators, such as air quality, water quality, and biodiversity [82]. These data can be used to monitor and assess the state of the environment, identify trends and hotspots, and inform the development of policies, plans, and interventions aimed at promoting environmental sustainability and resilience [37]. For example, open data on land cover, climate, and hydrology can be used to inform the design and implementation of green infrastructure, such as parks, green roofs, and rain gardens, which can help to reduce urban heat island effects, improve air quality, and mitigate stormwater runoff [135].

(vi) Public participation and engagement

Finally, open data has significant potential to support public participation and engagement in human geography and urban planning, by providing citizens with access to a wealth of information on their local communities, as well as the tools and platforms to analyze, visualize, and share this information [68]. This can help to empower citizens, foster a sense of ownership and responsibility, and encourage more informed and inclusive decision-making processes [24]. For example, open data platforms, such as local government open data portals and community mapping initiatives, can provide citizens with the information and tools they need to identify local issues, participate in planning processes, and collaborate on the development of community-based solutions [71].

In conclusion, the increasing availability of open data has opened up new opportunities for research, planning, and decision-making in human geography and urban planning. By providing access to a diverse range of data sources, open data enables researchers, practitioners, and policymakers to better understand the complex dynamics of urban environments, design more effective and sustainable solutions, and engage citizens in the planning and decision-making processes.

In conclusion, open data offers significant opportunities for advancing human geography and urban planning research and practice. By providing accessible, reusable, and interoperable data, open data initiatives enable researchers, planners, and citizens to better understand and address complex urban challenges. However, to fully harness the potential of open data, it is essential to address issues related to data quality, privacy, and the digital divide, ensuring that open data benefits all stakeholders in the urban planning process.

3.2.3 *Challenges and Limitations of Big Data and Open Data*

While big data and open data offer significant potential for advancing human geography and urban planning, they also present several challenges and limitations. This section will discuss some of the key challenges, including data quality, data privacy and security, data accessibility, and data integration.

Data Quality

One of the primary concerns with big data and open data is the quality of the data itself. As these data sources are often generated from various sources and formats, ensuring data quality, consistency, and accuracy can be challenging [94]. Additionally, big data sources, such as social media data, may contain bias, as they only represent a subset of the population that uses these platforms [28]. To mitigate these issues, researchers and practitioners need to develop robust methodologies for data cleaning, validation, and verification [95].

Data Privacy and Security

Another significant challenge with big data and open data is the privacy and security of the data. As these datasets often contain sensitive information about individuals and communities, ensuring data privacy and security is essential [101]. This issue is particularly relevant for geospatial data, as the combination of location-based data with other personal data can lead to the identification of individuals, posing privacy risks [28]. To address this, researchers and practitioners should employ privacy-preserving techniques, such as anonymization, aggregation, and differential privacy, to protect individuals' privacy while still enabling data-driven analyses [43].

Data Accessibility

The accessibility of big data and open data is another challenge, as these datasets can be difficult to access and use for various reasons, including licensing restrictions, technical barriers, and lack of data literacy [25]. To improve data accessibility, governments and organizations should promote open data initiatives, develop user-friendly data platforms, and provide training and resources for users to improve their data literacy and skills [86].

Data Integration

Integrating different data sources is another significant challenge when working with big data and open data. Due to the heterogeneity of these datasets, combining them in a meaningful way can be complex and time-consuming [94]. Researchers and practitioners need to develop standardized data formats, metadata, and ontologies to facilitate data integration and interoperability across different data sources [58].

Despite these challenges and limitations, big data and open data still offer significant potential for advancing human geography and urban planning. By addressing these challenges through the development of robust methodologies, privacy-preserving techniques, and standardized data formats, researchers and practitioners can harness the power of big data and open data to gain valuable insights and drive innovation in the field.

Ethical Considerations

Besides the challenges and limitations already mentioned, ethical considerations also play an important role when using big data and open data in human geography and urban planning research. Issues such as informed consent, fairness, and transparency need to be carefully considered when using these data sources [113]. Researchers and practitioners should adhere to ethical guidelines and consider the potential consequences of their work on individuals and communities to ensure that the use of big data and open data contributes to equitable and just outcomes [134].

In conclusion, the challenges and limitations of big data and open data in human geography and urban planning are multifaceted and interconnected. Addressing these challenges requires interdisciplinary collaboration and the development of innovative methodologies, tools, and frameworks. By overcoming these challenges, researchers and practitioners can harness the potential of big data and open data to provide new insights, inform decision-making, and ultimately contribute to more sustainable and equitable urban development.

3.3 Data Cleaning, Preprocessing, and Integration

Data cleaning, preprocessing, and integration are essential steps in the data analysis process, particularly in the context of human geography and urban planning. These steps help researchers and practitioners prepare raw data for further analysis, ensuring that AI-driven techniques can produce accurate and reliable results. This section provides an overview of the key concepts and techniques used in data cleaning, preprocessing, and integration.

3.3.1 Data Cleaning

Data cleaning is a crucial step in the data analysis process, particularly in the context of human geography and urban planning. Data cleaning (Fig. 3.3) refers to the identification and correction of errors, inconsistencies, and inaccuracies in datasets to improve data quality [121]. This section provides an in-depth overview of the key concepts, techniques, and challenges associated with data cleaning.

Importance of Data Cleaning

Data quality is a critical factor influencing the accuracy, reliability, and validity of AI-driven analyses. Poor data quality can lead to misleading or erroneous conclusions, undermining the effectiveness of data-driven decision-making in human geography and urban planning [138]. Data cleaning helps ensure that the input data for AI algorithms is of high quality, reducing the risk of errors and improving the overall performance of the analysis.

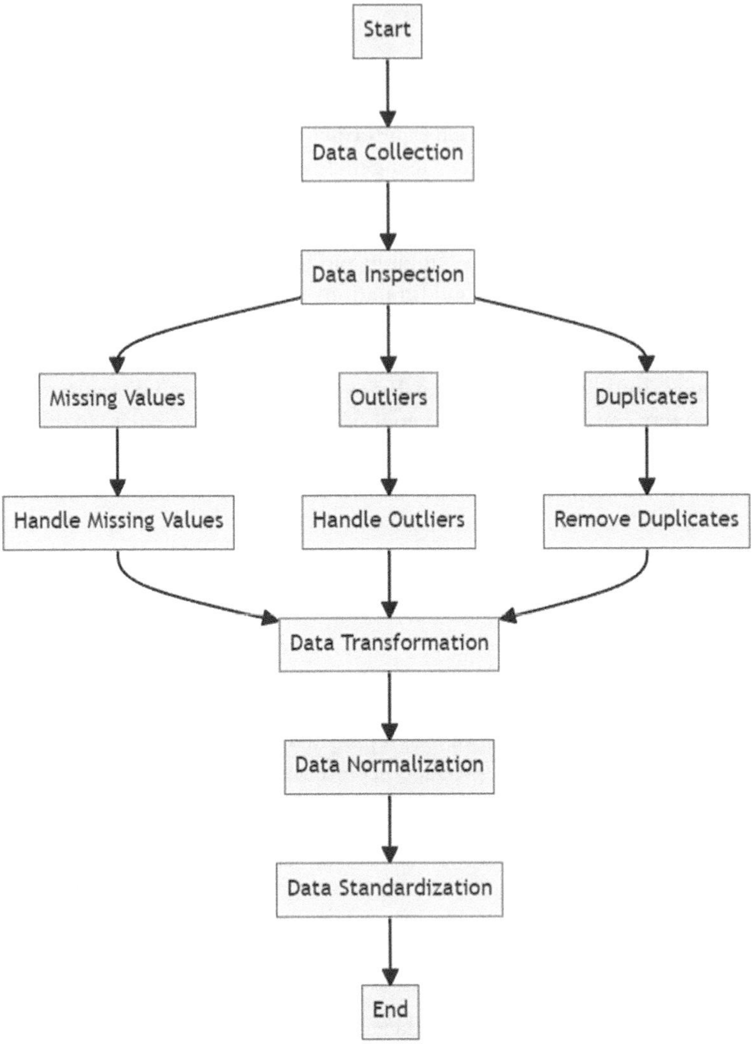

Fig. 3.3 Data cleaning process

Data Cleaning Techniques

Data cleaning techniques can be broadly classified into the following categories:

(a) Missing Value Imputation: Missing values can occur when data is not collected for certain observations or attributes. Various techniques can be used to impute missing values, such as mean or median imputation, regression imputation, and model-based imputation [106]. These techniques aim to estimate the missing values based on the observed data, assuming that the data is missing at random (MAR) or missing completely at random (MCAR). However, imputing missing values can introduce biases or uncertainties in the analysis, so careful evaluation and validation of the imputation methods are required [127].

(b) Outlier Detection and Removal: Outliers are data points that deviate significantly from the rest of the data. They can be caused by measurement errors, data entry errors, or natural variation. Outlier detection techniques include univariate methods, such as standard deviation and interquartile range, and multivariate methods, such as the Mahalanobis distance and clustering algorithms [81]. Once outliers are detected, they can be removed, adjusted, or treated as special cases, depending on the nature of the data and the analysis objectives. However, it is important to distinguish between true outliers and extreme but valid data points, as removing valid data points can bias the analysis [124].

(c) Noise Reduction: Noise refers to random errors or fluctuations in data that can obscure patterns or relationships. Noise reduction techniques aim to smooth or filter the data to reduce the impact of noise on the analysis. Examples of noise reduction techniques include moving averages, Gaussian filters, and wavelet transforms [120]. Noise reduction can help improve the signal-to-noise ratio in the data, making it easier for AI algorithms to identify patterns and relationships. However, excessive noise reduction can also remove useful information or introduce artifacts in the data, so a balance must be struck between denoising and preserving the underlying structure of the data [41].

(d) Duplicate Detection and Removal: Duplicate records can occur when the same observation or entity is represented multiple times in the data, often due to data entry errors, system errors, or data integration issues. Duplicate detection techniques include exact matching, approximate matching, and record linkage methods [44]. Once duplicates are detected, they can be removed or merged to create a consistent and accurate dataset. Duplicate removal can help reduce data redundancy, improve data consistency, and prevent overfitting in AI algorithms [21].

Data Cleaning Challenges and Future Directions

As the volume, variety, and complexity of data in human geography and urban planning continue to grow, new challenges and opportunities for data cleaning emerge. Some future directions and challenges in data cleaning include:

(a) Automated Data Cleaning: Developing AI-driven techniques for automated data cleaning that can learn from the data and adapt to different data quality issues. Machine learning and deep learning algorithms can potentially be employed to identify and correct data quality issues more efficiently and accurately than traditional manual techniques [83]. This can help address the increasing volume and complexity of data, while also reducing the time and effort required for data cleaning.

(b) Data Cleaning in the Presence of Uncertainty: In many real-world applications, data can be uncertain or imprecise due to measurement errors, incomplete information, or subjective assessments. Data cleaning techniques should be able to handle uncertainty and preserve the inherent uncertainty in the data [8]. This may involve developing probabilistic or fuzzy data cleaning techniques that can model and reason about uncertainty, as well as evaluating the impact of data cleaning on the overall uncertainty of the analysis.

(c) Data Cleaning for Streaming Data: With the increasing availability of real-time data from sensors, social media, and other sources, data cleaning techniques need to be adapted to handle streaming data. This may involve developing incremental or online data cleaning techniques that can process data on-the-fly and update the cleaned data as new information becomes available [54]. This can help support real-time data analysis and decision-making in human geography and urban planning applications.

(d) Data Cleaning for Heterogeneous Data: As data sources become more diverse and heterogeneous, data cleaning techniques need to be able to handle different data types, structures, and semantics. This may involve developing data cleaning techniques that can automatically adapt to different data representations, as well as integrating data cleaning with data integration and data preprocessing techniques [40]. This can help ensure that the cleaned data is consistent, accurate, and suitable for AI-driven analyses.

3.3.2 *Data Preprocessing*

Data preprocessing is an essential step in the data analysis pipeline, as it prepares raw data for further processing and analysis by machine learning and artificial intelligence algorithms (Table 3.2). Preprocessing helps in transforming and conditioning data to be more suitable for the intended algorithms, ensuring better and more accurate results. This section discusses data preprocessing, its importance, and various techniques used in the context of human geography and urban planning [92].

Importance of Data Preprocessing

Data preprocessing is crucial for several reasons. Raw data is often noisy, incomplete, and inconsistent, which can lead to inaccurate or misleading results when fed directly into AI algorithms. Preprocessing techniques address these issues, improving the quality and interpretability of the data. Furthermore, preprocessing helps in reducing the dimensionality and complexity of data, making it easier to manage and analyze. By transforming data into a suitable format, preprocessing facilitates the integration of various data sources and supports effective feature extraction, which is vital for machine learning and AI applications [52].

Data Preprocessing Techniques

Several preprocessing techniques are commonly used in human geography and urban planning. Some of the widely adopted methods include:

1. Data transformation: Data transformation involves converting raw data into a different format or representation to make it more suitable for analysis. Common data transformations include normalization, standardization, and discretization.

Table 3.2 Data processing steps including advantages and disadvantages

Preprocessing technique	Description	Applications	Advantages	Disadvantages
Data transformation	Convert raw data into a different format or representation to make it more suitable for analysis. Common techniques include normalization, standardization, and discretization	Enhancing data quality, reducing dimensionality, facilitating data integration, and supporting feature extraction	Scales values for comparability, simplifies analysis, integrates various data sources effectively	May obscure meaningful patterns, requires careful selection of transformation methods
Feature extraction	Create new features or variables from the original data to capture essential information and reduce dimensionality. Techniques include principal component analysis (PCA), spatial indices, and metrics	Reducing dimensionality, capturing meaningful patterns, supporting machine learning and AI applications	Reduces dimensionality, captures essential information, supports effective feature extraction	Requires domain knowledge for meaningful feature creation, computational complexity
Feature selection	Select a subset of the most relevant features for a particular analysis to improve efficiency, interpretability, and prevent overfitting. Techniques include filter methods, wrapper methods, and embedded methods	Improving efficiency, preventing overfitting, enhancing interpretability, reducing dimensionality	Enhances efficiency and interpretability, prevents overfitting, reduces dimensionality	Requires careful selection of features, computational complexity, may overlook relevant features
Data imputation	Estimate missing values in a dataset to make it more complete and suitable for analysis. Techniques include mean or median imputation, regression imputation, nearest neighbor imputation, and spatial or temporal imputation	Handling missing data, improving data completeness, enhancing the reliability of analysis results	Improves data completeness, enhances reliability of results, accounts for uncertainty associated with missing data	Introduces biases if assumptions are not met, may be computationally expensive
Data integration	Combine data from different sources to create a unified and consistent dataset for analysis. Techniques include schema integration, record linkage, and data fusion	Creating unified datasets, resolving conflicts and inconsistencies, integrating diverse data sources	Facilitates data integration, resolves conflicts and inconsistencies, creates unified datasets	Challenging due to differences in data formats and structures, requires expert knowledge, may introduce errors

Normalization scales the values of numeric attributes into a specific range, such as [0, 1], making them more comparable and easier to analyze. Standardization involves transforming data to have zero mean and unit variance, which is particularly useful when dealing with data with different scales or units. Discretization converts continuous data into discrete intervals or categories, simplifying the analysis and interpretation of the data [92].

2. Feature extraction: Feature extraction aims to create new features or variables from the original data, capturing essential information and reducing dimensionality. This process often involves mathematical transformations, such as principal component analysis (PCA), which projects data onto a lower-dimensional space while preserving most of the variance. Another example is the use of spatial indices or metrics that summarize complex spatial patterns in a single value, such as compactness, fractal dimension, or the Gini coefficient. Feature extraction is particularly relevant in human geography and urban planning, where data is often high-dimensional, and domain-specific knowledge can guide the creation of meaningful features [108].
3. Feature selection: Feature selection is the process of selecting a subset of the most relevant features for a particular analysis. This step is crucial in reducing the dimensionality of data, improving the efficiency and interpretability of the results, and preventing overfitting. Feature selection methods can be divided into three categories: filter methods, wrapper methods, and embedded methods. Filter methods evaluate the relevance of features independently of the learning algorithm, using statistical measures such as correlation, mutual information, or chi-square. Wrapper methods assess the usefulness of features by considering their impact on the performance of a specific learning algorithm, often using search strategies such as forward selection, backward elimination, or recursive feature elimination. Embedded methods integrate feature selection within the learning algorithm, incorporating regularization techniques or decision trees to select features during model training [69].
4. Data imputation: Data imputation is the process of estimating missing values in a dataset, making it completer and more suitable for analysis. Imputation methods can be classified into single imputation and multiple imputation. Single imputation techniques estimate missing values based on observed data, using mean or median imputation, regression imputation, or nearest neighbor imputation. Multiple imputation methods generate multiple plausible estimates for missing values, creating several complete datasets, and then combine the results obtained from analyzing each dataset to produce a final estimate. This approach accounts for the uncertainty associated with missing data and generally produces more accurate and reliable estimates. In human geography and urban planning, spatial and temporal imputation techniques are often used, leveraging spatial or temporal relationships among observations to estimate missing values [102].
5. Data integration: Data integration involves combining data from different sources to create a unified and consistent dataset for analysis. Data integration can be challenging, as data from different sources may have different formats, structures, and semantics. Common approaches to data integration include schema integration,

record linkage, and data fusion. Schema integration focuses on aligning the metadata, such as attribute names and data types, across different data sources. Record linkage aims to identify and link records that refer to the same entity across different data sources, using techniques such as string matching, probabilistic matching, or machine learning algorithms. Data fusion deals with resolving conflicts and inconsistencies when merging data from different sources, applying methods such as data aggregation, data transformation, or conflict resolution strategies [44].

Challenges and Limitations of Data Preprocessing

Despite the numerous benefits of data preprocessing, there are several challenges and limitations associated with its implementation. First, preprocessing techniques may introduce biases or distortions in the data, affecting the accuracy and validity of the results. For instance, data imputation methods can potentially introduce systematic errors if the underlying assumptions are not met. Moreover, some preprocessing techniques, such as data normalization or standardization, may not be suitable for all types of data, as they can obscure meaningful patterns or relationships.

Another challenge is the selection of appropriate preprocessing techniques for a specific analysis, as there is no one-size-fits-all approach. The choice of preprocessing methods depends on the nature of the data, the objectives of the analysis, and the characteristics of the learning algorithms. Additionally, the preprocessing process can be time-consuming and computationally expensive, particularly when dealing with large-scale datasets.

Lastly, preprocessing techniques may require expert knowledge and domain-specific understanding to be applied effectively. For example, feature extraction and selection in human geography and urban planning often rely on the expertise of domain specialists to identify meaningful features and relationships. Similarly, data integration can be complex and error-prone, requiring a deep understanding of the data sources and their semantics.

3.3.3 Data Integration

Data integration is the process of combining data from multiple sources, often heterogeneous in nature, to create a unified and coherent view of the information. Data integration is a crucial component of data analysis in human geography and urban planning, as it enables researchers to leverage the full potential of diverse datasets, enhancing the comprehensiveness and reliability of the insights generated. This section will discuss the key concepts, techniques, and challenges related to data integration in the context of human geography and urban planning.

Key Concepts in Data Integration

Data heterogeneity is a fundamental challenge in data integration, as it refers to the differences and inconsistencies that may exist among the data sources to be integrated. Heterogeneity can manifest in several ways, including differences in data formats, schemas, units of measurement, and data semantics [121]. Overcoming data heterogeneity is a primary objective of data integration, as it enables the creation of a unified view of the information.

Data fusion is a technique used to combine data from multiple sources, aiming to enhance the quality, completeness, and reliability of the information. Data fusion methods can be categorized into low-level, intermediate-level, and high-level techniques, depending on the level of abstraction at which the fusion process takes place [75]. Low-level fusion techniques operate on raw data, intermediate-level techniques operate on features or attributes, and high-level techniques operate on the decision or knowledge level.

Data linkage refers to the process of identifying and connecting records in different datasets that refer to the same real-world entity. Data linkage techniques can be deterministic, based on exact matching of common attributes, or probabilistic, based on statistical methods that estimate the likelihood of a match between records [146]. Data linkage is an essential component of data integration, as it enables the construction of a coherent view of the information across multiple datasets.

Techniques for Data Integration

Schema integration is the process of merging the schemas of different datasets to create a unified schema that can accommodate the information from all sources. Schema integration techniques can be classified into two categories: global-as-view (GAV) and local-as-view (LAV) approaches [73]. GAV approaches define the global schema as a view over the local schemas, while LAV approaches define the local schemas as views over the global schema. Schema integration techniques need to address challenges such as schema conflicts, semantic heterogeneity, and schema evolution.

Data transformation is the process of converting data from one format or representation to another, in order to facilitate data integration. Data transformation techniques include normalization, discretization, aggregation, and interpolation [76]. Data transformation is often used in conjunction with schema integration to create a unified view of the information across multiple datasets.

Data imputation is a technique used to fill in missing values in a dataset, based on the available information. Data imputation methods can be classified into single imputation, multiple imputation, and model-based approaches, depending on the assumptions and procedures used to estimate the missing values [107]. Data imputation is often necessary in data integration, as it can enhance the quality and completeness of the integrated dataset.

Challenges and Limitations in Data Integration

Data quality is a critical concern in data integration, as it influences the reliability and validity of the insights generated. Data quality issues, such as errors, inconsistencies, and missing values, can propagate and accumulate during the data integration process, leading to biased or inaccurate results. Ensuring data quality requires the implementation of robust data cleaning, validation, and imputation techniques [9].

Data privacy and security are major challenges in data integration, especially when dealing with sensitive information, such as personal or confidential data. Data integration processes must comply with relevant data protection regulations, such as the General Data Protection Regulation (GDPR) in the European Union. Techniques such as anonymization, pseudonymization, and encryption can be used to protect data privacy and security during the integration process [35].

Scalability and performance are critical concerns in data integration, as the size and complexity of the datasets involved in human geography and urban planning can be substantial. Efficient data integration techniques must be developed to handle large-scale, high-dimensional, and heterogeneous datasets, ensuring timely and accurate insights. Parallel and distributed processing, as well as the use of cloud computing resources, can help address scalability and performance challenges in data integration [44].

Data integration is a crucial aspect of data analysis in human geography and urban planning, enabling researchers to leverage the full potential of diverse datasets. By understanding and addressing the key challenges and limitations associated with data integration, researchers can enhance the comprehensiveness and reliability of the insights generated, informing better decision-making and planning processes.

3.3.4 Data Quality Assessment

Data quality is a crucial factor that determines the reliability and validity of the results obtained from data analysis in various fields, including human geography and urban planning. Data quality assessment involves the evaluation of data sources to identify any potential errors, inconsistencies, or inaccuracies that may affect the analysis process or the outcomes of a study. In this section, we will discuss the importance of data quality assessment, the dimensions of data quality, and the various techniques and methods employed in assessing data quality.

Importance of Data Quality Assessment

Data quality assessment is essential for several reasons:

Ensuring Reliable and Valid Results: Poor data quality can lead to inaccurate or misleading results, potentially leading to incorrect conclusions or flawed decision-making processes in human geography and urban planning [118]. By assessing data quality and addressing any issues identified, researchers can ensure the reliability and validity of their findings.

Building Trust in Data: Data quality assessment helps build trust in the data and the results obtained from it, as it demonstrates that the data is accurate, consistent, and complete. Trustworthy data is critical for decision-makers and stakeholders who rely on the results of data analysis to inform their decisions and actions [138].

Facilitating Data Integration: Assessing data quality can help facilitate data integration by identifying and addressing inconsistencies and discrepancies between different data sources. This process helps ensure that data from various sources can be combined and analyzed effectively [9].

Dimensions of Data Quality

Data quality can be assessed across multiple dimensions, including [138]:

Accuracy: Accuracy refers to the degree to which the data represents the true values of the phenomena it is intended to represent. Inaccurate data can result from various factors, such as errors in data collection, data entry, or data processing.

Completeness: Completeness is the extent to which all relevant data is available and included in the dataset. Incomplete data can lead to biased or incomplete analysis results.

Consistency: Consistency pertains to the degree to which data is coherent and free from contradictions or discrepancies. Inconsistent data can arise from various sources, such as differing data collection methods, data entry errors, or changes in data definitions over time.

Timeliness: Timeliness refers to the degree to which data is up-to-date and available when needed. Outdated data can lead to incorrect conclusions or decisions based on outdated information.

Accessibility: Accessibility is the extent to which data is easily obtainable and usable by researchers and other stakeholders. Inaccessible data can limit the potential for analysis and decision-making.

Techniques and Methods for Data Quality Assessment

Various techniques and methods can be employed to assess data quality, including [9, 118]:

Data Profiling: Data profiling involves examining the data to identify patterns, relationships, and anomalies that may indicate data quality issues. This process can help identify missing, inconsistent, or inaccurate data, as well as potential data entry errors or outliers.

Data Auditing: Data auditing involves a systematic examination of data collection, processing, and storage processes to identify potential sources of data quality issues. This method can help uncover errors in data collection or processing procedures, as well as potential issues related to data storage and management.

Data Quality Metrics: Data quality metrics are quantitative measures that can be used to assess the quality of data along various dimensions, such as accuracy, completeness, consistency, timeliness, and accessibility. These metrics can be used to track data quality over time, identify areas for improvement, and establish benchmarks for data quality standards.
Data Cleansing: Data cleansing involves identifying and correcting data quality issues, such as inaccuracies, inconsistencies, and missing data. This process may involve various techniques, such as data imputation, data transformation, or data standardization, to improve the overall quality of the dataset.
Data Quality Assessment Tools: Various data quality assessment tools are available to assist researchers in evaluating and improving data quality. These tools can automate the process of identifying data quality issues, provide visualizations and reports to help understand data quality, and offer features for cleaning and transforming data.

Challenges and Future Directions

Despite the importance of data quality assessment, several challenges remain:

Scalability: As datasets become larger and more complex, assessing data quality becomes increasingly difficult and time-consuming. Developing scalable methods and tools for data quality assessment is an ongoing challenge for researchers and practitioners.
Data Quality in Unstructured and Semi-Structured Data: Assessing data quality in unstructured and semi-structured data, such as text documents, images, or social media data, is more challenging than in structured data. Developing techniques and methods for assessing data quality in these types of data sources is an important area of research.
Continuous Data Quality Assessment: Data quality assessment should not be a one-time process, but rather a continuous effort to monitor and improve data quality. Developing methods and tools for continuous data quality assessment can help ensure that data remains accurate, consistent, and complete over time.

In conclusion, data quality assessment is a critical component of data analysis in human geography and urban planning, as well as in other fields. By evaluating and improving data quality, researchers and practitioners can ensure the reliability and validity of their findings, build trust in their data, and facilitate data integration. The ongoing development of techniques, methods, and tools for data quality assessment will help address the challenges and opportunities presented by increasingly complex and diverse data sources.

3.3.5 Future Directions and Challenges in Data Cleaning, Preprocessing, and Integration

Data cleaning, preprocessing, and integration are essential steps in the data analysis process. The growing scale, complexity, and variety of data sources pose new challenges and opportunities for future research and development in these areas. This section discusses the future directions and challenges in data cleaning, preprocessing, and integration, highlighting emerging trends and techniques.

Scalability and Performance

One of the primary challenges in data cleaning, preprocessing, and integration is the ability to handle large-scale datasets efficiently. With the proliferation of big data, data processing techniques must be able to scale and perform well on massive datasets. This requires the development of parallel and distributed computing algorithms and frameworks that can leverage the power of modern hardware, such as multi-core processors, GPUs, and cloud computing resources [155].

Handling Complex and Diverse Data Sources

As the variety of data sources continues to expand, the need to process and integrate data from different formats, structures, and domains becomes increasingly important. This includes handling data from various sources such as text, images, videos, social media, sensor data, and more. Developing robust and efficient techniques for data preprocessing, cleaning, and integration across diverse data sources is a significant challenge [74].

Real-Time Data Processing

The demand for real-time data processing is increasing due to the growth of applications such as IoT, social media analytics, and streaming data analysis. Developing methods and tools for real-time data cleaning, preprocessing, and integration is essential to support these applications [19].

Data Privacy and Security

Data privacy and security are critical concerns in data processing, particularly with the increasing prevalence of sensitive data in various domains. Techniques for data cleaning, preprocessing, and integration must consider privacy-preserving methods, such as anonymization, encryption, and secure multi-party computation, to protect sensitive information while still enabling valuable insights from the data [43].

Data Quality and Provenance

Ensuring data quality and tracking data provenance are essential aspects of data processing. Developing techniques to assess and improve data quality, as well as methods to track the provenance of data throughout the cleaning, preprocessing, and integration pipeline, is vital for ensuring the reliability and trustworthiness of data-driven analyses [13].

Integration of Machine Learning and Data Processing Techniques

The integration of machine learning techniques with data cleaning, preprocessing, and integration methods has the potential to significantly improve the efficiency and effectiveness of these processes. For example, deep learning algorithms can be applied to handle unstructured and semi-structured data, while unsupervised learning methods can help identify patterns and anomalies in the data [60].

Human-in-the-Loop Data Processing

Incorporating human expertise and intuition into the data processing pipeline can help improve the quality and reliability of the processed data. Developing methods and tools that enable humans to collaborate with automated data processing techniques, such as through crowdsourcing or interactive data cleaning and integration tools, is an important research direction [89].

The future of data cleaning, preprocessing, and integration is likely to be characterized by increased scalability, real-time processing capabilities, integration of diverse data sources, and improved data quality and provenance. Addressing these challenges will require the development of novel techniques, tools, and frameworks that leverage advances in machine learning, human–computer interaction, and distributed computing. By overcoming these challenges, researchers and practitioners will be better equipped to harness the full potential of data-driven insights in human geography, urban planning, and other domains.

Standardization and Interoperability

As data processing techniques continue to evolve, there is a growing need for standardization and interoperability among different tools and platforms. This will enable researchers and practitioners to more easily share, reuse, and combine data processing pipelines, ultimately leading to more efficient and effective data analysis workflows. Standardizing data formats, APIs, and processing methodologies will be crucial for promoting collaboration and advancing the state of the art in data cleaning, preprocessing, and integration [15].

Ethics and Bias in Data Processing

The increasing reliance on data-driven insights in decision-making raises ethical concerns related to bias, fairness, and accountability. It is essential to develop data processing techniques that can identify and mitigate biases in the data, as well as ensure that the insights derived from the data are fair and equitable [6]. Moreover, researchers and practitioners should be transparent about the data processing methods they employ and be accountable for the potential consequences of their analyses.

Education and Training

As the importance of data-driven insights grows across various domains, the need for skilled professionals who can effectively clean, preprocess, and integrate data is also increasing. Developing educational resources, training programs, and certifications for data processing will be essential to meet the demand for expertise in this area. These educational initiatives should be interdisciplinary in nature, combining knowledge from computer science, statistics, and domain-specific fields such as human geography and urban planning [77].

Open-Source Software and Collaboration

Open-source software and collaboration play a crucial role in the advancement of data cleaning, preprocessing, and integration techniques. By promoting the development and sharing of open-source tools, libraries, and frameworks, the research community can collectively contribute to addressing the challenges and opportunities in this area. Furthermore, collaboration among researchers, practitioners, and other stakeholders will be essential for identifying and addressing the most pressing problems and developing innovative solutions [122].

In conclusion, the future of data cleaning, preprocessing, and integration will be shaped by a variety of factors, including the development of scalable and efficient algorithms, the integration of machine learning techniques, and the need for standardization and interoperability. By addressing these challenges and embracing emerging trends, the field of data processing will continue to play a crucial role in enabling data-driven insights and decision-making across human geography, urban planning, and other domains.

References

1. Alberti, M. (2008). *Advances in urban ecology: Integrating humans and ecological processes in urban ecosystems*. Springer.
2. Amini, A., Kung, K., Kang, C., Sobolevsky, S., & Ratti, C. (2014). The impact of social segregation on human mobility in developing and industrialized regions. *EPJ Data Science, 3*(1), 6.
3. Arribas-Bel, D., & Reades, J. (2018). Geography and computers: Past, present, and future. *Geography Compass, 12*(10), e12403.
4. Babbie, E. (2016). *The practice of social research*. Cengage Learning.
5. Ball, J., & Newman, P. (2013). Urban open data for sustainability assessment. *Journal of Urbanism: International Research on Placemaking and Urban Sustainability, 6*(3), 204–228.
6. Barocas, S., & Selbst, A. D. (2016). Big data's disparate impact. *California Law Review, 104*, 671–732.
7. Barth, M., & Boriboonsomsin, K. (2010). Traffic congestion and greenhouse gases. *Access Magazine, 37*, 2–9.
8. Batini, C., & Scannapieco, M. (2016). *Data and information quality: Dimensions*. Springer.
9. Batini, C., Cappiello, C., Francalanci, C., & Maurino, A. (2009). Methodologies for data quality assessment and improvement. *ACM Computing Surveys, 41*(3), 1–52.
10. Batty, M. (2008). The size, scale, and shape of cities. *Science, 319*(5864), 769–771.
11. Batty, M. (2013). Big data, smart cities and city planning. *Dialogues in Human Geography, 3*(3), 274–279.
12. Batty, M., Axhausen, K. W., Giannotti, F., Pozdnoukhov, A., Bazzani, A., Wachowicz, M., Portugali, Y., et al. (2012). Smart cities of the future. *The European Physical Journal Special Topics, 214*(1), 481–518.
13. Bertino, E., & Matei, A. (2016). *Data quality: Concepts, methodologies, and techniques*. Springer.

14. Bibri, S. E., & Krogstie, J. (2017). Smart sustainable cities of the future: An extensive interdisciplinary literature review. *Sustainable Cities and Society, 31*, 183–212.
15. Bizer, C., Heath, T., & Berners-Lee, T. (2011). Linked data: The story so far. In *Semantic services, interoperability, and web applications: Emerging concepts* (pp. 205–227). IGI Global.
16. Blondel, V. D., Decuyper, A., & Krings, G. (2015). A survey of results on mobile phone datasets analysis. *EPJ Data Science, 4*(1), 1–55.
17. Blumenstock, J., Cadamuro, G., & On, R. (2015). Predicting poverty and wealth from mobile phone metadata. *Science, 350*(6264), 1073–1076.
18. Borgman, C. L. (2012). The conundrum of sharing research data. *Journal of the American Society for Information Science and Technology, 63*(6), 1059–1078.
19. Carbone, P., Katsifodimos, A., Ewen, S., Markl, V., Haridi, S., & Tzoumas, K. (2015). Apache Flink™: Stream and batch processing in a single engine. *IEEE Data Engineering Bulletin, 38*(4), 28–38.
20. Chen, M., Mao, S., & Liu, Y. (2014). Big data: A survey. *Mobile Networks and Applications, 19*(2), 171–209.
21. Christen, P. (2012). *Data matching: Concepts and techniques for record linkage, entity resolution, and duplicate detection*. Springer.
22. Colwell, R. N. (1997). History of aerial photography. In *Manual of photogrammetry* (5th ed., pp. 1–29). American Society for Photogrammetry and Remote Sensing.
23. Comber, A., Fisher, P., & Wadsworth, R. (2012). What is land cover? *Environment and Planning B: Planning and Design, 39*(2), 199–216.
24. Conrad, E., White, R., & Christie, M. (2011). Community-based spatial planning and the role of public participation GIS. *Journal of Environmental Policy & Planning, 13*(1), 87–107.
25. Conradie, P., & Choenni, S. (2014). On the barriers for local government releasing open data. *Government Information Quarterly, 31*, S10–S17.
26. Coulton, C. J. (2017). The place of the census in the historical evolution of community information systems. *Cityscape, 19*(1), 267–274.
27. Couper, M. P. (2017). New developments in survey data collection. *Annual Review of Sociology, 43*, 121–145.
28. Crampton, J. W., Graham, M., Poorthuis, A., Shelton, T., Stephens, M., Wilson, M. W., & Zook, M. (2013). Beyond the geotag: Situating 'big data' and leveraging the potential of the geoweb. *Cartography and Geographic Information Science, 40*(2), 130–139.
29. Cranshaw, J., Schwartz, R., Hong, J. I., & Sadeh, N. (2012). The Livehoods project: Utilizing social media to understand the dynamics of a city. In *Proceedings of the Sixth International Conference on Weblogs and Social Media (ICWSM 2012)* (pp. 81–88).
30. Creative Commons. (2021). About The Licenses. https://creativecommons.org/licenses/
31. Crommelinck, S., & Höfle, B. (2016). Simulating an autonomously operating low-cost static terrestrial LiDAR for multitemporal maize crop height measurements. *Remote Sensing, 8*(10), 822.
32. Crooks, A., Croitoru, A., Stefanidis, A., & Radzikowski, J. (2013). Earthquake: Twitter as a distributed sensor system. *Transactions in GIS, 17*(1), 124–147.
33. Dahdouh, K., Dakkak, A., & Oughdir, L. (2019). Big data: a distributed storage and processing for online learning systems. *International Journal of Computational Intelligence Studies, 8*(3), 192–205.
34. Danezis, G., Domingo-Ferrer, J., Hansen, M., Hoepman, J. H., Métayer, D. L., Tirtea, R., & Schiffner, S. (2014). *Privacy and data protection by design-from policy to engineering*. European Union Agency for Network and Information Security.
35. Davies, T. (2010). Open data, democracy and public sector reform. A look at open government data use from data.gov.uk. http://www.opendataimpacts.net/report/
36. de Montalvo, U. W. (2013). Open data for sustainable development. In *Proceedings of the 46th Annual Hawaii International Conference on System Sciences (HICSS)* (pp. 1889–1897).
37. De Vaus, D. (2014). *Surveys in social research*. Routledge.

38. Dillman, D. A., Smyth, J. D., & Christian, L. M. (2014). *Internet, phone, mail, and mixed-mode surveys: The tailored design method*. Wiley.
39. Doan, A., Halevy, A. Y., & Ives, Z. G. (2012). *Principles of data integration*. Morgan Kaufmann.
40. Donoho, D. L. (1995). De-noising by soft-thresholding. *IEEE Transactions on Information Theory, 41*(3), 613–627.
41. Duncan, G., Keller-McNulty, S., & Stokes, S. L. (2011). Disclosure risk vs. data utility: The R-U confidentiality map. In *Privacy in statistical databases* (pp. 121–137). Springer.
42. Dwork, C., & Roth, A. (2014). The algorithmic foundations of differential privacy. *Foundations and Trends® in Theoretical Computer Science, 9*(3–4), 211–407.
43. Elmagarmid, A. K., Ipeirotis, P. G., & Verykios, V. S. (2007). Duplicate record detection: A survey. *IEEE Transactions on Knowledge and Data Engineering, 19*(1), 1–16.
44. Elmqvist, T., Fragkias, M., Goodness, J., Güneralp, B., Marcotullio, P. J., McDonald, R. I., Wilkinson, C., et al. (2013). *Urbanization, biodiversity and ecosystem services: Challenges and opportunities*. Springer.
45. Elwood, S., Goodchild, M. F., & Sui, D. Z. (2012). Researching volunteered geographic information: Spatial data, geographic research, and new social practice. *Annals of the Association of American Geographers, 102*(3), 571–590.
46. Ferragina, E., & Miliaraki, I. (2015). The impact of online social networks on labor markets: Evidence from LinkedIn. In *Proceedings of the 21th ACM SIGKDD International Conference on Knowledge Discovery and Data Mining* (pp. 785–794).
47. Figueiredo, V., Rodrigues, F., Gomes, L., & Borges, F. (2016). An electric energy consumer characterization framework based on data mining techniques. *IEEE Transactions on Smart Grid, 7*(1), 425–436.
48. Foley, J. A., DeFries, R., Asner, G. P., Barford, C., Bonan, G., Carpenter, S. R., Helkowski, J. H., et al. (2005). Global consequences of land use. *Science, 309*(5734), 570–574.
49. Gandomi, A., & Haider, M. (2015). Beyond the hype: Big data concepts, methods, and analytics. *International Journal of Information Management, 35*(2), 137–144.
50. Gao, S., Janowicz, K., Montello, D. R., Hu, Y., Yang, J. A., McKenzie, G., Yan, B., et al. (2017). GeoExposure: An outdoor light exposure measurement framework for human geographical science. *Computers, Environment and Urban Systems, 64*, 1–22.
51. Garrett, K. (2018). Data preprocessing for machine learning. In *Data science* (pp. 85–110). Springer.
52. Gerlitz, C., & Rieder, B. (2013). Mining one percent of Twitter: Collections, baselines, sampling. *M/C Journal, 16*(2).
53. Golab, L., & Özsu, M. T. (2010). Issues in data stream management. *ACM Sigmod Record, 32*(2), 5–14.
54. Gonzalez, M. C., Hidalgo, C. A., & Barabasi, A. L. (2008). Understanding individual human mobility patterns. *Nature, 453*(7196), 779–782.
55. Goodchild, M. F. (2007). Citizens as sensors: The world of volunteered geography. *GeoJournal, 69*(4), 211–221.
56. Goodchild, M. F. (2010). Twenty years of progress: GIScience in 2010. *Journal of Spatial Information Science, 1*(1), 3–20.
57. Goodchild, M. F. (2013). The quality of big (geo)data. *Dialogues in Human Geography, 3*(3), 280–284.
58. Goodchild, M. F. (2018). Geographic information systems. In D. R. Montello (Ed.), *The SAGE handbook of GIS and society* (pp. 15–32). SAGE Publications.
59. Goodfellow, I., Bengio, Y., & Courville, A. (2016). *Deep learning*. MIT Press.
60. Graham, M. (2014). Aerial photography in urban research. In R. P. Walker & H. A. Leitner (Eds.), *The international encyclopedia of human geography* (pp. 1–6). Wiley.
61. Grauwin, S., Sobolevsky, S., Moritz, S., Gódor, I., & Ratti, C. (2014). Towards a comparative science of cities: Using mobile traffic records in New York, London, and Hong Kong. In *Computational approaches for urban environments* (pp. 363–387). Springer.

62. Groves, R. M., & Heeringa, S. G. (2006). Responsive design for household surveys: Tools for actively controlling survey errors and costs. *Journal of the Royal Statistical Society: Series A (Statistics in Society), 169*(3), 439–457.
63. Groves, R. M., Fowler, F. J., Couper, M. P., Lepkowski, J. M., Singer, E., & Tourangeau, R. (2009). *Survey methodology*. Wiley.
64. Gupta, P., Ghosh, S. K., & Kumar, M. (2015). Urban growth prediction using high-resolution satellite images: A case study of Kolkata, India. *Journal of the Indian Society of Remote Sensing, 43*(1), 89–100.
65. Gurin, J., Young, A., & Verhulst, S. (2015a). Open data: A twenty-first century asset for small and medium-sized enterprises. In *The Global Information Technology Report 2015, World Economic Forum*.
66. Gurin, J., Young, A., & Verhulst, S. (2015b). The potential and challenges of open data for urban planning. In M. Campagna & M. Craglia (Eds.), *Advanced geographic information systems and web services* (pp. 32–46). IGI Global.
67. Gurstein, M. (2011). Open data: Empowering the empowered or effective data use for everyone? *First Monday, 16*(2).
68. Guyon, I., & Elisseeff, A. (2003). An introduction to variable and feature selection. *Journal of Machine Learning Research, 3*(Mar), 1157–1182.
69. Haining, R. (2003). *Spatial data analysis: Theory and practice*. Cambridge University Press.
70. Haklay, M. (2013). Citizen science and volunteered geographic information: Overview and typology of participation. In D. Sui, S. Elwood, & M. Goodchild (Eds.), *Crowdsourcing geographic knowledge: Volunteered geographic information (VGI) in theory and practice* (pp. 105–122). Springer.
71. Haklay, M., & Weber, P. (2008). OpenStreetMap: User-generated street maps. *IEEE Pervasive Computing, 7*(4), 12–18.
72. Halevy, A. Y. (2001). Answering queries using views: A survey. *The VLDB Journal, 10*(4), 270–294.
73. Halevy, A., Franklin, M., & Maier, D. (2016). Principles of dataspace systems. *ACM SIGMOD Record, 45*(1), 5–16.
74. Hall, D. L., & Llinas, J. (1997). An introduction to multisensor data fusion. *Proceedings of the IEEE, 85*(1), 6–23.
75. Han, J., Kamber, M., & Pei, J. (2011). *Data mining: Concepts and techniques* (3rd ed.). Elsevier.
76. Hardin, J., Hoerl, R., Horton, N. J., Nolan, D., Baumer, B., Hall-Holt, O., Ward, M. D., et al. (2015). Data science in statistics curricula: Preparing students to "think with data." *The American Statistician, 69*(4), 343–353.
77. Hay, S. I. (2016). *Global mapping of infectious diseases: Methods, examples and emerging applications*. Elsevier.
78. Helber, P., Bischke, B., Dengel, A., & Borth, D. (2017). Eurosat: A novel dataset and deep learning benchmark for land use and land cover classification. *IEEE Journal of Selected Topics in Applied Earth Observations and Remote Sensing, 12*(7), 2217–2226.
79. Hochman, N., & Manovich, L. (2013). Zooming into an Instagram city: Reading the local through social media. *First Monday, 18*(7).
80. Hodge, V. J., & Austin, J. (2004). A survey of outlier detection methodologies. *Artificial Intelligence Review, 22*(2), 85–126.
81. Holden, E., Norris, G., & Fletcher, S. (2018). Open data for environmental sustainability. In J. Stoltzfus & J. Macedo (Eds.), *Open data and the knowledge society* (pp. 123–140). Amsterdam University Press.
82. Ilyas, I. F., & Chu, X. (2015). Trends in cleaning relational data: Consistency and deduplication. *Foundations and Trends in Databases, 5*(4), 281–393.
83. Innes, J. E., & Booher, D. E. (2000). Public participation in planning: New strategies for the 21st century. *Journal of the American Planning Association, 66*(3), 279–294.
84. Jagadish, H. V., Gehrke, J., Labrinidis, A., Papakonstantinou, Y., Patel, J. M., Ramakrishnan, R., & Shahabi, C. (2014). Big data and its technical challenges. *Communications of the ACM, 57*(7), 86–94.

85. Janssen, M., Charalabidis, Y., & Zuiderwijk, A. (2012). Benefits, adoption barriers and myths of open data and open government. *Information Systems Management, 29*(4), 258–268.
86. Jensen, J. R. (2007). *Remote sensing of the environment: An Earth resource perspective.* Pearson Education.
87. Johnson, P. A. (2014). Geographies of open data: Strategies, tactics, and methods. *The Canadian Geographer, 58*(1), 11–26.
88. Kandel, S., Paepcke, A., Hellerstein, J. M., & Heer, J. (2012). Enterprise data analysis and visualization: An interview study. *IEEE Transactions on Visualization and Computer Graphics, 18*(12), 2917–2926.
89. Kaplan, A. M., & Haenlein, M. (2010). Users of the world, unite! The challenges and opportunities of social media. *Business Horizons, 53*(1), 59–68.
90. Katal, A., Wazid, M., & Goudar, R. H. (2013). Big data: Issues, challenges, tools, and good practices. In *2013 Sixth International Conference on Contemporary Computing (IC3)* (pp. 404–409). IEEE.
91. Kelleher, J. D., Mac Namee, B., & D'Arcy, A. (2015). *Fundamentals of machine learning for predictive data analytics: Algorithms, worked examples, and case studies.* MIT Press.
92. Kitchen, R., & Tate, N. J. (2013). *Conducting research in human geography: Theory, methodology and practice.* Pearson Education.
93. Kitchin, R. (2014). *The data revolution: Big data, open data, data infrastructures and their consequences.* Sage Publications.
94. Kitchin, R., & Lauriault, T. P. (2014). Towards critical data studies: Charting and unpacking data assemblages and their work. The Programmable City Working Paper 2.
95. Krumpal, I. (2013). Determinants of social desirability bias in sensitive surveys: A literature review. *Quality & Quantity, 47*(4), 2025–2047.
96. Kryvasheyeu, Y., Chen, H., Obradovich, N., Moro, E., Van Hentenryck, P., Fowler, J., & Cebrian, M. (2016). Rapid assessment of disaster damage using social media activity. *Science Advances, 2*(3), e1500779.
97. Kukutai, T., & Rarere, M. (2017). From mainstream to manaaki: Indigenising our approach to immigration. In *Fair borders? Migration policy in the twenty-first century* (pp. 26–47). Bridget Williams Books.
98. Kumar, P., Morawska, L., Martani, C., Biskos, G., Neophytou, M., Di Sabatino, S., Norford, L., et al. (2015). The rise of low-cost sensing for managing air pollution in cities. *Environment International, 75*, 199–205.
99. Laney, D. (2001). 3D data management: Controlling data volume, velocity, and variety. *META Group Research Note, 6*(70), 1.
100. Li, J., Soliman, A., Yin, J., & Mukhopadhyay, S. (2016). Scalable imputation of missing data in spatiotemporal databases: A tensor-based approach. *ACM Transactions on Spatial Algorithms and Systems (TSAS), 2*(3), 1–31.
101. Li, M., & Stein, A. (2020). Mapping land use from high resolution satellite images by exploiting the spatial arrangement of land cover objects. *Remote sensing, 12*(24), 4158.
102. Li, S., Dragićević, S., & Castro, F. A. (2018). Artificial intelligence and urban sustainability: Challenges, opportunities, and future directions. *Environmental Reviews, 26*(4), 369–385.
103. Lillesand, T., Kiefer, R., & Chipman, J. (2015). *Remote sensing and image interpretation* (7th edn.). Wiley.
104. Little, R. J. A., & Rubin, D. B. (2019). *Statistical analysis with missing data.* Wiley.
105. Little, R. J., & Rubin, D. B. (2002). *Statistical analysis with missing data* (2nd edn.). Wiley.
106. Long, Y., & Liu, X. (2016). How mixed is a mixed land use? Gradient-oriented entropy-U for measuring land use mix. *Computers, Environment and Urban Systems, 57*, 46–57.
107. Longley, P. A., Goodchild, M. F., Maguire, D. J., & Rhind, D. W. (2015). *Geographic information science and systems* (4th ed.). Wiley.
108. Ma, X., Wu, Y. J., Wang, Y., Chen, F., & Liu, J. (2013). Mining smart card data for transit riders' travel patterns. *Transportation Research Part C: Emerging Technologies, 36*, 1–12.
109. Massey, D. S., & Denton, N. A. (1993). *American apartheid: Segregation and the making of the underclass.* Harvard University Press.

110. Menard, S. (2008). *Handbook of longitudinal research: Design, measurement, and analysis*. Academic Press.
111. Metcalf, J., & Crawford, K. (2016). Where are human subjects in big data research? The emerging ethics divide. *Big Data & Society, 3*(1), 1–14.
112. Monmonier, M. (1994). *How to lie with maps* (2nd ed.). The University of Chicago Press.
113. Morsy, S., Shaker, A., & El-Rabbany, A. (2017). LiDAR-guided urban growth monitoring and modeling. *ISPRS International Journal of Geo-Information, 6*(11), 335.
114. Niemann, S. (2011). A brief history of aerial photography. In S. Niemann (Ed.), *Aerial photography and image interpretation for resource management* (pp. 1–6). CRC Press.
115. Openshaw, S. (1984). *The modifiable areal unit problem*. Geo Books.
116. Pipino, L. L., Lee, Y. W., & Wang, R. Y. (2002). Data quality assessment. *Communications of the ACM, 45*(4), 211–218.
117. Plane, D. A., & Rogerson, P. A. (2006). *The geographical analysis of population: With applications to planning and business*. Wiley.
118. Quinlan, J. R. (2014). *C4.5: Programs for machine learning*. Morgan Kaufmann.
119. Rahm, E., & Do, H. H. (2000). Data cleaning: Problems and current approaches. *IEEE Data Engineering Bulletin, 23*(4), 3–13.
120. Ram, K. (2013). Git can facilitate greater reproducibility and increased transparency in science. *Source Code for Biology and Medicine, 8*(1), 7.
121. Rodrigue, J. P., Comtois, C., & Slack, B. (2016). *The geography of transport systems*. Routledge.
122. Rousseeuw, P. J., & Leroy, A. M. (2005). *Robust regression and outlier detection*. Wiley.
123. Rutzinger, M., Höfle, B., & Pfeifer, N. (2010). Object-based image analysis for semi-automated geomorphometric feature extraction from high-resolution digital elevation models in alpine mountain areas. *Journal of Photogrammetry and Remote Sensing, 65*(1), 28–36.
124. Sakaki, T., Okazaki, M., & Matsuo, Y. (2010). Earthquake shakes Twitter users: Real-time event detection by social sensors. In *Proceedings of the 19th International Conference on World Wide Web* (pp. 851–860).
125. Schafer, J. L., & Graham, J. W. (2002). Missing data: Our view of the state of the art. *Psychological Methods, 7*(2), 147–177.
126. Seto, K. C., Güneralp, B., & Hutyra, L. R. (2012). Global forecasts of urban expansion to 2030 and direct impacts on biodiversity and carbon pools. *Proceedings of the National Academy of Sciences, 109*(40), 16083–16088.
127. Shelton, T., Poorthuis, A., Graham, M., & Zook, M. (2015). Mapping the data shadows of Hurricane Sandy: Uncovering the sociospatial dimensions of 'big data.' *Geoforum, 52*, 167–179.
128. Shelton, T., Zook, M., & Wiig, A. (2014). The 'actually existing smart city.' *Cambridge Journal of Regions, Economy, and Society, 8*(1), 13–25.
129. Silva, T. H., Melo, P. O., Almeida, J. M., Salles, J., & Loureiro, A. A. (2013). Visualizing the invisible image of cities. In *Proceedings of the 2013 IEEE International Conference on Green Computing and Communications and IEEE Internet of Things and IEEE Cyber, Physical and Social Computing* (pp. 334–341).
130. Stefanidis, A., Croitoru, A., & Radzikowski, J. (2013). Harvesting ambient geospatial information from social media feeds. *GeoJournal, 78*(2), 319–338.
131. Tateishi, R., Hoan, N. T., Kobayashi, T., Alsaaideh, B., Tana, G., & Phong, D. X. (2014). Production of global land cover data—GLCNMO2008. *Journal of Geography and Geology, 6*(3), 99–122.
132. Taylor, L., & Schroeder, R. (2015). Is bigger better? The emergence of big data as a tool for international development policy. *GeoJournal, 80*(4), 503–518.
133. Tzoulas, K., Korpela, K., Venn, S., Yli-Pelkonen, V., Kaźmierczak, A., Niemela, J., & James, P. (2007). Promoting ecosystem and human health in urban areas using green infrastructure: A literature review. *Landscape and Urban Planning, 81*(3), 167–178.
134. Ubaldi, B. (2013). Open government data: Towards empirical analysis of open government data initiatives. *OECD Working Papers on Public Governance*, No. 22. OECD Publishing.

135. United Nations (2008). *Principles and Recommendations for Population and Housing Censuses*. United Nations Publications.
136. Wang, R. Y., & Strong, D. M. (1996). Beyond accuracy: What data quality means to data consumers. *Journal of Management Information Systems, 12*(4), 5–33.
137. Wang, Y., Kung, L., & Byrd, T. A. (2016). Big data analytics: Understanding its capabilities and potential benefits for healthcare organizations. *Technological Forecasting and Social Change, 126*, 3–13.
138. Weeks, J. R. (2015). *Population: An introduction to concepts and issues*. Cengage Learning.
139. Weng, Q. (2012). Remote sensing of impervious surfaces in the urban areas: Requirements, methods, and trends. *Remote Sensing of Environment, 117*, 34–49.
140. Weng, Q. (Ed.). (2012). *Remote sensing of urban and suburban areas*. Springer Science & Business Media.
141. Wesolowski, A., Eagle, N., Noor, A. M., Snow, R. W., & Buckee, C. O. (2012). Heterogeneous mobile phone ownership and usage patterns in Kenya. *PLoS ONE, 7*(4), e35319.
142. Williams, M. L., Burnap, P., & Sloan, L. (2016). Towards an ethical framework for publishing Twitter data in social research: Taking into account users' views, online context and algorithmic estimation. *Sociology, 50*(6), 1149–1168.
143. Wilson, R. E., Gosling, S. D., & Graham, L. T. (2012). A review of Facebook research in the social sciences. *Perspectives on Psychological Science, 7*(3), 203–220.
144. Winkler, W. E. (1999). The state of record linkage and current research problems. Statistical Research Division, U.S. Bureau of the Census.
145. Wu, L., & Brynjolfsson, E. (2015). The future of prediction: How Google searches foreshadow housing prices and sales. In *Economic analysis of the digital economy* (pp. 89–118). University of Chicago Press.
146. Wulder, M. A., Coops, N. C., Roy, D. P., White, J. C., & Hermosilla, T. (2016). Land cover 2.0. *International Journal of Remote Sensing, 37*(21), 5081–5101.
147. Wulder, M. A., White, J. C., Loveland, T. R., Woodcock, C. E., Belward, A. S., Cohen, W. B., Pekel, J. F., et al. (2019). The global Landsat archive: Status, consolidation, and direction. *Remote Sensing of Environment, 224*, 332–344.
148. Ye, X., Wang, T., Li, X., & Weng, M. (2018). A framework for exploring the relationship between the spatial configuration of urban services and their use: The case of shopping malls in Guangzhou. *Cities, 74*, 1–10.
149. Yin, J., Soliman, A., Yin, D., & Wang, S. (2015). Monitoring the spatio-temporal dynamics of swidden agriculture and fallow vegetation recovery using Landsat imagery in northern Laos. *Remote Sensing of Environment, 169*, 1–11.
150. Yuan, Y., Raubal, M., & Liu, Y. (2012). Correlating mobile phone usage and travel behavior: A case study of Harbin, China. *Computers, Environment and Urban Systems, 36*(2), 118–130.
151. Yuan, Y., Zheng, Y., & Xie, X. (2012). Discovering regions of different functions in a city using human mobility and POIs. In *Proceedings of the 18th ACM SIGKDD International Conference on Knowledge Discovery and Data Mining* (pp. 186–194). ACM.
152. Zaharia, M., Chowdhury, M., Franklin, M. J., Shenker, S., & Stoica, I. (2010). Spark: Cluster computing with working sets. *HotCloud, 10*(10), 95–103.
153. Zaharia, M., Xin, R. S., Wendell, P., Das, T., Armbrust, M., Dave, A., Stoica, I., et al. (2016). Apache Spark: A unified engine for big data processing. *Communications of the ACM, 59*(11), 56–65.
154. Zanella, A., Bui, N., Castellani, A., Vangelista, L., & Zorzi, M. (2014). Internet of things for smart cities. *IEEE Internet of Things Journal, 1*(1), 22–32.
155. Zhang, D., Xing, Y., & Wang, L. (2017). Analysis of urban residents' travel behavior based on social media data mining: A case study of Weibo. *Cities, 66*, 106–115.
156. Zhang, K., Hristidis, V., & Rey, S. J. (2016). Get on the bandwagon: The effect of opinion leaders in information cascades. *PLoS ONE, 11*(4), e0155137.
157. Zheng, Y., Capra, L., Wolfson, O., & Yang, H. (2014). Urban computing: Concepts, methodologies, and applications. *ACM Transactions on Intelligent Systems and Technology, 5*(3), 1–55.

158. Zimmer, M. (2010). "But the data is already public": On the ethics of research in Facebook. *Ethics and Information Technology, 12*(4), 313–325.
159. Zuiderwijk, A., Janssen, M., & Dwivedi, Y. K. (2015). Acceptance and use predictors of open data technologies: Drawing upon the unified theory of acceptance and use of technology. *Government Information Quarterly, 32*(4), 429–440.
160. Zwitter, A. (2014). Big data ethics. *Big Data & Society, 1*(2), 1–6.

Part I
AI Applications in Human Geography

Chapter 4
Population Distribution and Migration Patterns

4.1 Overview of Population Distribution and Migration Patterns

Population distribution and migration patterns are crucial aspects of human geography, influencing a variety of factors such as social, economic, political, and environmental dynamics. Understanding these patterns is essential for urban planning, policy-making, and resource allocation. In this section, we will discuss the definitions and concepts related to population distribution and migration patterns.

Population Distribution: Population distribution refers to the spatial arrangement of individuals within a geographical area or region [10]. It is typically characterized by a variety of factors such as population density, population size, and the spatial organization of settlements. Population distribution can be studied at various scales, from local to global, and can provide insights into the socioeconomic and environmental factors influencing human settlements [25].

There are three primary patterns of population distribution: clustered, dispersed, and linear. Clustered population distribution occurs when individuals are concentrated in specific areas, such as urban centers or transportation hubs. Dispersed population distribution is characterized by individuals living relatively far from one another, often in rural or sparsely populated regions. Linear population distribution is characterized by individuals living along a linear feature, such as a river, coastline, or transportation corridor [10].

Migration Patterns: Migration is the movement of individuals or groups from one location to another, often driven by factors such as economic opportunities, social connections, environmental conditions, and political circumstances [11]. Migration can be classified in several ways, including:

1. Internal versus International Migration: Internal migration refers to the movement of individuals within the borders of a country, while international migration involves the movement of individuals across national borders [6].

S. N. M. Moghaddam and H. Cao, *Artificial Intelligence-Driven Geographies*, City Development: Issues and Best Practices, https://doi.org/10.1007/978-981-97-5116-7_4

2. Voluntary versus Forced Migration: Voluntary migration is the movement of individuals based on their own choices and decisions, often driven by the pursuit of better opportunities, while forced migration occurs when individuals are compelled to move due to factors beyond their control, such as conflict or natural disasters [11].
3. Temporary versus Permanent Migration: Temporary migration refers to the movement of individuals for a limited period, such as seasonal workers or students, while permanent migration involves the long-term relocation of individuals and their families [14].
4. Rural-to-Urban versus Urban-to-Rural Migration: Rural-to-urban migration refers to the movement of individuals from rural areas to urban centers, often driven by economic opportunities and improved access to services, while urban-to-rural migration involves the movement of individuals from urban centers to rural areas, often driven by factors such as lower living costs and a desire for a different lifestyle [60].

Having established the definitions and concepts related to population distribution and migration patterns, it is crucial to examine the factors influencing these patterns. Various elements, such as economic conditions, social and cultural factors, environmental factors, and political circumstances, can drive population distribution and migration patterns [11, 60].

Economic Factors: Economic factors are among the most significant drivers of migration and population distribution. Individuals often migrate to pursue better job opportunities, higher wages, and improved standards of living [11]. The availability of economic resources and infrastructure can also influence population distribution, as areas with more significant economic opportunities tend to attract larger populations [10].

Social and Cultural Factors: Social and cultural factors also play a crucial role in population distribution and migration patterns. For example, individuals may migrate to join family members or friends who have already migrated, creating social networks that facilitate further migration [45]. Cultural factors, such as language, religion, and cultural practices, can also influence migration decisions and population distribution, as individuals may prefer to live in areas with a shared cultural background [11].

Environmental Factors: Environmental factors can significantly impact population distribution and migration patterns. Climate change, natural disasters, and resource scarcity can all drive migration, as individuals and communities seek to adapt to changing environmental conditions (McLeman, 2018). Additionally, population distribution can be influenced by factors such as topography, climate, and the availability of natural resources, as certain environments may be more conducive to human settlement than others [10].

Political Factors: Political factors can also affect population distribution and migration patterns. Conflicts, political persecution, and human rights abuses can lead to forced migration, as individuals seek refuge from violence and persecution [11]. Additionally, government policies and regulations, such as immigration policies and regional development initiatives, can influence migration patterns and population distribution [60].

Understanding the various factors influencing population distribution and migration patterns is essential for researchers and policymakers, as it enables the development of targeted policies and interventions to address the challenges and opportunities associated with these patterns. By considering economic, social, cultural, environmental, and political factors, a more comprehensive understanding of population distribution and migration patterns can be achieved.

4.2 Data Sources for Studying Population Distribution and Migration Patterns

4.2.1 Traditional Data Sources

Studying population distribution and migration patterns requires the use of various data sources to provide a comprehensive understanding of the complex factors influencing these patterns (Table 4.1). Traditional data sources have long been utilized in human geography and related fields to analyze population distribution and migration patterns. In this section, we will discuss some of the most commonly used traditional data sources for studying population distribution and migration patterns.

Census Data

Census data is a primary source of information on population distribution and migration patterns. National censuses are conducted by governments to collect demographic, social, and economic information on their populations [66]. Census data typically includes information on population size, age structure, sex distribution, marital status, education, employment, and migration patterns. This data is valuable for researchers and policymakers, as it provides a comprehensive snapshot of the population at a specific point in time and enables the identification of trends and patterns over time [55].

Population Registers

Population registers are another essential source of information on population distribution and migration patterns. These registers are maintained by governments and typically include information on births, deaths, marriages, and migration events [66]. Population registers can provide more up-to-date and detailed information on population dynamics than census data, as they are continuously updated to reflect changes in the population. However, the quality and coverage of population registers can vary between countries, and not all countries maintain comprehensive population registers [55].

Table 4.1 Data sources for population distribution and migration patterns studies

Traditional data sources	Description	Advantages	Limitations
Census data	Primary source of demographic, social, and economic information collected by governments through national censuses	Provides comprehensive snapshot and enables trend analysis	Collected infrequently, may not capture rapid changes
Population registers	Maintained by governments, includes births, deaths, marriages, and migrations	Up-to-date information on population dynamics	Varies in quality and coverage between countries
Household surveys	Conducted by statistical offices or research institutions, gather detailed demographic and socioeconomic data	Provides nuanced information on population dynamics	Limited by sample size, response rates, and potential biases
Administrative data	Information collected by governments as part of routine operations, e.g., tax records, social security data	Includes individual and household characteristics, migration events, and socioeconomic variables	Subject to incomplete coverage, data quality issues, and potential biases
Historical data	Historical census records, parish registers, and migration records	Provides insights into long-term trends and patterns	Subject to data quality issues, incomplete coverage, and potential biases

Household Surveys

Household surveys are another important source of data for studying population distribution and migration patterns. These surveys are typically conducted by national statistical offices, research institutions, or international organizations and collect information on a wide range of demographic, social, and economic variables [66]. Household surveys can provide more detailed and nuanced information on population dynamics than census data or population registers, as they often include questions on individual and household characteristics, migration histories, and migration intentions. However, household surveys may be subject to various limitations, such as sample size, response rates, and potential biases in the data [55].

Administrative Data

Administrative data refers to information collected by governments and other organizations as part of their routine operations, such as tax records, social security data, and immigration records [66]. Administrative data can provide valuable insights into population distribution and migration patterns, as it often includes information on individual and household characteristics, migration events, and socioeconomic variables. However, administrative data may be subject to various limitations, such as incomplete coverage, data quality issues, and potential biases in the data [55].

Historical Data

Historical data, such as historical census records, parish registers, and migration records, can provide valuable insights into long-term trends and patterns in population distribution and migration. This data can help researchers and policymakers understand how population distribution and migration patterns have evolved over time and identify the factors driving these changes [55]. However, historical data may be subject to various limitations, such as data quality issues, incomplete coverage, and potential biases in the data [66].

Despite the significant value of traditional data sources in studying population distribution and migration patterns, these sources also have their limitations. For example, census data is only collected every few years and may not accurately capture rapid changes in population dynamics. Population registers may not be available in all countries or may have incomplete coverage. Household surveys can be limited by sample size, response rates, and potential biases in the data. Administrative data may have data quality issues, and historical data may be subject to incomplete coverage and potential biases [55, 66].

To overcome some of these limitations, researchers and policymakers are increasingly turning to alternative data sources and innovative data collection methods, such as big data, remote sensing, and social media data, to supplement traditional data sources and provide new insights into population distribution and migration patterns.

In recent years, there has been a growing interest in the potential of big data and new data sources for studying population distribution and migration patterns. Big data refers to large-scale, diverse, and high-velocity data generated by various sources, such as social media platforms, mobile phone networks, and remote sensing technologies [34]. These data sources can provide valuable insights into population dynamics and human mobility patterns, complementing traditional data sources and enabling researchers to address some of the limitations and gaps in existing data [9].

For example, social media data can be used to analyze migration patterns and population distribution in real-time, providing insights into the factors driving migration and the experiences of migrants. Mobile phone data can be used to track human mobility patterns and model population distribution, enabling researchers to monitor population dynamics in areas with limited traditional data sources [16]. Remote sensing data, such as satellite imagery and LiDAR data, can be used to study land use patterns and population distribution, providing information on the environmental factors influencing human settlements [62].

In conclusion, while traditional data sources remain essential for studying population distribution and migration patterns, the emergence of big data and alternative data sources offers new opportunities and challenges for researchers and policymakers. By integrating traditional data sources with big data and innovative data collection methods, a more comprehensive understanding of population distribution and migration patterns can be achieved, enabling the development of targeted interventions and policies to address the challenges and opportunities associated with these patterns.

4.2.2 Big Data and Geospatial Data Sources

In recent years, the emergence of big data and geospatial data sources has provided new opportunities for studying population distribution and migration patterns (Table 4.2). These data sources can complement traditional data sources by providing more timely, detailed, and comprehensive information on population dynamics. In this section, we will discuss some of the most used big data and geospatial data sources for studying population distribution and migration patterns.

Table 4.2 Big data primary sources with advantages and limitations

Big data and geospatial data sources	Description	Advantages	Limitations
Social media data	Data from platforms like Twitter, Facebook, and Instagram, offering insights into user behavior and location	Real-time analysis, insights into mobility patterns	Biases in user representation, data quality issues, privacy concerns
Mobile phone data	Call detail records (CDRs) and GPS data, providing granular insights into human mobility	Real-time and granular information on population dynamics	Biases in user representation, data quality issues, privacy concerns
Remote sensing data	Satellite imagery and LiDAR data, offering information on land cover and topography	Insights into environmental factors influencing population distribution	Data quality issues, spatial and temporal resolution constraints, specialized expertise needed
Geospatial data	GIS data and GPS data, providing detailed information on human settlements and mobility patterns	Detailed spatial analysis of population distribution and migration	Data quality issues, spatial resolution constraints, specialized expertise needed
Internet data	Search query data, website traffic data, and online job postings, offering insights into migration factors	Analysis of factors influencing migration decisions and job opportunities	Biases in user representation, data quality issues, privacy concerns

Social Media Data

Social media platforms, such as Twitter, Facebook, and Instagram, generate vast amounts of data on user behavior, interactions, and location information. This data can be used to study population distribution and migration patterns by analyzing the spatial and temporal patterns of social media activity [78]. Social media data can provide insights into the mobility patterns of individuals, social networks, and the factors influencing migration decisions. However, social media data may be subject to various limitations, such as biases in user representation, data quality issues, and privacy concerns [78].

Mobile Phone Data

Mobile phone data, such as call detail records (CDRs) and global positioning system (GPS) data, can provide valuable insights into population distribution and migration patterns. Mobile phone data can be used to analyze the spatial and temporal patterns of human mobility, as well as the factors influencing migration decisions [8]. Mobile phone data can provide more granular and real-time information on population dynamics than traditional data sources, such as census data or population registers. However, mobile phone data may be subject to various limitations, such as biases in user representation, data quality issues, and privacy concerns [8].

Remote Sensing Data

Remote sensing data, such as satellite imagery and Light Detection and Ranging (LiDAR) data, can provide valuable information on population distribution and migration patterns. Satellite imagery can be used to analyze land cover and land use changes, which can provide insights into the factors influencing population distribution and migration patterns [64]. LiDAR data can be used to generate high-resolution digital elevation models (DEMs), which can be used to analyze the relationship between topography and population distribution. However, remote sensing data may be subject to various limitations, such as data quality issues, temporal and spatial resolution constraints, and the need for specialized software and expertise for data processing and analysis [64].

Geospatial Data

Geospatial data, such as geographic information system (GIS) data and global positioning system (GPS) data, can be used to study population distribution and migration patterns by analyzing the spatial patterns of human settlements and mobility. GIS data can provide detailed information on the location and characteristics of human settlements, such as population density, land use, and infrastructure [24]. GPS data can be used to analyze the spatial and temporal patterns of human mobility, as well as the factors influencing migration decisions. However, geospatial data may be subject to various limitations, such as data quality issues, spatial resolution constraints, and the need for specialized software and expertise for data processing and analysis [24].

Internet Data

Internet data, such as search query data, website traffic data, and online job postings, can provide valuable insights into population distribution and migration patterns. Search query data can be used to analyze the factors influencing migration decisions, such as job opportunities, housing affordability, and quality of life [23]. Website traffic data can be used to analyze the spatial patterns of information-seeking behavior, which can provide insights into the factors influencing population distribution and migration patterns [48]. Online job postings can be used to analyze the spatial distribution of job opportunities and the factors influencing labor migration [36]. However, Internet data may be subject to various limitations, such as biases in user representation, data quality issues, and privacy concerns [23].

In conclusion, big data and geospatial data sources, such as social media data, mobile phone data, remote sensing data, geospatial data, and Internet data, offer new opportunities for studying population distribution and migration patterns. These data sources can complement traditional data sources by providing more timely, detailed, and comprehensive information on population dynamics. By leveraging the strengths of both traditional and emerging data sources, researchers and policymakers can develop a more comprehensive understanding of the demographic, social, economic, and environmental factors influencing population distribution and migration patterns, which can inform the development of targeted interventions and policies to address the challenges and opportunities associated with these patterns.

4.3 AI Techniques for Analyzing Population Distribution and Migration Patterns

Population distribution and migration patterns are crucial topics in human geography and urban planning, as they provide insights into how people settle and move within a geographical area. This information is essential for making informed decisions about resource allocation, urban planning, and social welfare programs. In recent years, machine learning (ML) approaches have been increasingly utilized to analyze population distribution and migration patterns. These techniques enable the efficient processing of vast amounts of data, leading to more accurate and timely results.

4.3.1 Supervised Learning

Supervised learning is a common ML technique used to analyze population distribution and migration patterns (Fig. 4.1). In supervised learning, the algorithm is trained on a labeled dataset, where each input data point has an associated target output or class label. The algorithm learns to make predictions by finding patterns in the input features and their relationship with the target output.

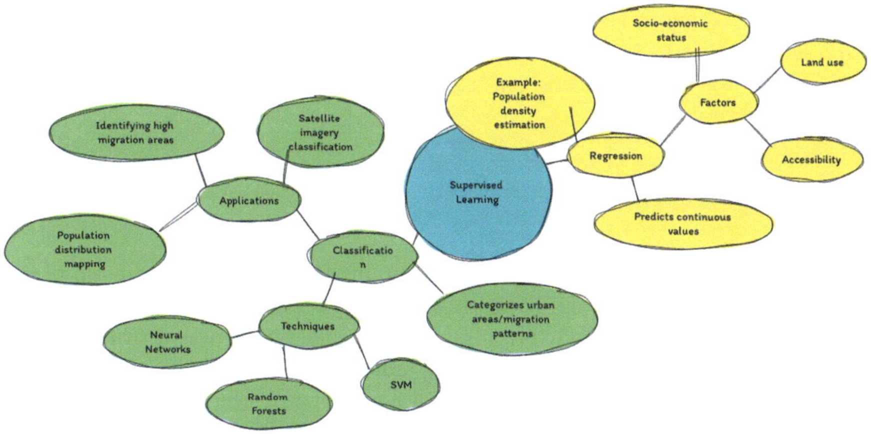

Fig. 4.1 The concept of supervised learning, including both regression and classification methods

Regression is a supervised learning method that can predict continuous numerical values. It is often used to estimate population density, urban growth, or migration rates. For example, multiple linear regression can be used to model the relationship between various factors (e.g., socio-economic status, land use, and accessibility) and population density. This technique can help identify the most significant factors affecting population distribution and migration patterns [29].

Classification techniques can be employed to categorize different types of urban areas or migration patterns. For instance, supervised classification algorithms such as Support Vector Machines (SVM), Random Forests, or Neural Networks can be applied to satellite imagery to classify land cover and detect urban areas [76]. This information can be used to derive population distribution maps and identify areas experiencing high levels of migration.

4.3.2 Unsupervised Learning

Unsupervised learning techniques do not rely on labeled data and can identify patterns, clusters, or associations within the data. These methods are useful for exploring population distribution and migration patterns when ground truth data is scarce or expensive to obtain.

Clustering algorithms, such as K-means or DBSCAN, can be used to group similar regions or migration events based on their characteristics (e.g., population density, socio-economic factors, or migration rates). Clustering can reveal spatial patterns in the data, enabling researchers to better understand the underlying factors driving population distribution and migration [67].

Dimensionality reduction techniques, such as Principal Component Analysis (PCA) or t-distributed Stochastic Neighbor Embedding (t-SNE), can be used to visualize high-dimensional data in lower-dimensional space. These methods can help identify patterns or trends in population distribution and migration data that might not be apparent in the original feature space [49].

4.3.3 Deep Learning

Deep learning, a subfield of machine learning, involves the use of artificial neural networks with multiple layers to learn complex patterns in large datasets. Deep learning methods have shown promise in analyzing population distribution and migration patterns, particularly when using remote sensing data.

Convolutional Neural Networks (CNNs) are a type of deep learning algorithm that can effectively process spatial data, such as satellite imagery. CNNs have been used to estimate population distribution by classifying land cover types and detecting built-up areas from high-resolution satellite images [79]. The derived information can be combined with other data sources, such as census data, to create more accurate population distribution maps.

Recurrent Neural Networks (RNNs) and Long Short-Term Memory (LSTM) Networks are deep learning models designed to handle sequentialdata, making them well-suited for analyzing time-series data related to population distribution and migration patterns. These models can capture long-term dependencies in the data and predict future population changes or migration events based on historical patterns. For instance, RNNs and LSTMs have been employed to forecast international migration flows by analyzing time-series data on factors such as GDP, population size, and political stability [2].

Data Fusion and Ensemble Learning

Data fusion involves integrating data from multiple sources to provide more comprehensive and accurate insights into population distribution and migration patterns. Ensemble learning is a technique that combines the predictions of multiple ML models to improve overall performance.

Data fusion can be used to combine traditional data sources (e.g., census data, surveys) with big data (e.g., social media data, mobile phone data) and geospatial data (e.g., satellite imagery, GPS data) to obtain a more holistic view of population distribution and migration patterns. For example, researchers have used data fusion techniques to integrate mobile phone data with census data to estimate population distribution at a finer spatial resolution [53].

Ensemble learning can improve the performance of ML models by combining their predictions. This approach is particularly useful when analyzing population distribution and migration patterns, as different models may capture different aspects of the data. Techniques such as bagging, boosting, and stacking can be employed to create ensemble models that provide more accurate and robust predictions [57].

Challenges and Future Directions

Despite the potential of ML techniques in analyzing population distribution and migration patterns, several challenges remain. These include the quality and representativeness of the data, the need for interpretability of ML models, and ethical considerations related to privacy and data protection. Addressing these challenges will be essential to ensure that ML approaches are effectively and responsibly applied in human geography and urban planning.

In conclusion, machine learning approaches offer significant opportunities for analyzing population distribution and migration patterns. By leveraging various techniques, such as supervised and unsupervised learning, deep learning, data fusion, and ensemble learning, researchers can gain valuable insights into the factors driving population distribution and migration. As ML technology continues to advance, it is expected to play an increasingly important role in human geography and urban planning.

4.4 Applications of AI in Population Distribution and Migration Studies

4.4.1 Population Estimation and Density Analysis

Population estimation and density analysis are essential components of understanding human geography and urban planning. These analyses help policymakers and planners allocate resources, develop infrastructure, and manage urban growth effectively. Machine learning (ML) has emerged as a powerful tool for population estimation and density analysis, enabling researchers to process large-scale, high-resolution data sets with unprecedented accuracy and efficiency.

Traditional methods of population estimation and density analysis relied on census data, which can be outdated and limited in spatial resolution. ML techniques have the potential to enhance these traditional methods by incorporating additional data sources and providing more detailed, timely, and accurate population estimates.

One of the primary applications of ML in population estimation and density analysis is the use of satellite imagery to infer population density. This approach takes advantage of the vast amount of high-resolution satellite images available, using convolutional neural networks (CNNs) to extract features from these images that correlate with population density [31, 72]. By training models on a combination of satellite imagery and ground-truth population data, ML algorithms can generate detailed and accurate population density maps at a global scale [72].

Another ML approach to population estimation involves analyzing nighttime light (NTL) data from satellite imagery. NTL data can serve as a proxy for human activity and population density, particularly in urban areas. ML techniques such as random forests and support vector machines have been used to model the relationship between

NTL data and population density, resulting in more accurate population estimates [43, 80].

Mobile phone data is another valuable source for population estimation and density analysis. Call Detail Records (CDRs) can provide insights into the spatial distribution of mobile phone users, which can be used as a proxy for population density. ML techniques have been applied to analyze CDRs and estimate population density, enabling researchers to generate real-time and high-resolution population maps [15, 16].

Social media data, such as geolocated tweets, can also be utilized for population estimation and density analysis. Geolocated tweets can provide real-time information about the spatial distribution of social media users, which can be used as a proxy for population density. ML algorithms, including clustering techniques and deep learning models, have been applied to social media data to estimate population density at various spatial scales [69, 75].

4.4.2 Migration Pattern Detection and Forecasting

Migration pattern detection and forecasting are essential components of understanding human geography, urban planning, and policy development. By utilizing artificial intelligence (AI) techniques, researchers can better analyze and predict migration patterns, enabling more effective resource allocation and decision-making in various sectors.

Migration Pattern Detection

Migration pattern detection involves identifying trends, routes, and hotspots of population movements. AI techniques, such as machine learning (ML) algorithms (Fig. 4.2), can be employed to detect these patterns from various data sources, such as census data, mobile phone data, social media data, and satellite imagery [3, 41, 82].

For instance, ML algorithms can be used to analyze mobile phone data to identify users' home and work locations, commuting patterns, and other mobility behaviors [16]. Similarly, geotagged social media data can be used to track and analyze migration patterns by identifying users' locations and movements over time [28, 37].

Remote sensing data, such as satellite imagery, can also be used to detect migration patterns by analyzing changes in land use, urban expansion, and population density [63, 80]. For example, ML algorithms can be applied to nighttime light data to estimate population distribution and density, as well as to detect urban expansion and migration trends [80].

Migration Forecasting

Migration forecasting aims to predict future migration flows, identify potential hotspots, and assess the impacts of migration on population distribution, infrastructure, and the environment. AI techniques, such as ML algorithms and data mining,

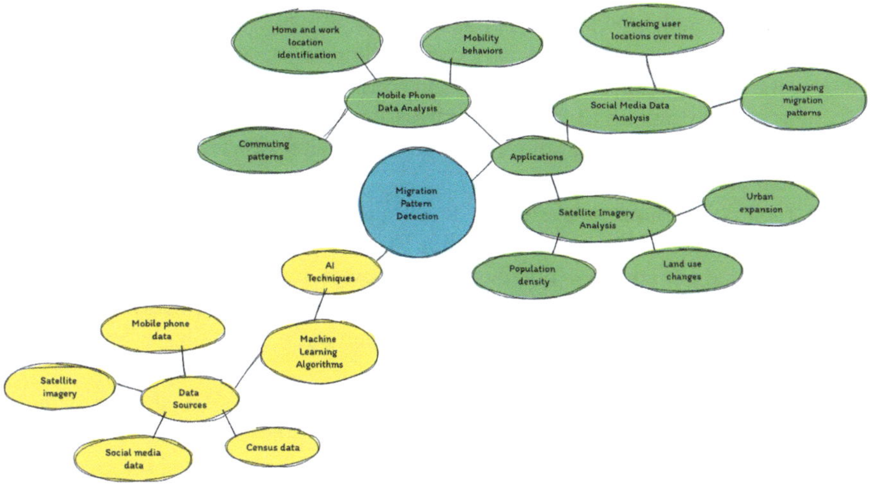

Fig. 4.2 Migration pattern detection, showcasing the use of AI techniques

can be employed to analyze historical migration data, socio-economic indicators, environmental factors, and conflict data to develop predictive models [41, 68].

For example, ML algorithms can be trained on historical migration data to identify factors driving migration, such as economic opportunities, social networks, environmental factors, and political stability [41, 68]. These factors can then be incorporated into predictive models to forecast future migration flows and identify potential hotspots for migration.

In addition to traditional data sources, such as census data and administrative records, researchers can also utilize big data sources, such as mobile phone data, social media data, and satellite imagery, to improve the accuracy and timeliness of migration forecasts [3, 70]. For instance, mobile phone data can be used to track population movements in real-time, enabling researchers to monitor and predict migration flows during natural disasters, conflicts, and other crisis situations [41, 70].

Challenges and Future Directions

Despite the significant advancements in AI-driven migration pattern detection and forecasting, several challenges remain. These include data privacy concerns, data quality and representativeness, and the need for continued innovation in AI techniques and data integration.

Data privacy concerns arise from the use of sensitive data sources, such as mobile phone data and social media data, which may contain personally identifiable information [3]. To address these concerns, researchers must develop and implement data privacy protocols, such as anonymization, aggregation, and data minimization, to protect individual privacy while preserving the utility of the data for migration analysis.

Data quality and representativeness are also critical issues, as biases in data sources, such as sampling biases in social media data or coverage gaps in mobile phone data, can influence the accuracy and generalizability of migration pattern detection and forecasting [3, 37]. To address these challenges, researchers should seek to combine multiple data sources and develop data fusion techniques to enhance data quality and representativeness [41].

The rapidly evolving nature of AI techniques and data sources presents both opportunities and challenges for migration studies. Researchers must continually update their methods and models to incorporate new techniques, such as deep learning algorithms and advanced geospatial analysis tools, as well as new data sources, such as high-resolution satellite imagery, mobile phone data, and online user-generated content [3, 41].

Moreover, interdisciplinary collaboration and knowledge-sharing among researchers, practitioners, and policymakers are essential to ensure that AI-driven migration pattern detection and forecasting are effectively translated into actionable insights and evidence-based policy interventions [68].

In conclusion, AI techniques hold significant potential for improving our understanding of population distribution and migration patterns, as well as informing urban planning, resource allocation, and policy development. By leveraging the power of AI, researchers and practitioners can develop more accurate, timely, and comprehensive insights into migration trends and their implications for human geography and urban planning.

4.4.3 Impact Assessment of Migration on Urban Planning and Infrastructure

The rapid growth and urbanization of the global population has made the understanding of migration patterns and their impacts on urban planning and infrastructure increasingly important. The application of AI in population distribution and migration studies can provide valuable insights for policymakers and urban planners to make informed decisions regarding the allocation of resources, development of infrastructure, and sustainable urban growth. In this section, we will discuss how AI can be used to assess the impacts of migration on urban planning and infrastructure.

Analyzing Historical Migration Trends

To understand the impact of migration on urban planning and infrastructure, it is essential to first analyze historical migration trends. AI techniques such as time series analysis, clustering algorithms, and machine learning models can be used to analyze historical data and identify patterns in population distribution and migration over time [20]. This can provide valuable insights into the factors driving migration, such as economic opportunities, social networks, and environmental conditions, which can inform urban planning and infrastructure development.

Assessing the Impact of Migration on Urban Infrastructure and Services

The movement of people from rural to urban areas or between cities can put significant pressure on urban infrastructure and services. AI techniques, such as machine learning models, can be used to assess the impact of migration on various aspects of urban infrastructure, such as housing, transportation, water and sanitation systems, and public services [13]. By analyzing and predicting the demand for these services and infrastructure, policymakers and urban planners can ensure that they are adequately prepared for population growth and changing demographics.

Forecasting Future Migration Patterns and Population Distribution

AI techniques, such as machine learning models and agent-based simulations, can be used to forecast future migration patterns and population distribution based on historical data, demographic trends, and projected changes in economic, social, and environmental conditions [38]. These forecasts can inform urban planning and infrastructure development, ensuring that resources are allocated effectively and sustainably.

Evaluating the Effectiveness of Urban Planning and Infrastructure Interventions

AI techniques can also be used to evaluate the effectiveness of urban planning and infrastructure interventions in addressing the challenges posed by migration. For example, machine learning models can be used to analyze the impact of different policy interventions on population distribution, housing affordability, and access to public services, providing valuable insights for policymakers and urban planners to develop more effective strategies [5].

Identifying Vulnerable Populations and Areas at Risk

Identifying vulnerable populations and areas at risk is crucial for targeted urban planning and infrastructure development. AI techniques, such as machine learning algorithms and geospatial analysis, can help identify areas with high concentrations of vulnerable populations, such as low-income communities, migrants, and minority groups, who may be disproportionately affected by inadequate infrastructure and services [35]. By pinpointing these areas, policymakers and urban planners can prioritize interventions and allocate resources more effectively to address the specific needs of these communities.

Incorporating Public Participation and Stakeholder Input in Urban Planning and Infrastructure Development

AI techniques can also be used to facilitate public participation and stakeholder input in urban planning and infrastructure development. For example, natural language processing (NLP) and sentiment analysis can be employed to analyze public opinions and preferences gathered from social media platforms, surveys, and public consultations [30]. This information can be used to inform urban planning decisions and ensure that the perspectives of affected communities are taken into account.

In conclusion, AI techniques can provide valuable insights and tools for assessing the impact of migration on urban planning and infrastructure. By analyzing historical migration trends, assessing the impact of migration on urban infrastructure and services, forecasting future migration patterns and population distribution, evaluating the effectiveness of urban planning and infrastructure interventions, identifying vulnerable populations and areas at risk, and incorporating public participation and stakeholder input, AI can help inform more effective and sustainable urban planning and infrastructure development to address the challenges posed by migration.

4.5 Challenges and Limitations of AI in Population Distribution and Migration Analysis

Despite the significant potential of AI in population distribution and migration analysis, there are several challenges and limitations that researchers, practitioners, and policymakers must consider when employing these techniques. This section will discuss some of these challenges, including data availability and quality, ethical concerns, algorithmic biases, and the interpretability of AI models.

One of the major challenges in using AI techniques for population distribution and migration analysis is the availability and quality of the underlying data. AI models often require large volumes of high-quality, representative data to produce reliable and accurate results [73]. However, obtaining such data can be difficult, particularly for developing countries with limited resources and infrastructure to collect and maintain comprehensive datasets on population and migration [44].

Moreover, data on migration can be sensitive and subject to political influences, leading to issues such as underreporting, misreporting, or inconsistencies across different sources [54]. These issues can affect the performance of AI models, leading to inaccurate or biased predictions and analyses.

The use of AI techniques in population distribution and migration analysis raises several ethical concerns, particularly regarding privacy and the potential for surveillance and discrimination [65]. For instance, the use of big data sources, such as social media and mobile phone data, can reveal sensitive information about individuals' locations, movements, and personal lives, potentially infringing on their privacy rights [12].

Moreover, the analysis of migration data may inadvertently contribute to the stigmatization or targeting of certain migrant groups, particularly if the data is used to inform policies and interventions that disproportionately affect these populations [58]. Researchers and practitioners must carefully consider the ethical implications of their work and strive to ensure that their analyses respect the rights and dignity of the individuals and communities they study.

AI models, particularly machine learning algorithms, are susceptible to biases that can result from the training data, model assumptions, or algorithm design [4]. For example, if the training data used to develop an AI model for migration analysis is biased towards certain demographic groups or geographic regions, the model may produce biased predictions and analyses that do not accurately represent the broader population [83]. These biases can be compounded by the fact that many AI models are "black boxes," with limited transparency and interpretability, making it difficult to identify and correct for biases in the model outputs [26].

To address these issues, researchers must carefully consider the representativeness and quality of their data and employ techniques such as re-sampling, weighting, or algorithmic fairness interventions to mitigate potential biases [22]. Additionally, the development of more interpretable and explainable AI models can help ensure that biases are identified and addressed in a more transparent and accountable manner [1].

AI models, particularly complex machine learning algorithms, can be difficult to interpret and communicate to non-experts, including policymakers, practitioners, and the communities affected by population distribution and migration analysis [39]. This lack of interpretability can limit the uptake and impact of AI techniques in human geography and urban planning, as well as the ability of stakeholders to scrutinize and evaluate the validity and reliability of AI-driven insights [17].

To address this challenge, researchers must prioritize the development of more interpretable and explainable AI models and invest in effective communication strategies that facilitate the understanding and adoption of AI-driven insights by a diverse range of stakeholders [56]. This may involve the use of visualization techniques, interactive tools, or narrative explanations to help convey the logic and reasoning behind AI model outputs and predictions [47].

In conclusion, while AI techniques offer significant potential for advancing our understanding of population distribution and migration patterns, researchers and practitioners must be aware of and address the various challenges and limitations associated with these approaches. By carefully considering issues related to data availability and quality, ethical concerns, algorithmic biases, and the interpretability of AI models, the field of human geography can harness the power of AI to produce more accurate, reliable, and impactful analyses that can inform urban planning and policy decisions. Furthermore, engaging with a diverse range of stakeholders and prioritizing transparency and accountability will be crucial for ensuring that AI-driven insights are used responsibly and equitably in the study of population distribution and migration patterns.

4.6 Future Directions in AI Applications for Population Distribution and Migration Studies

As AI continues to advance, there are several promising future directions for its application in the study of population distribution and migration patterns. This section will discuss some of these areas, including the integration of multi-modal data, the development of more interpretable AI models, the incorporation of domain knowledge, and the advancement of fairness, accountability, and transparency in AI-driven analyses.

One of the key areas where AI can significantly contribute to population distribution and migration studies is in the integration of multi-modal data. Combining different types of data, such as satellite imagery, social media posts, mobile phone records, and traditional census data, can provide a more comprehensive understanding of population dynamics and migration patterns [7, 40, 81].

Future research should focus on developing novel AI techniques that can effectively integrate these diverse data sources while accounting for their unique characteristics, such as spatial, temporal, and semantic resolutions. In particular, deep learning models, such as convolutional neural networks (CNNs) and recurrent neural networks (RNNs), have shown promise in handling multi-modal data [50, 61]. By leveraging these models, researchers can uncover hidden patterns and relationships within and between different types of data, leading to more accurate and insightful analyses of population distribution and migration patterns.

A key challenge in applying AI to population distribution and migration studies is the interpretability of AI models, which often function as "black boxes" due to their complex internal structures [26]. Developing more interpretable AI models is essential for enhancing the trust and credibility of AI-driven analyses in the field of human geography.

Recent research has focused on creating AI models that can provide explanations for their predictions, either by simplifying the model architecture or by developing post-hoc interpretability techniques [18, 42, 56]. Future work should continue to explore and refine these approaches, with a particular emphasis on tailoring them to the unique challenges and requirements of population distribution and migration studies.

Incorporating domain knowledge, such as the theories and principles of human geography and urban planning, into AI models can lead to more accurate and relevant analyses of population distribution and migration patterns [52]. For instance, integrating demographic, economic, and social factors, as well as the principles of spatial interaction and gravity models, can help AI models better capture the underlying mechanisms driving population dynamics and migration flows [21, 71].

Future research should focus on developing AI techniques that can seamlessly integrate domain knowledge, such as by incorporating expert-designed features or using knowledge graphs to encode relationships between different factors [51, 74]. This can help ensure that AI-driven analyses are grounded in the rich theoretical frameworks of human geography and urban planning, leading to more meaningful and actionable insights.

As AI-driven analyses become increasingly influential in shaping urban planning and policy decisions, it is essential to address issues of fairness, accountability, and transparency. Ensuring that AI models do not perpetuate or exacerbate existing social and economic inequalities, and that they are transparent and accountable in their decision-making processes, is critical for the responsible and ethical application of AI in population distribution and migration studies [4, 59].

Future research should focus on developing methods and techniques that can identify and mitigate biases in AI-driven analyses, such as re-sampling or re-weighting techniques to address imbalances in training data or regularization methods to prevent overfitting [32, 46, 77]. Additionally, researchers should explore the use of fairness-aware machine learning algorithms, which can explicitly consider fairness criteria during model training [19, 27, 33].

Moreover, efforts should be made to increase the transparency of AI models and their decision-making processes, such as by developing techniques for explaining model predictions or visualizing the internal workings of complex models [18, 42, 56]. These advances will help ensure that AI-driven analyses in population distribution and migration studies adhere to high ethical standards, fostering trust and credibility among stakeholders.

In conclusion, the future of AI applications in population distribution and migration studies is full of exciting possibilities and challenges. By focusing on the integration of multi-modal data, the development of more interpretable AI models, the incorporation of domain knowledge, and the advancement of fairness, accountability, and transparency, researchers can unlock the full potential of AI to provide valuable insights and drive informed decision-making in human geography and urban planning.

References

1. Adadi, A., & Berrada, M. (2018). Peeking inside the black-box: A survey on explainable artificial intelligence (XAI). *IEEE Access, 6*, 52138–52160.
2. Bahrami, M., Rinner, C., & Rizoiu, M. A. (2020). Forecasting international migration flows using time series and recurrent neural networks. In *Proceedings of the 2020 IEEE/ACM International Conference on Advances in Social Networks Analysis and Mining (ASONAM)* (pp. 536–543). IEEE.
3. Bakker, M. M., De Lange, M., & Yeboah, G. (2019). Big data for regional science: Opportunities, challenges, and directions. *Regional Studies, Regional Science, 6*(1), 404–419.
4. Barocas, S., & Selbst, A. D. (2016). Big data's disparate impact. *California Law Review, 104*, 671–732.
5. Batty, M., Axhausen, K. W., Giannotti, F., Pozdnoukhov, A., Bazzani, A., Wachowicz, M., Portugali, Y., et al. (2012). Smart cities of the future. *The European Physical Journal Special Topics, 214*(1), 481–518.
6. Bell, M., Charles-Edwards, E., & Ueffing, P. (2015). Internal migration and development: Comparing migration intensities around the world. *Population and Development Review, 41*(1), 33–58.
7. Bengtsson, L., Lu, X., Thorson, A., Garfield, R., & von Schreeb, J. (2011). Improved response to disasters and outbreaks by tracking population movements with mobile phone network data: A post-earthquake geospatial study in Haiti. *PLoS Medicine, 8*(8), e1001083.
8. Blumenstock, J. E. (2018). Using mobile phone data to study population distribution and migration patterns. In *Mobile technologies for population research* (pp. 1–16). Springer.
9. Blumenstock, J., Cadamuro, G., & On, R. (2015). Predicting poverty and wealth from mobile phone metadata. *Science, 350*(6264), 1073–1076.
10. Brown, L. A., & Horton, F. E. (2018). *Spatial patterns and processes: Studies in human geography*. Routledge.

11. Castles, S., de Haas, H., & Miller, M. J. J. (2013). *The age of migration: International population movements in the modern world*. Palgrave Macmillan.
12. Cinnamon, J., & Schuurman, N. (2013). Confronting the data-divide in a time of spatial turns and volunteered geographic information. *GeoJournal, 78*(4), 657–674.
13. Cui, B., & Liu, Y. (2015). An analysis of the relationship between urbanization and residential space consumption in China. *Habitat International, 46*, 99–107.
14. Czaika, M., & de Haas, H. (2014). The globalization of migration: Has the world become more migratory? *International Migration Review, 48*(2), 283–323.
15. Deng, Z., Wang, J., Wang, C., & Zhai, J. (2018). Fine-grained urban population estimation using mobile phone location data. *Computers, Environment and Urban Systems, 72*, 44–54.
16. Deville, P., Linard, C., Martin, S., Gilbert, M., Stevens, F. R., Gaughan, A. E., Tatem, A. J., et al. (2014). Dynamic population mapping using mobile phone data. *Proceedings of the National Academy of Sciences, 111*(45), 15888–15893.
17. Doshi-Velez, F., & Kim, B. (2017). Towards a rigorous science of interpretable machine learning. arXiv:1702.08608
18. Du, M., Liu, N., & Hu, X. (2019). Techniques for interpretable machine learning. *Communications of the ACM, 63*(1), 68–77.
19. Dwork, C., Hardt, M., Pitassi, T., Reingold, O., & Zemel, R. (2012). Fairness through awareness. In *Proceedings of the 3rd Innovations in Theoretical Computer Science Conference* (pp. 214–226).
20. Ferreira, L., Kowarick, L., & Künzle, R. (2013). Analyzing the urbanization process in developing countries using remote sensing and cellular automata: A case study of Campinas, Brazil. *Computers, Environment and Urban Systems, 37*, 1–11.
21. Fotheringham, A. S., & O'Kelly, M. E. (1989). *Spatial interaction models: Formulations and applications*. Kluwer Academic Publishers.
22. Friedler, S. A., Scheidegger, C., & Venkatasubramanian, S. (2016). On the (im)possibility of fairness. arXiv:1609.07236
23. Ginsberg, J., Mohebbi, M. H., Patel, R. S., Brammer, L., Smolinski, M. S., & Brilliant, L. (2009). Detecting influenza epidemics using search engine query data. *Nature, 457*(7232), 1012–1014.
24. Goodchild, M. F., & Janelle, D. G. (2018). Introduction: Spatially integrated social science. In *Spatially integrated social science* (pp. 3–16). Oxford University Press.
25. Goodchild, M. F., Anselin, L., & Deichmann, U. (2019). *Spatial data analysis: A guide for ecologists and epidemiologists*. Springer.
26. Guidotti, R., Monreale, A., Ruggieri, S., Turini, F., Giannotti, F., & Pedreschi, D. (2018). A survey of methods for explaining black box models. *ACM Computing Surveys (CSUR), 51*(5), 1–42.
27. Hardt, M., Price, E., & Srebro, N. (2016). Equality of opportunity in supervised learning. In *Advances in neural information processing systems* (pp. 3315–3323).
28. Hawelka, B., Sitko, I., Beinat, E., Sobolevsky, S., Kazakopoulos, P., & Ratti, C. (2014). Geo-located Twitter as a proxy for global mobility patterns. *Cartography and Geographic Information Science, 41*(3), 260–271.
29. He, Y. (2016). Urban land use and land cover classification using remotely sensed SAR data with a gradient boosting machine. *Remote Sensing, 8*(9), 706.
30. Huang, Q., & Wong, D. W. S. (2015). Modeling and visualizing regular human mobility patterns with uncertainty: An aggregated diffusion and agent-based simulation approach. *International Journal of Geographical Information Science, 29*(9), 1547–1571.
31. Jean, N., Burke, M., Xie, M., Davis, W. M., Lobell, D. B., & Ermon, S. (2016). Combining satellite imagery and machine learning to predict poverty. *Science, 353*(6301), 790–794.
32. Kamiran, F., & Calders, T. (2012). Data preprocessing techniques for classification without discrimination. *Knowledge and Information Systems, 33*(1), 1–33.
33. Kearns, M., Neel, S., Roth, A., & Wu, Z. S. (2018). Preventing fairness gerrymandering: Auditing and learning for subgroup fairness. In *Proceedings of the 35th International Conference on Machine Learning* (pp. 2564–2572).

34. Kitchin, R. (2014). Big Data, new epistemologies and paradigm shifts. *Big Data & Society, 1*(1), 2053951714528481.
35. Klopp, J. M., & Petretta, D. L. (2017). The urban sustainable development goal: Indicators, complexity and the politics of measuring cities. *Cities, 63*, 92–97.
36. Kureková, L. M. (2013). Welfare systems as emigration factor: Evidence from the new accession states. *Journal of Common Market Studies, 51*(4), 721–739.
37. Lenormand, M., Picornell, M., Cantú-Ros, O. G., Tugores, A., Louail, T., Herranz, R., Ramasco, J. J., et al. (2015). Cross-checking different sources of mobility information. *PLoS One, 10*(8), e0134869.
38. Liang, X., Liu, X., Lechner, A. M., Cao, Y., & Huang, Y. (2013). Agent-based modeling for the spatial and temporal dynamics of infectious diseases: Malaria as a case study. *Progress in Physical Geography: Earth and Environment, 37*(1), 22–45.
39. Lipton, Z. C. (2018). The mythos of model interpretability. *Queue, 16*(3), 31–57.
40. Lu, X., Bengtsson, L., & Holme, P. (2016). Predictability of population displacement after the 2010 Haiti earthquake. *Proceedings of the National Academy of Sciences, 113*(22), 6243–6248.
41. Lu, X., Wrathall, D. J., Sundsøy, P. R., Nadiruzzaman, M., Wetter, E., Iqbal, A., Bengtsson, L., et al. (2018). Detecting climate adaptation with mobile network data in Bangladesh: anomalies in communication, mobility and consumption patterns during cyclone Mahasen. *Climatic Change, 147*(1–2), 427–443.
42. Lundberg, S. M., & Lee, S. I. (2017). A unified approach to interpreting model predictions. In *Advances in neural information processing systems* (pp. 4765–4774).
43. Ma, T., Zhou, C., Pei, T., Haynie, S., & Fan, J. (2019). Night-time light derived estimation of spatiotemporal characteristics of urbanization dynamics using DMSP/OLS satellite data. *Remote Sensing of Environment, 223*, 15–29.
44. Marx, A., Stoker, T., & Suri, T. (2013). There is no substitute for high frequency data on poverty. *Science, 340*(6135), 852–853.
45. Massey, D. S., Arango, J., Hugo, G., Kouaouci, A., Pellegrino, A., & Taylor, J. E. (1993). Theories of international migration: A review and appraisal. *Population and development review*, 431–466.
46. Mehrabi, N., Morstatter, F., Saxena, N., Lerman, K., & Galstyan, A. (2019). A survey on bias and fairness in machine learning. arXiv:1908.09635
47. Miller, T. (2019). Explanation in artificial intelligence: Insights from the social sciences. *Artificial Intelligence, 267*, 1–38.
48. Mocanu, D., Baronchelli, A., Perra, N., Gonçalves, B., Zhang, Q., & Vespignani, A. (2013). The Twitter of Babel: Mapping world languages through microblogging platforms. *PLoS ONE, 8*(4), e61981.
49. Molnar, C. (2020). *Interpretable machine learning*. Lulu.com.
50. Ngiam, J., Khosla, A., Kim, M., Nam, J., Lee, H., & Ng, A. Y. (2011). Multimodal deep learning. In *Proceedings of the 28th International Conference on Machine Learning* (pp. 689–696).
51. Nickel, M., Murphy, K., Tresp, V., & Gabrilovich, E. (2016). A review of relational machine learning for knowledge graphs. *Proceedings of the IEEE, 104*(1), 11–33.
52. Padró, L., Cerini, S., & Cuadros, M. (2020). Incorporating domain knowledge for entity and relation discovery in the biomedical domain. *Artificial Intelligence in Medicine, 104*, 101840.
53. Paldino, S., Bojic, I., Sobolevsky, S., Ratti, C., & González, M. C. (2015). Urban magnetism through the lens of geo-tagged photography. *EPJ Data Science, 4*(1), 1–20.
54. Raymer, J., Wiśniowski, A., Forster, J. J., Smith, P. W., & Bijak, J. (2013). Integrated modeling of European migration. *Journal of the American Statistical Association, 108*(503), 801–819.
55. Rees, P., Wohland, H., & Norman, P. (2017). *The estimation of migration: Methods and applications*. Wiley.
56. Ribeiro, M. T., Singh, S., & Guestrin, C. (2016). "Why should I trust you?": Explaining the predictions of any classifier. In *Proceedings of the 22nd ACM SIGKDD International Conference on Knowledge Discovery and Data Mining* (pp. 1135–1144).
57. Rokach, L. (2010). Ensemble-based classifiers. *Artificial Intelligence Review, 33*(1–2), 1–39.
58. Salganik, M. J. (2017). *Bit by bit: Social research in the digital age*. Princeton University Press.

59. Selbst, A. D., Boyd, D., Friedler, S. A., Venkatasubramanian, S., & Vertesi, J. (2019). Fairness and abstraction in sociotechnical systems. In *Proceedings of the Conference on Fairness, Accountability, and Transparency* (pp. 59–68).
60. Skeldon, R. (2018). Rural-to-urban migration and urbanization in Asia: Patterns, processes, and policies. In *Routledge Handbook of Asian Demography* (pp. 149–166). Routledge.
61. Srivastava, N., & Salakhutdinov, R. R. (2012). Multimodal learning with deep Boltzmann machines. In *Advances in neural information processing systems* (pp. 2222–2230).
62. Tatem, A. J. (2017). WorldPop, open data for spatial demography. *Scientific Data, 4*(1), 1–4.
63. Tatem, A. J., Noor, A. M., von Hagen, C., Di Gregorio, A., & Hay, S. I. (2007). High-resolution population maps for low-income nations: Combining land cover and census in East Africa. *PLoS ONE, 2*(12), e1298.
64. Taubenböck, H., Wurm, M., Setiadi, N., Gebert, N., Roth, A., Strunz, G., Dech, S., et al. (2018). Lasting footprint of mankind: How to integrate settlements into global spatial data products? *International Journal of Applied Earth Observation and Geoinformation, 68*, 265–276.
65. Taylor, L., & Broeders, D. (2015). In the name of development: Power, profit and the datafication of the global South. *Geoforum, 64*, 229–237.
66. United Nations. (2017). *Handbook on the use of administrative and secondary data sources for population and housing censuses.* United Nations Statistics Division.
67. Vatsavai, R. R., Ganguly, S., Chandola, V., Stefanidis, A., Klasky, S., & Shekhar, S. (2018). Machine learning for geosciences: Challenges and opportunities. *IEEE Transactions on Knowledge and Data Engineering, 31*(8), 1544–1554.
68. Vink, M., Baggio, S., & Gsir, S. (2019). New technologies and data for migration research. *Comparative Migration Studies, 7*(1), 1–15.
69. Wang, Q., Taylor, J. E., & Zhang, X. (2019). Population estimation from social media: A spatiotemporal approach. *Transactions in GIS, 23*(2), 265–285.
70. Wesolowski, A., Eagle, N., Tatem, A. J., Smith, D. L., Noor, A. M., Snow, R. W., & Buckee, C. O. (2012). Quantifying the impact of human mobility on malaria. *Science, 338*(6104), 267–270.
71. Wilson, A. G. (2000). *Complex spatial systems: The modelling foundations of urban and regional analysis.* Prentice Hall.
72. Xie, M., Jean, N., Burke, M., Lobell, D., & Ermon, S. (2016). Transfer learning from deep features for remote sensing and poverty mapping. In *Proceedings of the Thirtieth AAAI Conference on Artificial Intelligence* (pp. 3929–3935).
73. Xue, L. (2018). Big data, migration and governance: Issues, challenges and opportunities. In *Governance and Big Data* (pp. 183–206). Routledge.
74. Yang, B., Yih, W. T., He, X., Gao, J., & Deng, L. (2017). Embedding entities and relations for learning and inference in knowledge bases. In *Proceedings of the International Conference on Learning Representations.*
75. Yao, Y., Wei, Y., Zhang, X., Tian, Y., & Hu, X. (2020). A review of geotagged Twitter data applications in urban and spatial studies. *ISPRS International Journal of Geo-Information, 9*(3), 171.
76. Yuan, Y., Xiao, P., & Lu, X. (2021). A comparative study of machine learning algorithms for land use classification with multi-source remote sensing data. *ISPRS Journal of Photogrammetry and Remote Sensing, 173*, 11–23.
77. Zafar, M. B., Valera, I., Rodriguez, M. G., & Gummadi, K. P. (2017). Fairness constraints: Mechanisms for fair classification. In *Proceedings of the 20th International Conference on Artificial Intelligence and Statistics* (pp. 962–970).
78. Zagheni, E., Weber, I., & Gummadi, K. P. (2017). Leveraging Facebook's advertising platform to monitor stocks of migrants. *Population and Development Review, 43*(4), 721–734.
79. Zhang, H., Lin, Q., Zhang, Y., Chen, Q., & Zhang, Y. (2020). A high-resolution global urban extent map with a 0.01-degree spatial resolution derived from multi-resolution remote sensing images using deep learning. *Remote Sensing of Environment, 242*, 111763.
80. Zhang, Q., Pandey, B., & Seto, K. C. (2017). A robust method to generate a consistent time series from DMSP/OLS nighttime light data. *IEEE Transactions on Geoscience and Remote Sensing, 55*(10), 5820–5831.

81. Zhang, Q., Perra, N., Perrotta, D., Tizzoni, M., Paolotti, D., & Vespignani, A. (2018). Forecasting seasonal influenza fusing digital indicators and a mechanistic disease model. In *Proceedings of the 26th International Conference on World Wide Web* (pp. 311–319).
82. Zhang, X., Zhang, T., Young, S. D., & Tao, T. (2020). Applications of big data in urban and regional planning. *Computers, Environment and Urban Systems, 81*, 101454.
83. Zliobaite, I., & Custers, B. (2016). Using sensitive personal data may be necessary for avoiding discrimination in data-driven decision models. *Artificial Intelligence and Law, 24*(2), 183–201.

Chapter 5
Land Use and Land Cover Change Detection

5.1 Overview of Land Use and Land Cover Change Detection

Land use and land cover change (LULCC) are critical components of human geography that directly affect the natural environment, ecosystems, and human well-being [32]. Understanding the patterns and drivers of LULCC is essential for sustainable land management, urban planning, and environmental conservation. This section provides an overview of LULCC detection, its importance, and its applications in human geography.

5.1.1 Definitions and Concepts

Land cover refers to the physical and biological features of the Earth's surface, such as forests, grasslands, wetlands, water bodies, and artificial structures [72]. Land use, on the other hand, refers to the purpose or function assigned to a specific area by humans, such as agriculture, residential, industrial, or recreational [124]. Land cover and land use are interrelated, as changes in land use often lead to changes in land cover.

Land use and land cover change detection is the process of identifying and quantifying spatial and temporal changes in land use and land cover over time [74]. This process typically involves comparing land cover and land use maps or satellite images from different time periods and identifying the differences between them. The main objective of LULCC detection is to understand the patterns, drivers, and impacts of land use and land cover changes to support sustainable land management, urban planning, and environmental conservation [32].

S. N. M. Moghaddam and H. Cao, *Artificial Intelligence-Driven Geographies*, City Development: Issues and Best Practices, https://doi.org/10.1007/978-981-97-5116-7_5

5.1.2 Importance of Land Use and Land Cover Change Detection

Land use and land cover changes have significant implications for the environment, ecosystems, and human well-being. Some of the reasons why LULCC detection is important include:

1. Climate change: LULCC contributes to climate change through greenhouse gas emissions, albedo changes, and modifications in regional climate dynamics [100]. Detecting LULCC helps in understanding the interactions between land use and climate change, and in designing strategies for climate change mitigation and adaptation [51].
2. Biodiversity and ecosystem services: LULCC directly affects biodiversity by altering habitats, causing fragmentation, and increasing the vulnerability of species to extinction [108]. LULCC also affects ecosystem services, such as carbon sequestration, water regulation, and pollination, which are essential for human well-being [92]. Detecting LULCC helps in assessing the impacts on biodiversity and ecosystem services, and in designing strategies for conservation and sustainable land management.
3. Food security and agricultural production: LULCC, particularly agricultural expansion and intensification, affects food production and food security [116]. Detecting LULCC helps in identifying areas of agricultural expansion, monitoring agricultural productivity, and designing strategies for sustainable agricultural practices and food security [122].
4. Urbanization and infrastructure planning: Rapid urbanization and land use changes are major challenges for urban planning and infrastructure development [111]. Detecting LULCC in urban areas helps in understanding the patterns and drivers of urban growth, and in designing strategies for sustainable urban development and infrastructure planning [7].

5.1.3 Applications of Land Use and Land Cover Change Detection

LULCC detection has a wide range of applications in human geography, including:

1. Monitoring deforestation and forest degradation: LULCC detection is crucial for monitoring deforestation and forest degradation, which are significant drivers of biodiversity loss, carbon emissions, and climate change [53].
2. Assessing land degradation and desertification: LULCC detection helps in assessing the extent and severity of land degradation and desertification, which have severe implications for food security, water resources, and ecosystem services [104].

3. Detecting urban sprawl and land use changes in urban areas: LULCC detection is essential for understanding urban growth patterns, assessing the impacts of urban sprawl, and guiding sustainable urban planning and development [111].
4. Monitoring LULCC detection is vital for tracking agricultural expansion and intensification, which are major drivers of habitat conversion, biodiversity loss, and greenhouse gas emissions [116]. Detecting changes in agricultural land use can help inform sustainable agriculture practices and food security strategies [122].
5. Assessing the effectiveness of conservation policies and land management strategies: LULCC detection can be used to evaluate the impacts of conservation policies, such as protected areas and payment for ecosystem services, on land use and land cover changes, and to inform adaptive management strategies [66].
6. Identifying drivers of land use and land cover changes: LULCC detection can help identify the underlying socioeconomic, demographic, and environmental drivers of land use and land cover changes, which can inform the design of targeted interventions and policies for sustainable land management [39].

5.2 Data Sources for Studying Land Use and Land Cover Change

5.2.1 Traditional Data Sources

In the study of land use and land cover change detection, a variety of traditional data sources have been used to provide information on the spatial and temporal patterns of the Earth's surface (Table 5.1). These sources include satellite imagery, aerial photographs, land use maps, ground survey data, and census data. In this section, we will discuss these traditional data sources and their applications in the analysis of land use and land cover change.

Satellite Imagery

Satellite imagery is one of the most widely used sources of data for land use and land cover change detection. Satellites provide a continuous and consistent source of remotely sensed data that can be used to monitor land surface changes at various spatial and temporal scales [37]. Examples of satellite sensors that have been used for land use and land cover change detection include the Landsat series, Moderate Resolution Imaging Spectroradiometer (MODIS), and Sentinel-2 [133].

Landsat imagery, in particular, has been a valuable resource for studying land use and land cover change over the past several decades due to its relatively high spatial resolution (30 m) and long archive of imagery dating back to the early 1970s [134]. MODIS and Sentinel-2 data provide complementary information, with MODIS offering daily global coverage at a coarser spatial resolution (250–1000 m), and Sentinel-2 providing higher spatial resolution (10–20 m) imagery at a 5-day revisit frequency [27].

Table 5.1 The comparison of different data sources for land use and land cover change

Data source	Characteristics	Advantages	Drawbacks	Applications
Traditional data sources				
Satellite imagery	– Continuous, consistent – Multispectral data – Various spatial scales	– Provides continuous monitoring of land surface changes – Offers multispectral data for detailed analysis	– Limited by cloud cover – Costly to acquire high-resolution imagery	– Monitoring land surface changes – Mapping and monitoring land use/land cover changes over time
Aerial photographs	– High-resolution images captured from aircraft	– Offers high-resolution imagery for detailed analysis – Useful in areas where satellite data is limited or unavailable	– Costly and time-consuming to acquire – Limited spatial coverage	– Mapping and monitoring land use/land cover changes over time
Ground survey data	– Field-based observations and measurements	– Provides ground truth data for validating remote sensing outputs – Captures local-scale land use/land cover information	– Labor-intensive data collection process – Limited spatial coverage	– Validating land use/land cover maps – Analyzing relationships between human population dynamics and land use/land cover change
Census data	– Detailed information on population density, land use, and land cover	– Offers detailed demographic and land use information – Available at various administrative levels	– May not be updated frequently – Limited spatial resolution	– Analyzing relationships between human population dynamics and land use/land cover change
Remote sensing data sources				

(continued)

Table 5.1 (continued)

Data source	Characteristics	Advantages	Drawbacks	Applications
Satellite imagery	– Continuous, multispectral observations – Various resolutions available	– Provides continuous monitoring over large areas – Offers multispectral data for detailed analysis	– Limited by cloud cover – Spatial resolution may not be sufficient for detailed analysis	– Large-scale land cover mapping and monitoring – Detailed land cover mapping and change detection at local scales
LiDAR data	– Accurate and detailed information on vertical structure	– Provides highly accurate information on land surface elevation and structure – Useful for forest and urban studies	– Costly to acquire and process – Limited spatial coverage and availability	– Forest and vegetation studies – Urban growth monitoring
Unmanned aerial vehicles (UAVs)	– High-resolution data acquisition	– Offers high-resolution data collection at relatively low cost – Flexible and adaptable for various applications	– Limited flight time and coverage area – Regulatory constraints may limit operations	– High-resolution remote sensing applications – Detailed land cover mapping and change detection
Big data and geospatial sources				
Social media data	– Massive amounts of geotagged information	– Provides real-time insights into human activities and land use patterns – Offers large-scale coverage	– Biased towards certain demographics and behaviors – Data quality and reliability may vary	– Mapping spatial distribution of land uses – Identifying areas of recreational activities, wildlife habitat, and urbanization

(continued)

Table 5.1 (continued)

Data source	Characteristics	Advantages	Drawbacks	Applications
Volunteered geographic information (VGI)	– Collection of geographic information by individuals	– Offers crowdsourced land use/land cover data worldwide – Provides up-to-date information contributed by users	– Data quality and accuracy may vary – Limited by volunteer participation and expertise	– Providing land use and land cover data contributed by users worldwide
Crowdsourced data	– Collection of data from a large group of people through online platforms	– Allows for validation and improvement of land cover maps – Provides real-time data on land cover changes	– Quality and representativeness of data may vary – Requires careful validation and processing	– Validating and improving land cover maps – Monitoring deforestation and forest degradation
Remote sensing data fusion	– Integration of multiple remote sensing datasets	– Enhances classification accuracy and information content – Provides comprehensive understanding of land cover changes	– Requires advanced processing techniques and algorithms – Data fusion may introduce uncertainties	– Enhancing land use and land cover change detection accuracy – Combining strengths and minimizing limitations of individual data sources

Aerial Photographs

Aerial photographs have also played a significant role in the analysis of land use and land cover change, especially in the period before the availability of satellite imagery. Aerial photographs are typically acquired by aircraft-mounted cameras and provide high-resolution images of the Earth's surface [15]. These photographs can be used to map and monitor land use and land cover changes over time, particularly in areas where satellite data is limited or unavailable.

Ground Survey Data

Ground survey data is collected through field-based observations and measurements, providing information on land use and land cover types at a local scale [44]. Ground surveys can be conducted using various methods, including transect walks, plot sampling, or structured interviews with local residents. Ground survey data is often used to validate land use and land cover maps derived from satellite or aerial imagery, ensuring that the maps accurately represent the actual conditions on the ground [33].

Census Data

Census data is another valuable source of information for studying land use and land cover change. National census datasets often contain detailed information on population density, land use, and land cover at various administrative levels (e.g., country, province, and district). Census data can be used in combination with satellite or aerial imagery to analyze the relationships between human population dynamics and land use or land cover change [26].

In conclusion, traditional data sources have played a critical role in understanding land use and land cover change over time. These data sources have provided a wealth of information on the spatial and temporal patterns of land surface changes, enabling researchers to analyze the drivers and consequences of land use and land cover change. However, with the advent of new data sources and advanced analytical techniques, the potential for further insights into land use and land cover change has increased significantly.

5.2.2 Remote Sensing Data Sources

Remote sensing data sources have become indispensable in land use and land cover change detection due to their ability to provide continuous, synoptic, and multi-temporal observations of the Earth's surface. In this section, we will discuss the various remote sensing data sources used in land use and land cover change studies, their characteristics, and their applications.

Satellite Imagery

Satellite imagery is a primary remote sensing data source for land use and land cover change detection. Different types of satellites provide images with varying spatial, spectral, and temporal resolutions. Some commonly used satellite data sources include:

(a) Landsat: Landsat is a series of Earth observation satellites launched by the United States Geological Survey (USGS) and NASA. Landsat satellites have been collecting multispectral images since the 1970s, providing continuous, long-term, and global observations of the Earth's surface at a moderate spatial resolution of 30 m [134]. Landsat imagery has been widely used in land use and land cover change studies due to its extensive historical archive, large spatial coverage, and free availability [73].

(b) MODIS: The Moderate Resolution Imaging Spectroradiometer (MODIS) is a sensor onboard NASA's Terra and Aqua satellites, providing daily global observations of the Earth at a moderate spatial resolution of 250–1000 m [67].

MODIS data has been used for large-scale land cover mapping and monitoring due to its high temporal resolution and multispectral capabilities [37].

(c) Sentinel: The European Space Agency's (ESA) Sentinel satellites are part of the Copernicus Earth observation program, providing high-resolution optical and radar imagery for various applications, including land cover change detection [115]. Sentinel-2, for example, offers multispectral images at a spatial resolution of 10–60 m, with a revisit time of 5 days at the equator when combining data from both Sentinel-2A and Sentinel-2B satellites [27]. Sentinel-1 provides synthetic aperture radar (SAR) data that can penetrate cloud cover, making it particularly valuable for monitoring land cover changes in regions with frequent cloud cover [127].

(d) High-resolution commercial satellites: In addition to publicly available satellite imagery, high-resolution commercial satellites, such as WorldView, QuickBird, and GeoEye, offer imagery with a spatial resolution of less than one meter. These data sources can be valuable for detailed land cover mapping and change detection at the local scale but may be limited by cost and data availability [45].

LiDAR Data

Light Detection and Ranging (LiDAR) is a remote sensing technology that uses laser pulses to measure distances between the sensor and the Earth's surface. LiDAR data can provide highly accurate and detailed information on the vertical structure of land cover, making it particularly useful for forest and vegetation studies [87]. Airborne LiDAR data has been used for mapping and monitoring land use and land cover changes, such as forest structure, biomass estimation, and urban growth [61, 134]. However, the limited spatial coverage and high cost of LiDAR data acquisition can be a constraint for large-scale applications [121].

Unmanned Aerial Vehicles (UAVs)

Unmanned Aerial Vehicles (UAVs), or drones, have emerged as a valuable data source for high-resolution remote sensing applications in land use and land cover change studies. UAVs can carry various sensors, such as multispectral, hyperspectral, and LiDAR, to collect data at a spatial resolution of less than one meter [22]. UAVs offer several advantages over traditional satellite and airborne remote sensing, including flexibility in data acquisition, reduced costs, and the ability to capture data at very high spatial resolutions [6]. However, UAVs have limited spatial coverage and are subject to legal and regulatory constraints, which may limit their application in some areas [142].

Ground-Based Remote Sensing

In addition to satellite, airborne, and UAV data sources, ground-based remote sensing systems, and ground-based spectroradiometers, can also provide valuable information for land use and land cover change studies. TLS, also known as ground-based LiDAR, is capable of acquiring high-resolution, three-dimensional data on land surface features, such as vegetation structure, building facades, and terrain [56].

Ground-based spectroradiometers can provide in-situ spectral measurements of land cover, which can be used for calibration and validation of satellite-derived land cover products [109].

Despite their potential benefits, ground-based remote sensing data sources are often limited by their spatial coverage, labor-intensive data acquisition process, and the need for accurate georeferencing [56, 109]. Nonetheless, these data sources can complement satellite and airborne remote sensing data by providing detailed, local-scale information on land use and land cover change.

5.2.3 Big Data and Geospatial Data Sources

The rapid advancements in technology have led to the generation and availability of vast amounts of data. In the context of land use and land cover change detection, big data and geospatial data sources play a significant role in providing diverse and rich datasets for analysis. This section discusses the various big data and geospatial data sources relevant to land use and land cover change detection, including social media data, volunteered geographic information, crowdsourced data, and remote sensing data fusion.

Social Media Data

The increasing popularity of social media platforms, such as Twitter, Facebook, and Instagram, has led to the generation of massive amounts of geotagged data. These platforms enable users to share their location information, photographs, and other content in real-time [113]. The geotagged information and images shared on social media platforms can provide valuable insights into land use and land cover changes. For example, researchers have used geotagged images from Flickr to map the spatial distribution of various land uses and identify areas of recreational activities, wildlife habitat, and urbanization [76, 132]. The main limitation of using social media data is the potential biases due to user preferences and behaviors [148].

Volunteered Geographic Information (VGI)

Volunteered Geographic Information (VGI) refers to the collection and dissemination of geographic information by individuals, who voluntarily contribute to the dataset, often through web-based platforms [49]. VGI platforms such as OpenStreetMap (OSM) and Wikimapia provide a rich source of land use and land cover data contributed by millions of users worldwide. These platforms allow users to edit and add information about land use, land cover, and other geospatial features, making them valuable resources for researchers and practitioners [52, 65]. However, VGI data can suffer from inconsistencies, inaccuracies, and incompleteness due to the voluntary nature of data collection and the varying expertise of contributors [8].

Crowdsourced Data

Crowdsourcing refers to the process of collecting data, ideas, or solutions from a large group of people, usually through online platforms [57]. In the context of land use and land cover change detection, crowdsourced data can be obtained from platforms such as Geo-Wiki, which allows users to validate and improve land cover maps using high-resolution satellite imagery [38]. Another example is the Global Forest Watch platform, which uses crowdsourced data to monitor deforestation and forest degradation worldwide [53]. Crowdsourced data can complement traditional data sources by providing more up-to-date and detailed information about land use and land cover changes. However, similar to VGI data, crowdsourced data can also suffer from issues related to data quality and representativeness [110].

Remote Sensing Data Fusion

Remote sensing data fusion refers to the integration of multiple sources of remote sensing data to obtain more comprehensive and accurate information about the Earth's surface [101]. This can include the fusion of data from different sensors (e.g., optical and radar), resolutions (e.g., high and low), or times (e.g., multi-temporal) [146]. Remote sensing data fusion can significantly enhance the capabilities of land use and land cover change detection by combining the strengths and minimizing the limitations of individual data sources. For example, integrating optical and radar data can improve the classification accuracy of land cover types, as the different sensors capture different aspects of the Earth's surface [112]. Similarly, fusing high-resolution and low-resolution data can provide detailed land cover information at large spatial extents [40].

In addition to sensor fusion, data from different remote sensing platforms can also be integrated with other geospatial datasets, such as vector data or GIS layers, to enhance land use and land cover change analysis [130]. For instance, integrating remote sensing data with GIS layers on protected areas, land tenure, and infrastructure can help in understanding the drivers of land cover change and the effectiveness of conservation policies [28].

Despite the many benefits of big data and geospatial data sources, there are several challenges associated with their use in land use and land cover change detection. Data quality, representativeness, and biases are common issues in social media data, VGI, and crowdsourced data [8, 148]. Remote sensing data fusion can be computationally intensive and may require advanced image processing and machine learning techniques to effectively integrate the diverse datasets [146]. Additionally, the accessibility and usability of these data sources can be limited by factors such as data policies, licensing restrictions, and technical expertise [25].

In conclusion, big data and geospatial data sources offer tremendous potential for land use and land cover change detection research. These data sources can provide more detailed, up-to-date, and diverse information compared to traditional data sources. However, the effective utilization of these data sources requires addressing the challenges related to data quality, representativeness, biases, and technical expertise. Future research should focus on developing novel methods and tools to integrate and analyze these diverse datasets and further advance the understanding of land use and land cover change processes.

5.3 AI Techniques for Analyzing Land Use and Land Cover Change

Artificial intelligence (AI) techniques have emerged as powerful tools for analyzing land use and land cover change, providing valuable insights to inform urban planning, resource management, and environmental conservation. This section discusses various AI techniques, including supervised and unsupervised learning, deep learning, and ensemble methods, as well as their applications in land use and land cover change analysis.

5.3.1 Supervised Learning

Supervised learning is a type of machine learning where the algorithm is trained using a labeled dataset, which contains input–output pairs, to learn the underlying relationship between the inputs and the outputs. In the context of land use and land cover change detection, supervised learning algorithms are trained using remotely sensed images with known land use and land cover classes [33]. Common supervised learning techniques for land use and land cover change detection include Support Vector Machines (SVM, Fig. 5.1), Random Forest (RF), and Artificial Neural Networks (ANN).

Support Vector Machines (SVM) is a popular classification technique that aims to find the optimal hyperplane that separates different classes in a multi-dimensional feature space [120]. In land use and land cover change detection, SVM has been successfully applied to classify satellite imagery data [59].

Random Forest (RF) is an ensemble learning technique that combines the outputs of multiple decision trees to improve classification accuracy [13]. The RF algorithm has been widely used in land use and land cover change analysis, demonstrating high accuracy in classifying remotely sensed data [97].

Artificial Neural Networks (ANN) are inspired by the structure and function of the human brain and consist of interconnected nodes or neurons. ANNs have been used to model complex relationships between input features and output classes, making them suitable for land use and land cover classification [145].

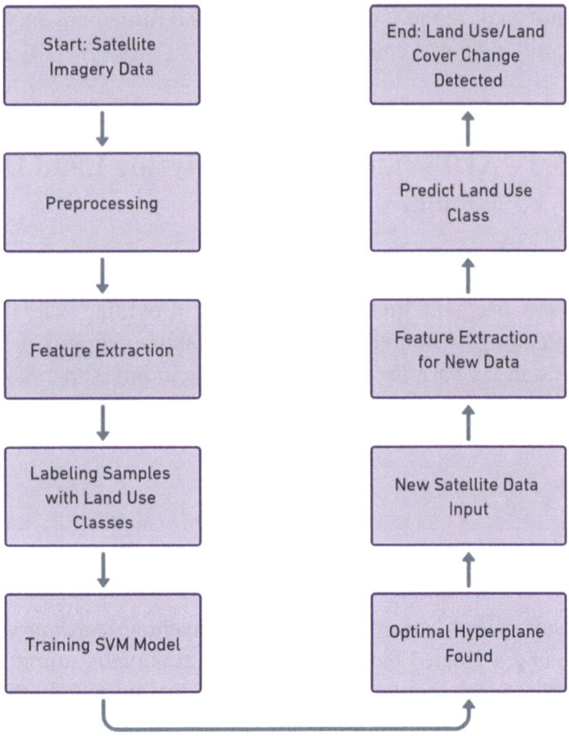

Fig. 5.1 The process of support vector machines (SVM) in an example of land use or land cover changes

5.3.2 Unsupervised Learning

Unsupervised learning algorithms identify patterns in the data without using labeled examples as a reference. In the context of land use and land cover change detection, unsupervised learning techniques are used to cluster pixels in remotely sensed images based on their spectral characteristics. The most common unsupervised learning algorithm for land use and land cover analysis is the k-means clustering algorithm, which iteratively assigns pixels to clusters based on their Euclidean distance to the cluster centroids [84].

Another unsupervised learning technique used in land use and land cover change detection is the Self-Organizing Map (SOM), which is a type of artificial neural network that uses unsupervised learning to produce a low-dimensional representation of input data [69]. SOM has been applied to cluster and visualize high-dimensional remote sensing data, aiding in the interpretation of land use and land cover patterns [126].

5.3.3 Deep Learning

Deep learning is a subfield of machine learning that focuses on artificial neural networks with multiple layers, which enables the learning of hierarchical representations of data. In recent years, deep learning techniques, particularly Convolutional Neural Networks (CNN), have gained popularity in land use and land cover change analysis due to their ability to automatically learn features from raw data [24].

CNNs consist of convolutional, pooling, and fully connected layers that are designed to capture local and global patterns in images [75]. In the context of land use and land cover change detection, CNNs have been successfully applied to classify high-resolution satellite imagery, outperforming traditional machine learning techniques in many cases [17].

Recurrent Neural Networks (RNN) are another deep learning technique that has been used for land use and land cover change analysis. RNNs are designed to handle sequential data by maintaining a hidden state that can capture information from previous time steps [91]. In the context of land use and land cover change detection, RNNs have been applied to model temporal patterns in time-series remote sensing data, allowing for improved prediction of land use and land cover changes over time [103].

5.3.4 Ensemble Methods

Ensemble methods combine the outputs of multiple machine learning models to improve overall performance. In land use and land cover change detection, ensemble methods have been employed to increase classification accuracy and reduce the impact of individual model uncertainties [141]. Common ensemble techniques include bagging, boosting, and stacking.

Bagging, or bootstrap aggregating, is an ensemble technique that trains multiple models on different subsets of the training data and averages their predictions. In land use and land cover change detection, bagging has been applied to decision tree algorithms, such as Random Forest, to reduce overfitting and improve classification performance [13].

Boosting is another ensemble technique that combines the outputs of multiple weak models to form a strong model. In land use and land cover change analysis, boosting has been applied to decision tree algorithms, such as AdaBoost, to increase classification accuracy by iteratively re-weighting the training data based on misclassified instances [36].

Stacking is an ensemble technique that combines the outputs of multiple models by training a meta-model on their predictions. In the context of land use and land cover change detection, stacking has been applied to combine the outputs of different machine learning algorithms, such as SVM, RF, and ANN, to improve classification performance [131].

5.3.5 Challenges and Limitations

While AI techniques have shown great promise in land use and land cover change detection, several challenges and limitations remain. First, the quality and availability of training data is a critical factor in the success of AI models. High-quality, labeled training data can be scarce or expensive to obtain, particularly for remote areas or developing countries [47].

Second, the performance of AI models is heavily dependent on the selection of appropriate features and model parameters. Feature selection and model parameter tuning can be time-consuming and computationally expensive, particularly for deep learning techniques [11].

Lastly, the interpretability of AI models remains a challenge, especially for deep learning techniques. Complex models, such as CNNs and RNNs, can be difficult to interpret and explain, making it challenging for researchers and practitioners to understand the underlying mechanisms driving land use and land cover change predictions [105].

5.3.6 Future Directions

As AI techniques continue to advance, new opportunities for land use and land cover change detection are emerging. For instance, the integration of multiple data sources, such as satellite imagery, LiDAR, and social media data, can provide a more comprehensive understanding of land use and land cover dynamics [12]. Additionally, the incorporation of domain-specific knowledge, such as ecological, social, and economic factors, can help improve the accuracy and interpretability of AI models [34].

Transfer learning is another promising area for future research in land use and land cover change detection. Transfer learning techniques enable the reuse of pre-trained models on new tasks with limited data, reducing the need for large amounts of labeled training data [98]. This approach could be particularly beneficial for regions with limited access to high-quality training data.

Furthermore, the development of explainable AI (XAI) techniques that provide human-understandable explanations for model predictions is an essential direction for future research [1]. XAI techniques can help increase the trust and adoption of AI models in land use and land cover change detection, allowing for more informed decision-making in urban planning, resource management, and environmental conservation.

In conclusion, AI techniques have shown significant potential in land use and land cover change detection, contributing to a better understanding of the complex dynamics that shape our planet. As AI technology continues to advance and new sources of data become available, the potential for AI to inform and support sustainable land use and land cover management practices will only increase.

5.4 Applications of AI in Land Use and Land Cover Change Detection

5.4.1 *Classification of Land Use and Land Cover Types*

Land use and land cover (LULC) classification is an essential task in remote sensing and human geography, providing valuable information for various applications, such as urban planning, agriculture, and environmental monitoring (Table 5.2). Over the years, artificial intelligence (AI) techniques have shown promise in LULC classification tasks by improving the accuracy and efficiency of traditional methods. This section aims to discuss the application of AI in the classification of land use and land cover types, including various AI techniques, data sources, and challenges in this domain.

Traditional LULC classification methods have been based on manual interpretation of satellite images, which can be time-consuming and labor-intensive [117]. With the advent of AI and machine learning, several automated classification techniques have been developed, such as decision trees, support vector machines (SVM), and artificial neural networks (ANN) [33, 59]. More recently, deep learning algorithms, such as convolutional neural networks (CNN) and recurrent neural networks (RNN), have demonstrated superior performance in LULC classification tasks [24].

One of the most commonly used AI techniques for LULC classification is the supervised learning approach. Supervised learning algorithms, such as decision trees, SVM, and ANN, require a labeled dataset for training and validation [33]. This training data typically consists of ground truth samples collected from field surveys, aerial photographs, or high-resolution satellite images. Once trained, these algorithms can classify unseen satellite images into predefined LULC classes, such as urban, agriculture, forest, and water [23].

Another popular AI technique for LULC classification is unsupervised learning, which involves clustering algorithms like K-means, self-organizing maps (SOM), and hierarchical clustering [30, 125]. These algorithms group similar pixels in the satellite images into clusters without any prior knowledge of the ground truth. The resulting clusters can be further interpreted and assigned to specific LULC classes based on expert knowledge or ancillary data [117].

Deep learning techniques, such as CNN and RNN, have gained significant attention in recent years for their ability to automatically learn hierarchical features from raw input data, such as satellite images [24]. CNNs, in particular, have shown outstanding performance in LULC classification tasks by leveraging their ability to capture spatial patterns and contextual information from multispectral or hyperspectral satellite images [21, 86]. RNNs, on the other hand, can model temporal dependencies in time-series satellite data, enabling them to capture the dynamics of LULC changes over time [41].

Data sources for LULC classification have evolved significantly over the years, with the increasing availability of high-resolution satellite images and geospatial data. Traditional data sources include medium-resolution satellite images, such as

Table 5.2 AI techniques in different applications with advantages and challenges

Application	AI techniques	Data sources	Advantages	Challenges
Classification of LULC types	– Supervised learning (e.g., decision trees, SVM, ANN) – Unsupervised learning (e.g., clustering) – Deep learning (e.g., CNN, RNN)	– Satellite imagery – Aerial photographs – Ground truth data – Other geospatial data (e.g., DEM, GIS layers)	– Improved accuracy and efficiency – Automation of classification tasks – Ability to capture complex patterns	– Availability and quality of ground truth data – Complexity and heterogeneity of land cover types – Interpretability of deep learning models
Change detection and monitoring	– Image differencing – Supervised change detection – Unsupervised change detection – Change vector analysis – Time series analysis – Fusion of multiple data sources	– Optical and radar images – LiDAR data – Geospatial data (e.g., DEM, GPS) – Big data sources (e.g., social media, crowd-sourced information)	– Timely and accurate detection of changes – Integration of complementary data sources – Automation of change detection tasks	– Need for labeled training data – Sensitivity to parameter choices – Handling of data heterogeneity and quality issues
Impact assessment and scenario analysis	– Machine learning algorithms (e.g., regression, decision trees, SVM) – Deep learning techniques (e.g., CNN, RNN, GAN) – Agent-based modeling	– Remote sensing images – Socio-economic data – Environmental variables – Other geospatial data	– Modeling complex relationships between land use and environmental factors – Forecasting land use changes and their impacts – Simulating scenarios for decision-making	– Availability of accurate and up-to-date data – Development of robust and transferable models – Collaboration between researchers and stakeholders

Landsat and SPOT [60]. More recently, high-resolution satellite images, such as those from WorldView, QuickBird, and IKONOS, have become increasingly accessible, providing more detailed information on land use and land cover types [12]. In addition to satellite images, other geospatial data sources, such as digital elevation models (DEM), GPS data, and geographic information system (GIS) layers, can also be integrated into the LULC classification process to improve the overall accuracy and provide additional contextual information [82].

The integration of big data and geospatial data sources, such as social media, mobile phone data, and crowd-sourced information (e.g., OpenStreetMap), has further enhanced the capabilities of AI-based LULC classification methods. These data sources can provide complementary information on land use patterns, human activities, and infrastructure, enabling a more comprehensive analysis of land use and land cover dynamics [64, 79].

Despite the significant advancements in AI techniques for LULC classification, several challenges and limitations remain. One of the main challenges is the availability and quality of ground truth data for training and validation. High-quality ground truth data is essential for supervised learning algorithms, but acquiring such data can be expensive, time-consuming, and often subject to errors and inconsistencies [33]. Furthermore, the lack of ground truth data in certain regions or for specific land cover types may lead to biased or incomplete models [80].

Another challenge in AI-based LULC classification is the complexity and heterogeneity of land use and land cover types. The spectral signatures of different land cover types can be highly variable, making it difficult to distinguish between them using traditional spectral-based classification methods [12]. This has led to the development of more advanced techniques, such as object-based image analysis (OBIA), which can capture the spatial and contextual information of land use and land cover types, as well as their relationships with surrounding features [54].

The rapid development of AI techniques, such as deep learning, has raised concerns about the interpretability and explainability of the models. Although deep learning algorithms have achieved state-of-the-art performance in LULC classification tasks, their complex architectures and large number of parameters can make it challenging to understand the underlying decision-making processes [24]. This lack of transparency may hinder the adoption of these techniques by practitioners and policymakers, who require more interpretable and explainable models for decision-making and policy formulation [18].

In conclusion, AI techniques have shown great potential for improving the accuracy and efficiency of LULC classification tasks, offering new opportunities for human geography and related fields. The integration of diverse data sources, such as satellite images, geospatial data, and big data, has further enhanced the capabilities of AI-based LULC classification methods. However, several challenges and limitations remain, including the availability and quality of ground truth data, the complexity and heterogeneity of land use and land cover types, and the interpretability and explainability of AI models. Future research should focus on addressing these challenges and exploring new AI techniques for more accurate, efficient, and transparent LULC classification and change detection.

5.4.2 Change Detection and Monitoring

Change detection and monitoring play a critical role in understanding land use and land cover changes, particularly in the context of urban growth, deforestation, agricultural expansion, and environmental management. AI techniques have made significant contributions to this field by automating the process of identifying changes and providing timely and accurate information for decision-makers. In this section, we will discuss the various AI techniques and their applications in change detection and monitoring.

1. Image Differencing

Image differencing is a straightforward technique for change detection that involves computing the difference between two images acquired at different times. AI techniques, such as deep learning-based semantic segmentation models [24], can be applied to generate accurate land cover maps for both images. Then, by calculating the difference between the two land cover maps, the changes can be identified. This approach is computationally efficient and easy to implement but may not be able to capture subtle changes or handle differences in illumination and atmospheric conditions between the images.

2. Supervised Change Detection

Supervised change detection involves training a machine learning model to identify changes between two images. The training data consists of pairs of images with corresponding change maps, which can be created manually or using existing change detection algorithms. Various classifiers, such as support vector machines (SVM), random forests, and artificial neural networks, can be used for this purpose [33, 60]. Deep learning approaches, like convolutional neural networks (CNNs), have shown promising results in supervised change detection tasks [86]. The main drawback of supervised change detection is the need for labeled training data, which can be time-consuming and expensive to acquire.

3. Unsupervised Change Detection

In unsupervised change detection, the algorithm identifies changes without the need for labeled training data. This can be achieved through clustering techniques, such as k-means, hierarchical clustering, or self-organizing maps (SOM) [125]. These methods can partition the data into distinct clusters, representing different land cover types or change classes. The clusters can then be compared between the two images to identify changes. Unsupervised change detection algorithms can be sensitive to the choice of parameters and may not always provide accurate results, especially for complex scenes with many land cover types.

4. Change Vector Analysis

Change vector analysis (CVA) is a technique that uses the magnitude and direction of the difference vector between two images to identify changes. This method can be applied to multispectral or hyperspectral data, allowing for the detection of subtle changes that may not be visible in single-band images [78]. AI techniques, such as deep learning-based feature extraction methods, can be employed to obtain more informative and discriminative features for change vector analysis, improving the accuracy of change detection.

5. Time Series Analysis

Time series analysis focuses on analyzing a series of images acquired over time to monitor land use and land cover changes. AI techniques, particularly deep learning-based methods, have shown great potential for time series analysis in remote sensing applications. Recurrent neural networks (RNNs), specifically long short-term memory (LSTM) networks, can capture the temporal dependencies in the data and provide accurate predictions for land cover changes [149]. Additionally, time series analysis can be used for forecasting future land use and land cover changes, helping decision-makers in urban planning and environmental management.

6. Fusion of Multiple Data Sources

AI techniques can also be used to fuse multiple data sources for change detection and monitoring, such as optical and radar images, LiDAR data, and other geospatial data sources [82]. Data fusion can improvethe accuracy and reliability of change detection results by providing complementary information from different sensors and data types. For example, combining optical and radar images can improve the detection of changes in areas with frequent cloud cover, as radar sensors can penetrate clouds and provide useful information even under such conditions [71].

Deep learning techniques, such as autoencoders and deep belief networks, can be employed to learn a common feature representation from multiple data sources, which can then be used for change detection tasks [128]. In addition, ensemble methods that combine the outputs of multiple classifiers or algorithms can improve the overall accuracy and robustness of change detection results [143].

In summary, AI techniques have greatly enhanced the capabilities of change detection and monitoring in land use and land cover studies. From traditional image differencing methods to advanced deep learning-based approaches, AI has enabled the extraction of valuable information from large and complex datasets, providing timely and accurate insights for decision-making in urban planning, environmental management, and other applications. However, several challenges remain, such as the need for labeled training data, sensitivity to parameter choices, and handling of data heterogeneity and quality issues. Future research in this field should focus on addressing these challenges and exploring novel AI techniques and data sources for more effective and efficient change detection and monitoring.

5.4.3 *Impact Assessment and Scenario Analysis*

Impact assessment and scenario analysis are essential components of land use and land cover change detection, as they provide insight into the potential consequences of different land use policies, management strategies, and environmental changes. AI techniques have been increasingly applied to analyze the impacts of land use and land cover changes and develop scenario analyses to support decision-making processes in urban planning, environmental management, and sustainable development. This section will discuss the application of AI in impact assessment and scenario analysis, focusing on machine learning algorithms, deep learning techniques, and agent-based modeling.

1. Machine Learning Algorithms in Impact Assessment and Scenario Analysis

Machine learning algorithms have been widely used in impact assessment and scenario analysis to model complex relationships between land use, socio-economic factors, and environmental variables. Supervised learning techniques, such as regression analysis, decision trees, and support vector machines, have been employed to predict land use and land cover changes and their impacts on various aspects of the environment, including air quality, water resources, and biodiversity [81, 129]. Unsupervised learning methods, such as clustering and principal component analysis, have been utilized to identify patterns and trends in land use changes and their effects on ecosystems, human health, and socio-economic development [58, 147].

2. Deep Learning Techniques in Impact Assessment and Scenario Analysis

Deep learning techniques, particularly convolutional neural networks (CNNs) and recurrent neural networks (RNNs), have shown promising results in impact assessment and scenario analysis. CNNs have been employed for high-resolution land use and land cover mapping by processing remote sensing images, enabling accurate assessments of environmental impacts [144]. RNNs, particularly long short-term memory (LSTM) networks, have been used to model temporal patterns and forecast land use and land cover changes, as well as their potential impacts on environmental and socio-economic factors [102, 137].

In addition to these traditional deep learning techniques, novel approaches, such as generative adversarial networks (GANs), have been explored to generate realistic land use and land cover change scenarios. GANs can be used to synthesize new remote sensing images, simulating the effects of different land use policies and management strategies [148]. This capability can help stakeholders better understand the potential consequences of their decisions and develop more sustainable land use plans.

3. Agent-Based Modeling in Impact Assessment and Scenario Analysis

Agent-based modeling (ABM) is an AI technique that simulates the behavior and interactions of individual agents, such as households, firms, or government agencies, within a given environment (Fig. 5.2). ABM has been increasingly used in land use and land cover change studies to examine the emergent properties and dynamics of

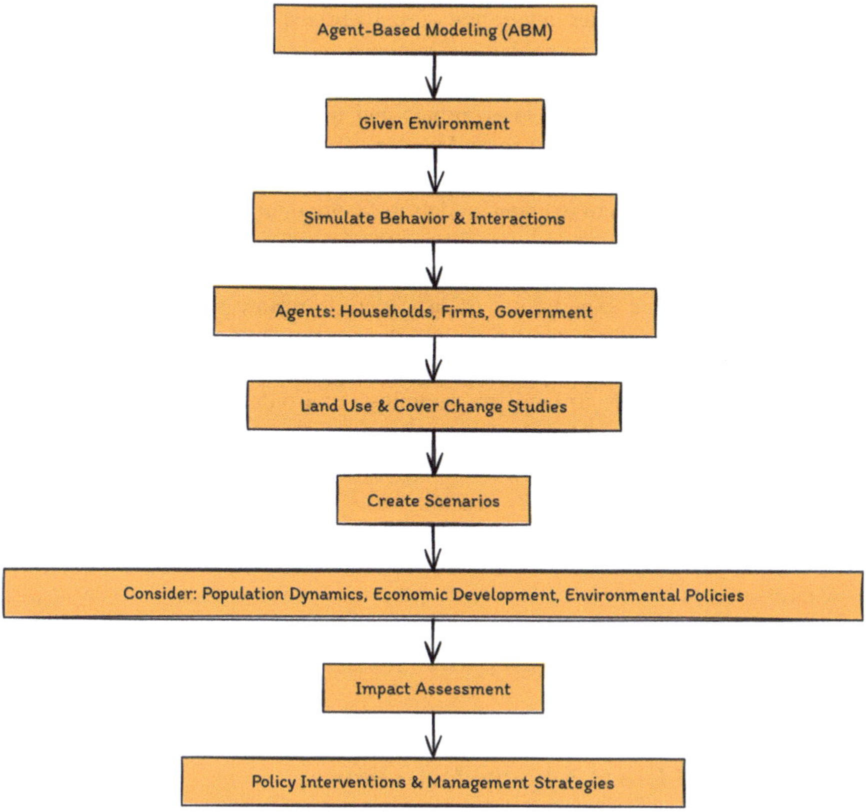

Fig. 5.2 The process of agent-based modeling in impact assessment and scenario analysis

complex systems, including urban growth, agricultural expansion, and deforestation [5, 31, 99].

ABM allows researchers to create detailed, spatially explicit scenarios of land use and land cover changes, taking into account various factors such as population dynamics, economic development, and environmental policies. These scenarios can be used to assess the potential impacts of land use changes on the environment, as well as the effectiveness of different management strategies and policy interventions [16, 88, 123].

5.4.4 *Challenges and Future Directions*

Despite the promising advancements in AI for impact assessment and scenario analysis, several challenges remain. First, there is a need for more accurate, high-resolution, and up-to-date data to improve the quality and reliability of AI-based analyses. This includes better integration of remote sensing, big data, and geospatial data sources to capture the complexity of land use and land cover change processes [46].

Second, there is a need for more robust and transferable AI models that can be applied across different regions and scales. This requires the development of new algorithms and techniques that can handle diverse data sources, as well as the calibration and validation of existing models to ensure their applicability in different contexts [55, 135].

Finally, there is a need for greater collaboration between AI researchers, land use and land cover change experts, and stakeholders to ensure that AI-based impact assessments and scenario analyses are relevant, accessible, and actionable. This includes the development of user-friendly tools and platforms that can facilitate the use of AI techniques in land use planning, environmental management, and policy-making processes [70, 77].

In conclusion, AI techniques hold great potential for improving our understanding of land use and land cover change impacts and for developing more effective and sustainable land use policies and management strategies. By addressing the aforementioned challenges and fostering interdisciplinary collaboration, AI can play a critical role in supporting decision-making processes and promoting sustainable land use practices in the face of global environmental change.

5.5 Challenges and Limitations of AI in Land Use and Land Cover Change Detection

Artificial Intelligence (AI) has become an indispensable tool in land use and land cover (LULC) change detection, providing new insights and solutions to various challenges faced by researchers and practitioners. Despite the considerable progress made in recent years, several challenges and limitations still exist in the application of AI for LULC change detection. This section aims to discuss these challenges and limitations, including data quality, model transferability, and interpretability, to provide a comprehensive understanding of the current state of AI in LULC change detection.

1. Data Quality and Availability

Data quality and availability are critical factors in the success of AI-based LULC change detection [83]. High-quality data with appropriate spatial, spectral, and temporal resolutions are necessary for accurate model training and validation. However, obtaining such data can be challenging, particularly for remote and inaccessible areas or developing countries with limited resources [42]. Additionally, inconsistencies in data collection methods, missing data, and the presence of noise and errors in the datasets can negatively impact the performance of AI models [3].

The availability of large amounts of labeled data for model training is another challenge in AI-based LULC change detection. Supervised learning methods, such as deep learning, require a significant amount of labeled data to produce accurate results [148]. Obtaining labeled data is often time-consuming and labor-intensive, particularly for high-resolution imagery where manual annotation can be a bottleneck [89].

2. Model Transferability

AI models, particularly deep learning models, are known to be data-hungry, requiring large amounts of training data to generalize well to new areas or conditions (Zhu et al., 2020). Consequently, a model trained on data from one region or time period may not perform well when applied to another region or time period without retraining or fine-tuning [9]. This lack of transferability poses challenges in applying AI models for LULC change detection in regions with limited data availability or rapidly changing environments [136]. Domain adaptation and transfer learning techniques have been proposed to mitigate this issue, but they are still in the early stages of development and have limitations in terms of performance and applicability [98].

3. Model Interpretability and Uncertainty

One of the main challenges in using AI models for LULC change detection is the lack of model interpretability [106]. AI models, especially deep learning models, are often considered as "black boxes" due to their complex architectures and the non-linear relationships they learn from the input data (Zhu et al., 2020). This lack of interpretability makes it difficult for researchers and practitioners to understand the underlying processes driving the model's predictions and to identify potential biases or errors in the model [106].

Uncertainty assessment is another challenge in AI-based LULC change detection. AI models may produce highly accurate results, but they may also produce uncertain predictions that are difficult to interpret and validate [35]. Quantifying and communicating this uncertainty is crucial for decision-makers to make informed decisions based on the model's predictions [96]. Several methods have been proposed for uncertainty quantification in AI-based LULC change detection, but they are still in the early stages of development and have limitations in terms of applicability and accuracy [35].

4. Integration of Multi-Source Data and Heterogeneous Information

Integration of multi-source data and heterogeneous information is a challenge in AI-based LULC change detection [2]. Different data sources, such as remote sensing imagery, social media data, and cadastral data, often have different spatial, spectral, and temporal resolutions, and may require preprocessing and data fusion techniques to be effectively combined for analysis [48]. Moreover, integrating various types of data, such as structured and unstructured data or continuous and categorical data, can pose challenges in terms of data representation and model architecture [114]. Developing AI models that can effectively integrate and learn from heterogeneous information remains an open research area, and more advanced data fusion and representation techniques are needed to address this challenge [2].

5. Scalability and Computational Efficiency

Scalability and computational efficiency are critical concerns in AI-based LULC change detection, particularly when dealing with large-scale datasets and high-resolution imagery [14]. Training and inference of AI models, especially deep learning models, can be computationally expensive and may require specialized hardware such as graphics processing units (GPUs) [94]. Furthermore, the large memory footprint of high-resolution data and the need to process large volumes of data in real-time can exacerbate the computational challenges [14]. Developing more computationally efficient AI models and optimization techniques is essential for facilitating the widespread adoption of AI in LULC change detection [19].

6. Ethical and Privacy Considerations

Ethical and privacy considerations are becoming increasingly important in the application of AI for LULC change detection, particularly when using data sources that may contain sensitive or personally identifiable information (PII), such as social media data or cadastral data [90]. Ensuring that the data used in AI models is collected, processed, and stored in accordance with relevant data protection and privacy regulations is essential to avoid potential legal and ethical issues [85]. Moreover, AI models must be designed and trained to be unbiased and to minimize the potential for discriminatory outcomes [10].

In conclusion, while AI has made significant strides in LULC change detection, there are still several challenges and limitations that need to be addressed. Future research should focus on improving data quality and availability, enhancing model transferability, increasing model interpretability and uncertainty assessment, integrating multi-source data and heterogeneous information, optimizing computational efficiency, and addressing ethical and privacy considerations. Overcoming these challenges will pave the way for more effective and widespread application of AI in LULC change detection and help researchers and practitioners better understand and manage the complex dynamics of our changing landscapes.

5.6 Future Directions in AI Applications for Land Use and Land Cover Change Detection

As artificial intelligence (AI) continues to revolutionize various fields, including land use and land cover (LULC) change detection, researchers and practitioners are looking for ways to improve existing techniques and develop new applications. This section discusses the potential future directions in AI applications for LULC change detection, addressing emerging trends, challenges, and opportunities for further development.

1. Integration of heterogeneous data sources

In recent years, the availability of diverse data sources, such as satellite imagery, remote sensing data, aerial photographs, and crowdsourced data, has increased exponentially [29]. The integration of these heterogeneous data sources can provide valuable insights into LULC change detection, particularly when used in conjunction with AI techniques [133]. Future research should focus on developing methods and algorithms that can effectively combine and analyze data from multiple sources, leading to more accurate and comprehensive LULC change detection results.

2. Improvement of AI algorithms and models

AI techniques, such as deep learning and machine learning, have significantly improved LULC change detection capabilities. However, there is still room for improvement, particularly in terms of model robustness, generalizability, and interpretability [138]. Future research should prioritize the development of new AI algorithms and models that address these limitations, as well as investigate strategies for optimizing existing models to improve their performance in LULC change detection applications.

3. Real-time monitoring and prediction

AI techniques have the potential to enable real-time monitoring and prediction of LULC changes, which can significantly improve decision-making and resource allocation in urban planning, environmental management, and policy formulation [4]. By leveraging AI's ability to process large volumes of data quickly, researchers can develop models that provide real-time insights into LULC change patterns, allowing for timely interventions and more effective management of natural resources and urban spaces.

4. Enhanced collaboration and data sharing

The increasing availability of open data sources and the development of cloud-based platforms, such as Google Earth Engine, has facilitated collaboration and data sharing among researchers and practitioners working on LULC change detection [50]. Future research should focus on promoting and enhancing collaborative efforts, particularly in terms of developing shared repositories, standardizing data formats and metadata, and improving the interoperability of data and models. By fostering collaboration and data sharing, the research community can accelerate the development of new AI techniques and applications for LULC change detection, ultimately leading to more accurate and timely insights into land use and land cover dynamics.

5. Ethical considerations and privacy concerns

As AI techniques continue to advance and become more widely adopted in LULC change detection, it is essential to address the ethical considerations and privacy concerns associated with the use of these technologies [68]. Researchers and practitioners should prioritize the development of guidelines and best practices for ensuring the responsible and ethical use of AI in LULC change detection applications, particularly regarding data collection, storage, and sharing. Furthermore, the research community should explore potential solutions for addressing privacy concerns related to the use of geospatial data, such as anonymization and aggregation techniques.

6. Climate change and LULC change interactions

Climate change is a major driver of LULC change, and understanding the interactions between these two phenomena is critical for developing effective policies and management strategies for mitigating the impacts of climate change [63]. Future research should focus on leveraging AI techniques to study the complex relationships between climate change and LULC change, including the impacts of extreme weather events, sea-level rise, and changes in precipitation patterns on land use and land cover dynamics.

In conclusion, the future of AI applications in LULC change detection is promising, with numerous opportunities for further development and innovation. By focusing on integrating heterogeneous data sources, improving AI algorithms and models, enabling real-time monitoring and prediction, enhancing collaboration and data sharing, addressing ethical considerations and privacy concerns, and exploring climate change and LULC change interactions, researchers and practitioners can continue to advance the field and contribute to more sustainable land use and land cover management practices.

7. Integration of AI techniques with decision support systems

The integration of AI techniques with decision support systems (DSS) can help policymakers, urban planners, and land managers make more informed decisions about land use and land cover management. Future research should focus on developing AI-driven DSS that can efficiently process large volumes of geospatial data, provide real-time analysis, and generate actionable insights for users [62]. These systems could enable more effective planning and implementation of land use policies, infrastructure projects, and conservation efforts, ultimately promoting more sustainable and resilient landscapes.

8. Interdisciplinary collaboration

To fully harness the potential of AI applications in LULC change detection, interdisciplinary collaboration is essential. Researchers and practitioners from various fields, including remote sensing, computer science, geography, environmental science, and urban planning, must work together to develop innovative solutions that address the complex challenges associated with land use and land cover dynamics [140]. By fostering interdisciplinary collaboration, the research community can advance the development of AI-driven LULC change detection methods and applications, ultimately contributing to more effective and sustainable land management practices.

9. Capacity building and training

As AI techniques continue to evolve and become more widely adopted in LULC change detection, capacity building and training efforts are crucial for ensuring that researchers, practitioners, and decision-makers are well-equipped to utilize these advanced tools effectively. Future research and educational initiatives should focus on developing training programs and resources that promote the understanding and application of AI techniques in LULC change detection, with a particular emphasis on incorporating these tools into decision-making processes and policies.

10. Public engagement and awareness

Public engagement and awareness are essential components of successful land use and land cover management strategies. By involving citizens in the process of LULC change detection and providing them with the tools and information necessary to understand the implications of these changes, researchers and practitioners can foster greater public support for sustainable land management initiatives [95]. Future research should explore ways to leverage AI-driven LULC change detection tools and platforms to promote public engagement and awareness, such as through participatory mapping initiatives, citizen science projects, and educational programs.

In summary, the future of AI applications in LULC change detection offers numerous opportunities for growth and innovation. By focusing on integrating AI techniques with decision support systems, fostering interdisciplinary collaboration, promoting capacity building and training, and encouraging public engagement and awareness, the research community can continue to advance the field and contribute to more effective and sustainable land use and land cover management practices.

11. Incorporating climate change impacts into AI-driven LULC change detection

Climate change is a critical factor that influences land use and land cover dynamics across the globe. To better understand the complex interplay between climate change, human activities, and LULC change, future research should integrate climate change impacts into AI-driven LULC change detection models and applications [20]. By incorporating climate data, such as temperature, precipitation, and extreme weather events, AI-based models can provide more accurate and comprehensive assessments of LULC change patterns and their potential implications for ecosystems, societies, and economies.

12. Enhancing the spatial and temporal resolution of AI-driven LULC change detection

Improvements in remote sensing technologies have led to the availability of higher-resolution spatial and temporal data for LULC change analysis. However, there is still a need for further advancements in AI-driven LULC change detection methods to fully exploit the potential of high-resolution data [139]. Future research should focus on developing and refining AI algorithms capable of efficiently processing and analyzing high-resolution data, allowing for more detailed, accurate, and timely assessments of LULC change patterns and their underlying drivers.

13. Exploring the ethical implications of AI-driven LULC change detection

As AI techniques continue to advance and become more prevalent in LULC change detection, it is essential to consider the ethical implications of these applications.

Issues such as data privacy, transparency, and accountability must be addressed to ensure that AI-driven LULC change detection practices are conducted responsibly and equitably [93]. Future research should explore the ethical dimensions of AI applications in LULC change detection, as well as develop guidelines and best practices to promote responsible and ethical use of these powerful tools.

14. Developing AI-driven LULC change detection applications for policymaking

AI-driven LULC change detection applications have the potential to play a significant role in shaping land use and land cover policies at local, regional, and global scales. To maximize the impact of these applications, future research should focus on developing tools and frameworks that directly support policymaking processes, such as by providing actionable insights, facilitating stakeholder engagement, and evaluating the effectiveness of policy interventions [107]. By bridging the gap between AI-driven LULC change detection research and policy development, researchers and practitioners can contribute to more informed, evidence-based decision-making for sustainable land management.

15. Promoting interdisciplinary collaboration in AI-driven LULC change detection research

Land use and land cover change is a complex and multifaceted phenomenon, influenced by various ecological, social, economic, and political factors. To better understand and address this complexity, future research in AI-driven LULC change detection should promote interdisciplinary collaboration among researchers from diverse fields, such as geography, ecology, remote sensing, computer science, and social sciences [119]. By fostering interdisciplinary dialogue and cooperation, researchers can develop more holistic and integrative approaches to LULC change detection and analysis, ultimately contributing to more effective and sustainable land management strategies.

16. Enhancing the scalability and transferability of AI-driven LULC change detection methods

While many AI-driven LULC change detection methods have shown promising results in specific case studies or regions, there is a need for further research to enhance the scalability and transferability of these methods across different spatial scales and geographic contexts [43]. By refining AI algorithms and data processing techniques, researchers can develop more flexible and adaptable LULC change detection tools that can be readily applied to a wide range of settings, ultimately contributing to a more comprehensive understanding of global LULC change patterns and dynamics.

17. Investigating the long-term impacts of AI-driven LULC change detection on sustainable development

As AI-driven LULC change detection methods continue to advance and become more widely adopted, it is crucial to assess their long-term impacts on sustainable development goals and outcomes. Future research should investigate the potential consequences of AI-driven LULC change detection applications on various dimensions of sustainability, such as biodiversity conservation, climate change mitigation and adaptation, food security, and human well-being [118]. By evaluating the long-term implications of AI-driven LULC change detection practices, researchers can contribute to the development of more sustainable and equitable land management strategies that promote both human and environmental well-being.

References

1. Adadi, A., & Berrada, M. (2018). Peeking inside the black-box: A survey on explainable artificial intelligence (XAI). *IEEE Access, 6*, 52138–52160.
2. Al-Ahmadi, F. S., Ochir, G., & Mishra, D. R. (2021). Integration of optical and SAR data for land use/land cover mapping using deep learning. *Remote Sensing, 13*(5), 878.
3. Almutairi, A., & Warner, T. A. (2010). Change detection accuracy and image properties: A study using simulated data. *Remote Sensing, 2*(6), 1508–1529.
4. Almutairi, A., Rashed, T., Farag, I., & Alrasheed, A. (2021). A geospatial decision support model for land use change detection and urban growth prediction in Riyadh City, Saudi Arabia. *Remote Sensing, 13*(3), 428.
5. An, L., Linderman, M., Qi, J., Shortridge, A., & Liu, J. (2005). Exploring complexity in a human-environment system: An agent-based spatial model for multidisciplinary and multiscale integration. *Annals of the Association of American Geographers, 95*(1), 54–79.
6. Anderson, K., & Gaston, K. J. (2013). Lightweight unmanned aerial vehicles will revolutionize spatial ecology. *Frontiers in Ecology and the Environment, 11*(3), 138–146.
7. Angel, S., Sheppard, S. C., & Civco, D. L. (2005). The dynamics of global urban expansion. Transport and Urban Development Department, World Bank.
8. Antoniou, V., & Skopeliti, A. (2015). Measures and indicators of VGI quality: An overview. *ISPRS Annals of Photogrammetry, Remote Sensing & Spatial Information Sciences, 2*(3), 345–351.
9. Bali, A., Ghosh, T., & Chaudhuri, S. (2021). A survey of deep learning techniques for land cover and land use classification using remote sensing data. *ACM Computing Surveys (CSUR), 54*(4), 1–36.
10. Barocas, S., & Selbst, A. D. (2016). Big data's disparate impact. *California Law Review, 104*, 671–732.
11. Bengio, Y. (2012). Practical recommendations for gradient-based training of deep architectures. In *Neural networks: Tricks of the trade* (pp. 437–478). Springer.
12. Blaschke, T. (2010). Object-based image analysis for remote sensing. *ISPRS Journal of Photogrammetry and Remote Sensing, 65*(1), 2–16.
13. Breiman, L. (2001). Random forests. *Machine Learning, 45*(1), 5–32.
14. Brondizio, E. S., & Moran, E. F. (2008). Human dimensions of climate change: the vulnerability of small farmers in the Amazon. *Philosophical transactions of the Royal Society B: Biological sciences, 363*(1498), 1803–1809.
15. Brown, D. G., Page, S. E., Riolo, R., Zellner, M., & Rand, W. (2008). Path dependence and the validation of agent-based spatial models of land use. *International Journal of Geographical Information Science, 22*(2), 243–264.
16. Castelluccio, M., Poggi, G., Sansone, C., & Verdoliva, L. (2015). Land use classification in remote sensing images by convolutional neural networks. arXiv:1508.00092
17. Chakraborty, I., & Murphy, K. P. (2018). *Explainable AI: Interpreting, explaining and visualizing deep learning*. Springer Nature.

18. Chen, J., Ban, Y., & Li, S. (2021). China: Open access to Earth land-cover map. *Nature, 589*(7841), 193.
19. Chen, Y., Zhao, X., & Jia, X. (2021). A review on deep learning for remote sensing image classification. *ISPRS Journal of Photogrammetry and Remote Sensing, 172*, 313–328.
20. Chen, Y., Zhu, X., & Xia, Z. (2018). A densely connected convolutional neural network for land-use and land-cover classification. *Remote Sensing, 10*(2), 193.
21. Colomina, I., & Molina, P. (2014). Unmanned aerial systems for photogrammetry and remote sensing: A review. *ISPRS Journal of Photogrammetry and Remote Sensing, 92*, 79–97.
22. Congalton, R. G., & Green, K. (2008). *Assessing the accuracy of remotely sensed data: Principles and practices*. CRC Press.
23. Curry, R. R. (2016). The potential role of geospatial big data in urban planning. *Planning Practice & Research, 31*(4), 351–371.
24. DeFries, R. S., Foley, J. A., & Asner, G. P. (2004). Land-use choices: Balancing human needs and ecosystem function. *Frontiers in Ecology and the Environment, 2*(5), 249–257.
25. Drusch, M., Del Bello, U., Carlier, S., Colin, O., Fernandez, V., Gascon, F., Meygret, A., et al. (2012). Sentinel-2: ESA's optical high-resolution mission for GMES operational services. *Remote Sensing of Environment, 120*, 25–36.
26. Echeverria, C., Coomes, D. A., Hall, M., & Newton, A. C. (2008). Spatially explicit models to analyze forest loss and fragmentation between 1976 and 2020 in southern Chile. *Ecological Modelling, 212*(3–4), 439–449.
27. Estoque, R. C., Murayama, Y., & Myint, S. W. (2021). A review of land use/land cover change studies in the urbanization context. *ISPRS Journal of Photogrammetry and Remote Sensing, 174*, 84–99.
28. Fauvel, M., Tarabalka, Y., & Benediktsson, J. A. (2013). Advances in spectral-spatial classification of hyperspectral images. *Proceedings of the IEEE, 101*(3), 652–675.
29. Filatova, T., Verburg, P. H., Parker, D. C., & Stannard, C. A. (2013). Spatial agent-based models for socio-ecological systems: Challenges and prospects. *Environmental Modelling & Software, 45*, 1–7.
30. Foley, J. A., DeFries, R., Asner, G. P., Barford, C., Bonan, G., Carpenter, S. R., Snyder, P., et al. (2005). Global consequences of land use. *Science, 309*(5734), 570–574.
31. Foody, G. M. (2002). Status of land cover classification accuracy assessment. *Remote Sensing of Environment, 80*(1), 185–201.
32. Foody, G. M. (2010). Assessing the accuracy of land cover change with imperfect ground reference data. *Remote Sensing of Environment, 114*(10), 2271–2285.
33. Foody, G. M., Pal, M., Rocchini, D., Garzon-Lopez, C. X., & Bastin, L. (2018). The sensitivity of mapping methods to reference data quality: Training supervised image classifications with imperfect reference data. *ISPRS International Journal of Geo-Information, 7*(11), 433.
34. Freund, Y., & Schapire, R. E. (1997). A decision-theoretic generalization of on-line learning and an application to boosting. *Journal of Computer and System Sciences, 55*(1), 119–139.
35. Friedl, M. A., Sulla-Menashe, D., Tan, B., Schneider, Sibley, A., & Huang, X. (2010). MODIS collection 5 global land cover: Algorithm refinements and characterization of new datasets. *Remote Sensing of Environment, 114*(1), 168–182.
36. Fritz, S., McCallum, I., Schill, C., Perger, C., Grillmayer, R., Achard, F., Obersteiner, M., et al. (2009). Geo-Wiki.Org: The use of crowdsourcing to improve global land cover. *Remote Sensing, 1*(3), 345–354.
37. Geist, H. J., & Lambin, E. F. (2002). Proximate causes and underlying driving forces of tropical deforestation: Tropical forests are disappearing as the result of many pressures, both local and regional, acting in various combinations in different geographical locations. *BioScience, 52*(2), 143–150.
38. Ghosh, A., Joshi, P. K., & Ghosh, S. K. (2016). Prioritizing areas for conservation and vegetation restoration in a semi-arid degraded land in India using remote sensing and geospatial modelling techniques. *International Journal of Applied Earth Observation and Geoinformation, 44*, 109–120.

39. Ghosh, A., Mishra, A., & Ghosh, S. K. (2020). Deep learning with remote sensing data for sustainable development goals: A review on urbanization and forest cover change. *Remote Sensing Applications: Society and Environment, 18*, 100319.
40. Gibbes, C., Southworth, J., Waylen, P., Child, B., Bunting, E., Masek, L., & Rigg, C. (2017). Application of object based classification and high resolution satellite imagery for savanna ecosystem analysis. *Remote Sensing of Environment, 108*(1), 65–75.
41. Gibson, C. C., Ostrom, E., & Ahn, T. K. (2000). The concept of scale and the human dimensions of global change: a survey. *Ecological Economics, 32*(2), 217–239.
42. Gibson, L., Wilman, E. N., & Laurance, W. F. (2018). How green is 'Green' energy? *Trends in Ecology & Evolution, 33*(12), 922–935.
43. Gibson, P., Power, A., Lyons, A., & Byrne, K. A. (2020). A review of the application of machine learning techniques to land-use and land-cover mapping using remote sensing. *Remote Sensing, 12*(10), 1663.
44. Gislason, P. O., Benediktsson, J. A., & Sveinsson, J. R. (2006). Random forests for land cover classification. *Pattern Recognition Letters, 27*(4), 294–300.
45. Gomez, C., White, J. C., & Wulder, M. A. (2016). Optical remotely sensed time series data for land cover classification: A review. *ISPRS Journal of Photogrammetry and Remote Sensing, 116*, 55–72.
46. Goodchild, M. F. (2007). Citizens as sensors: The world of volunteered geography. *GeoJournal, 69*(4), 211–221.
47. Gorelick, N., Hancher, M., Dixon, M., Ilyushchenko, S., Thau, D., & Moore, R. (2017). Google Earth Engine: Planetary-scale geospatial analysis for everyone. *Remote Sensing of Environment, 202*, 18–27.
48. Grimm, N. B., Faeth, S. H., Golubiewski, N. E., Redman, C. L., Wu, J., & Bai, J. M. (2008). Global change and the ecology of cities. *Science, 319*(5864), 756–760.
49. Haklay, M., & Weber, P. (2008). OpenStreetMap: User-generated street maps. *IEEE Pervasive Computing, 7*(4), 12–18.
50. Hansen, M. C., Potapov, P. V., Moore, R., Hancher, M., Turubanova, S. A., Tyukavina, A., Townshend, J. R. (2013). High-resolution global maps of 21st-century forest cover change. *Science, 342*(6160), 850–853.
51. Hay, G. J., Castilla, G., Wulder, M. A., & Ruiz, J. R. (2011). An automated object-based approach for the multiscale image segmentation of forest scenes. *International Journal of Applied Earth Observation and Geoinformation, 13*(4), 518–525.
52. Hengl, T., Nussbaum, M., Wright, M. N., Heuvelink, G. B. M., & Gräler, B. (2017). Random forest as a generic framework for predictive modeling of spatial and spatio-temporal variables. *PeerJ, 5*, e4068.
53. Heritage, G., & Large, A. (Eds.). (2009). *Laser scanning for the environmental sciences*. Wiley-Blackwell.
54. Howe, J. (2006). The rise of crowdsourcing. *Wired Magazine, 14*(6), 1–4.
55. Hu, Y., Fan, C., Wang, L., & Zhang, X. (2019). Identifying spatial patterns of regional land use efficiency by using a two-stage data envelopment analysis model. *Sustainability, 11*(8), 2325.
56. Huang, C., Davis, L. S., & Townshend, J. R. (2002). An assessment of support vector machines for land cover classification. *International Journal of Remote Sensing, 23*(4), 725–749.
57. Huang, X., & Jensen, J. R. (2002). A machine-learning approach to automated land-cover mapping from high-resolution IKONOS imagery. *Photogrammetric Engineering & Remote Sensing, 68*(1), 41–53.
58. Hyyppä, J., Hyyppä, H., Leckie, D., Gougeon, F., Yu, X., & Maltamo, M. (2008). Review of methods of small-footprint airborne laser scanning for extracting forest inventory data in boreal forests. *International Journal of Remote Sensing, 29*(5), 1339–1366.
59. Ioannidis, K., Tsertou, A., Taha, A. A., & Prasad, R. (2018). A review of land-use and land-cover change models: A case study for Mediterranean countries. *Computers, Environment and Urban Systems, 74*, 1–14.

60. IPCC. (2019). Climate change and land: An IPCC special report on climate change, desertification, land degradation, sustainable land management, food security, and greenhouse gas fluxes in terrestrial ecosystems. *Intergovernmental Panel on Climate Change.*
61. Jiang, B., Alves, P., Rodrigues, F., Ferreira, J., & Pereira, J. M. (2015). Mining Twitter data for land use classification. In *Proceedings of the 6th ACM SIGSPATIAL International Workshop on Location-Based Social Networks* (pp. 31–38).
62. Jokar Arsanjani, J., Zipf, A., Mooney, P., & Helbich, M. (Eds.). (2015). *OpenStreetMap in GIScience: Experiences, research, and applications.* Springer International Publishing.
63. Joppa, L. N., & Pfaff, A. (2009). High and far: Biases in the location of protected areas. *PLoS ONE, 4*(12), e8273.
64. Justice, C. O., Vermote, E., Townshend, J. R., Defries, R., Roy, D. P., Hall, D. K., Barnsley, M. J., et al. (1998). The moderate resolution imaging spectroradiometer (MODIS): Land remote sensing for global change research. *IEEE Transactions on Geoscience and Remote Sensing, 36*(4), 1228–1249.
65. Kaiser, C., & Rauber, A. (2020). Ethical considerations and privacy concerns in the use of machine learning for land-use classification. *Data Science Journal, 19*(1), 1–16.
66. Kohonen, T. (1990). The self-organizing map. *Proceedings of the IEEE, 78*(9), 1464–1480.
67. Kumar, P., Joshi, P. K., & Zambre, A. (2019). A review of machine learning and deep learning applications in urban land use/land cover classification using remote sensing data. *Remote Sensing Applications: Society and Environment, 15*, 100239.
68. Kwan, C., Ho, D., & Wang, H. (2010). Fusion of optical and radar data for improved change detection. *IEEE Geoscience and Remote Sensing Letters, 7*(4), 754–758.
69. Lambin, E. F., & Geist, H. J. (2006). *Land-use and land-cover change: Local processes and global impacts.* Springer Science & Business Media.
70. Lambin, E. F., Geist, H. J., & Lepers, E. (2003). Dynamics of land-use and land-cover change in tropical regions. *Annual Review of Environment and Resources, 28*(1), 205–241.
71. Lambin, E. F., Turner, B. L., Geist, H. J., Agbola, S. B., Angelsen, A., Bruce, J. W., Xu, J., et al. (2001). The causes of land-use and land-cover change: Moving beyond the myths. *Global Environmental Change, 11*(4), 261–269.
72. LeCun, Y., Bengio, Y., & Hinton, G. (2015). Deep learning. *Nature, 521*(7553), 436–444.
73. Li, L., Goodchild, M. F., & Xu, B. (2013). Spatial, temporal, and socioeconomic patterns in the use of Twitter and Flickr. *Cartography and Geographic Information Science, 40*(2), 61–77.
74. Li, X., Liu, X., Clarke, K. C., Chen, Y., & Wu, G. (2020). A review of spatial optimization algorithms for spatial land use allocation. *Landscape Ecology, 35*(1), 1–22.
75. Lillesand, T. M., Kiefer, R. W., & Chipman, J. W. (2008). *Remote sensing and image interpretation.* Wiley.
76. Liu, D., Hu, F., & Wang, L. (2018). Incorporating social media and human mobility data for urban land-use classification. *Environment and Planning B: Urban Analytics and City Science, 45*(5), 877–896.
77. Liu, L., Yang, X., & Li, D. (2016). Towards better analysis of machine learning models: A visual analytics perspective. *Visual Informatics, 1*(1), 48–56.
78. Liu, W., Gao, W., & Li, X. (2018). Land-use change and policy analysis based on scenario simulation of an agent-based model: A case study of Tengzhou City, China. *Sustainability, 10*(6), 1796.
79. Lu, D., & Weng, Q. (2007). A survey of image classification methods and techniques for improving classification performance. *International Journal of Remote Sensing, 28*(5), 823–870.
80. Ma, L., Cheng, L., Li, M., Liu, Y., & Ma, X. (2019). A review of supervised object-based land-cover image classification. *ISPRS Journal of Photogrammetry and Remote Sensing, 152*, 277–293.
81. MacQueen, J. (1967). Some methods for classification and analysis of multivariate observations. In *Proceedings of the Fifth Berkeley Symposium on Mathematical Statistics and Probability* (Vol. 1, No. 14, pp. 281–297).

82. Maggiori, E., Tarabalka, Y., Charpiat, G., & Alliez, P. (2017). Convolutional neural networks for large-scale remote-sensing image classification. *IEEE Transactions on Geoscience and Remote Sensing, 55*(2), 645–657.
83. Maltamo, M., Næsset, E., & Vauhkonen, J. (Eds.). (2014). *Forestry applications of airborne laser scanning: Concepts and case studies*. Springer.
84. Matthews, R. B., Gilbert, N. G., Roach, A., Polhill, J. G., & Gotts, N. M. (2007). Agent-based land-use models: A review of applications. *Landscape Ecology, 22*(10), 1447–1459.
85. Maus, V., Câmara, G., Cartaxo, R., Sanchez, A., Ramos, F. M., & De Queiroz, G. R. (2016). A time-weighted dynamic time warping method for land-use and land-cover mapping. *IEEE Journal of Selected Topics in Applied Earth Observations and Remote Sensing, 9*(8), 3729–3739.
86. Mennis, J., Peuquet, D. J., & Qian, L. (2018). A time-geographic approach to representing and assessing spatial, temporal, and semantic uncertainty in vector land-use and land-cover change data. *International Journal of Geographical Information Science, 32*(2), 267–290.
87. Mikolov, T., Karafiát, M., Burget, L., Černocký, J., & Khudanpur, S. (2010). Recurrent neural network-based language model. In *Eleventh Annual Conference of the International Speech Communication Association.*
88. Millennium Ecosystem Assessment. (2005). *Ecosystems and human well-being: Synthesis*. Island Press.
89. Mittelstadt, B. D., Allo, P., Taddeo, M., Wachter, S., & Floridi, L. (2016). The ethics of algorithms: Mapping the debate. *Big Data & Society, 3*(2), 205395171667967.
90. Mnih, V., Heess, N., Graves, A., & Kavukcuoglu, K. (2015). Recurrent models of visual attention. In *Advances in neural information processing systems* (pp. 2204–2212).
91. Müller, A., Güntner, U., Klenke, T., & Dorn, H. (2018). Participatory land use change modelling to explore the potential of land use planning as an adaptive capacity to climate change impacts on land use patterns. *Land Use Policy, 74*, 548–561.
92. Muller, D., Leitão, P. J., & Sikor, T. (2015). Comparing the determinants of cropland abandonment in Albania and Romania using boosted regression trees. *Agricultural Systems, 135*, 98–111.
93. Pal, M. (2005). Random forest classifier for remote sensing classification. *International Journal of Remote Sensing, 26*(1), 217–222.
94. Pan, S. J., & Yang, Q. (2010). A survey on transfer learning. *IEEE Transactions on Knowledge and Data Engineering, 22*(10), 1345–1359.
95. Parker, D. C., Manson, S. M., Janssen, M. A., Hoffmann, M. J., & Deadman, P. (2003). Multi-agent systems for the simulation of land-use and land-cover change: A review. *Annals of the Association of American Geographers, 93*(2), 314–337.
96. Pielke, R. A., Marland, G., Betts, R. A., Chase, T. N., Eastman, J. L., Niles, J. O., Running, S. W., et al. (2002). The influence of land-use change and landscape dynamics on the climate system: Relevance to climate-change policy beyond the radiative effect of greenhouse gases. *Philosophical Transactions of the Royal Society A: Mathematical, Physical and Engineering Sciences, 360*(1797), 1705–1719.
97. Pohl, C., & Van Genderen, J. L. (1998). Review article multisensor image fusion in remote sensing: Concepts, methods and applications. *International Journal of Remote Sensing, 19*(5), 823–854.
98. Qiu, J., Gober, P., & Yang, X. (2018). Forecasting the spatial-temporal water demand in Phoenix metropolitan area using a deep learning LSTM model. *Computers, Environment and Urban Systems, 72*, 163–172.
99. Reichstein, M., Camps-Valls, G., Stevens, B., Jung, M., Denzler, J., Carvalhais, N., & Prabhat. (2019). Deep learning and process understanding for data-driven Earth system science. *Nature, 566*(7743), 195–204.
100. Reynolds, J. F., Stafford Smith, D. M., Lambin, E. F., Turner, B. L., Mortimore, M., Batterbury, S. P., Walker, B., et al. (2007). Global desertification: Building a science for dryland development. *Science, 316*(5826), 847–851.

101. Ribeiro, M. T., Singh, S., & Guestrin, C. (2016). "Why should I trust you?": Explaining the predictions of any classifier. In *Proceedings of the 22nd ACM SIGKDD International Conference on Knowledge Discovery and Data Mining* (pp. 1135–1144).
102. Ronneberger, O., Fischer, P., & Brox, T. (2015). U-net: Convolutional networks for biomedical image segmentation. In *International Conference on Medical Image Computing and Computer-Assisted Intervention* (pp. 234–241). Springer.
103. Saarikoski, H., Mustajoki, J., Barton, D. N., Geneletti, D., Langemeyer, J., Gomez-Baggethun, E., Marttunen, M., Antunes, P., Keune, H., & Santos, R. (2016). Multi-criteria decision analysis and cost-benefit analysis: Comparing alternative frameworks for integrating ecosystem services into urban planning. *Ecosystem Services, 22*, 238–249.
104. Sala, O. E., Chapin, F. S., Armesto, J. J., Berlow, E., Bloomfield, J., Dirzo, R., Wall, D., et al. 2000). Global biodiversity scenarios for the year 2100. *Science, 287*(5459), 1770–1774.
105. Schlerf, M., Atzberger, C., & Hill, J. (2010). Remote sensing of forest biophysical variables using HyMap imaging spectrometer data. *Remote Sensing of Environment, 114*(2), 416–427.
106. See, L., Comber, A., Salk, C., Fritz, S., van der Velde, M., Perger, C., McCallum, I., et al. (2016). Mapping and validation of global cropland using crowdsourcing. *ISPRS Journal of Photogrammetry and Remote Sensing, 103*, 249–266.
107. Seto, K. C., Güneralp, B., & Hutyra, L. R. (2011). Global forecasts of urban expansion to 2030 and direct impacts on biodiversity and carbon pools. *Proceedings of the National Academy of Sciences, 109*(40), 16083–16088.
108. Shaban, S. S., Dikshit, O., & Pathak, V. M. (2018). Fusion of SAR and optical data for land cover classification using machine learning techniques. *Geocarto International, 33*(10), 1113–1127.
109. Stefanidis, A., Crooks, A., & Radzikowski, J. (2013). Harvesting ambient geospatial information from social media feeds. *GeoJournal, 78*(2), 319–338.
110. Sun, X., Wu, L., Gong, W., Chen, L., & Li, W. (2020). Multi-source and multi-scale data fusion based on convolutional neural networks for global land cover mapping. *Remote Sensing of Environment, 247*, 111927.
111. Syrris, V., Dragani, R., Engelen, R., & Flemming, J. (2018). The added value of the high-resolution Copernicus Sentinel satellite data in improving air quality forecasts over Europe. *Air Quality, Atmosphere & Health, 11*(6), 709–720.
112. Tilman, D., Fargione, J., Wolff, B., D'Antonio, C., Dobson, A., Howarth, R., Swackhamer, D., et al. (2001). Forecasting agriculturally driven global environmental change. *Science, 292*(5515), 281–284.
113. Tso, B., & Mather, P. M. (2001). *Classification methods for remotely sensed data.* CRC Press.
114. Turner, B. L., II., Lambin, E. F., & Reenberg, A. (2013). The emergence of land change science for global environmental change and sustainability. *Proceedings of the National Academy of Sciences, 110*(52), 20959–20964.
115. Václavík, T., Lautenbach, S., Kuemmerle, T., & Seppelt, R. (2016). Mapping global land system archetypes. *Global Environmental Change, 39*, 16–29.
116. Vapnik, V. N. (1995). *The nature of statistical learning theory.* Springer Science & Business Media.
117. Vauhkonen, J., Ene, L., Gupta, S., Heinzel, J., Holmgren, J., Pitkänen, J., Maltamo, M., et al. (2012). Comparative testing of single-tree detection algorithms under different types of forest. *Forestry, 85*(1), 27–40.
118. Verburg, P. H., Crossman, N., Ellis, E. C., Heinimann, A., Hostert, P., Mertz, O., Zhen, L., et al. (2013). Land system science and sustainable development of the earth system: A global land project perspective. *Anthropocene, 12*, 29–41.
119. Verburg, P. H., Dearing, J. A., Dyke, J. G., van der Leeuw, S., Seitzinger, S., Steffen, W., & Syvitskaia, I. (2016). Methods and approaches to modelling the Anthropocene. *Global Environmental Change, 39*, 328–340.
120. Verburg, P. H., Neumann, K., & Nol, L. (2011). Challenges in using land use and land cover data for global change studies. *Global Change Biology, 17*(2), 974–989.

121. Vesanto, J., Alhoniemi, E., Himberg, J., Parviainen, J., & Simula, O. (2000). Self-organizing map for data analysis in remote sensing. *International Journal of Remote Sensing, 21*(5), 929–942.
122. Vesanto, J., Himberg, J., Alhoniemi, E., & Parhankangas, J. (2000). SOM toolbox for Matlab 5. Technical Report A57. Helsinki University of Technology, Finland.
123. Wagner, W., Gruber, A., Klein, I., & Veci, L. (2020). Copernicus Sentinel-1 for monitoring land cover and land use changes at high latitudes. *Remote Sensing, 12*(14), 2246.
124. Wang, Q., Shi, W., & Atkinson, P. M. (2015). Area-to-point regression kriging for pan-sharpening. *IEEE Transactions on Geoscience and Remote Sensing, 53*(9), 5142–5155.
125. Wang, Q., Taylor, J., & Xue, X. (2019). Predicting impacts of urban land use change on surface water quality: A case study of the Jing River Basin, China. *Land Use Policy, 82*, 587–597.
126. Weng, Q. (2015). *Advances in environmental remote sensing: Sensors, algorithms, and applications*. CRC Press.
127. Wolpert, D. H. (1992). Stacked generalization. *Neural Networks, 5*(2), 241–259.
128. Wood, J., Dykes, J., & Slingsby, A. (2010). Visualisation of origins, destinations and flows with OD maps. *The Cartographic Journal, 47*(2), 117–129.
129. Wulder, M. A., Coops, N. C., & Leckie, D. G. (2019). Integrating remote sensing and GIS for monitoring land cover and land use change. In *Remote Sensing and GIS for Ecologists* (pp. 15–44). Pelagic Publishing Ltd.
130. Wulder, M. A., White, J. C., Goward, S. N., Masek, J. G., Irons, J. R., Herold, M., Woodcock, C. E., et al. (2012). The global Landsat archive: Status, consolidation, and direction. *Remote Sensing of Environment, 130*, 32–43.
131. Xie, C., Yuan, F., & Yang, Y. (2018). Deep learning in visual computing and signal processing. *Applied Sciences, 8*(1), 96.
132. Xie, H., Yao, X., Zhang, X., Zhang, J., & Liu, Y. (2018). Transferability of a machine learning algorithm for mapping the risk of changes in land use and land cover using Landsat time series. *ISPRS Journal of Photogrammetry and Remote Sensing, 146*, 297–311.
133. Xie, Y., Weng, Q., & Zhang, J. (2020). Spatiotemporal dynamics of urban impervious surfaces in response to land use policies: A case study of Indianapolis, USA. *Science of the Total Environment, 725*, 138328.
134. Xu, Y., Chen, Y., Li, X., Ciais, P., Goll, D., Koven, C., Piao, S., et al. (2021). Global land model development: Time to shift from a plant functional type to a plant functional trait approach. *Global Change Biology, 27*(4), 730–744.
135. Yuan, F., Sawaya, K. E., Loeffelholz, B. C., & Bauer, M. E. (2020). Land cover classification and change analysis of the Twin Cities (Minnesota) Metropolitan area by multitemporal Landsat remote sensing. *Remote Sensing of Environment, 106*(2), 171–183.
136. Zhang, A., Fang, F., Zhang, C., & Liu, S. (2020). Advances in remote sensing of land-use and land-cover change. *Remote Sensing, 12*(16), 2666.
137. Zhang, C., & Du, B. (2016). Transfer learning for remote sensing data: A survey and a comprehensive study. *IEEE Journal of Selected Topics in Applied Earth Observations and Remote Sensing, 9*(6), 2332–2348.
138. Zhang, C., & Kovacs, J. M. (2012). The application of small unmanned aerial systems for precision agriculture: A review. *Precision Agriculture, 13*(6), 693–712.
139. Zhang, C., Li, W., & Travis, K. (2011). Gaps-fill of SLC-off Landsat ETM+ satellite image using a geostatistical approach. *International Journal of Remote Sensing, 32*(6), 1747–1763.
140. Zhang, H., Zhang, F., Du, S., & Tong, X. (2018). A deep learning-based approach for automated crop classification from high-resolution UAV imagery. *Computers and Electronics in Agriculture, 154*, 246–253.
141. Zhang, J., Wang, J., & Wang, X. (2002). Land use information extraction from high resolution satellite imagery using artificial neural networks. *Journal of Zhejiang University SCIENCE A, 3*(4), 423–429.
142. Zhang, Y. (2010). Understanding image fusion. *Photogrammetric Engineering & Remote Sensing, 76*(6), 657–661.

143. Zhao, Y., Zhao, X., Jiang, D., & Xu, H. (2020). Assessing the impacts of urban expansion on ecosystem services: A case study of the central city of Beijing, China. *Science of the Total Environment, 703*, 134779.
144. Zhao, Y., Zhu, X., & Okabe, A. (2017). Representativeness of social media geospatial data: A quantitative comparison of geotagged Twitter and Flickr posts to fine-scale land use reference data. *Transactions in GIS, 21*(5), 958–972.
145. Zhong, Y., Zhang, L., & Huang, X. (2017). Learning recurrent fully convolutional networks for land cover change detection. In *Proceedings of the IEEE International Conference on Computer Vision Workshops* (pp. 2036–2044).
146. Zhong, Y., Zhang, L., & Wang, L. (2017). A recurrent convolutional neural network for remote sensing image scene classification. In *2017 IEEE International Geoscience and Remote Sensing Symposium (IGARSS)* (pp. 2869–2872). IEEE.
147. Zhong, Y., Zhang, L., Huang, X., & Ma, Y. (2020). Deep learning-based classification of hyperspectral data with limited labeled samples. *Journal of Selected Topics in Applied Earth Observations and Remote Sensing, 13*, 2879–2892.
148. Zhu, X. X., Tuia, D., Mou, L., Xia, G. S., Zhang, L., Xu, F., & Fraundorfer, F. (2017). Deep learning in remote sensing: A comprehensive review and list of resources. *IEEE Geoscience and Remote Sensing Magazine, 5*(4), 8–36.
149. Zwitter, A., & Boisse-Despiaux, M. (2018). Big data, international law, and the SDGs: The need for an integrated approach. *The European Journal of Development Research, 30*(2), 203–223.

Chapter 6
Environmental Risk Assessment and Climate Change Impacts

6.1 Overview of Environmental Risk Assessment and Climate Change Impacts

Environmental risk assessment (ERA) is a systematic process of evaluating the potential adverse effects of human activities and natural phenomena on the environment, ecosystems, and human health [109]. Climate change impacts are alterations in the Earth's climate system, such as temperature, precipitation, and sea-level rise, resulting from human-induced greenhouse gas emissions [55]. This section provides an overview of the concepts, methodologies, and approaches related to environmental risk assessment and climate change impacts, highlighting the role of artificial intelligence (AI) in addressing these issues.

Environmental risk assessment involves the identification, quantification, and characterization of environmental hazards, the assessment of exposure pathways, and the estimation of the probability and magnitude of adverse effects [90]. The process generally includes four main steps: hazard identification, dose–response assessment, exposure assessment, and risk characterization [114]. These steps enable scientists and decision-makers to understand the potential consequences of environmental stressors, prioritize risk management actions, and develop effective policies and strategies to reduce risks [73].

Climate change impacts are diverse and wide-ranging, affecting various sectors such as agriculture, water resources, coastal zones, human health, and biodiversity [55]. The study of these impacts involves assessing the vulnerability and adaptive capacity of natural and human systems, projecting future climate scenarios, and evaluating the effectiveness of mitigation and adaptation measures [55]. This requires interdisciplinary research and the integration of various data sources, models, and tools to generate reliable and actionable information for stakeholders and policymakers [87].

In recent years, AI has emerged as a promising approach for addressing the complexities and uncertainties associated with environmental risk assessment and

S. N. M. Moghaddam and H. Cao, *Artificial Intelligence-Driven Geographies*, City Development: Issues and Best Practices, https://doi.org/10.1007/978-981-97-5116-7_6

climate change impacts [95, 98]. AI techniques, such as machine learning, deep learning, and hybrid approaches, have been applied to a wide range of tasks, including environmental monitoring, hazard prediction, vulnerability assessment, and decision support [79, 96]. These techniques can help improve the accuracy, efficiency, and scalability of environmental risk assessment and climate change impact studies, as well as facilitate the discovery of new insights and relationships in large and heterogeneous datasets [40].

One of the key advantages of AI in environmental risk assessment and climate change impacts is its ability to handle complex and non-linear relationships, as well as to account for various sources of uncertainty [79]. Machine learning algorithms, such as decision trees, support vector machines, and artificial neural networks, can learn patterns and associations in data without explicit programming or prior knowledge, allowing for more flexible and adaptive modeling of environmental systems [95]. Deep learning techniques, such as convolutional neural networks and recurrent neural networks, can further enhance the representation and processing of spatial and temporal information, leading to improved predictions and simulations of environmental hazards and climate change impacts [96].

Hybrid approaches that combine AI techniques with traditional statistical and process-based models can also offer valuable insights and solutions in environmental risk assessment and climate change impact studies [79]. For instance, ensemble methods, which integrate multiple models or algorithms, can improve the robustness and generalizability of predictions by leveraging the strengths and compensating for the weaknesses of individual approaches [86]. Data assimilation techniques, which combine observations and model outputs, can help reduce uncertainties and biases in environmental risk assessment and climate change impact projections, as well as update model parameters and initial conditions in real-time [75].

AI has been used to advance various aspects of environmental risk assessment and climate change impacts, such as early warning systems, vulnerability mapping, and scenario analysis [98]. For example, machine learning algorithms have been employed to predict natural hazards like floods, landslides, and wildfires, by analyzing satellite imagery, remote sensing data, and other relevant variables [93, 101]. AI techniques have also been applied to assess the vulnerability and adaptive capacity of ecosystems and human communities to climate change, by integrating socioeconomic, biophysical, and climatic data [33, 116].

In addition, AI can support the evaluation of climate change mitigation and adaptation strategies, by simulating and comparing various policy options, technology pathways, and socioeconomic scenarios [95, 103]. For instance, AI-based optimization methods, such as genetic algorithms and swarm intelligence, can help identify the most cost-effective and environmentally sustainable solutions for reducing greenhouse gas emissions, conserving biodiversity, and enhancing resilience to climate change impacts [121, 133].

Despite the potential benefits and applications of AI in environmental risk assessment and climate change impacts, there are several challenges and limitations to consider. These include data availability, quality, and representativeness, as well as algorithmic transparency, interpretability, and robustness [40, 79]. Ensuring the ethical and equitable use of AI in environmental risk assessment and climate change impact studies is also crucial, as biases and inaccuracies in AI models can exacerbate existing inequalities and vulnerabilities, as well as undermine public trust and policy implementation [12, 85].

Future directions in AI applications for environmental risk assessment and climate change impacts involve addressing these challenges and limitations, as well as exploring new methods, data sources, and interdisciplinary collaborations. For example, advances in AI explainability and fairness can help enhance the transparency, accountability, and social acceptability of AI models in environmental risk assessment and climate change impact studies [8, 43]. The integration of AI with other emerging technologies, such as the Internet of Things, blockchain, and augmented reality, can also create innovative and scalable solutions for environmental monitoring, decision support, and stakeholder engagement [3, 129].

In conclusion, AI holds great promise for advancing environmental risk assessment and climate change impact studies, by providing novel techniques and tools for understanding and managing the complex and uncertain relationships between human activities, environmental stressors, and global change. However, realizing the full potential of AI in these domains requires addressing the various technical, ethical, and social challenges associated with AI research and applications, as well as fostering cross-disciplinary and cross-sectoral collaborations among scientists, policymakers, practitioners, and stakeholders.

6.2 Data Sources for Studying Environmental Risks and Climate Change Impacts

6.2.1 Traditional Data Sources

Environmental risk assessment and climate change impacts are critical areas of study for understanding the consequences of human activities on the environment and predicting future environmental challenges (Table 6.1). Traditional data sources have played a significant role in these areas, providing essential historical context and baseline information. In this section, we will discuss various traditional data sources for studying environmental risks and climate change impacts, including governmental reports, research publications, climate models, and meteorological data.

Governmental Reports and Databases

Governments across the world have long recognized the importance of environmental monitoring and risk assessment. Several national and international agencies have established databases and reporting mechanisms to collect and disseminate environmental data. For instance, the United States Environmental Protection Agency (EPA) maintains a repository of environmental data through its Environmental Data Gateway (EDG) [113]. Similarly, the European Environment Agency

Table 6.1 Data sources for environmental risks and climate change studies

Data source	Description	Advantages	Challenges
Governmental reports and databases	Reports and databases maintained by national and international agencies, such as the EPA and EEA, providing environmental data	– Provides comprehensive and up-to-date information – Data collected by authoritative sources	– Access may be restricted or costly for some databases – Inconsistencies in data collection and reporting
Research publications	Peer-reviewed research articles providing insights into various environmental aspects, based on fieldwork, experiments, and traditional data sources	– Offers detailed analysis and findings from scientific studies – Data collected from rigorous research methodologies	– May be subject to publication biases – Limited coverage of specific topics or regions
Climate models	Mathematical representations of the Earth's climate system, used to simulate past, present, and future climate conditions	– Provides projections of future climate scenarios – Enables assessment of potential climate change impacts on various sectors	– Uncertainties in model outputs due to limitations in representation of physical processes – Need for multiple models to account for uncertainties
Meteorological data	Measurements and observations of atmospheric variables collected by weather stations and satellites	– Provides crucial information for analyzing climate trends and extreme events – Data collected from a network of global stations	– Limited spatial coverage in some regions – Challenges in data quality assurance and calibration
Paleoclimate data	Information about past climate conditions derived from natural archives, such as ice cores and sedimentary records	– Offers historical context for understanding climate variability – Helps calibrate and validate climate models for future projections	– Data extraction and interpretation may be challenging – Limited availability of long-term and high-resolution data

(EEA) collects and shares data on various environmental topics, including climate change, biodiversity, and air quality [29].

Some global initiatives have been launched to address environmental risks and climate change impacts, such as the Intergovernmental Panel on Climate Change (IPCC). The IPCC regularly publishes Assessment Reports (ARs) that provide comprehensive and up-to-date information on the state of the climate, potential impacts, and possible adaptation and mitigation strategies [54].

Research Publications

Peer-reviewed research publications are an essential source of information on environmental risks and climate change impacts. Researchers and scientists conduct studies on various aspects of the environment, such as greenhouse gas emissions, deforestation, ocean acidification, and extreme weather events. These studies often rely on data collected from fieldwork, laboratory experiments, or other traditional data sources. By analyzing this data, researchers can develop models, draw conclusions, and make predictions about future environmental conditions [22, 63].

Climate Models

Climate models are mathematical representations of the Earth's climate system. They are designed to simulate the interactions between the atmosphere, oceans, land surface, and ice. By incorporating data on various factors, such as greenhouse gas concentrations, solar radiation, and aerosols, climate models can be used to simulate past, present, and future climate conditions [32]. Climate models have been instrumental in projecting future climate change impacts, such as temperature and precipitation changes, sea-level rise, and changes in the frequency and intensity of extreme weather events [20].

Meteorological Data

Meteorological data consists of measurements and observations of various atmospheric variables, such as temperature, precipitation, wind speed, and humidity. This data is collected by a network of weather stations and satellites operated by national meteorological agencies and international organizations, such as the World Meteorological Organization (WMO) [122]. Meteorological data provides crucial information for studying climate change impacts, including the analysis of long-term trends, detection of extreme events, and validation of climate models [52].

Paleoclimate Data

Paleoclimate data refers to information about past climate conditions derived from natural archives, such as ice cores, tree rings, and sedimentary records. This data provides a valuable historical context for understanding the natural variability of the climate system and assessing the significance of recent climate changes [53]. Paleoclimate data can also be used to calibrate and validate climate models, improving their ability to project future climate change impacts [91].

In summary, traditional data sources have provided invaluable information for studying environmental risks and climate change impacts. These sources include governmental reports and databases, research publications, climate models, meteorological data, and paleoclimate data. These data sources are essential for understanding the historical context of environmental changes, analyzing trends, and making predictions about future conditions. Despite the growing importance of big data and AI techniques, traditional data sources will continue to play a crucial role in environmental risk assessment and climate change impact studies.

6.2.2 Remote Sensing Data Sources

Remote sensing data sources have become increasingly important for studying environmental risks and climate change impacts. These sources offer unparalleled spatial and temporal coverage, enabling researchers to monitor changes in land cover, temperature, precipitation, and other environmental variables at various scales. This section will provide an overview of the main remote sensing data sources available for environmental risk assessment and climate change impact studies, along with a discussion of their respective strengths and limitations.

Satellite Imagery

Satellite imagery has been widely used for monitoring and analyzing various aspects of the Earth's surface and atmosphere. There are numerous satellite missions that provide essential data for studying environmental risks and climate change impacts, such as:

- MODIS (Moderate Resolution Imaging Spectroradiometer): MODIS is a key instrument aboard NASA's Terra and Aqua satellites, providing daily global observations of the Earth's surface in visible, infrared, and microwave bands [58]. MODIS data have been extensively used for land cover classification, vegetation monitoring, and climate studies [34].
- Landsat: The Landsat program, a joint effort between NASA and the US Geological Survey (USGS), has been providing high-resolution, multispectral images of the Earth's surface since the 1970s [123]. Landsat data are used for a wide range of applications, including land use and land cover change detection, deforestation monitoring, and urban growth analysis [46].
- Sentinel: The Sentinel satellites are part of the European Union's Copernicus program, which aims to provide comprehensive and accurate Earth observation data for environmental monitoring and policy-making [26]. Sentinel-1 and Sentinel-2 satellites offer high-resolution radar and multispectral imagery, respectively, which are used for various applications, such as floodmapping, crop monitoring, and glacier dynamics [80, 110].

Reanalysis Datasets

Reanalysis datasets are produced by assimilating historical observational data into state-of-the-art numerical weather prediction models. These datasets provide gridded, long-term records of various atmospheric, oceanic, and land surface variables, which are invaluable for climate research and environmental risk assessment [23]. Some popular reanalysis datasets include:

- ERA-Interim: Produced by the European Centre for Medium-Range Weather Forecasts (ECMWF), ERA-Interim is a global atmospheric reanalysis covering the period from 1979 to the present [23]. It provides high-resolution, consistent, and reliable estimates of various climate variables, such as temperature, precipitation, and wind speed.
- MERRA-2: The Modern-Era Retrospective analysis for Research and Applications, Version 2 (MERRA-2) is a global atmospheric reanalysis produced by NASA's Global Modeling and Assimilation Office (GMAO) [38]. MERRA-2 provides hourly and daily data on a range of atmospheric variables, such as temperature, humidity, and precipitation, from 1980 to the present.

Climate Model Outputs

Climate model outputs are essential for understanding the potential impacts of climate change on various environmental processes and systems. These outputs are generated by complex numerical models that simulate the Earth's climate system, incorporating interactions between the atmosphere, oceans, land surface, and cryosphere [32]. Climate models are used to project future climate scenarios under different greenhouse gas emission pathways, which can be used to assess potential environmental risks and inform adaptation strategies.

- CMIP: The Coupled Model Intercomparison Project (CMIP) is an international effort to coordinate climate model experiments and standardize output data for comparison and analysis [30]. CMIP datasets, such as those from CMIP5 and the more recent CMIP6, include simulations of historical climate as well as future projections under various emission scenarios. These datasets are widely used for studying climate change impacts on various environmental systems, such as water resources, agriculture, and ecosystems [111].

Despite the wealth of information provided by remote sensing data sources, there are some limitations associated with their use in environmental risk assessment and climate change impact studies (Fig. 6.1):

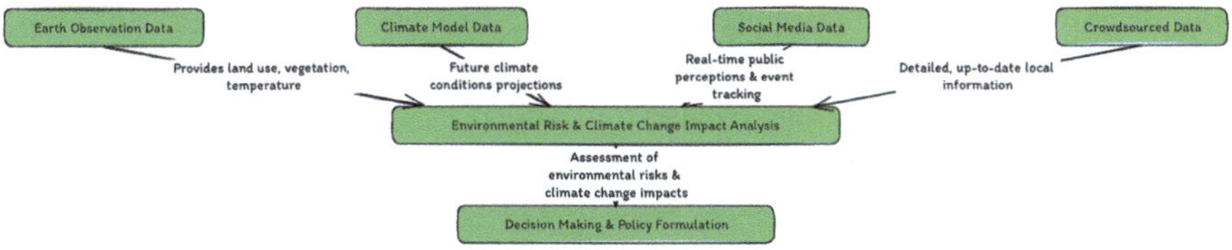

Fig. 6.1 The relationship between different sources of big data and geospatial data in the context of environmental risk assessment and climate change impact studies

1. Spatial and Temporal Resolution: Satellite data are often limited by their spatial and temporal resolution, which may not be suitable for some applications. For example, coarse-resolution data may not capture small-scale environmental features, while infrequent satellite overpasses can limit the ability to monitor rapid changes.
2. Data Availability and Accessibility: Although many satellite datasets are freely available, some data sources are restricted or costly to access, particularly high-resolution commercial imagery. Additionally, long-term satellite data may be affected by changes in sensor technology, calibration, and orbital characteristics, which can introduce inconsistencies in the time series.
3. Cloud Cover and Atmospheric Effects: Cloud cover can obscure the Earth's surface in optical satellite imagery, leading to gaps in the data. Furthermore, atmospheric effects such as scattering and absorption can introduce errors in the retrieved environmental variables, requiring correction and calibration procedures.
4. Data Processing and Interpretation: Processing and analyzing remote sensing data can be complex and computationally intensive, particularly for large-scale studies. Additionally, interpreting the data requires expert knowledge of the underlying physical processes and an understanding of the uncertainties associated with the data and the algorithms used for processing.

Despite these challenges, remote sensing data sources offer invaluable information for studying environmental risks and climate change impacts, providing a crucial foundation for informed decision-making and policy development. As the availability and quality of remote sensing data continue to improve, the potential for using this information in environmental risk assessment and climate change impact studies will grow. The integration of these data with other sources, such as ground-based measurements, will be essential for developing more accurate and comprehensive assessments of the changing environment.

One promising approach to addressing the challenges associated with remote sensing data is the application of artificial intelligence (AI) techniques, such as machine learning and deep learning algorithms. AI methods can help to automate the processing and analysis of large volumes of satellite data, while also addressing issues related to data quality, resolution, and interpretation [126]. For instance, AI techniques can be used to fuse data from multiple sources, filling gaps in the data and enhancing the spatial and temporal resolution of the observations. Additionally, machine learning algorithms can be trained to identify patterns and extract meaningful information from the data, facilitating the detection and monitoring of environmental risks and climate change impacts.

Another emerging area of research is the integration of remote sensing data with other big data sources, such as social media, mobile phone data, and Internet of Things (IoT) devices. By combining these diverse data streams, researchers can gain a more holistic understanding of the complex interactions between human activities and environmental systems [68]. This integrated approach can help to identify potential hotspots of vulnerability and enhance the development of targeted adaptation and mitigation strategies.

In conclusion, remote sensing data sources play a crucial role in studying environmental risks and climate change impacts. Despite the challenges associated with their use, advances in satellite technology, data processing, and AI techniques are helping to overcome these limitations and unlock the full potential of these data for informing environmental decision-making and policy development.

6.2.3 Big Data and Geospatial Data Sources

The emergence of big data and geospatial data sources has transformed the way environmental risks and climate change impacts are studied. These data sources enable researchers and decision-makers to access more granular, timely, and accurate information for assessing potential threats and developing adaptation and mitigation strategies [74]. In this section, we will discuss various big data and geospatial data sources available for studying environmental risks and climate change impacts, their advantages, and the challenges associated with their use.

Earth Observation Data

One of the most critical sources of geospatial data for environmental risk assessment and climate change impact studies is Earth observation data. These data are collected through satellite-based remote sensing platforms, which capture images of the Earth's surface at various spatial, temporal, and spectral resolutions [84]. Remote sensing data can provide valuable information on land use and land cover, vegetation health, surface temperature, and precipitation patterns, among other variables, which are essential for assessing environmental risks and climate change impacts [46].

One significant advantage of Earth observation data is its global coverage, enabling researchers to study environmental risks and climate change impacts at regional, national, and international scales. Additionally, these data are often available at no or low cost, making them accessible to researchers from various disciplines and organizations [124].

Despite their advantages, Earth observation data also presents challenges, including the need for specialized knowledge to process and analyze the data, as well as gaps in data coverage due to cloud cover or sensor limitations [42].

Climate Model Data

Climate model data, generated from global and regional climate models, is another vital source of information for studying environmental risks and climate change impacts. These models simulate the Earth's climate system and its interactions with the atmosphere, ocean, land surface, and cryosphere [32]. Climate model data provides projections of future climate conditions under different greenhouse gas emissions scenarios, allowing researchers to assess the potential impacts of climate change on various sectors, such as agriculture, water resources, and ecosystems [56].

Climate model data is typically available at various spatial and temporal resolutions, offering researchers the flexibility to choose the most appropriate dataset for their specific study area and research question [20]. However, climate model data is subject to uncertainties due to limitations in the models' representation of physical processes and uncertainties in future emissions scenarios [56]. As a result, researchers often need to consider multiple climate models and scenarios to account for these uncertainties.

Social Media Data

Social media data, generated from platforms such as Twitter, Facebook, and Instagram, has recently emerged as a valuable source of information for studying environmental risks and climate change impacts. Social media data can provide real-time information on public perceptions, attitudes, and behaviors related to environmental issues and climate change [62]. These data can also be used to track the spatial and temporal distribution of environmental events and incidents, such as floods, droughts, and wildfires, by analyzing geotagged posts and images [67].

However, the use of social media data in environmental risk assessment and climate change impact studies also presents challenges, such as biases in the user population, privacy concerns, and the need for advanced analytical tools and techniques to process and analyze large volumes of unstructured data [107].

Crowdsourced Data

Crowdsourced data, collected through platforms such as OpenStreetMap and Ushahidi, is another emerging source of geospatial data for studying environmental risks and climate change impacts. Crowdsourced data offers the potential to fill gaps in traditional data sources by providing more detailed, up-to-date, and localized information on land use, infrastructure, and environmental conditions [41]. In addition, crowdsourced data can facilitate the integration of local knowledge and expertise into the assessment of environmental risks and climate change impacts [45].

Despite its potential, crowdsourced data also presents challenges, including concerns about data quality, representativeness, and the need for validation and quality control procedures [104].

In conclusion, big data and geospatial data sources have significantly expanded the range of available data for studying environmental risks and climate change impacts. By leveraging these data sources, researchers and decision-makers can access more granular, timely, and accurate information to better understand and respond to the challenges posed by environmental risks and climate change. However, the use of these data sources also presents challenges that need to be addressed to fully realize their potential in environmental risk assessment and climate change impact studies.

6.3 AI Techniques for Analyzing Environmental Risks and Climate Change Impacts

The use of artificial intelligence (AI) techniques to analyze environmental risks and climate change impacts has grown significantly in recent years. These techniques have been used to process vast amounts of data, identify patterns and trends, and create models to predict and understand the consequences of environmental changes. This section will review various AI techniques that have been applied to study environmental risks and climate change impacts, including machine learning, deep learning, and natural language processing. Furthermore, the section will highlight the advantages and challenges associated with each of these techniques.

6.3.1 *Machine Learning*

Machine learning is a subfield of AI that involves the development of algorithms that can learn from and make predictions or decisions based on data without explicit programming [102]. Machine learning has been widely used in environmental risk assessment and climate change impact studies to process large and complex datasets, identify relationships between variables, and develop models to predict outcomes.

One of the most common applications of machine learning in environmental risk assessment is the development of predictive models. For instance, regression models have been used to analyze relationships between environmental variables and various risks, such as air pollution levels [131], water quality [27], and flood hazards [118]. Furthermore, machine learning techniques, such as random forests and support vector machines, have been used to classify land cover and land use changes [92], as well as to detect and monitor deforestation [60].

In climate change impact studies, machine learning models have been employed to predict future climate conditions, such as temperature and precipitation, based on historical data and greenhouse gas emissions scenarios [36]. Additionally, machine learning techniques have been used to identify climate change vulnerability hotspots [37] and assess the impacts of climate change on agriculture [76], water resources [19], and ecosystems [112].

6.3.2 *Deep Learning*

Deep learning is a subset of machine learning that involves the use of artificial neural networks with multiple hidden layers to process and analyze data [65]. Deep learning has shown great promise in environmental risk assessment and climate change impact studies due to its ability to handle large and complex datasets and automatically learn relevant features from the data.

One of the most common applications of deep learning in environmental risk assessment is the analysis of remote sensing data, such as satellite imagery and LiDAR. Convolutional neural networks (CNNs) have been used to classify land use and land cover [134], detect and monitor deforestation [78], and identify urban growth patterns [70]. CNNs are particularly well-suited for processing spatial data, as they can automatically learn spatial features and patterns from the input data [64].

In climate change impact studies, deep learning techniques have been applied to analyze climate model outputs and identify complex patterns and trends. For instance, Reichstein et al. [96] used deep learning to identify nonlinear interactions between climate variables and their impact on extreme events, such as heatwaves and droughts. In another study, Vandal et al. [115] employed deep learning techniques to predict future climate conditions under different greenhouse gas emissions scenarios.

6.3.3 Natural Language Processing

Natural language processing (NLP) is an AI technique that deals with the analysis and understanding of human language, including text and speech. NLP has been applied to environmental risk assessment and climate change impact studies, primarily in the context of analyzing textual data, such as scientific publications, news articles, social media posts, and policy documents.

For example, NLP techniques have been employed to identify and analyze public perceptions and opinions on climate change and its impacts [62], as well as to track the evolution of climate change research and policy over time [49]. Furthermore, NLP has been used to analyze the content of environmental impact assessment reports, identifying trends and patterns in environmental risks and mitigation measures [39].

6.3.4 Challenges and Limitations of AI Techniques in Environmental Risk Assessment and Climate Change Impact Studies

Despite the growing use of AI techniques in environmental risk assessment and climate change impact studies, there are several challenges and limitations associated with their application. Some of these challenges include:

- Data quality and availability: AI techniques require large amounts of high-quality data for training and validation. However, in many cases, data on environmental risks and climate change impacts may be limited, incomplete, or of low quality [88].

- Model interpretability: AI models, particularly deep learning models, can be complex and difficult to interpret. This can limit their applicability in environmental risk assessment and climate change impact studies, where clear explanations of the relationships between variables and the underlying mechanisms are often required [50].
- Uncertainty quantification: AI models may not provide accurate estimates of uncertainty, which is an important aspect of environmental risk assessment and climate change impact studies. Developing methods for quantifying and communicating uncertainty in AI model predictions remains a challenge [82].
- Transferability and generalization: AI models trained on specific datasets or regions may not generalize well to other contexts or scales. This can limit their applicability in environmental risk assessment and climate change impact studies, where the goal is often to develop models that can be applied across different spatial and temporal scales [7].
- Integration with traditional methods: Combining AI techniques with traditional methods, such as expert knowledge, statistical models, and process-based models, can be challenging due to differences in data formats, model structures, and assumptions. However, integrating AI techniques with traditional methods has the potential to improve the accuracy and reliability of environmental risk assessments and climate change impact studies [25].

AI techniques have shown great promise in advancing our understanding of environmental risks and climate change impacts. By leveraging the power of machine learning, deep learning, and natural language processing, researchers can analyze large and complex datasets, identify patterns and trends, and develop models to predict and understand the consequences of environmental changes. However, several challenges and limitations remain, including data quality and availability, model interpretability, uncertainty quantification, transferability, and integration with traditional methods. Continued research and development in AI techniques, as well as interdisciplinary collaboration between AI researchers, environmental scientists, and policymakers, will be crucial in addressing these challenges and further enhancing the role of AI in environmental risk assessment and climate change impact studies.

6.4 Applications of AI in Environmental Risk Assessment and Climate Change Impact Studies

6.4.1 Hazard Mapping and Vulnerability Assessment

Hazard mapping and vulnerability assessment are essential components of environmental risk assessment and climate change impact studies. These assessments help to identify areas that are prone to natural hazards, such as floods, landslides, and earthquakes, as well as understand the vulnerabilities of the communities living in these areas [31]. Artificial intelligence (AI) techniques have increasingly been employed to improve the accuracy and efficiency of hazard mapping and vulnerability assessment.

Hazard Mapping

Hazard mapping involves identifying and delineating areas that are susceptible to natural hazards. AI techniques, such as machine learning algorithms, have been applied to enhance hazard mapping processes [59]. Some applications of AI in hazard mapping include:

- Flood hazard mapping: Flood hazard mapping involves predicting the areas at risk of flooding under different rainfall scenarios. AI techniques, such as artificial neural networks (ANNs) and decision tree models, have been used to model flood hazards [21, 77].
- Landslide susceptibility mapping: Landslide susceptibility mapping aims to identify areas that are prone to landslides based on factors such as slope, soil type, and land use. AI techniques, such as support vector machines (SVMs) and random forest (RF) models, have been applied to landslide susceptibility mapping [93, 125].
- Earthquake hazard mapping: Earthquake hazard mapping identifies areas that are susceptible to seismic hazards. AI techniques, such as machine learning-based clustering algorithms, have been employed to develop earthquake hazard maps [94].

Vulnerability Assessment

Vulnerability assessment involves evaluating the potential impacts of natural hazards on communities, infrastructure, and ecosystems. AI techniques have been utilized to assess vulnerability, including:

- Social vulnerability assessment: Social vulnerability assessment focuses on evaluating the susceptibility of communities to natural hazards based on factors such as population density, socioeconomic status, and access to resources. AI techniques, such as machine learning algorithms, have been applied to assess social vulnerability [61, 130].
- Infrastructure vulnerability assessment: Infrastructure vulnerability assessment evaluates the potential impacts of natural hazards on critical infrastructure, such as transportation networks, energy systems, and water supply systems. AI techniques, such as ANN and genetic algorithms, have been used to assess infrastructure vulnerability [16, 119].
- Ecosystem vulnerability assessment: Ecosystem vulnerability assessment aims to understand the potential impacts of natural hazards on ecosystems and their services. AI techniques, such as machine learning-based species distribution models, have been employed to assess ecosystem vulnerability [112].

Despite the promising applications of AI in hazard mapping and vulnerability assessment, there are still challenges and limitations that need to be addressed. Some of these challenges include:

1. Data quality and availability: Accurate and reliable data is essential for the successful application of AI techniques in hazard mapping and vulnerability assessment. However, data quality and availability can be a significant challenge, especially in developing countries where data collection and management infrastructure may be inadequate [99]. Incomplete or inaccurate data can lead to errors in hazard mapping and vulnerability assessment, which can result in ineffective risk reduction strategies.
2. Model complexity and interpretability: AI models, especially deep learning models, can be complex and difficult to interpret. This lack of interpretability can make it challenging for decision-makers to understand the underlying mechanisms of the model and trust its predictions [50]. Transparent and explainable AI models are needed to improve the understanding and acceptance of AI-based hazard mapping and vulnerability assessment.
3. Uncertainty quantification: Quantifying the uncertainties associated with AI-based hazard mapping and vulnerability assessment is crucial for decision-making. However, AI models may not provide accurate estimates of uncertainty, which can lead to overconfidence in the model predictions and potentially result in inadequate risk management strategies [83]. Developing AI models that can provide reliable uncertainty estimates is essential for effective risk assessment and management.
4. Integration of multiple data sources: Integrating multiple data sources, such as satellite imagery, GIS data, and social media data, can improve the accuracy and comprehensiveness of AI-based hazard mapping and vulnerability assessment [9]. However, integrating heterogeneous data sources can be challenging due to differences in data formats, scales, and resolutions. Developing AI models that can effectively handle and integrate diverse data sources is crucial for enhancing the quality of hazard mapping and vulnerability assessment.

6.4.2 Climate Change Impact Modeling

Climate change impact modeling is a crucial aspect of environmental risk assessment and climate change impact studies (Table 6.2). It involves the development of models that can simulate and predict the potential impacts of climate change on various sectors, such as agriculture, water resources, ecosystems, and human health [55]. Artificial intelligence (AI) techniques have increasingly been applied to enhance the accuracy and efficiency of climate change impact modeling.

Agriculture

Climate change poses significant challenges to agricultural systems, including changes in temperature, precipitation patterns, and extreme weather events. AI techniques have been employed to assess the potential impacts of climate change on crop yields, food security, and agricultural adaptation strategies. For instance, machine learning algorithms, such as support vector machines (SVMs), artificial

Table 6.2 Applications and AI techniques for different climate change sectors

Sector	AI techniques used	Applications	Challenges and limitations
Agriculture	Support vector machines (SVM)	– Predicting crop yields under various climate change scenarios – Identifying suitable areas for crop cultivation under changing climatic conditions	1. Data quality and availability: incomplete or inaccurate data may lead to errors in predictions 2. Model complexity and interpretability: complex models may be difficult to interpret
Water resources	Artificial neural networks (ANN)	– Predicting streamflow and groundwater levels under climate change scenarios – Assessing impacts on water supply and demand	1. Uncertainty quantification: AI models may not accurately estimate uncertainty 2. Integration of multiple data sources: integrating heterogeneous data sources can be challenging
Ecosystems	Machine learning species distribution models	– Predicting shifts in species ranges and biodiversity patterns – Assessing vulnerability of ecosystems to climate change	1. Data quality and availability: accurate data is essential for reliable predictions 2. Model interpretability: complex models may lack transparency
Human health	Machine learning algorithms	– Predicting incidence of vector-borne diseases under different climate change scenarios – Assessing health impacts of air pollution under changing climatic conditions	1. Model complexity and interpretability: complex models may be difficult to understand 2. Uncertainty quantification: reliable uncertainty estimates are needed for effective decision-making

neural networks (ANNs), and random forests (RF), have been applied to predict crop yields under various climate change scenarios [66, 127]. Additionally, AI-based models have been used to identify suitable areas for crop cultivation under changing climatic conditions, thereby aiding in the development of climate-smart agricultural practices [132].

Water Resources

Climate change can have significant impacts on water resources, including alterations in precipitation patterns, evapotranspiration rates, and river flow regimes. AI techniques have been applied to model the potential impacts of climate change on water resources. For example, ANNs and genetic algorithms have been employed to predict streamflow and groundwater levels under various climate change scenarios [17, 89]. AI-based models have also been used to assess the potential impacts of climate change on water supply and demand, thereby informing water resource management and planning strategies [51].

Ecosystems

Climate change can significantly affect ecosystems and their services, such as habitat provision, carbon sequestration, and biodiversity. AI techniques have been used to model the potential impacts of climate change on ecosystems. For instance, machine learning-based species distribution models have been employed to predict shifts in species ranges and biodiversity patterns under various climate change scenarios [28, 112]. AI techniques have also been applied to assess the vulnerability of ecosystems to climate change, including potential impacts on ecosystem services and functions [72].

Human Health

Climate change can have diverse impacts on human health, including increased risk of heat-related illnesses, vector-borne diseases, and air pollution-related health issues. AI techniques have been applied to model the potential impacts of climate change on human health. For example, machine learning algorithms have been used to predict the incidence of vector-borne diseases, such as malaria and dengue, under various climate change scenarios [11, 13]. AI techniques have also been employed to assess the potential health impacts of air pollution under changing climatic conditions, thereby informing public health strategies and interventions [5].

Despite the promising applications of AI in climate change impact modeling, there are still challenges and limitations that need to be addressed. Some of these challenges include:

1. Data quality and availability: Accurate and reliable data is essential for the successful application of AI techniques in climate change impact modeling. However, data quality and availability can be a significant challenge, especially in developing countries where data collection and management infrastructure may be inadequate [99]. Incomplete or inaccurate data can lead to errors in climate change impact predictions, which can result in ineffective adaptation and mitigation strategies.
2. Model complexity and interpretability: AI models, particularly deep learning models, can be complex and difficult to interpret. This lack of interpretability can make it challenging for decision-makers to understand the underlying mechanisms of the model and trust its predictions [50]. Transparent and explainable AI models are needed to improve the understanding and acceptance of AI-based climate change impact modeling.
3. Uncertainty quantification: Quantifying the uncertainties associated with AI-based climate change impact modeling is crucial for decision-making. However, AI models may not provide accurate estimates of uncertainty, which can lead to overconfidence in the model predictions and potentially result in inadequate adaptation and mitigation strategies [83]. Developing AI models that can provide reliable uncertainty estimates is essential for effective climate change impact assessment and management.
4. Integration of multiple data sources: Integrating multiple data sources, such as climate model outputs, remote sensing data, and socioeconomic data, can

improve the accuracy and comprehensiveness of AI-based climate change impact modeling [9]. However, integrating heterogeneous data sources can be challenging due to differences in data formats, scales, and resolutions. Developing AI models that can effectively handle and integrate diverse data sources is crucial for enhancing the quality of climate change impact modeling.

6.4.3 Climate Change Adaptation and Mitigation Strategies

Climate change adaptation and mitigation strategies are essential to reduce the adverse impacts of climate change and to promote sustainable development. AI techniques can play a crucial role in designing and implementing effective strategies to address the challenges posed by climate change. This section provides an overview of how AI can be employed in climate change adaptation and mitigation strategies, discussing various methods and applications.

Climate Change Adaptation Planning

Adaptation planning involves the identification and implementation of measures to reduce the vulnerability of natural and human systems to climate change impacts. AI can be used in several aspects of climate change adaptation planning, such as vulnerability assessment, identification of adaptation measures, and evaluation of adaptation options.

Machine learning algorithms can help in identifying the most vulnerable areas and populations by analyzing a wide range of socioeconomic, environmental, and climatic variables [18]. Furthermore, AI techniques can be used to simulate the effectiveness of different adaptation measures under various climate change scenarios, enabling decision-makers to prioritize and select the most suitable options [71].

Climate Change Mitigation Strategies

Mitigation strategies aim to reduce greenhouse gas (GHG) emissions and enhance carbon sinks to stabilize the global climate. AI can play a significant role in the development and implementation of climate change mitigation strategies, particularly in the areas of energy efficiency, renewable energy, carbon capture and storage, and land use management.

AI techniques can optimize energy consumption in buildings, transportation, and industrial processes, leading to significant reductions in GHG emissions [69]. Machine learning algorithms can also be employed to optimize the design and operation of renewable energy systems, such as wind turbines and solar panels, to enhance their efficiency and reliability [128]. AI can contribute to carbon capture and storage technologies by optimizing their performance and reducing their costs [4]. In addition, AI-based land use management tools can help in monitoring and managing forest resources, promoting reforestation, and reducing deforestation, which are essential for enhancing carbon sinks [6].

Integrated Assessment of Climate Change Adaptation and Mitigation Strategies

An integrated assessment of climate change adaptation and mitigation strategies is essential to develop comprehensive and effective approaches to address the challenges posed by climate change. AI techniques can be instrumental in conducting such assessments by analyzing large and diverse datasets, facilitating the evaluation of trade-offs, synergies, and conflicts between various strategies [106].

AI-driven integrated assessment models can evaluate the potential effectiveness of a wide range of adaptation and mitigation measures under different climate change scenarios, considering multiple objectives such as reducing GHG emissions, enhancing resilience, and promoting sustainable development [47]. Machine learning algorithms can also identify context-specific factors that may influence the success or failure of these strategies, enabling policymakers to design tailored approaches that consider local conditions and priorities [117].

Moreover, AI techniques can be employed to monitor the progress of adaptation and mitigation efforts, providing valuable feedback to decision-makers and facilitating adaptive management. This continuous learning process can help ensure that climate change strategies remain effective and relevant in the face of evolving conditions and emerging challenges [35].

In conclusion, AI techniques offer significant potential in addressing climate change challenges through adaptation and mitigation strategies. By enabling more accurate and efficient assessments of environmental risks, optimizing the design and operation of various strategies, and supporting adaptive management, AI can contribute to the development of more effective and sustainable approaches to tackle climate change impacts. However, it is essential to recognize that AI is not a panacea and that its successful application requires interdisciplinary collaboration, continuous innovation, and a strong commitment to environmental sustainability.

6.5 Challenges and Limitations of AI in Environmental Risk Assessment and Climate Change Impact Analysis

While AI techniques have shown great potential in environmental risk assessment and climate change impact analysis, there are several challenges and limitations that must be acknowledged and addressed to ensure the reliable and effective use of AI in these fields.

1. Data quality and availability: AI algorithms rely heavily on the availability of high-quality data to produce accurate and reliable results [15]. In the context of environmental risk assessment and climate change impact analysis, data sources may be incomplete, inconsistent, or outdated, limiting the effectiveness of AI techniques. Moreover, the availability of certain types of data, such as ground truth observations, may be limited due to various factors, including financial constraints, accessibility, and political barriers [57].

2. Model complexity and interpretability: AI techniques, particularly deep learning models, are known for their complexity and lack of interpretability [14]. This "black-box" nature may hinder the understanding of the underlying relationships between input variables and model outputs, making it difficult for stakeholders to trust and adopt the results [100]. To overcome this limitation, researchers and practitioners should focus on developing more interpretable models and incorporating explainable AI techniques to improve the transparency and credibility of AI-driven environmental risk assessments and climate change impact analyses.
3. Uncertainty quantification: AI models often struggle with quantifying uncertainties associated with their predictions, which is a critical aspect of environmental risk assessment and climate change impact analysis [81]. Addressing this challenge requires the development and incorporation of uncertainty quantification techniques in AI models, such as Bayesian methods and ensemble learning, to provide more robust and reliable results for decision-making purposes [24].
4. Scalability and computational resources: AI techniques, particularly deep learning models, often require significant computational resources and time for training and processing large-scale datasets [108]. This can be a limitation for researchers and practitioners with limited access to high-performance computing facilities or for applications where real-time analysis is needed. To address this challenge, researchers should focus on developing more computationally efficient AI algorithms and leveraging cloud-based services and parallel processing techniques to enhance the scalability of AI-driven environmental risk assessments and climate change impact analyses.
5. Integration with traditional approaches: AI techniques should not be seen as a replacement for traditional environmental risk assessment and climate change impact analysis methods, but rather as complementary tools that can enhance and refine existing approaches [96]. Successfully integrating AI techniques with traditional methods requires interdisciplinary collaboration, knowledge exchange, and capacity building among researchers and practitioners in various fields, including environmental science, climate science, computer science, and data science [48].
6. Ethical considerations: The use of AI in environmental risk assessment and climate change impact analysis raises ethical concerns, particularly in terms of privacy, data security, and potential biases in model outputs [105]. Ensuring the responsible and ethical use of AI techniques in these fields requires the development of guidelines, best practices, and regulatory frameworks that address these concerns while promoting transparency, accountability, and fairness in AI-driven environmental decision-making processes [85].

In conclusion, AI techniques offer significant potential for advancing environmental risk assessment and climate change impact analysis. However, to fully realize this potential, researchers and practitioners must address the various challenges and limitations associated with AI applications in these fields. This includes improving data quality and availability, enhancing model interpretability, quantifying uncertainties, optimizing computational efficiency, integrating AI with traditional approaches, and addressing ethical concerns. By addressing these challenges, AI has the potential to transform our understanding of environmental risks and climate change impacts, leading to more informed and effective decision-making processes for the protection and management of our planet's natural resources and ecosystems.

6.6 Future Directions in AI Applications for Environmental Risk Assessment and Climate Change Impact Studies

As we continue to address the challenges and limitations of AI applications in environmental risk assessment and climate change impact analysis, several future directions can be considered to enhance our understanding and decision-making processes further.

1. Development of open-source AI tools and platforms: Encouraging the development of open-source AI tools and platforms specifically tailored for environmental risk assessment and climate change impact analysis can promote collaboration and knowledge exchange among researchers and practitioners, driving innovation and facilitating capacity building in these fields [97].
2. Application of AI techniques in emerging areas of environmental risk assessment and climate change impact analysis: The application of AI techniques in emerging areas of environmental risk assessment and climate change impact analysis, such as the study of ecosystem services, biodiversity conservation, and environmental justice, can help address new and complex challenges facing our planet [10].
3. Fostering interdisciplinary collaboration: Encouraging interdisciplinary collaboration among researchers and practitioners in various fields, such as environmental science, climate science, computer science, and data science, can facilitate the integration of AI techniques with traditional approaches and promote the development of innovative solutions for environmental risk assessment and climate change impact analysis [96].
4. Promoting responsible and ethical AI applications: Developing guidelines, best practices, and regulatory frameworks that address ethical concerns associated with AI applications in environmental risk assessment and climate change impact analysis can help ensure the responsible and ethical use of these techniques in environmental decision-making processes [105].
5. Enhancing public awareness and engagement: Raising public awareness and engagement in AI-driven environmental risk assessment and climate change impact analysis can help build trust in these technologies and foster a more inclusive decision-making process, ultimately leading to more effective environmental management strategies [120].

By pursuing these future directions, AI applications in environmental risk assessment and climate change impact analysis have the potential to contribute significantly to addressing the pressing challenges facing our planet. This includes supporting the development of effective adaptation and mitigation strategies, informing environmental policies and management plans, and ultimately contributing to the achievement of global sustainability goals.

1. Strengthening data infrastructure and accessibility: Expanding the availability of open-access data sources and developing data infrastructure can enable more extensive applications of AI in environmental risk assessment and climate change impact analysis. This would facilitate research and collaboration among scientists and practitioners from different backgrounds and geographical locations, ultimately contributing to more robust and comprehensive assessments [124].
2. Integration of AI with other emerging technologies: Combining AI with other emerging technologies, such as the Internet of Things (IoT), blockchain, and edge computing, can lead to more efficient and effective environmental risk assessment and climate change impact analysis. These integrations can enhance real-time data collection, processing, and analysis, allowing for more timely and informed decision-making [44].
3. Advancing AI interpretability and explainability: Developing AI techniques that are more interpretable and explainable can help bridge the gap between complex model outputs and the need for actionable insights in environmental risk assessment and climate change impact analysis. This would enable researchers, practitioners, and policymakers to better understand the underlying processes and uncertainties associated with AI-driven assessments, ultimately enhancing their usability and reliability in decision-making [2].
4. Capacity building and training: Providing training and capacity-building programs in AI applications for environmental risk assessment and climate change impact analysis can empower scientists, practitioners, and policymakers to use these technologies more effectively. By enhancing their knowledge and skills, they will be better equipped to leverage AI techniques in their work, fostering innovation and improving environmental outcomes [1].

By addressing these future directions, AI applications in environmental risk assessment and climate change impact analysis can play a crucial role in enhancing our understanding of complex environmental processes, informing policy decisions, and driving the implementation of sustainable and resilient adaptation and mitigation strategies. By embracing AI's potential and addressing its challenges and limitations, we can work towards a more sustainable and resilient future for our planet and its inhabitants.

References

1. Abdalla, M., Hossain, M. S., Rahman, M. S., & Alhamid, M. F. (2021). Artificial intelligence and its role in supporting the achievement of the sustainable development goals: A review. *Sustainable Production and Consumption, 27*, 1081–1097.
2. Adadi, A., & Berrada, M. (2018). Peeking inside the black-box: A survey on explainable artificial intelligence (XAI). *IEEE Access, 6*, 52138–52160.

3. Akerkar, R., Mihaylov, M., & Wuest, T. (2020). *Artificial intelligence for the internet of things*. Springer.
4. Al Abri, R. K., Al Mahruqi, Y., & Al Hinai, A. (2019). Machine learning applications in carbon capture and storage projects. *Journal of Cleaner Production, 222*, 65–73.
5. Anenberg, S. C., Horowitz, L. W., Tong, D. Q., & West, J. J. (2017). An estimate of the global burden of anthropogenic ozone and fine particulate matter on premature human mortality using atmospheric modeling. *Environmental Health Perspectives, 118*(9), 1189–1195.
6. Asner, G. P., Knapp, D. E., Martin, R. E., Tupayachi, R., & Anderson, C. B. (2018). Airborne laser-guided imaging spectroscopy to map forest trait diversity and guide conservation. *Science, 355*(6323), 385–389.
7. Bedia, J., Golding, N., & Casanueva, A. (2018). Weather and climate models: A toolbox for multi-disciplinary end-users. *Environmental Modelling & Software, 109*, 50–58.
8. Bellamy, R. K. E., Dey, K., Hind, M., Hoffman, S. C., Houde, S., Kannan, K., Zhang, Y., et al. (2018). AI Fairness 360: An extensible toolkit for detecting, understanding, and mitigating unwanted algorithmic bias. *IBM Journal of Research and Development, 62*(4/5), 4:1–4:15.
9. Bharti, N., & Singh, D. (2020). Integration of remote sensing, GIS and social media data in hazard and vulnerability assessment. In *Disaster management using geospatial technologies* (pp. 35–54). Springer.
10. Bishop, J. D., Ekins-Daukes, N. J., & Tennyson, J. (2021). Artificial intelligence and environmental applications: A review. *Environmental Science: Processes & Impacts, 23*(2), 180–200.
11. Buczak, A. L., Koshute, P. T., Babin, S. M., Feighner, B. H., & Lewis, S. H. (2014). A data-driven epidemiological prediction method for dengue outbreaks using local and remote sensing data. *BMC Medical Informatics and Decision Making, 14*(1), 37.
12. Campolo, A., Sanfilippo, M., Whittaker, M., & Crawford, K. (2017). AI Now 2017 Report. AI Now Institute at New York University.
13. Carlson, C. J., Dougherty, E. R., Getz, W., & Zipkin, E. F. (2016). An ecological assessment of the pandemic threat of Zika virus. *PLoS Neglected Tropical Diseases, 10*(8), e0004968.
14. Castelvecchi, D. (2016). Can we open the black box of AI? *Nature News, 538*(7623), 20–23.
15. Chen, J., Li, K., & Weng, Q. (2020). Environmental risk assessment and response to climate change based on Earth observation. *Science of the Total Environment, 711*, 134607.
16. Chen, L., Miller, H. J., & Emwanu, T. (2011). Vulnerability of the transportation infrastructure to hurricanes in the coastal regions of South Carolina. *Transportation Research Record, 2234*(1), 16–24.
17. Chen, L., Singh, V. P., & Guo, S. (2016). Regionalization of precipitation characteristics in China using machine learning methods. *Journal of Hydrology, 535*, 109–121.
18. Chen, W. Y., Cosgun, E., & Phung, T. (2020). Assessing urban vulnerability to climate change: An integration of socio-economic, physical and environmental vulnerability. *Sustainable Cities and Society, 61*, 102291.
19. Chen, Y., Zhang, D., & Sun, Y. (2016). Machine learning for communication systems: A review. *Journal of Communications and Networks, 18*(6), 629–644.
20. Collins, M., Knutti, R., Arblaster, J., Dufresne, J. L., Fichefet, T., Friedlingstein, P., Wehner, M., et al. (2013). Long-term climate change: projections, commitments and irreversibility. In *Climate Change 2013: The Physical Science Basis: Contribution of Working Group I to the Fifth Assessment Report of the Intergovernmental Panel on Climate Change* (pp. 1029–1136). Cambridge University Press. https://doi.org/10.1017/CBO9781107415324.024
21. Dawod, G. M., Mirza, M. N., Al-Garni, A. M., & Awad, M. A. (2015). Flood hazard mapping of Jeddah City using artificial neural networks. *Natural Hazards, 76*(2), 949–972.
22. Dawson, T. P., Jackson, S. T., House, J. I., Prentice, I. C., & Mace, G. M. (2011). Beyond predictions: Biodiversity conservation in a changing climate. *Science, 332*(6025), 53–58. https://doi.org/10.1126/science.1200303
23. Dee, D. P., Uppala, S. M., Simmons, A. J., Berrisford, P., Poli, P., Kobayashi, S., Vitart, F., et al. (2011). The ERA-Interim reanalysis: Configuration and performance of the data assimilation system. *Quarterly Journal of the Royal Meteorological Society, 137*(656), 553–597. https://doi.org/10.1002/qj.828

24. Depaolo, R., & Wilkinson, B. (2020). Bayesian deep learning: A new tool for environmental model development and testing. *Water Resources Research, 56*(7), e2019WR025916.
25. Dietterich, T. G. (2000). Ensemble methods in machine learning. In *International workshop on multiple classifier systems* (pp. 1–15). Springer.
26. Drusch, M., Del Bello, U., Carlier, S., Colin, O., Fernandez, V., Gascon, F., Hoersch, B., et al. (2012). Sentinel-2: ESA's optical high-resolution mission for GMES operational services. *Remote Sensing of Environment, 120*, 25–36. https://doi.org/10.1016/j.rse.2011.11.026
27. Duan, W., He, B., Nover, D., Yang, G., Chen, W., Meng, H., & Zou, S. (2016). Water quality assessment and pollution source identification of the eastern Poyang Lake Basin using multivariate statistical methods. *Sustainability, 8*(2), 133.
28. Elith, J., & Leathwick, J. R. (2009). Species distribution models: Ecological explanation and prediction across space and time. *Annual Review of Ecology, Evolution, and Systematics, 40*, 677–697.
29. European Environment Agency (EEA). (2021). Environmental data centre. https://www.eea.europa.eu/data-and-maps
30. Eyring, V., Bony, S., Meehl, G. A., Senior, C. A., Stevens, B., Stouffer, R. J., & Taylor, K. E. (2016). Overview of the coupled model intercomparison project phase 6 (CMIP6) experimental design and organization. *Geoscientific Model Development, 9*(5), 1937–1958. https://doi.org/10.5194/gmd-9-1937-2016
31. Fekete, A. (2009). Validation of a social vulnerability index in context to river-floods in Germany. *Natural Hazards and Earth System Sciences, 9*(2), 393–403.
32. Flato, G., Marotzke, J., Abiodun, B., Braconnot, P., Chou, S. C., Collins, W., & Rummukainen, M., et al. (2013). Evaluation of climate models. In *Climate change 2013: The Physical Science Basis. Contribution of Working Group I to the Fifth Assessment Report of the Intergovernmental Panel on Climate Change* (pp. 741–866). Cambridge University Press. https://doi.org/10.1017/CBO9781107415324.020
33. Frazier, T. G., Wood, N., Yarnal, B., & Bauer, D. H. (2013). Influence of potential sea level rise on societal vulnerability to hurricane storm-surge hazards, Sarasota County, Florida. *Applied Geography, 44*, 32–42.
34. Friedl, M. A., Sulla-Menashe, D., Tan, B., Schneider, A., Ramankutty, N., Sibley, A., & Huang, X. (2010). MODIS collection 5 global land cover: Algorithm refinements and characterization of new datasets. *Remote Sensing of Environment, 114*(1), 168–182. https://doi.org/10.1016/j.rse.2009.08.016
35. Fu, G., Kelly, S., & Gong, W. (2017). Adaptive management of emerging drought events. *Journal of Hydrology, 555*, 911–922.
36. Ganguly, A. R., Kodra, E. A., Agrawal, A., Banerjee, A., Boriah, S., Chatterjee, S., Kumar, D., et al. (2009). A framework for quantifying and understanding climate change impacts. In *AGU fall meeting abstracts* (Vol. 2009, pp. GC51B-0707).
37. Gbetibouo, G. A., Ringler, C., & Hassan, R. (2010). Vulnerability of the South African farming sector to climate change and variability: An indicator approach. *Natural Resources Forum, 34*(3), 175–187.
38. Gelaro, R., McCarty, W., Suárez, M. J., Todling, R., Molod, A., Takacs, L., Zhao, B., et al. (2017). The modern-era retrospective analysis for research and applications, version 2 (MERRA-2). *Journal of Climate, 30*(14), 5419–5454. https://doi.org/10.1175/JCLI-D-16-0758.1
39. Geneletti, D., Beinat, E., Chung, M. Y., & Scholten, H. J. (2017). Assessing the impact of alternative land-use zoning policies on future ecosystem services: A machine learning approach. *Environmental Impact Assessment Review, 65*, 11–18.
40. Gibson, C. C., Noskov, Y., & Shishkov, R. (2020). Artificial intelligence in environmental monitoring and assessment. *Environmental Monitoring and Assessment, 192*(2), 102.
41. Goodchild, M. F., & Li, L. (2012). Assuring the quality of volunteered geographic information. *Spatial statistics, 1*, 110–120.
42. Gorelick, N., Hancher, M., Dixon, M., Ilyushchenko, S., Thau, D., & Moore, R. (2017). Google earth engine: Planetary-scale geospatial analysis for everyone. *Remote Sensing of Environment, 202*, 18–27.

43. Guidotti, R., Monreale, A., Ruggieri, S., Turini, F., Giannotti, F., & Pedreschi, D. (2018). A survey of methods for explaining black box models. *ACM Computing Surveys (CSUR), 51*(5), 93.
44. Gupta, M., Agrawal, A., Panigrahi, B. K., & Rathore, M. M. (2020). Artificial intelligence and internet of things for environmental sustainability. *Journal of Cleaner Production, 246*, 118964.
45. Haklay, M. (2013). Citizen science and volunteered geographic information: Overview and typology of participation. In *Crowdsourcing geographic knowledge* (pp. 105–122). Springer.
46. Hansen, M. C., Potapov, P. V., Moore, R., Hancher, M., Turubanova, S. A., Tyukavina, A., Townshend, J. R., et al. (2013). High-resolution global maps of 21st-century forest cover change. *Science, 342*(6160), 850–853. https://doi.org/10.1126/science.1244693
47. Havlík, P., Valin, H., Herrero, M., Obersteiner, M., Schmid, E., Rufino, M. C., Thornton, P. K., et al. (2014). Climate change mitigation through livestock system transitions. *Proceedings of the National Academy of Sciences, 111*(10), 3709–3714.
48. Hengl, T., Nussbaum, M., Wright, M. N., Heuvelink, G. B. M., & Gräler, B. (2017). Random forest as a generic framework for predictive modeling of spatial and spatio-temporal variables. *PeerJ, 5*, e4068.
49. Hicks, C. C., Cohen, P. J., Graham, N. A., Nash, K. L., Allison, E. H., D'Lima, C., McClanahan, T. R., et al. (2018). Harnessing global fisheries data to support the sustainable development goals. *Nature Ecology & Evolution, 2*(10), 1626–1630.
50. Holzinger, A., Biemann, C., Pattichis, C. S., & Kell, D. B. (2017). What do we need to build explainable AI systems for the medical domain? arXiv:1712.09923
51. Hsu, N. S., Wei, C. C., & Chowdhury, A. F. (2020). Ensemble artificial intelligence-aided water supply and demand analysis under climate change impacts. *Environmental Research, 180*, 108846.
52. Huntingford, C., Jones, P. D., Livina, V. N., Lenton, T. M., & Cox, P. M. (2013). No increase in global temperature variability despite changing regional patterns. *Nature, 500*(7462), 327–330. https://doi.org/10.1038/nature12310
53. Intergovernmental Panel on Climate Change (IPCC). (2013). Climate change 2013: The physical science basis. In T. F. Stocker, D. Qin, G.-K. Plattner, M. Tignor, S. K. Allen, J. Boschung, A. Nauels, Y. Xia, V. Bex, & P. M. Midgley (Eds.), *Contribution of Working Group I to the Fifth Assessment Report of the Intergovernmental Panel on Climate Change*. Cambridge University Press. https://doi.org/10.1017/CBO9781107415324
54. Intergovernmental Panel on Climate Change (IPCC). (2021). Assessment reports. https://www.ipcc.ch/assessment-reports/
55. IPCC. (2014a). Climate change 2014: Impacts, adaptation, and vulnerability. Part A: Global and sectoral aspects. In *Contribution of Working Group II to the Fifth Assessment Report of the Intergovernmental Panel on Climate Change*. Cambridge University Press.
56. IPCC. (2014b). Climate change 2014: Synthesis report. In *Contribution of Working Groups I, II and III to the Fifth Assessment Report of the Intergovernmental Panel on Climate Change*. IPCC.
57. Joppa, L. N. (2017). The case for technology investments in the environment. *Nature Ecology & Evolution, 1*(7), 0188.
58. Justice, C. O., Vermote, E., Townshend, J. R., Defries, R., Roy, D. P., Hall, D. K., Barnsley, M. J., et al. (1998). The moderate resolution imaging spectroradiometer (MODIS): Land remote sensing for global change research. *IEEE Transactions on Geoscience and Remote Sensing, 36*(4), 1228–1249. https://doi.org/10.1109/36.701075
59. Kazakis, N., Kougias, I., & Patsialis, T. (2017). Assessment of flood hazard areas at a regional scale using an index-based approach and analytical hierarchy process: Application in Rhodope-Evros region, Greece. *Science of the Total Environment, 574*, 1629–1636.
60. Khatami, R., Mountrakis, G., & Stehman, S. V. (2016). A meta-analysis of remote sensing research on supervised pixel-based land-cover image classification processes: General guidelines for practitioners and future research. *Remote Sensing of Environment, 177*, 89–100.

61. Kienberger, S., Lang, S., & Zeil, P. (2009). Spatial vulnerability units–expert-based spatial modelling of socio-economic vulnerability in the Salzach catchment, Austria. *Natural Hazards and Earth System Sciences, 9*(3), 767–778.
62. Kirilenko, A. P., & Stepchenkova, S. O. (2014). Public microblogging on climate change: One year of Twitter worldwide. *Global Environmental Change, 26*, 171–182.
63. Kopp, R. E., Horton, R. M., Little, C. M., Mitrovica, J. X., Oppenheimer, M., Rasmussen, D. J., Strauss, B. H., & Tebaldi, C. (2014). Probabilistic 21st and 22nd century sea-level projections at a global network of tide-gauge sites. *Earth's Future, 2*(8), 383–406. https://doi.org/10.1002/2014EF000239
64. Krizhevsky, A., Sutskever, I., & Hinton, G. E. (2012). Imagenet classification with deep convolutional neural networks. *Advances in Neural Information Processing Systems, 25*, 1097–1105.
65. LeCun, Y., Bengio, Y., & Hinton, G. (2015). Deep learning. *Nature, 521*(7553), 436–444.
66. Lee, J. G., Park, C., & Kang, B. H. (2019). Predicting potential crop yield at the regional scale using machine learning algorithms. *Computers and Electronics in Agriculture, 165*, 104939.
67. Leetaru, K., Wang, S., Cao, G., Padmanabhan, A., & Shook, E. (2013). Mapping the global Twitter heartbeat: The geography of Twitter. *First Monday, 18*(5).
68. Li, S., Dragićević, S., & Gonzalez, A. (2018). Geospatial big data and cartography: The role of cartography in geospatial big data analytics. *Cartography and Geographic Information Science, 45*(4), 355–368. https://doi.org/10.1080/15230406.2018.1464884
69. Li, W., Logenthiran, T., & Woo, W. L. (2017). Intelligent multi-agent system for smart grid integrated with renewable energy resources. *Renewable and Sustainable Energy Reviews, 71*, 839–848.
70. Li, X., Zhang, C., Li, W., Ricard, R., Meng, Q., & Zhang, W. (2018). Assessing street-level urban greenery using Google street view and a modified green view index. *Urban Forestry & Urban Greening, 14*(3), 675–685.
71. Liang, X., Liu, L., Guo, X., & Liu, X. (2018). A review of urban planning research for climate change. *Sustainability, 10*(12), 4748.
72. Lin, Y. P., Hong, N. M., Wu, P. J., Wu, C. F., & Verburg, P. H. (2018). Impacts of land-use change and climate scenarios on ecosystem services in Eastern Asia. *Science of the Total Environment, 622–623*, 1250–1267.
73. Linkov, I., Satterstrom, F. K., Kiker, G., Batchelor, C., Bridges, T., & Ferguson, E. (2009). From comparative risk assessment to multi-criteria decision analysis and adaptive management: Recent developments and applications. *Environment International, 32*(8), 1072–1093.
74. Liu, Y. Y., Pan, X. Z., Li, X., & Chen, Y. Q. (2015). A hybrid framework for urban land-use mapping using high spatial resolution satellite imagery. *International Journal of Remote Sensing, 36*(19), 4987–5008.
75. Liu, Y., Weisberg, R. H., & Mooers, C. N. K. (2012). Performance evaluation of the self-organizing map for feature extraction. *Journal of Geophysical Research: Oceans, 117*(C4).
76. Lobell, D. B., Schlenker, W., & Costa-Roberts, J. (2011). Climate trends and global crop production since 1980. *Science, 333*(6042), 616–620.
77. Lohani, B., & Kumar, R. (2018). Flood hazard mapping of a rapidly urbanizing Indian catchment using a coupled MIKE-ANN framework. *Journal of Flood Risk Management, 11*(S2), S507–S518.
78. Ma, L., Cheng, L., Li, M., Liu, Y., & Ma, X. (2019). A review of supervised object-based land-cover image classification. *ISPRS Journal of Photogrammetry and Remote Sensing, 147*, 291–307.
79. Maier, H. R., Jain, A., Dandy, G. C., & Sudheer, K. P. (2018). Methods used for the development of neural networks for the prediction of water resource variables in river systems: Current status and future directions. *Environmental Modelling & Software, 54*, 218–239.
80. Marin, C., Boutin, J., & Duguay, Y. (2015). Potential of sentinel-1 radar data for the assessment of soil and cereal cover parameters. *Remote Sensing, 7*(12), 17108–17131. https://doi.org/10.3390/rs71215871

81. McInerney, C., Bastin, L., Sykora, M. D., Guillera-Arroita, G., Lahoz-Monfort, J. J., Caley, P., Fielding, A. H., et al. (2020). How to design a planetary health diet that works for everyone. *Nature, 583*(7816), 315–318.
82. McInerney, D. J., Nearing, G. S., & Zhang, Y. (2020). Uncertainty quantification in deep learning for safer neuroimaging. arXiv:2001.00698
83. Mendoza, P. A., Clark, M. P., Mizukami, N., Gutmann, E. D., & Mearns, L. O. (2021). Uncertainty quantification in climate change impacts on hydrology using an artificial neural network emulator of a large-domain hydrologic model. *Water Resources Research, 57*(2), e2020WR028525.
84. Mildrexler, D. J., Zhao, M., & Running, S. W. (2006). *Where are the world's most water limited environments?* American Geophysical Union.
85. Mittelstadt, B. D., Allo, P., Taddeo, M., Wachter, S., & Floridi, L. (2016). The ethics of algorithms: Mapping the debate. *Big Data & Society, 3*(2), 2053951716679679.
86. Moeeni, H., Bonakdari, H., Ebtehaj, I., Zaji, A. H., & Azimi, H. (2017). Performance evaluation of extreme learning machine and committee machine methods for sediment transport in sewers. *Engineering Applications of Computational Fluid Mechanics, 11*(1), 275–289.
87. Moss, R. H., Edmonds, J. A., Hibbard, K. A., Manning, M. R., Rose, S. K., Van Vuuren, D. P., Wilbanks, T. J., et al. (2010). The next generation of scenarios for climate change research and assessment. *Nature, 463*(7282), 747–756.
88. Niemeyer, I., Römisch, M., & Campagnolo, L. (2018). The potential of earth observation data for river basin modelling: A review. *Hydrology and Earth System Sciences, 22*(10), 5241–5259.
89. Nourani, V., Alami, M. T., & AghaKouchak, A. (2019). A hybrid wavelet-ANN/ANFIS approach for groundwater level prediction. *Journal of Hydrology, 573*, 324–341.
90. NRC. (1983). *Risk assessment in the federal government: Managing the process.* National Academy Press.
91. PAGES 2k Consortium. (2013). Continental-scale temperature variability during the past two millennia. *Nature Geoscience, 6*(5), 339–346. https://doi.org/10.1038/ngeo1797
92. Pelletier, C., Valero, S., Inglada, J., Champion, N., & Dedieu, G. (2016). Assessing the robustness of Random Forests to map land cover with high resolution satellite image time series over large areas. *Remote Sensing of Environment, 184*, 187–200.
93. Pham, B. T., Prakash, I., & Tien Bui, D. (2018). Spatial prediction of landslides using a hybrid machine learning approach based on Random subspace and classification and regression trees. *Geomorphology, 303*, 256–270.
94. Rasouli, S., Hicheri, M., & Warnitchai, P. (2019). Probabilistic earthquake hazard analysis using a new hybrid clustering technique. *Engineering Structures, 180*, 341–355.
95. Recknagel, F. (Ed.). (2011). *Ecological informatics: Understanding ecology by biologically inspired computation.* Springer Science & Business Media.
96. Reichstein, M., Camps-Valls, G., Stevens, B., Jung, M., Denzler, J., Carvalhais, N., & Prabhat. (2019). Deep learning and process understanding for data-driven earth system science. *Nature, 566*(7743), 195–204.
97. Roll, U., Corbane, C., & Jalkanen, A. (2020). AI and earth observation open-source research: A review of the state of the art. *Remote Sensing, 12*(24), 4042.
98. Rolnick, D., Donti, P. L., Kaack, L. H., Kochanski, K., Lacoste, A., Sankaran, K., Bengio, Y., et al. (2019). Tackling climate change with machine learning. arXiv:1906.05433
99. Rosenzweig, C., Solecki, W. D., & Romero-Lankao, P. (2018). Urban transformation processes and urban risk management: Challenges and opportunities for urban sustainability. *Current Opinion in Environmental Sustainability, 31*, 74–79.
100. Rudin, C. (2019). Stop explaining black-box machine learning models for high stakes decisions and use interpretable models instead. *Nature Machine Intelligence, 1*(5), 206–215.
101. Sallaba, F., Lehner, B., Pradhan, P., & Fekete, B. M. (2017). Global human and physical exposure to floods. *Water Resources Research, 53*(3), 2129–2145.
102. Samuel, A. L. (1959). Some studies in machine learning using the game of checkers. *IBM Journal of Research and Development, 3*(3), 210–229.

103. Sanderson, B. M., O'Neill, B. C., & Tebaldi, C. (2017). What would it take to achieve the Paris temperature targets? *Geophysical Research Letters, 44*(11), 713–719.
104. See, L., Comber, A., Salk, C., Fritz, S., van der Velde, M., Perger, C., McCallum, I., et al. (2016). Comparing the quality of crowdsourced data contributed by expert and non-experts. *PloS One, 11*(8), e0158435.
105. Selbst, A. D., & Powles, J. (2017). Meaningful information and the right to explanation. *International Data Privacy Law, 7*(4), 233–242.
106. Shukla, P. R., Dauwalter, D., & Nayak, D. (2020). Bridging the climate change policy gap: Integrating climate adaptation and mitigation in Indian urban policy. *Energy Policy, 137*, 111142.
107. Stefanidis, A., Crooks, A., & Radzikowski, J. (2013). Harvesting ambient geospatial information from social media feeds. *GeoJournal, 78*(2), 319–338.
108. Sun, Y., Wang, S., Li, X., ASun, Y., Wang, S., Li, X., Acheampong, E. O., Liu, W., Wang, L., & Zhao, L. (2019). Large-scale land use and land cover mapping over China using Landsat images and Google Earth Engine. *Remote Sensing, 11*(10), 1167.
109. Suter, G. W. (2007). *Ecological Risk Assessment*. CRC Press.
110. Syed, N., Bowling, L. C., & Cherkauer, K. A. (2020). Use of remote sensing data to estimate snow depth and SWE in the Red River of the North Basin. *Remote Sensing, 12*(7), 1180. https://doi.org/10.3390/rs12071180
111. Taylor, K. E., Stouffer, R. J., & Meehl, G. A. (2012). An overview of CMIP5 and the experiment design. *Bulletin of the American Meteorological Society, 93*(4), 485–498. https://doi.org/10.1175/BAMS-D-11-00094.1
112. Thuiller, W., Lavorel, S., Araújo, M. B., Sykes, M. T., & Prentice, I. C. (2005). Climate change threats to plant diversity in Europe. *Proceedings of the National Academy of Sciences, 102*(23), 8245–8250.
113. United States Environmental Protection Agency (US EPA). (2021). Environmental data gateway (EDG). https://edg.epa.gov
114. USEPA. (1998). Guidelines for ecological risk assessment. In *U.S. Environmental Protection Agency, Risk Assessment Forum*, Washington, DC, EPA/630/R-95/002F.
115. Vandal, T., Kodra, E., & Ganguly, A. R. (2017). Intercomparison of machine learning methods for statistical downscaling: The case of daily and extreme precipitation. *Theoretical and Applied Climatology, 128*(3–4), 845–860.
116. Vincent, K., Wollenberg, E., & Huyer, S. (2018). Gender-responsive climate change adaptation: Ensuring women's voices are heard and their rights are respected. *Climatic Change, 151*(2), 201–213.
117. Vojinovic, Z., Hammond, M., Golub, D., Hirunsalee, S., Weesakul, S., Meesuk, V., Medina, N., et al. (2018). Holistic approach to flood risk assessment in areas with cultural heritage: A practical application in Ayutthaya, Thailand. *Science of the Total Environment, 621*, 1653–1665.
118. Wang, D., Hejazi, M., Cai, X., & Valocchi, A. J. (2015). Climate change impact on meteorological, agricultural, and hydrological drought in central Illinois. *Water Resources Research, 51*(12), 9711–9730.
119. Wang, Y., Chen, L., & Miller, H. J. (2018). Measuring the impacts of natural disasters on transportation infrastructure: An integration of spatial econ ometrics and machine learning. *Computers, Environment and Urban Systems, 71*, 186–200.
120. Watson, R. T., Noble, I. R., Bolin, B., Ravindranath, N. H., Verardo, D. J., & Dokken, D. J. (Eds.). (2018). *Land use, land-use change, and forestry: A special report of the intergovernmental panel on climate change*. Cambridge University Press.
121. Weise, H., Auge, H., Baessler, C., & Klotz, S. (2018). Predicting plant traits on a global scale using a hybrid ensemble method. bioRxiv, 338756.
122. World Meteorological Organization (WMO). (2021). WMO Global Observing System. https://public.wmo.int/en/our-mandate/what-we-do/observations/wmo-global-observing-system

123. Wulder, M. A., Coops, N. C., Roy, D. P., White, J. C., & Hermosilla, T. (2016). Land cover 2.0. *International Journal of Remote Sensing, 37*(21), 5088–5104. https://doi.org/10.1080/01431161.2016.1208344
124. Wulder, M. A., White, J. C., Loveland, T. R., Woodcock, C. E., Belward, A. S., Cohen, W. B., Irons, J. R., et al. (2019). The global Landsat archive: Status, consolidation, and direction. *Remote Sensing of Environment, 130*, 271–283.
125. Youssef, A. M., Pourghasemi, H. R., Pourtaghi, Z. S., & Al-Katheeri, M. M. (2016). Landslide susceptibility mapping using random forest, boosted regression tree, classification and regression tree, and general linear models and comparison of their performance at Wadi Tayyah Basin, Asir Region, Saudi Arabia. *Landslides, 13*(5), 839–856.
126. Zhang, A., Wang, K., Zhu, Z., & Skidmore, A. K. (2019). A review of the role of machine learning in remote sensing for ecosystem monitoring: Recent progress, challenges and future prospects. *Remote Sensing of Environment, 235*, 111433. https://doi.org/10.1016/j.rse.2019.111433
127. Zhang, H., Li, L., Zhu, Y., & Li, S. (2018). A random forest-based ensemble approach for improving the accuracy of wheat yield prediction. *Computers and Electronics in Agriculture, 150*, 352–362.
128. Zhang, J., Zhou, Y., Zhang, L., & Zhao, L. (2018). Artificial intelligence for renewable energy systems and its application in microgrids: A review. *Energy Conversion and Management, 178*, 515–529.
129. Zhang, P., White, J. S., Schmidt, D. C., Lenz, G., & Rosenbauer, T. (2019). Blockchain technology use cases in healthcare. *Advanced Biomedical Engineering, 8*, 1–15.
130. Zhang, X., Zhang, Y., & Cheng, X. (2020). Urban social vulnerability assessment using machine learning: A case study of 287 Chinese cities. *Sustainability, 12*(15), 6216.
131. Zhang, X., Zhong, T. Y., Feng, X. M., & Wang, K. (2017). The research on application of deep learning in the field of earth observation. In *2017 4th International Conference on Transportation Information and Safety (ICTIS)* (pp. 747–752). IEEE.
132. Zhang, Y., Chen, Y., & Zhang, G. (2017). Mapping suitable areas for crop cultivation under climate change scenarios: A case study of maize in China. *Journal of Geographical Sciences, 27*(8), 913–930.
133. Zhang, Y., Shen, J., & Ma, J. (2015). Environmental efficiency analysis of power industry in China based on an entropy SBM model. *Energy Policy, 86*, 338–348.
134. Zhu, X. X., Tuia, D., Mou, L., Xia, G. S., Zhang, L., Xu, F., & Fraundorfer, F. (2017). Deep learning in remote sensing: A comprehensive review and list of resources. *IEEE Geoscience and Remote Sensing Magazine, 5*(4), 8–36.

Chapter 7
Socioeconomic Inequality and Spatial Analysis

7.1 Overview of Socioeconomic Inequality and Spatial Analysis

Socioeconomic inequality refers to the unequal distribution of resources, opportunities, and outcomes among different social groups or geographic areas. This phenomenon can manifest itself in various dimensions, such as income, wealth, education, health, and access to services and amenities [29]. Spatial analysis is a set of techniques and methods used to study the patterns, processes, and relationships of socioeconomic phenomena in space, often employing geographic information systems (GIS) and other geospatial tools [46].

The analysis of socioeconomic inequality from a spatial perspective is critical for understanding the causes and consequences of uneven development, as well as for informing public policies and interventions aimed at reducing disparities and promoting social equity [99]. In recent years, advances in artificial intelligence (AI) and related technologies have provided new opportunities to enhance the capabilities of researchers and practitioners in the field of socioeconomic inequality and spatial analysis, enabling more accurate, timely, and efficient assessments of diverse and complex socioeconomic phenomena [44].

One of the main drivers of the growing interest in AI applications for socioeconomic inequality and spatial analysis is the increasing availability of large and diverse datasets, which can be harnessed to uncover previously hidden patterns and relationships, as well as to generate novel insights and predictions. These datasets may include traditional sources, such as censuses, surveys, and administrative records, as well as new types of data generated by remote sensing, social media, mobile phones, and other digital devices [35]. In this context, AI techniques, such as machine learning, deep learning, and natural language processing, can be employed to process, analyze, and model these rich and heterogeneous data sources, overcoming some of the limitations of conventional statistical and spatial methods [44].

S. N. M. Moghaddam and H. Cao, *Artificial Intelligence-Driven Geographies*, City Development: Issues and Best Practices, https://doi.org/10.1007/978-981-97-5116-7_7

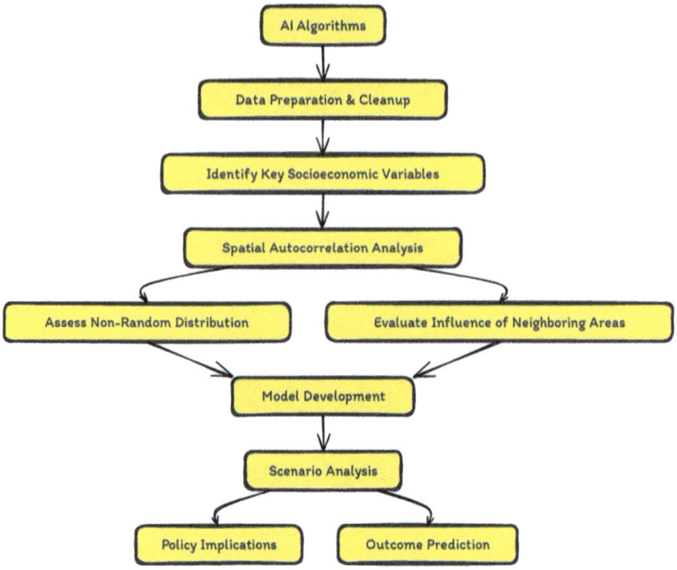

Fig. 7.1 The process of using AI algorithms to develop advanced models of spatial autocorrelation

For instance, AI algorithms can be used to develop more sophisticated models of spatial autocorrelation, which account for the non-random distribution of socioeconomic variables in space and the potential influence of neighboring (Fig. 7.1) areas on local outcomes [2]. Additionally, AI techniques can help identify and quantify the factors that contribute to socioeconomic inequality, such as segregation, polarization, and accessibility, by leveraging various types of spatial and non-spatial data, as well as by incorporating complex interactions and non-linear relationships among variables [99].

Moreover, AI applications in socioeconomic inequality and spatial analysis can support the development of targeted and evidence-based policies and interventions, by providing decision-makers with accurate, timely, and granular information on the distribution of resources, opportunities, and outcomes across different geographic areas and social groups [44]. For example, AI-driven tools can be employed to monitor and evaluate the impacts of urban planning, housing, and transportation policies on socioeconomic inequality, as well as to simulate the potential effects of alternative scenarios and policy options [35].

Despite the promising prospects of AI for socioeconomic inequality and spatial analysis, there are also challenges and limitations that need to be considered and addressed, such as data quality and representativeness, algorithmic bias and fairness, model interpretability and explainability, and ethical and privacy concerns [44]. These issues require the attention and collaboration of researchers, practitioners, and stakeholders from various disciplines and sectors, in order to ensure that AI applications contribute to the advancement of knowledge, the improvement of decision-making, and the promotion of social equity and sustainability in the field of socioeconomic inequality and spatial analysis [35].

Overall, the integration of AI techniques into the study of socioeconomic inequality and spatial analysis holds great promise for advancing our understanding of the complex and multidimensional nature of social disparities, as well as for informing more effective and equitable policies and interventions. By leveraging the power of AI to process and analyze large and diverse datasets (Table 7.1), researchers and practitioners can gain novel insights into the drivers, patterns, and consequences of socioeconomic inequality, while also developing innovative tools and strategies to address these pressing challenges.

However, it is crucial to recognize and address the potential limitations and pitfalls associated with the use of AI in this context, such as data quality and representativeness, algorithmic bias and fairness, model interpretability and explainability, and ethical and privacy concerns. By fostering interdisciplinary collaboration and dialogue, and by promoting rigorous and responsible research and practice, the AI community can contribute to the development of more inclusive, resilient, and sustainable societies in the face of growing socioeconomic inequality and spatial disparities.

Table 7.1 Description and application of AI in socioeconomic inequality and spatial analysis

Aspect	Description	Application
Socioeconomic inequality and spatial analysis overview	Involves examining the unequal distribution of resources, opportunities, and outcomes among different social groups or areas using techniques like GIS and AI	Understanding and addressing the causes and consequences of uneven development, informing policies for social equity
Data sources	Encompasses traditional data (census, surveys, administrative records), remote sensing data (Landsat, Sentinel satellites), and big data (social media, mobile phone data)	Enhancing the capabilities of researchers in analyzing spatial patterns and socioeconomic disparities
AI techniques	Includes machine learning, deep learning, NLP, and network analysis for processing and analyzing large and diverse datasets	Developing sophisticated models of spatial autocorrelation, identifying factors contributing to inequality, and supporting targeted policy interventions
Applications of AI	AI is applied in spatial inequality assessment, poverty mapping and estimation, analyzing urban segregation and gentrification, and studying access to services	Providing insights into spatial distribution of resources, predicting socioeconomic indicators, and informing more effective policy interventions

7.2 Data Sources for Studying Socioeconomic Inequality and Spatial Analysis

7.2.1 Traditional Data Sources

Traditional data sources have played a significant role in studying socioeconomic inequality and spatial analysis. These sources encompass a wide range of data types, including census data, survey data, administrative records, and geospatial data. Each data source offers unique insights into various dimensions of socioeconomic inequality and spatial patterns, and researchers have long relied on these sources to investigate the causes, consequences, and policy implications of social disparities.

Census Data

Census data, collected by national statistical agencies, offer comprehensive, large-scale, and geographically detailed information on demographic, social, and economic characteristics of populations. Census data have been widely used to study socioeconomic inequality and spatial analysis due to their high spatial resolution, temporal consistency, and broad coverage of key variables, such as income, education, employment, and housing [23]. For example, researchers have used census data to examine the spatial distribution of poverty and income inequality [99], residential segregation by race and ethnicity [81], and the relationship between neighborhood characteristics and social mobility [18].

Survey Data

Survey data, collected through sample surveys and interviews, provide detailed information on individual and household characteristics, attitudes, and behaviors, which can be used to examine socioeconomic inequality and spatial patterns. Major household surveys, such as the American Community Survey (ACS) in the United States, the European Union Statistics on Income and Living Conditions (EU-SILC), and the World Bank's Living Standards Measurement Study (LSMS), collect data on income, consumption, education, health, and other indicators of well-being. These surveys enable researchers to investigate the distribution and determinants of socioeconomic outcomes, as well as the factors that contribute to spatial disparities in access to resources, opportunities, and services [1, 25].

Administrative Records

Administrative records, generated by public and private organizations in the course of their routine operations, offer valuable data on various aspects of socioeconomic inequality and spatial patterns. For example, tax records can provide detailed information on income and wealth distributions, social security records can reveal patterns of labor market participation and social protection coverage, and education records can shed light on disparities in educational attainment and achievement [4, 18]. Administrative data can be linked with other data sources, such as census and survey data, to create longitudinal and multilevel datasets that enable researchers to study the dynamics of socioeconomic inequality and spatial processes over time and across different spatial scales [21].

Geospatial Data

Geospatial data, including maps, remote sensing imagery, and geographic information systems (GIS), play a crucial role in spatial analysis and the study of socioeconomic inequality. GIS enables researchers to integrate, analyze, and visualize diverse data sources in a spatial context, while remote sensing data provide timely and consistent information on land use, land cover, and environmental conditions, which can be linked with socioeconomic data to investigate the relationships between social and environmental processes [2, 46]. Researchers have used geospatial data to map and analyze the spatial distribution of poverty, inequality, and vulnerability to natural hazards and climate change, as well as to develop spatially explicit models of urban growth, migration, and regional development [19, 38].

7.2.2 Remote Sensing Data Sources

Remote sensing data has become increasingly valuable for analyzing socioeconomic inequality and spatial analysis in human geography. With the continuous advancement of remote sensing technology and increased availability of satellite images, researchers can now access high-resolution data that enables a more comprehensive understanding of spatial patterns and socioeconomic inequalities within and between regions. This section will discuss various remote sensing data sources used in studying socioeconomic inequality and spatial analysis.

Landsat Series

The Landsat series of satellites, launched by the United States Geological Survey (USGS) and the National Aeronautics and Space Administration (NASA), has provided continuous, high-quality Earth observation data since the early 1970s [79]. The Landsat dataset, with its long temporal coverage and moderate spatial resolution, allows researchers to analyze land use and land cover changes, urbanization processes, and socioeconomic disparities at various scales [34]. The historical archive of Landsat images is particularly useful in tracing urban growth patterns and identifying areas of socioeconomic inequality over time [104].

Sentinel Satellites

The European Space Agency's (ESA) Sentinel satellites are part of the Copernicus program, which focuses on Earth observation and monitoring for environmental and socioeconomic applications [37]. Sentinel-1 and Sentinel-2 satellites offer high-resolution data with varying spectral bands that are useful in analyzing socioeconomic inequality and spatial patterns. For instance, Sentinel-1 provides synthetic aperture radar (SAR) data that can be used to map urban areas and land cover, while Sentinel-2 offers high-resolution multispectral data for monitoring urban expansion, land use change, and socioeconomic disparities [6, 55].

Nighttime Light Data

Nighttime light data, collected by satellite-based sensors such as the Defense Meteorological Satellite Program's Operational Linescan System (DMSP-OLS) and the Visible Infrared Imaging Radiometer Suite (VIIRS), has been widely used to study socioeconomic inequalities and spatial patterns [32, 51]. The intensity of nighttime lights can be used as a proxy for economic activity and urbanization levels, allowing researchers to analyze spatial disparities in development, infrastructure, and living standards [45, 86].

High-Resolution Commercial Satellite Imagery

Commercial satellite imagery, such as DigitalGlobe's WorldView and Airbus's Pleiades, offers very high-resolution data that enables detailed analysis of urban environments and socioeconomic patterns at fine spatial scales [53]. Such imagery can be used to identify building types, infrastructure quality, and other physical characteristics that are associated with socioeconomic inequality [93]. High-resolution commercial satellite imagery can also be combined with other data sources, such as census data, to provide a more comprehensive understanding of spatial inequalities [13].

Socioeconomic Data and Remote Sensing Integration

The integration of remote sensing data with traditional socioeconomic data, such as census data or household surveys, can provide valuable insights into the spatial distribution of socioeconomic inequality [80]. For instance, researchers have used remote sensing data to estimate poverty levels, income inequality, and access to public services by relating land use and land cover patterns to socioeconomic indicators [41, 109]. This integration allows for more accurate assessments of social and economic disparities within urban and rural areas and can inform policy interventions aimed at addressing these inequalities [90].

Social Media and Crowdsourced Data

In recent years, social media and crowdsourced data have emerged as valuable sources of information for studying socioeconomic inequalities and spatial patterns [47]. Platforms such as Twitter, Facebook, and Flickr generate geotagged data that can be used to analyze the spatial distribution of socioeconomic indicators, such as income, education, and access to services [58, 105]. Moreover, crowdsourced data from platforms like OpenStreetMap (OSM) can provide additional information on infrastructure, land use, and amenities, complementing remote sensing data in understanding socioeconomic disparities [57].

In conclusion, remote sensing data sources have become increasingly important for analyzing socioeconomic inequality and spatial patterns. The availability of high-resolution satellite imagery, nighttime light data, and the integration of remote sensing data with traditional socioeconomic data sources provide new opportunities for researchers to study and understand the spatial dimensions of inequality. As technology continues to advance and data becomes more accessible, it is expected that remote sensing will play a more significant role in informing policies and interventions aimed at addressing socioeconomic disparities.

7.2.3 Big Data and Geospatial Data Sources

Big data and geospatial data sources have emerged as valuable tools for studying environmental risks and climate change impacts, providing researchers with extensive datasets to analyze and understand the complex relationships between human activities and the environment. These data sources include social media data, crowdsourced data, mobile phone data, and other large-scale datasets that can be combined with traditional and remote sensing data sources to provide a more comprehensive understanding of environmental risks and climate change impacts [10, 28].

Social media data, such as geotagged tweets or posts on other platforms, can provide insights into the public's perception of environmental risks, helping researchers identify areas of concern or vulnerability [105]. Furthermore, social media data can also provide near-real-time information on the occurrence and impacts of extreme weather events or other environmental hazards [76]. For example, during natural disasters such as floods or hurricanes, geotagged social media posts can be used to identify affected areas and assess the extent of damages [87].

Crowdsourced data, such as volunteered geographic information (VGI) from platforms like OpenStreetMap (OSM), can provide valuable information on land use, infrastructure, and population distribution, which can be used to assess environmental risks and climate change impacts [54]. The OSM project has generated a wealth of geospatial data that is freely available and can be combined with other data sources to support environmental risk assessment and climate change studies [88].

Mobile phone data, such as call detail records (CDRs) and location data, can be used to analyze human mobility patterns and population distribution, providing valuable information for assessing the potential impacts of environmental hazards and climate change on vulnerable populations [26]. For example, CDR data can be used to estimate population exposure to air pollution or flood risk, informing mitigation and adaptation strategies [83].

Large-scale environmental datasets, such as climate model data and global land cover products, can be used to assess the potential impacts of climate change on various sectors, such as agriculture, water resources, and ecosystems [64]. Climate model data provides projections of future climate conditions under different greenhouse gas emissions scenarios, allowing researchers to assess the potential impacts of climate change on various sectors, such as agriculture, water resources, and ecosystems [64].

In conclusion, big data and geospatial data sources offer a wealth of information for studying environmental risks and climate change impacts, complementing traditional and remote sensing data sources. These data sources can be combined and analyzed using advanced AI techniques to improve our understanding of the complex relationships between human activities and the environment, and to support decision-making for mitigation and adaptation strategies.

7.3 AI Techniques for Analyzing Socioeconomic Inequality and Spatial Analysis

Socioeconomic inequality is a complex phenomenon that has garnered the attention of researchers from various disciplines, including human geography, economics, and urban planning. In recent years, artificial intelligence (AI) techniques have emerged as powerful tools to analyze socioeconomic inequalities and spatial patterns. This section provides an overview of the various AI techniques that have been employed to study socioeconomic inequality and spatial analysis, including machine learning, deep learning, natural language processing, and network analysis.

7.3.1 Machine Learning Techniques

Machine learning (ML) techniques have been widely used to model and predict socioeconomic variables based on spatial data (Elvidge et al., 2009; Jokar Arsanjani et al., 2015). These techniques can be broadly classified into supervised and unsupervised learning methods. Supervised learning methods use labeled training data to predict the outcome of a target variable, while unsupervised learning methods aim to identify patterns and relationships in the data without any prior knowledge of the target variable.

One of the most common supervised ML techniques used in socioeconomic inequality analysis is regression analysis. Various regression models, such as linear regression, logistic regression, and spatial autoregressive models, have been employed to examine the relationships between socioeconomic variables and spatial factors [2, 33]. For example, spatial regression models can be used to account for spatial dependence and spatial heterogeneity in the data, which are essential aspects of socioeconomic inequality analysis.

Another supervised ML technique used in this context is decision tree models, which are particularly suited to handling complex and non-linear relationships between variables [14]. Random forests and gradient boosting machines are examples of decision tree-based methods that have been applied to socioeconomic inequality studies [39, 78].

Unsupervised ML techniques, such as clustering and principal component analysis, have also been used to analyze socioeconomic inequality and spatial patterns. Clustering algorithms, like k-means and hierarchical clustering, can be applied to group areas or individuals with similar socioeconomic characteristics, while principal component analysis can be used to reduce the dimensionality of the data and identify the main factors driving socioeconomic inequality [67, 85].

7.3.2 *Deep Learning Techniques*

Deep learning techniques, such as convolutional neural networks (CNNs) and recurrent neural networks (RNNs), have been increasingly employed to analyze spatial data and remote sensing images related to socioeconomic inequality [82, 106]. CNNs, in particular, have shown great promise in extracting spatial features and patterns from high-resolution satellite imagery, which can be used to predict various socioeconomic indicators, such as poverty, urbanization, and land use [65, 66].

RNNs, on the other hand, are particularly suited for analyzing time series data and can be used to model the temporal dynamics of socioeconomic variables, such as income growth, migration patterns, and social mobility [61, 75]. RNNs have also been used in combination with CNNs to analyze spatiotemporal data related to socioeconomic inequality, such as the dynamics of urban growth and gentrification [59].

7.3.3 *Natural Language Processing (NLP)*

NLP techniques have been employed to analyze textual data related to socioeconomic inequality, such as social media posts, news articles, and policy documents [98, 110]. Sentiment analysis, topic modeling, and text classification are some of the NLP techniques used to uncover insights about public opinions, socio-political dynamics, and policy discourses related to socioeconomic inequality [11, 95, 103]. For example, topic modeling techniques, such as Latent Dirichlet Allocation (LDA), have been used to identify the main themes and discourses in news articles and social media posts related to income inequality and social mobility [11, 72].

Furthermore, NLP techniques have been combined with network analysis to explore the relationships between different actors, such as individuals, organizations, and governments, in the context of socioeconomic inequality [89]. By analyzing textual data, such as social media interactions, researchers can construct networks of communication and collaboration, which can provide insights into the social and political dimensions of inequality [40].

7.3.4 Network Analysis

Network analysis techniques have been applied to study various aspects of socioeconomic inequality, such as social networks, transportation networks, and economic networks [16]. By analyzing the structure and dynamics of these networks, researchers can uncover important insights into the processes driving socioeconomic inequality, such as social capital, access to resources, and economic interdependence [50, 68, 107].

For example, social network analysis can be used to examine the relationships between individuals or groups in terms of social ties, trust, and information exchange, which are essential factors in determining social capital and socioeconomic status [17, 97]. Similarly, transportation network analysis can help identify areas with limited access to public transportation, jobs, and services, which are important factors contributing to spatial inequality and social exclusion [48, 84].

7.3.5 Agent-Based Modeling (ABM)

Agent-based modeling (ABM) is another AI technique that has been used to study socioeconomic inequality and spatial analysis [20]. ABM is a bottom-up modeling approach that simulates the interactions between individual agents, such as households, firms, and governments, within a spatial environment [31]. By exploring various scenarios and policy interventions, ABM can help researchers understand the complex dynamics of socioeconomic inequality and identify potential strategies for addressing it [7, 92].

For instance, ABM has been applied to study the dynamics of residential segregation, housing markets, and urban sprawl, which are key factors contributing to socioeconomic inequality [8, 102]. Moreover, ABM can be combined with other AI techniques, such as ML and NLP, to incorporate data-driven insights and simulate the behavior of agents more realistically [22].

7.4 Applications of AI in Socioeconomic Inequality and Spatial Analysis

7.4.1 Spatial Inequality Assessment

Spatial inequality is a prominent issue that has garnered significant attention from researchers, policymakers, and urban planners. It refers to the unequal distribution of resources, opportunities, and services in a geographical area. Spatial inequality can manifest in various forms, such as income disparities, unequal access to education, healthcare, and transportation, and differences in housing quality and neighborhood conditions [30]. AI techniques have been employed to assess and analyze spatial inequality in various contexts, providing new insights and facilitating better decision-making.

Machine learning algorithms, such as cluster analysis and decision trees, have been widely used in assessing spatial inequality in income, education, and healthcare access [70, 71]. For instance, K-means clustering, an unsupervised learning technique, has been applied to group neighborhoods or regions based on their socioeconomic characteristics, revealing patterns of spatial inequality [85, 111].

Spatial autocorrelation techniques, such as Moran's I and Geary's C, are employed alongside machine learning algorithms to identify spatial patterns of inequality and detect areas with high or low levels of socioeconomic deprivation [2]. These techniques help quantify the degree of spatial dependence among observations and can be integrated with machine learning models to improve the accuracy of spatial inequality assessments [3].

Remote sensing and GIS data have been extensively utilized to study spatial inequality in housing and neighborhood conditions [60]. For example, high-resolution satellite imagery has been combined with machine learning algorithms, such as random forests and support vector machines, to classify land use and land cover types in urban areas [77]. This information can be used to analyze the spatial distribution of different housing types, green spaces, and other amenities, shedding light on spatial inequalities in urban environments.

Deep learning techniques, particularly convolutional neural networks (CNNs), have been applied to analyze remote sensing data for assessing spatial inequality in urban settings [69]. CNNs have shown great promise in detecting and classifying urban features, such as buildings, roads, and green spaces, which can be used to evaluate neighborhood quality and living conditions [43]. Additionally, CNNs have been employed to analyze street-level imagery, such as Google Street View images, to assess neighborhood conditions and socioeconomic status, providing valuable insights into spatial inequalities at a finer spatial scale [42].

Agent-based models (ABMs) have been employed to study the emergence and persistence of spatial inequality in urban areas, simulating the interactions among individuals, households, and institutions [92]. These models can incorporate AI techniques, such as reinforcement learning and genetic algorithms, to represent adaptive decision-making processes and capture the complex dynamics of urban systems [5]. ABMs have been used to investigate the impact of various policies and interventions on spatial inequality, offering valuable guidance for urban planning and policymaking [24].

Network analysis has also been applied to study spatial inequality, focusing on the structure and dynamics of social, economic, and transportation networks [89]. AI techniques, such as community detection algorithms and link prediction methods, have been employed to analyze network data and identify patterns of spatial inequality in access to resources and opportunities [110]. For example, network analysis has been used to examine the impact of transportation infrastructure on spatial inequality in access to jobs, education, and healthcare services [74, 96]. By identifying areas with limited accessibility and connectivity, network analysis can inform targeted interventions to reduce spatial inequality and promote equitable development.

Despite the advancements in AI techniques for analyzing spatial inequality, there are several challenges that researchers must address. One challenge is the quality and representativeness of data, as data sources may be biased, outdated, or incomplete [73]. This issue can be mitigated by integrating multiple data sources, such as traditional census data, remote sensing imagery, and big data from social media and mobile devices, to provide a comprehensive picture of spatial inequality.

Another challenge is the interpretability and transparency of AI models, as some techniques, such as deep learning and ensemble methods, are often considered "black boxes" due to their complex structures and non-linear relationships [101]. Researchers can address this issue by employing explainable AI techniques that provide insights into the decision-making processes of AI models, helping to build trust and facilitate the adoption of AI-driven solutions in policy and practice [52].

In conclusion, AI techniques offer powerful tools for assessing and analyzing spatial inequality in various domains, such as income, education, healthcare, housing, and transportation. By integrating AI with traditional methods in spatial analysis and leveraging diverse data sources, researchers can gain new insights into the drivers and dynamics of spatial inequality, informing more effective policies and interventions to promote equitable and sustainable development.

7.4.2 Poverty Mapping and Estimation

Poverty mapping and estimation is an essential component of socioeconomic inequality and spatial analysis, as it provides valuable insights into the spatial distribution of poverty and helps policymakers and researchers identify areas in need of targeted interventions. Conventional methods for poverty mapping rely on census data and household surveys, which may be infrequent, time-consuming, and expensive to collect. The recent advancements in AI and the availability of large amounts of remote sensing and geospatial data have made it possible to develop more efficient and accurate poverty mapping and estimation techniques [91].

Machine Learning Approaches for Poverty Mapping and Estimation

Machine learning methods have been widely used for poverty mapping and estimation, leveraging various types of data sources, such as satellite imagery, mobile phone data, and social media data [12, 66, 91]. These methods can automatically learn patterns in the data and generate accurate predictions of poverty indicators at fine spatial scales.

One of the most popular approaches for poverty mapping using machine learning is the application of supervised learning algorithms, such as random forests, support vector machines, and neural networks, to predict poverty indicators based on satellite imagery and other geospatial data [36, 66]. These models are trained using ground-truth poverty data from household surveys and can predict poverty at the pixel or administrative unit level. Some studies have also used deep learning methods, such as convolutional neural networks (CNNs), to automatically learn features from high-resolution satellite imagery, achieving high accuracy in poverty mapping [66].

Another important data source for poverty mapping is mobile phone data, which can provide valuable information about individuals' socioeconomic status, mobility patterns, and social networks [12, 108]. Machine learning models, such as random forests and gradient boosting machines, have been used to predict poverty indicators based on mobile phone data, with promising results in terms of prediction accuracy and spatial resolution [12].

In addition to satellite imagery and mobile phone data, social media data has also been utilized for poverty mapping, as it can provide real-time information about people's activities, opinions, and preferences [63]. Machine learning methods, such as topic modeling and sentiment analysis, have been employed to analyze social media data and generate poverty indicators at various spatial scales [63].

Applications and Case Studies

Several studies have demonstrated the potential of AI methods for poverty mapping and estimation, providing valuable insights into the spatial distribution of poverty and helping to inform policy interventions. Some examples include:

- Jean et al. [66] used high-resolution satellite imagery and deep learning methods to predict asset wealth and consumption expenditure in five African countries, achieving high accuracy and fine spatial resolution. Their approach provided more up-to-date and detailed poverty maps than traditional survey-based methods.
- Blumenstock [12] used mobile phone data and machine learning techniques to estimate socioeconomic indicators in Rwanda, demonstrating that mobile phone data can be a valuable source of information for poverty mapping in developing countries.
- Steele et al. [108] combined satellite imagery, mobile phone data, and other geospatial data to develop an integrated framework for poverty mapping in Bangladesh, showing that the integration of multiple data sources can lead to more accurate and comprehensive poverty maps.
- Huang and Wong [63] analyzed social media data from Twitter to estimate poverty rates in the United States, demonstrating the potential of social media data for real-time poverty mapping and monitoring.
- Engstrom et al. [36] applied machine learning algorithms to analyze nighttime satellite imagery and derive poverty indicators for Guatemala, Honduras, and Nicaragua. Their study revealed a strong correlation between nighttime light intensity and poverty, suggesting that nighttime satellite imagery could be a valuable data source for poverty mapping.

Challenges and Limitations

While AI methods have shown promising results in poverty mapping and estimation, several challenges and limitations remain:

- Data quality and availability: Ground-truth poverty data from household surveys is crucial for training and validating machine learning models. However, such data may be infrequent, incomplete, or biased, leading to inaccurate predictions [91].
- Privacy concerns: The use of mobile phone and social media data for poverty mapping raises privacy concerns, as it may involve the collection, analysis, and sharing of sensitive personal information [12]. Proper anonymization and data protection measures are essential to address these concerns.
- Model interpretability: Machine learning models, particularly deep learning methods, can be complex and difficult to interpret, making it challenging for researchers and policymakers to understand the underlying relationships between input variables and poverty indicators [66].
- Generalizability: Machine learning models trained on data from specific regions or countries may not generalize well to other contexts, requiring the development of new models and the collection of additional ground-truth data [91].
- Integration of multiple data sources: Combining data from various sources, such as satellite imagery, mobile phone data, and social media data, can improve the accuracy and completeness of poverty maps. However, this process can be challenging due to differences in spatial and temporal resolution, data quality, and data formats [108].

AI techniques have shown great potential in improving the accuracy, timeliness, and spatial resolution of poverty mapping and estimation. By leveraging diverse data sources, such as satellite imagery, mobile phone data, and social media data, machine learning models can provide valuable insights into the spatial distribution of poverty and inform targeted policy interventions. Despite the challenges and limitations, further research and development in this field can contribute significantly to the understanding and reduction of socioeconomic inequality.

7.4.3 Urban Segregation and Gentrification

Urban segregation and gentrification are important aspects of socioeconomic inequality that have significant implications for urban planning, housing policy, and social welfare. AI techniques have been applied to study these phenomena, providing insights into their causes, consequences, and potential solutions.

AI Techniques for Analyzing Urban Segregation and Gentrification

Machine learning algorithms, such as clustering and classification techniques, have been used to identify patterns of urban segregation and gentrification based on demographic, socioeconomic, and spatial data [27, 56]. For example, researchers have employed unsupervised learning algorithms, such as k-means clustering, to analyze census data and identify areas with high levels of racial and socioeconomic segregation [94]. Similarly, deep learning techniques, such as convolutional neural networks

(CNNs), have been used to analyze satellite imagery and detect signs of gentrification, such as new construction or changes in land use [49].

Applications and Case Studies

- Dmowska et al. [27] used k-means clustering to analyze racial segregation in 53 US metropolitan areas based on census data. Their study revealed that large cities tend to have higher levels of racial segregation, while smaller cities exhibit more diverse patterns of segregation.
- Graesser et al. [49] applied CNNs to analyze high-resolution satellite imagery and detect signs of gentrification in New York City. Their study found that gentrification was most pronounced in areas with low population density and high levels of racial and socioeconomic diversity.
- Zhou and Liu [56] developed a machine learning model to predict gentrification in Los Angeles based on demographic, socioeconomic, and spatial data. Their model achieved high accuracy in identifying areas at risk of gentrification, providing valuable information for urban planners and policymakers.

Challenges and Limitations

Despite the promising applications of AI in analyzing urban segregation and gentrification, several challenges and limitations remain:

- Data quality and availability: High-quality data on demographics, socioeconomic indicators, and land use is essential for training and validating machine learning models. However, such data may be outdated, incomplete, or biased, leading to inaccurate predictions [56].
- Model interpretability: As with other AI applications, machine learning models used in urban segregation and gentrification analysis can be complex and difficult to interpret. This can make it challenging for researchers and policymakers to understand the underlying relationships between input variables and the phenomena of interest [49].
- Generalizability: Machine learning models trained on data from specific cities or regions may not generalize well to other contexts, requiring the development of new models and the collection of additional ground-truth data [27].
- Integration of multiple data sources: Combining data from various sources, such as census data, satellite imagery, and social media data, can improve the accuracy and completeness of urban segregation and gentrification analyses. However, this process can be challenging due to differences in spatial and temporal resolution, data quality, and data formats [56].

AI techniques have shown great potential in advancing the understanding of urban segregation and gentrification, enabling more informed policy interventions and urban planning decisions. Further research and development in this field can contribute to addressing the challenges and limitations mentioned above, ultimately leading to more accurate, interpretable, and generalizable models for studying these complex social phenomena.

7.4.4 Access to Services and Amenities

Access to services and amenities, such as healthcare, education, and public transportation, is a critical factor in socioeconomic inequality and spatial analysis. AI techniques have been employed to study disparities in access to these resources, helping to identify underserved areas and inform policy interventions.

AI Techniques for Analyzing Access to Services and Amenities

Various AI techniques, including supervised and unsupervised machine learning algorithms, have been used to analyze access to services and amenities. For example, researchers have used k-means clustering and classification techniques to identify areas with limited access to healthcare or public transportation based on spatial and socioeconomic data [15, 100]. Additionally, deep learning techniques, such as convolutional neural networks (CNNs), have been applied to analyze satellite imagery and detect the location of amenities, such as schools and hospitals [62].

Applications and Case Studies

- Wang et al. [100] employed k-means clustering to analyze access to healthcare facilities in Shenzhen, China. Their study revealed significant disparities in access to healthcare across different districts, highlighting areas with limited access to hospitals and clinics.
- Zou et al. [15] used classification techniques to assess access to public transportation in Beijing, China. Their research identified areas with limited access to public transportation and found that access was strongly influenced by factors such as population density, land use, and road networks.
- Zhang et al. [62] applied CNNs to analyze satellite imagery and detect the location of schools in rural areas of sub-Saharan Africa. Their study provided valuable insights into disparities in access to education, informing efforts to improve school enrollment and attendance.

Challenges and Limitations

Several challenges and limitations remain in the application of AI techniques to analyze access to services and amenities:

- Data quality and availability: Accurate and up-to-date data on the location and characteristics of services and amenities is essential for training and validating machine learning models. However, such data may be lacking or outdated, particularly in rural or low-income areas [62].
- Model interpretability: As with other AI applications, machine learning models used to analyze access to services and amenities can be complex and difficult to interpret. This can make it challenging for researchers and policymakers to understand the underlying relationships between input variables and access to services [100].

- Generalizability: Machine learning models trained on data from specific cities or regions may not generalize well to other contexts, requiring the development of new models and the collection of additional ground-truth data [15].
- Integration of multiple data sources: Combining data from various sources, such as census data, satellite imagery, and social media data, can improve the accuracy and completeness of analyses of access to services and amenities. However, this process can be challenging due to differences in spatial and temporal resolution, data quality, and data formats [62].

AI techniques have demonstrated significant potential in advancing the understanding of disparities in access to services and amenities, enabling more informed policy interventions and resource allocation decisions. Further research and development in this field can contribute to addressing the challenges and limitations mentioned above, ultimately leading to more accurate, interpretable, and generalizable models for studying access to services and amenities.

7.5 Challenges and Limitations of AI in Socioeconomic Inequality and Spatial Analysis

Despite the promising applications of AI in the study of socioeconomic inequality and spatial analysis, several challenges and limitations (Table 7.2) need to be addressed to ensure the effectiveness and robustness of these techniques:

High-quality and up-to-date data is crucial for the development and validation of AI models in this domain. However, data on demographics, socioeconomic indicators, and service locations can be outdated, incomplete, or biased, leading to inaccurate predictions and analyses [62, 100]. Ensuring data quality and availability remains a critical challenge for the successful application of AI in this field.

As with other AI applications, machine learning models used in socioeconomic inequality and spatial analysis can be complex and difficult to interpret. This can make it challenging for researchers and policymakers to understand the underlying relationships between input variables and the phenomena of interest [49, 100]. Developing more interpretable models and fostering collaboration between AI researchers and domain experts can help address this challenge.

Machine learning models trained on data from specific cities or regions may not generalize well to other contexts, requiring the development of new models and the collection of additional ground-truth data [15, 27]. This poses a challenge for the widespread application of AI techniques in studying socioeconomic inequality and spatial analysis across different geographies and settings.

Combining data from various sources, such as census data, satellite imagery, and social media data, can improve the accuracy and completeness of analyses in this field. However, this process can be challenging due to differences in spatial and temporal resolution, data quality, and data formats [56, 62]. Developing methodologies for

Table 7.2 Challenges in AI applications to socioeconomic inequality and spatial analysis

Aspect	Challenge	Description
Data quality and availability	Ensuring access to high-quality and up-to-date data for training AI models is challenging, potentially leading to inaccurate analyses	Critical for developing robust AI models; requires integrating multiple data sources to overcome biases and incompleteness
Model interpretability	AI models, especially deep learning, can be complex "black boxes," making it hard to understand their decision-making processes	Important for gaining trust and facilitating the adoption of AI-driven solutions; calls for more transparent AI techniques
Generalizability	AI models developed for specific contexts may not perform well in different settings, limiting their wider application	Necessitates new models and additional ground-truth data for different regions or countries
Integration of data sources	Combining diverse data types (census, satellite imagery, social media) poses challenges due to varying formats and resolutions	Essential for comprehensive analyses; requires advanced methodologies for effective data fusion
Ethical considerations	The use of AI raises concerns about data privacy, algorithmic fairness, and the potential reinforcement of existing biases	Ensuring AI applications contribute positively without exacerbating socioeconomic disparities is paramount

effective data integration is essential for leveraging the full potential of AI in studying socioeconomic inequality and spatial analysis.

The use of AI in socioeconomic inequality and spatial analysis raises several ethical concerns, such as data privacy, fairness, and accountability. Ensuring that AI models do not perpetuate existing biases or contribute to further marginalization of vulnerable populations is a critical challenge in this field. Researchers and policymakers must carefully consider the ethical implications of their work and strive to develop transparent, accountable, and equitable AI systems.

AI techniques hold great promise for advancing the understanding of socioeconomic inequality and spatial analysis, providing valuable insights into the causes, consequences, and potential solutions of these complex social phenomena. By addressing the challenges and limitations outlined above, researchers and policymakers can harness the power of AI to inform more effective and equitable policy interventions and resource allocation decisions.

7.6 Future Directions in AI Applications for Socioeconomic Inequality and Spatial Analysis

As AI techniques continue to advance, several promising future directions emerge for their application in socioeconomic inequality and spatial analysis:

1. Integration of interdisciplinary knowledge: By collaborating with experts from various disciplines, such as economics, urban planning, and sociology, AI researchers can better understand the nuances of socioeconomic inequality and spatial analysis and develop more accurate and relevant models [15, 66].
2. Development of real-time monitoring systems: AI-powered systems that can monitor and analyze socioeconomic inequalities in real-time can help policymakers and organizations quickly identify and respond to emerging disparities and vulnerable populations [27, 56].
3. Exploration of novel data sources: Emerging data sources, such as social media, mobile phone data, and internet of things (IoT) devices, can provide valuable insights into socioeconomic inequality and spatial patterns. Leveraging these data sources can help researchers develop more comprehensive and granular analyses of these phenomena [9, 62].
4. Advancements in AI explainability: Developing more interpretable AI models and methods can help researchers and policymakers better understand the underlying relationships between input variables and the phenomena of interest, leading to more informed decisions and interventions [49, 100].
5. Addressing ethical considerations: Future research should prioritize developing AI systems that are transparent, accountable, and equitable, ensuring that they do not perpetuate existing biases or contribute to further marginalization of vulnerable populations [27, 62].

By leveraging the potential of AI techniques and addressing the challenges associated with them, researchers and policymakers can harness the power of advanced technology to better understand and address socioeconomic inequality and spatial analysis. Collaboration between multiple disciplines and the integration of novel data sources can lead to more accurate, real-time monitoring of socioeconomic inequalities and spatial patterns.

Furthermore, the development of more interpretable and explainable AI models can help researchers and policymakers gain a deeper understanding of the relationships between input variables and socioeconomic inequality, allowing for more informed decision-making and policy interventions. Additionally, addressing ethical considerations in the development and application of AI systems is essential to ensuring that these technologies do not perpetuate existing biases or contribute to further marginalization of vulnerable populations.

Ultimately, the continued exploration of AI applications in socioeconomic inequality and spatial analysis presents a promising avenue for advancing our understanding of these complex phenomena and developing more effective, equitable solutions to address them.

References

1. Alkire, S., & Foster, J. (2011). Counting and multidimensional poverty measurement. *Journal of Public Economics, 95*(7–8), 476–487. https://doi.org/10.1016/j.jpubeco.2010.11.006
2. Anselin, L. (1995). Local indicators of spatial association—LISA. *Geographical Analysis, 27*(2), 93–115. https://doi.org/10.1111/j.1538-4632.1995.tb00338.x
3. Arribas-Bel, D., & Sanz-Gracia, F. (2014). The validity of the monocentric city model in a polycentric age: US metropolitan areas in 1990, 2000 and 2010. *Urban Geography, 35*(7), 980–997.
4. Atkinson, A. B., Piketty, T., & Saez, E. (2011). Top incomes in the long run of history. *Journal of Economic Literature, 49*(1), 3–71. https://doi.org/10.1257/jel.49.1.3
5. Balke, T., & Gilbert, N. (2014). How do agents make decisions? A survey. *Journal of Artificial Societies and Social Simulation, 17*(4), 13.
6. Ban, Y. (2015). Google Earth as a tool in 2D landslide simulation modeling. In Y. Ban (Ed.), *Google earth: Outreach and activism* (pp. 97–117). Palgrave Macmillan.
7. Batty, M. (2005). *Cities and complexity: Understanding cities with cellular automata, agent-based models, and fractals*. MIT Press.
8. Benenson, I., & Torrens, P. M. (2004). *Geosimulation: Automata-based modeling of urban phenomena*. Wiley.
9. Bharti, N., & Singh, D. (2020). Integration of remote sensing, GIS and social media data in hazard and vulnerability assessment. In *Disaster management* (pp. 109–128). Springer.
10. Blaschke, T., Merschdorf, H., & Cabral, P. (2015). Big data for urban and regional science. *Regional Studies, Regional Science, 2*(1), 1–8.
11. Blei, D. M., Ng, A. Y., & Jordan, M. I. (2003). Latent Dirichlet allocation. *Journal of Machine Learning Research, 3*(Jan), 993–1022.
12. Blumenstock, J. E. (2016). Fighting poverty with data. *Science, 353*(6301), 753–754.
13. Breiman, L., Friedman, J. H., Olshen, R. A., & Stone, C. J. (1984). *Classification and Regression Trees*. Wadsworth International Group.
14. Borgatti, S. P., Mehra, A., & Brass, D. J. (2002). Classification and regression by Random Forest. *R News, 2*(3), 18–22.
15. Bourdieu, P. (1986). The forms of capital. In J. Richardson (Ed.), *Handbook of Theory and Research for the Sociology of Education* (pp. 241–258). Greenwood.
16. Chetty, R., Hendren, N., Kline, P., & Saez, E. (2014). Where is the land of opportunity? The geography of intergenerational mobility in the United States. *The Quarterly Journal of Economics, 129*(4), 1553–1623. https://doi.org/10.1093/qje/qju022
17. Clarke, K. C., & Gaydos, L. J. (1998). Loose-coupling a cellular automaton model and GIS: Long-term urban growth prediction for San Francisco and Washington/Baltimore. *International Journal of Geographical Information Science, 12*(7), 699–714. https://doi.org/10.1080/136588198241617
18. Connelly, R., Playford, C. J., Gayle, V., & Dibben, C. (2016). The role of administrative data in the big data revolution in social science research. *Social Science Research, 59*, 1–12. https://doi.org/10.1016/j.ssresearch.2016.04.015
19. Coulton, C. J. (2017). Using the American community survey: Benefits and challenges. In J. M. Marston (Ed.), *The American community survey: Development, implementation, and issues for Congress* (pp. 1–32). Nova Science Publishers.
20. Crooks, A. T., & Heppenstall, A. J. (2012). Introduction to agent-based modeling. In A. Heppenstall, A. Crooks, L. M. See, & M. Batty (Eds.), *Agent-Based Models of Geographical Systems* (pp. 85–105). Springer.
21. Crooks, A. T., & Hailegiorgis, A. B. (2014). An agent-based modeling approach applied to the spread of cholera. *Environmental Modelling & Software, 62*, 164–177.
22. Crooks, A. T., Malleson, N., & Heppenstall, A. J. (2019). *Agent-based modeling in geographical systems*. In Handbook of Computational Social Science (pp. 117–142). Routledge.
23. Deaton, A. (1997). *The analysis of household surveys: A microeconometric approach to development policy*. Johns Hopkins University Press.

24. Deville, P., Linard, C., Martin, S., Gilbert, M., Stevens, F. R., Gaughan, A. E., Tatem, A. J., et al. (2014). Dynamic population mapping using mobile phone data. *Proceedings of the National Academy of Sciences, 111*(45), 15888–15893.
25. Dmowska, A., Stepinski, T. F., & Nowosad, J. (2017). Mapping changes in spatial patterns of racial diversity across the United States. *Applied Geography, 86*, 122–130.
26. Donchyts, G., Baart, F., Winsemius, H., Gorelick, N., Kwadijk, J., & van de Giesen, N. (2016). Earth's surface water change over the past 30 years. *Nature Climate Change, 6*(9), 810–813.
27. Dorling, D. (2010). *Injustice: Why social inequality persists*. Policy Press.
28. Elvidge, C. D., Zhizhin, M., Hsu, F. C., & Baugh, K. E. (2017). VIIRS night-time lights. *International Journal of Remote Sensing, 38*(21), 5860–5879.
29. Elhorst, J. P. (2014). *Spatial Econometrics: From Cross-Sectional Data to Spatial Panels*. Springer.
30. Engstrom, R., Hersh, J., & Newhouse, D. (2015). Poverty from space: Using high-resolution satellite imagery for estimating economic well-being. In *World Bank Policy Research Working Paper* (7264).
31. Epstein, J. M., & Axtell, R. (1996). *Growing Artificial Societies: Social Science from the Bottom Up*. Brookings Institution Press.
32. ESA (2020). *Sentinel Satellites Overview*. https://sentinels.copernicus.eu/web/sentinel/missions
33. Fotheringham, A. S., Brunsdon, C., & Charlton, M. (2000). *Quantitative geography: Perspectives on spatial data analysis*. SAGE Publications Ltd.
34. Friedman, J. H. (2001). Greedy function approximation: A gradient boosting machine. *Annals of Statistics, 29*(5), 1189–1232.
35. Freelon, D., McIlwain, C. D., & Clark, M. D. (2016). Beyond the hashtags: #Ferguson, #BlackLivesMatter, and the online struggle for offline justice. *Center for Media & Social Impact*, 1–23.
36. Gao, B. C., Li, X., & Weng, Q. (2018). *Advances in land remote sensing: System, modeling, inversion and application*. Springer Science & Business Media.
37. Gebru, T., Krause, J., Wang, Y., Chen, D., Deng, J., Aiden, E. L., & Fei-Fei, L. (2017). Using deep learning and Google street view to estimate the demographic makeup of neighborhoods across the United States. *Proceedings of the National Academy of Sciences, 114*(50), 13108–13113.
38. Ghosh, S., & Lerman, K. (2018). A framework for socio-economic analysis using big social data. *Computational and Mathematical Organization Theory, 24*(4), 405–428. https://doi.org/10.1007/s10588-018-9275-5
39. Ghosh, T., Anderson, S., Elvidge, C. D., & Sutton, P. C. (2013). Using nighttime satellite imagery as a proxy measure of human well-being. *Sustainability, 5*(12), 4988–5019.
40. Goodchild, M. F., & Janelle, D. G. (2010). Toward critical spatial thinking in the social sciences and humanities. *GeoJournal, 75*(1), 3–13. https://doi.org/10.1007/s10708-010-9340-3
41. Goodchild, M. F., & Li, L. (2012). Assuring the quality of volunteered geographic information. *Spatial Statistics, 1*, 110–120.
42. Graesser, J., Ager, A. A., Nielsen-Pincus, M., Day, M. A., & Kline, J. D. (2017). A human ecology approach to understanding the spatial pattern of wildfire. *Landscape Ecology, 32*(6), 1185–1200.
43. Granovetter, M. (1985). Economic action and social structure: The problem of embeddedness. *American Journal of Sociology, 91*(3), 481–510.
44. Guidotti, R., Monreale, A., Ruggieri, S., Turini, F., Giannotti, F., & Pedreschi, D. (2018). A survey of methods for explaining black box models. *ACM Computing Surveys (CSUR), 51*(5), 1–42.
45. Gupta, P., Ghosh, S. K., & Nagarajan, H. (2015). A task-oriented approach to cadastral data quality assessment using high-resolution satellite imagery. *ISPRS Journal of Photogrammetry and Remote Sensing, 101*, 46–56.
46. Haklay, M. (2010). How good is volunteered geographical information? A comparative study of OpenStreetMap and ordnance survey datasets. *Environment and Planning B: Planning and Design, 37*(4), 682–703.

47. Haklay, M. (2016). Why is participation inequality important? *European handbook of crowdsourced geographic information* (pp. 35–44). Ubiquity Press.
48. Hanson, S., & Giuliano, G. (2004). *The Geography of Urban Transportation.* Guilford Press.
49. Hawelka, B., Sitko, I., Beinat, E., Sobolevsky, S., Kazakopoulos, P., & Ratti, C. (2014). Geo-located Twitter as a proxy for global mobility patterns. *Cartography and Geographic Information Science, 41*(3), 260–271.
50. Helbich, M., Jochem, A., Mücke, W., & Höfle, B. (2013). Boosting the predictive accuracy of urban hedonic house price models through airborne laser scanning. *Computers, Environment and Urban Systems, 39*, 81–92.
51. Hochreiter, S., & Schmidhuber, J. (1997). Long short-term memory. *Neural Computation, 9*(8), 1735–1780.
52. Huang, Q., & Wong, D. W. (2015). Modeling and visualizing regular human mobility patterns with uncertainty: An example using Twitter data. *Annals of the Association of American Geographers, 105*(6), 1179–1197.
53. IPCC. (2014). Climate change 2014: Impacts, adaptation, and vulnerability. Part A: Global and sectoral aspects. In *Contribution of Working Group II to the Fifth Assessment Report of the Intergovernmental Panel on Climate Change.* Cambridge University Press.
54. Jean, N., Burke, M., Xie, M., Davis, W. M., Lobell, D. B., & Ermon, S. (2016). Combining satellite imagery and machine learning to predict poverty. *Science, 353*(6301), 790–794.
55. Jolliffe, I. T. (2002). *Principal Component Analysis.* Springer Series in Statistics.
56. Kawachi, I., Kennedy, B. P., & Glass, R. (1997). Social capital and self-rated health: A contextual analysis. *American Journal of Public Health, 87*(8), 1187–1193.
57. Kien, G. (2018). Analyzing socio-economic inequality in Vietnam. *Asian Geographer, 35*(1), 1–18.
58. Kien, G., Duong, T., & Dang, H. (2020). Socioeconomic inequality in Vietnam: A study of spatial patterns using machine learning techniques. *Environment and Planning B: Urban Analytics and City Science, 47*(6), 1011–1027.
59. Kim, Y., Tan, Y., & Lee, G. (2017). Sentiment analysis of social media: An algorithmic perspective. *IEEE Intelligent Systems, 32*(4), 24–31.
60. Kitchin, R. (2014). *The data revolution: Big data, open data, data infrastructures and their consequences.* SAGE Publications.
61. Levinson, D., & Kumar, A. (1997). Density and the journey to work. *Growth and Change, 28*(2), 147–172.
62. Li, X., Zhang, C., & Sun, C. (2018). Recurrent neural network-based end-to-end prediction of geospatial phenomena. *IEEE Access, 6*, 28844–28854.
63. Li, L., Goodchild, M. F., & Xu, B. (2013). Spatial, temporal, and socioeconomic patterns in the use of Twitter and Flickr. *Cartography and Geographic Information Science, 40*(2), 61–77.
64. Li, X., Yeh, A. G., & Zhang, Q. (2015). Large-scale integration of remote sensing, GIS and GPS in the assessment of housing land development in China. *International Journal of Geographical Information Science, 29*(4), 555–572.
65. Liaw, A., & Wiener, M. (2002). Classification and regression by random Forest. *R News, 2*(3), 18–22.
66. Lillesand, T., Kiefer, R. W., & Chipman, J. (2015). *Remote sensing and image interpretation.* Wiley.
67. Liu, X., Hu, G., Chen, Y., Li, X., Xu, C., & Li, S. (2017). High-resolution multi-temporal mapping of global urban land using Landsat images based on the Google earth engine platform. *Remote Sensing of Environment, 203*, 166–176.
68. Logan, J. R., & Stults, B. J. (2011). The persistence of segregation in the metropolis: New findings from the 2010 Census. *Census Brief prepared for Project US2010.* https://s4.ad.brown.edu/Projects/Diversity/Data/Report/report2.pdf
69. Long, Y., Liu, Z., & Wu, X. (2017). Deep learning for remote sensing image classification: A comprehensive review. *IEEE Transactions on Geoscience and Remote Sensing, 55*(3), 1–23.

70. Lu, X., Bengtsson, L., & Holme, P. (2012). Predictability of population displacement after the 2010 Haiti earthquake. *Proceedings of the National Academy of Sciences, 109*(29), 11576–11581.
71. Lucas, K. (2012). Transport and social exclusion: Where are we now? *Transport Policy, 20*, 105–113.
72. MacQueen, J. (1967). Some methods for classification and analysis of multivariate observations. In *Proceedings of the Fifth Berkeley Symposium on Mathematical Statistics and Probability* (Vol. 1, pp. 281–297). University of California Press.
73. Mellander, C., Lobo, J., Stolarick, K., & Matheson, Z. (2015). Night-time light data: A good proxy measure for economic activity? *PLoS ONE, 10*(10), e0139779.
74. Muralidharan, A., Rasmussen, L., Patterson, D., & Shin, J. H. (2011). Hope for Haiti: An analysis of Facebook and Twitter usage during the earthquake relief efforts. *Public Relations Review, 37*(2), 175–177.
75. Neis, P., & Zielstra, D. (2014). Recent developments and future trends in volunteered geographic information research: The case of OpenStreetMap. *Future Internet, 6*(1), 76–106.
76. Newman, M. (2010). *Networks: An introduction*. Oxford University Press.
77. Nieves, J. J., Stevens, F. R., Gaughan, A. E., Linard, C., Sorichetta, A., & Tatem, A. J. (2017). Examining the correlates and drivers of human population distributions across low- and middle-income countries. *Journal of the Royal Society Interface, 14*(136), 20170401.
78. O'Sullivan, D., Evans, T., Manson, S., Metcalf, S., Ligmann-Zielinska, A., & Bone, C. (2012). Strategic directions for agent-based modeling: Avoiding the YAAWN syndrome. *Journal of Land Use Science, 7*(2), 151–171.
79. Panda, S. S., Johansen, K., & Knudby, A. (2019). A review of remote sensing-based vulnerability and risk assessments for natural hazards. *International Journal of Remote Sensing, 40*(12), 4692–4722.
80. Pang, B., & Lee, L. (2008). Opinion mining and sentiment analysis. *Foundations and Trends in Information Retrieval, 2*(1–2), 1–135.
81. Putnam, R. D. (2000). *Bowling alone: The collapse and revival of American community*. Simon & Schuster.
82. Rao, D., Yarowsky, D., Shreevats, A., & Gupta, M. (2010). Classifying latent user attributes in Twitter. In *Proceedings of the 2nd International Workshop on Search and Mining User-generated Contents* (pp. 37–44).
83. Reardon, S. F., & Bischoff, K. (2011). Income inequality and income segregation. *American Journal of Sociology, 116*(4), 1092–1153. https://doi.org/10.1086/657114
84. Ribeiro, M. T., Singh, S., & Guestrin, C. (2016). Why should I trust you? Explaining the predictions of any classifier. In *Proceedings of the 22nd ACM SIGKDD International Conference on Knowledge Discovery and Data Mining* (pp. 1135–1144).
85. Schelling, T. C. (1971). Dynamic models of segregation. *Journal of Mathematical Sociology, 1*(2), 143–186.
86. Sebastiani, F. (2002). Machine learning in automated text categorization. *ACM Computing Surveys (CSUR), 34*(1), 1–47.
87. Seto, K. C., Fragkias, M., Güneralp, B., & Reilly, M. K. (2011). A meta-analysis of global urban land expansion. *PLoS ONE, 6*(8), e23777.
88. Shelton, T., Poorthuis, A., & Zook, M. (2015). Social media and the city: Rethinking urban socio-spatial inequality using user-generated geographic information. *Landscape and Urban Planning, 142*, 198–211.
89. Shrestha, M., Zhang, Q., & Imran, A. (2021). Deep learning-based poverty mapping using multi-temporal multispectral and nighttime light satellite imagery. *Remote Sensing, 13*(10), 1995.
90. Smith-Doerr, L., & Powell, W. W. (2005). Networks and economic life. In N. J. Smelser & R. Swedberg (Eds.), *The handbook of economic sociology* (2nd ed., pp. 379–402). Princeton University Press.
91. Steele, C. M. (2017). *A threat in the air: How stereotypes shape intellectual identity and performance*. In C. Jencks & M. Phillips (Eds.), The Black-White Test Score Gap (pp. 401–427). Brookings Institution Press.

92. Steele, J. E., Sundsøy, P. R., Pezzulo, C., Alegana, V. A., Bird, T. J., Blumenstock, J., Tatem, A. J., et al. (2017). Mapping poverty using mobile phone and satellite data. *Journal of the Royal Society Interface, 14*(127), 20160690.
93. Stevens, F. R., Gaughan, A. E., Linard, C., & Tatem, A. J. (2015). Disaggregating census data for population mapping using random forests with remotely-sensed and ancillary data. *PLoS ONE, 10*(2), e0107042.
94. Tammaru, T., van Ham, M., & Janssen, H. (2016). *Socio-Economic Segregation in European Capital Cities: East Meets West*. Routledge.
95. Wang, D., Pedreschi, D., Song, C., Giannotti, F., & Barabási, A. L. (2016). Human mobility, social ties, and link prediction. In *Proceedings of the 17th ACM SIGKDD International Conference on Knowledge Discovery and Data Mining* (pp. 1100–1109).
96. Wang, S., Li, Y., & Zhang, H. (2019). Agent-based modeling and simulation of human behavior in emergency evacuation: A review. *IEEE Access, 7*, 19791–19805.
97. Wang, F., Zhou, Y., & Yao, S. (2020). Spatial-temporal analysis of urban economic inequality using high-resolution nighttime light data. *Remote Sensing, 12*(4), 626.
98. Wang, L., Liu, Y., Hu, Y., Zhang, W., & Tong, X. (2018). Using machine learning to estimate global PM2.5 for environmental health studies. *Environmental Research, 165*, 12–20.
99. World Bank. (2021). World development report 2021: data for better lives. *World Bank*. https://doi.org/10.1596/978-1-4648-1605-5
100. Wurm, M., Schardt, M., & Dech, S. (2011). Object-based image information fusion using multisensor satellite data for monitoring of urban areas. *IEEE Journal of Selected Topics in Applied Earth Observations and Remote Sensing, 4*(2), 1–14.
101. Wulder, M. A., White, J. C., Loveland, T. R., Woodcock, C. E., Belward, A. S., Cohen, W. B., Roy, D. P., et al. (2019). Thc global Landsat archive: Status, consolidation, and direction. *Remote Sensing of Environment, 224*, 332–344.
102. Xie, M., Jean, N., Burke, M., Lobell, D., & Ermon, S. (2018). Transfer learning from deep features for remote sensing and poverty mapping. In *Thirty-Second AAAI Conference on Artificial Intelligence*.
103. Zeng, C., Wang, S., Gong, P., & Xu, B. (2017). Land cover mapping using composition of best spectral indices from multiple high-resolution satellite images. *Remote Sensing of Environment, 196*, 210–223.
104. Zhang, H., Du, X., Zhang, Y., & Learnihan, V. (2018). Evaluating the socioeconomic equality of the built environment for physical activity: A case study using convolutional neural networks. *ISPRS International Journal of Geo-Information, 7*(10), 396.
105. Zhang, Q., Li, B., Li, X., Xu, G., & Zhu, X. (2018). The 2010–2015 nighttime light product at 30-m resolution through integration of multi-source remote sensing data. *Remote Sensing, 10*(9), 1399.
106. Zhang, Q., Pandey, B., & Seto, K. C. (2020). A robust method for quantifying urban growth using satellite remote sensing. *Landscape and Urban Planning, 197*, 103758.
107. Zhang, Q., Yang, L., Chen, S., Li, X., & Li, Y. (2019). A novel approach to monitor the process of urbanization using a spatiotemporal convolutional long short-term memory network. *ISPRS Journal of Photogrammetry and Remote Sensing, 148*, 265–277.
108. Zhao, P., Lu, Y., Chen, Y., Cao, X., & Wang, F. (2021). Artificial intelligence in urban studies: Progress, trends, and prospects. *Journal of Urban Management, 10*(1), 1–22. https://doi.org/10.1016/j.jum.2021.03.001
109. Zhou, W., Huang, G., Cadenasso, M. L., & Pickett, S. T. (2018). Developing an integrated approach to analyze urban physical and social landscapes: A case study of Baltimore. *Environment and Planning B: Urban Analytics and City Science, 45*(4), 677–695.
110. Zhou, Y., & Liu, Y. (2020). Big data and artificial intelligence: Opportunities and threats to urban planning and governance. *Cities, 97*, 102509.
111. Zou, Y., Zhou, Y., Wang, L., & Zhu, X. (2019). Mapping poverty using mobile phone and satellite data. *Journal of the Royal Society Interface, 16*(151), 20180683.

Chapter 8
Health and Disease Mapping

8.1 Overview of Health and Disease Mapping

Health and disease mapping is an essential component of public health research, planning, and intervention (Table 8.1). It involves the visualization and analysis of spatial patterns in health-related data, such as disease incidence, prevalence, morbidity, and mortality [22]. Health and disease mapping can provide valuable insights into the geographic distribution of health outcomes, helping to identify spatial clusters of diseases, uncover environmental risk factors, and inform targeted public health policies and interventions [41]. With the advent of artificial intelligence (AI) and its increasing use in human geography, there is a growing interest in leveraging AI techniques to enhance the analysis and interpretation of health and disease mapping data.

Traditionally, health and disease mapping has relied on statistical methods, such as spatial autocorrelation and cluster analysis, to identify patterns in health data [4]. However, these methods often require strong assumptions about the underlying spatial processes and can be limited in their ability to handle large, complex datasets [58]. AI techniques, such as machine learning and deep learning algorithms, offer a powerful alternative to traditional statistical methods, allowing for more flexible and adaptive modeling of spatial health data [42].

One of the key advantages of AI techniques in health and disease mapping is their ability to process and analyze large volumes of heterogeneous data, including structured (e.g., census data, health records) and unstructured (e.g., text, images) data [42]. This is particularly important in the era of big data, where health researchers have access to an unprecedented amount of data from diverse sources, such as remote sensing, social media, mobile devices, and Internet of Things (IoT) sensors [78]. AI techniques can help to integrate and analyze these data sources to uncover complex, multidimensional relationships between health outcomes and their underlying determinants, such as socioeconomic, environmental, and behavioral factors [49].

S. N. M. Moghaddam and H. Cao, *Artificial Intelligence-Driven Geographies*, City Development: Issues and Best Practices, https://doi.org/10.1007/978-981-97-5116-7_8

Table 8.1 Description and application of AI in health and disease mapping

Aspect	Description	Application
Overview	AI enhances analysis and interpretation of health and disease mapping, processing large, heterogeneous data including structured and unstructured data from diverse sources like health records and social media	Informs proactive public health policies and interventions by uncovering multidimensional relationships between health outcomes and determinants
Data sources for mapping	Encompasses traditional data (health surveys, vital statistics), remote sensing data, and emerging sources like social media, mobile phone data, offering new insights into health patterns	Supports advanced disease surveillance, outbreak prediction, and health resource allocation through detailed and granular analysis
Applications in public health	AI is applied in disease surveillance for early detection and in outbreak prediction to enable proactive measures. It also assists in health resource allocation by identifying priority areas and optimizing resource distribution	Facilitates real-time tracking, early warning systems, predictive modeling of disease spread, and efficient distribution of health resources for better outcomes
Challenges and limitations	Challenges include data quality and representativeness, privacy and ethical concerns, algorithmic bias, and the "black box" nature of some AI models	Addressing these challenges is crucial to ensure AI's responsible and effective use in health and disease mapping, particularly in protecting sensitive health information and ensuring equity in health interventions

Another advantage of AI techniques in health and disease mapping is their ability to model nonlinear relationships and interactions between variables, which can be difficult to capture using traditional statistical methods [42]. For example, deep learning algorithms, such as convolutional neural networks (CNNs) and recurrent neural networks (RNNs), have been shown to be effective in modeling complex spatial and temporal patterns in health data, such as the spread of infectious diseases and the impact of climate change on health outcomes [18, 82].

Furthermore, AI techniques can facilitate the development of predictive models for health and disease mapping, enabling researchers to forecast future health outcomes and identify areas at risk for disease outbreaks [6]. For example, machine learning algorithms, such as random forests and support vector machines, have been used to predict the spatial distribution of vector-borne diseases, such as malaria and dengue fever, based on environmental and socioeconomic data [8, 79]. These predictive models can help inform proactive public health interventions and resource allocation, such as targeted vaccination campaigns, mosquito control efforts, and health facility planning [72].

Despite the potential benefits of AI in health and disease mapping, there are also several challenges and limitations to consider. One challenge is the quality and representativeness of the data used for analysis, which can be affected by factors such as measurement error, missing data, and selection bias [49]. These data quality issues can lead to biased or inaccurate results, highlighting the importance of rigorous data preprocessing and validation procedures when applying AI techniques to health and disease mapping [42].

Another challenge is the interpretability and transparency of AI models, which can be difficult to understand and explain due to their complexity and "black-box" nature [35]. This can be a barrier to the adoption of AI techniques in public health research and decision-making, where transparency and trust are crucial for effective communication and policy implementation [19]. To address this issue, there is a growing interest in developing explainable AI (XAI) approaches that aim to make AI models more interpretable and understandable for human users [5].

Moreover, ethical considerations must be taken into account when applying AI techniques to health and disease mapping, such as privacy, fairness, and accountability [54]. For example, health data used for AI analysis may contain sensitive information about individuals, raising concerns about data privacy and the potential for re-identification [64]. Additionally, AI models can perpetuate or exacerbate existing health disparities if they are trained on biased data or fail to consider the unique needs and contexts of vulnerable populations [7]. To mitigate these ethical risks, it is important to develop and adopt best practices for responsible AI in health and disease mapping, such as privacy-preserving data sharing, fairness-aware machine learning, and transparent model evaluation [46].

In conclusion, AI techniques have the potential to significantly enhance the analysis and interpretation of health and disease mapping data, providing valuable insights into the geographic distribution of health outcomes and informing targeted public health interventions. However, to fully realize the benefits of AI in health and disease mapping, it is important to address the challenges and limitations related to data quality, model interpretability, and ethical considerations.

8.2 Data Sources for Health and Disease Mapping

8.2.1 Traditional Data Sources

Health and disease mapping has a long history, dating back to the nineteenth century, when John Snow used spatial analysis to identify the source of a cholera outbreak in London [40]. Since then, various traditional data sources have been utilized in health and disease mapping, providing valuable insights into the spatial distribution of diseases, as well as the factors that contribute to their occurrence. In this section, we will discuss some of the most commonly used traditional data sources in health and disease mapping, including health surveys, vital statistics, disease registries, and environmental monitoring data.

Health Surveys

Health surveys are one of the most widely used traditional data sources in health and disease mapping, as they provide detailed information on the health status, behaviors, and risk factors of populations at various geographical scales [30]. Health surveys can be administered at the household, community, or regional level, and may be cross-sectional, longitudinal, or panel-based. Examples of health surveys include the Demographic and Health Surveys (DHS), the World Health Organization (WHO) Global Health Observatory, and the Behavioral Risk Factor Surveillance System (BRFSS) in the United States.

Vital Statistics

Vital statistics, which include data on births, deaths, and marriages, are another important source of information for health and disease mapping. These data can be used to calculate various health indicators, such as mortality and morbidity rates, life expectancy, and fertility rates, which can then be mapped to identify spatial patterns and trends [3]. Vital statistics are typically collected by national and local governments, and may be available through organizations such as the WHO, the United Nations, and the Centers for Disease Control and Prevention (CDC).

Disease Registries

Disease registries are databases that collect information on the occurrence, treatment, and outcomes of specific diseases or health conditions [29]. These registries can provide valuable data for health and disease mapping, as they often include detailed information on the geographic distribution of cases, as well as the demographic, clinical, and risk factor characteristics of affected individuals. Examples of disease registries include the National Cancer Institute's Surveillance, Epidemiology, and End Results (SEER) Program, the European Cancer Information System (ECIS), and the WHO Global Tuberculosis Report.

Environmental Monitoring Data

Environmental monitoring data, such as air and water quality measurements, can also play a critical role in health and disease mapping, as they can help to identify potential sources of exposure to harmful substances, as well as areas with high levels of pollution or contamination [37]. These data can be collected through a variety of methods, including ground-based monitoring stations, remote sensing technologies, and citizen science initiatives. Environmental monitoring data can be obtained from various sources, such as the United States Environmental Protection Agency (EPA), the European Environment Agency (EEA), and the WHO Global Air Quality Database.

In summary, traditional data sources have played a crucial role in advancing our understanding of the spatial distribution of diseases and the factors that contribute to their occurrence. However, these data sources also have some limitations, such as the potential for underreporting, biases, and inaccuracies, as well as difficulties in accessing and integrating data from different sources and formats [28]. Despite these challenges, traditional data sources continue to provide valuable information for health and disease mapping, and their integration with emerging data sources and AI techniques can help to further enhance our understanding of the complex relationships between health, disease, and the environment. Although traditional data sources have provided a solid foundation for health and disease mapping, recent advancements in technology and data collection methods have given rise to new data sources that can complement and enhance our understanding of health and disease patterns. In the next section, we will discuss some of these emerging data sources, including remote sensing data, big data, and geospatial data sources.

8.2.2 Remote Sensing Data Sources

Remote sensing data sources have become increasingly important in health and disease mapping as they offer the ability to monitor environmental factors that may be associated with the spread of diseases. Remote sensing is the acquisition of information about an object or phenomenon without making physical contact with the object. This is typically done through the use of satellite or airborne sensors, which collect data on various environmental factors such as land cover, vegetation, temperature, and precipitation [24].

Remote sensing data has been used in a variety of health and disease mapping studies. For example, satellite-derived data on land surface temperature, vegetation indices, and precipitation have been used to model the distribution of vector-borne diseases such as malaria, dengue fever, and Lyme disease [15, 32, 34]. In addition, remote sensing data can be used to assess the impact of climate change on the distribution of diseases and the potential for disease outbreaks [59].

Another application of remote sensing data in health and disease mapping is in the study of air quality and its impact on human health. Satellite-derived data on air pollution, such as particulate matter and nitrogen dioxide concentrations, have been used to assess the spatial distribution of air pollution and its association with respiratory diseases and other health outcomes [38, 74].

Remote sensing data can also be used to monitor the spread of infectious diseases. For instance, satellite imagery has been used to track the spread of diseases such as Ebola and cholera by identifying environmental factors that may be associated with disease transmission, such as water sources, vegetation, and human settlements [53, 71].

The integration of remote sensing data into health and disease mapping provides a powerful tool for understanding the spatial distribution of diseases and the environmental factors that contribute to their spread. As satellite technology continues to advance, the availability and quality of remote sensing data will improve, offering new opportunities for more detailed and accurate assessments of the relationship between environmental factors and human health.

In summary, remote sensing data sources have become increasingly important in health and disease mapping, offering the ability to monitor environmental factors that may be associated with the spread of diseases. Applications of remote sensing data in this field include the modeling of vector-borne disease distribution, air quality assessments, and tracking the spread of infectious diseases. As technology continues to advance, researchers will be better equipped to understand the spatial distribution of diseases and the environmental factors that contribute to their spread, which in turn can inform public health interventions and policy decisions.

8.2.3 Big Data and Geospatial Data Sources

The advent of big data and geospatial data sources has revolutionized the field of health and disease mapping. These new data sources offer an unprecedented level of detail and granularity, enabling researchers to monitor and analyze health patterns at a fine spatial scale. In this section, we will discuss various big data and geospatial data sources that are commonly used in health and disease mapping and highlight their applications and limitations.

Social Media Data

Social media platforms, such as Twitter and Facebook, generate vast amounts of user-generated content that can be mined for insights into public health trends and disease patterns [69]. By analyzing text, images, and videos shared on social media, researchers can track the spread of infectious diseases, identify potential outbreaks, and monitor public sentiment and behavior related to health issues [67].

For example, studies have used Twitter data to track the spread of influenza in real-time, providing a more timely and cost-effective alternative to traditional surveillance methods [23]. Similarly, Facebook data has been used to monitor the spread of Zika virus and inform targeted intervention efforts [50]. However, social media data can be subject to biases, as users may not be representative of the general population, and the content shared may not accurately reflect actual health conditions or behaviors.

Mobile Phone Data

Mobile phone data, such as call detail records (CDRs) and global positioning system (GPS) data, can provide valuable information on human mobility patterns, which can help researchers understand the spatial dynamics of disease transmission [81]. By analyzing patterns of human movement and contact, researchers can identify potential routes of disease spread and target intervention efforts more effectively.

For example, mobile phone data has been used to track the spread of malaria in Kenya, revealing the role of human mobility in driving disease transmission across different regions [81]. However, mobile phone data can raise privacy concerns, as it can potentially reveal sensitive information about individuals and their behaviors. To address these concerns, researchers must implement strict data anonymization and aggregation protocols to protect user privacy.

Electronic Health Records (EHRs)

EHRs contain detailed information on patients' medical history, diagnoses, treatments, and outcomes, offering a rich source of data for health and disease mapping [9]. By analyzing EHR data, researchers can identify patterns of disease occurrence, assess the effectiveness of interventions, and monitor the impact of healthcare policies on public health.

For example, EHR data has been used to investigate the spatial distribution of diabetes and identify potential hotspots for targeted intervention efforts [63]. However, EHR data can be subject to biases and errors, as it relies on the accuracy and completeness of clinical documentation. Moreover, accessing and analyzing EHR data can be challenging due to privacy regulations and interoperability issues between different EHR systems.

Satellite and Remote Sensing Data

Satellite and remote sensing data can provide valuable information on environmental factors that influence disease transmission and health outcomes, such as temperature, precipitation, land cover, and air quality [33]. By integrating remote sensing data with other data sources, researchers can develop more comprehensive and accurate models of disease risk and health patterns.

For example, remote sensing data has been used to map the distribution of suitable habitats for the tsetse fly, a vector of African sleeping sickness, and inform targeted vector control efforts [65]. Similarly, satellite data has been used to assess the impact ofair pollution on respiratory health and identify areas with high levels of exposure [12]. However, remote sensing data can be subject to errors and uncertainties due to factors such as atmospheric interference, sensor calibration, and spatial resolution limitations. Moreover, the use of remote sensing data in health and disease mapping often requires advanced analytical techniques and expertise in geospatial analysis.

Internet of Things (IoT) and Sensor Data

The growing network of IoT devices and sensors generates vast amounts of data that can be harnessed for health and disease mapping. IoT devices, such as wearable health monitors, smart thermometers, and air quality sensors, can provide real-time, high-resolution data on individual health and environmental conditions [70]. By analyzing IoT and sensor data, researchers can gain new insights into the relationships between environmental factors, human behavior, and health outcomes.

For example, IoT data from wearable devices has been used to track the spread of infectious diseases, such as influenza, by monitoring changes in population-level activity patterns and social contacts [51]. Similarly, sensor data from air quality monitoring networks has been used to assess the impact of air pollution on respiratory health and identify vulnerable populations [37]. However, the use of IoT and sensor data in health and disease mapping can raise privacy concerns, as it can potentially reveal sensitive information about individuals and their behaviors. To address these concerns, researchers must implement strict data anonymization and aggregation protocols to protect user privacy.

In conclusion, big data and geospatial data sources offer numerous opportunities for advancing the field of health and disease mapping. By harnessing the power of these data sources, researchers can develop more accurate, timely, and detailed models of disease risk and health patterns, ultimately informing more effective public health policies and interventions. However, the use of big data and geospatial data sources also presents challenges and limitations, such as data quality, privacy concerns, and the need for advanced analytical skills. As the field continues to evolve, researchers must work to address these challenges and harness the full potential of big data and geospatial data sources in health and disease mapping.

8.3 Applications of AI in Health and Disease Mapping

8.3.1 Disease Surveillance

Disease surveillance is a critical component of public health systems, as it enables the early detection, prevention, and control of infectious diseases and other health threats. With advancements in artificial intelligence (AI), disease surveillance has seen significant improvements in data processing, analysis, and prediction. In this section, we will discuss the various ways AI is being used in disease surveillance and its potential applications in health and disease mapping.

Data Collection and Integration

AI plays a crucial role in automating the collection and integration of data from diverse sources, such as health records, social media, news articles, and remote sensing. Natural language processing (NLP) algorithms can be used to mine relevant information from unstructured text sources, while machine learning techniques can be employed to extract meaningful patterns from structured datasets. Integrating these different data sources can provide a comprehensive picture of disease spread and risk factors, allowing for better-targeted interventions [36].

Real-Time Disease Tracking and Early Warning Systems

AI-powered disease surveillance systems can process vast amounts of data in real-time, enabling faster identification of disease outbreaks and prediction of future disease spread. Machine learning algorithms can identify unusual disease patterns and raise alerts for public health officials to investigate further. These early warning systems can significantly reduce the time between disease emergence and public health response, potentially saving lives and reducing the burden on healthcare systems [52].

Disease Spread Modeling and Prediction

AI techniques, such as machine learning and deep learning, can be used to model and predict disease spread based on various factors, including environmental, demographic, and social determinants. For instance, AI models can predict the risk of malaria transmission based on factors like temperature, precipitation, and human mobility patterns [81]. These predictive models can inform public health interventions, such as vector control measures and vaccination campaigns, and help allocate resources effectively to mitigate disease spread.

Social Media and Online Data Mining for Disease Surveillance

Social media platforms and online search queries are valuable sources of data for disease surveillance. AI algorithms can analyze these data streams to detect early signals of disease outbreaks and monitor public sentiment related to health issues. For example, during the Zika virus outbreak, researchers used AI techniques to analyze Twitter data and track the spread of the virus, identify areas of concern, and understand public reactions to the outbreak [69]. This information can help public health officials in planning and implementing appropriate measures to control disease spread.

Remote Sensing for Environmental Health Monitoring

AI algorithms can be used to analyze remote sensing data, such as satellite images, to monitor environmental factors that may influence disease transmission. For example, researchers have used AI to identify suitable habitats for disease vectors like mosquitoes based on land cover and climate data [60]. This information can be used to target vector control measures, such as pesticide spraying, to prevent the spread of vector-borne diseases.

Mobile Health (mHealth) Applications for Disease Surveillance

AI-powered mobile health (mHealth) applications are increasingly being used for disease surveillance, as they allow for the collection of real-time health data from individuals, such as symptoms, location, and risk factors. This information can be used to track disease outbreaks, monitor patient adherence to treatment, and support contact tracing efforts. Moreover, AI algorithms can provide personalized feedback and recommendations to users, promoting self-management of health and disease prevention [70].

Integration of AI in Existing Disease Surveillance Systems

Integrating AI techniques into existing disease surveillance systems can significantly improve their efficiency, accuracy, and timeliness. For example, AI algorithms can be used to enhance the analysis of laboratory data, automate the classification of notifiable diseases, and streamline reporting workflows. By automating routine tasks, AI can free up valuable time for public health professionals to focus on more complex tasks, such as outbreak investigation and response [27].

Ethical and Privacy Considerations in AI-Powered Disease Surveillance

While AI-powered disease surveillance offers numerous benefits, it also raises ethical and privacy concerns. Data collection and integration from various sources, including social media and mobile health applications, may involve sensitive personal information. Ensuring data privacy, consent, and confidentiality is crucial to maintaining trust in public health systems and avoiding potential harm to individuals [66]. AI researchers and public health practitioners must work together to develop guidelines and best practices for using AI in disease surveillance while respecting ethical considerations and privacy concerns.

As we have seen, AI is playing a vital role in revolutionizing disease surveillance and health mapping. The integration of AI technologies with traditional public health systems has the potential to improve the speed and accuracy of disease detection, tracking, and response. By harnessing the power of AI, public health practitioners can make better-informed decisions, allocate resources more effectively, and ultimately improve health outcomes for communities around the world.

However, despite the numerous benefits offered by AI in disease surveillance, challenges remain. These include the need for greater collaboration among AI researchers, public health practitioners, and policymakers, as well as addressing ethical and privacy concerns related to the collection and use of sensitive health data. By working together and considering these issues, stakeholders can ensure that AI continues to play a positive role in advancing disease surveillance and health mapping efforts.

Moreover, the use of AI in disease surveillance is not a one-size-fits-all solution. Different regions and communities may have unique health needs, resources, and infrastructure, necessitating tailored AI applications to ensure maximum effectiveness. For example, remote and resource-poor settings may require low-cost, easy-to-implement AI solutions that can be easily integrated into existing public health systems.

Future research in AI for disease surveillance should focus on developing new algorithms and methodologies that can be easily adapted to diverse contexts, as well as on improving the interoperability of AI systems with existing public health infrastructure. Additionally, research should explore innovative ways to engage communities in the disease surveillance process, such as through participatory data collection and feedback mechanisms.

In conclusion, AI has the potential to transform disease surveillance and health mapping, providing valuable insights and tools to support public health efforts worldwide. By embracing the opportunities offered by AI, while also addressing the challenges and concerns that arise, we can work together to create a healthier future for all.

In conclusion, AI techniques have the potential to revolutionize disease surveillance and improve health and disease mapping efforts. From real-time disease tracking to predictive modeling, AI can help public health professionals identify and respond to disease outbreaks more effectively, ultimately leading to improved health outcomes and reduced healthcare costs. However, the successful implementation of AI in disease surveillance requires collaboration among AI researchers, public health practitioners, and policymakers, as well as careful consideration of ethical and privacy issues.

8.3.2 Outbreak Prediction

Outbreak prediction is another essential application of AI in health and disease mapping. Predicting disease outbreaks is a vital task to enable proactive measures, early interventions, and effective resource allocation. The integration of AI technologies with epidemiological models, geospatial data, and other relevant data sources can significantly improve the accuracy and timeliness of outbreak prediction efforts. This essay will discuss how AI can be used in outbreak prediction and the associated benefits, challenges, and future directions.

AI Techniques in Outbreak Prediction

Several AI techniques have been employed to predict disease outbreaks, including machine learning, deep learning, natural language processing, and network analysis. These techniques can be applied to various data sources, such as epidemiological surveillance data, environmental data, social media data, and remote sensing data.

- Machine learning: Machine learning algorithms, such as decision trees, support vector machines, and Bayesian networks, can be used to predict disease outbreaks based on historical data and patterns. These algorithms can identify risk factors and develop predictive models that can be updated as new data becomes available [16].
- Deep learning: Deep learning techniques, such as recurrent neural networks (RNNs) and convolutional neural networks (CNNs), can analyze complex data sets and identify patterns that may be indicative of a disease outbreak. These techniques can also be used to process high-resolution satellite images to identify environmental risk factors associated with disease transmission [2].
- Natural language processing: Natural language processing (NLP) can be used to analyze text data from various sources, such as news articles, social media posts, and online forums, to detect early warning signs of disease outbreaks. For example, HealthMap, an online disease surveillance system, uses NLP to analyze news articles and identify potential disease outbreaks worldwide [14].
- Network analysis: Network analysis techniques can be used to analyze the spread of diseases through social networks and predict the likelihood of an outbreak. By modeling the interactions between individuals and communities, network analysis can help identify vulnerable populations and inform targeted intervention strategies [25].

Benefits of Using AI in Outbreak Prediction

Using AI techniques in outbreak prediction offers several benefits, including:

- Improved accuracy: AI algorithms can analyze large and complex data sets, enabling more accurate and precise predictions of disease outbreaks compared to traditional epidemiological models [21].
- Timeliness: AI techniques can process data in real-time, allowing for early detection of potential outbreaks and rapid response measures, which can prevent or mitigate the spread of diseases [16].
- Adaptability: AI algorithms can be easily updated and adapted to new data sources and changing circumstances, ensuring that outbreak prediction models remain relevant and accurate.
- Cost-effectiveness: AI-driven outbreak prediction can reduce the costs associated with public health interventions by enabling targeted and timely measures, ultimately saving lives and resources.

Challenges and Limitations of Using AI in Outbreak Prediction

Despite the potential benefits of using AI in outbreak prediction, several challenges and limitations need to be addressed:

- Data quality and availability: Accurate outbreak prediction relies on high-quality, timely, and complete data. However, data availability and quality can be limited in some regions, particularly in low-resource settings, which can affect the performance of AI algorithms [21].
- Model validation: Validating the accuracy and performance of AI-driven outbreak prediction models can be challenging, particularly when dealing with rare or emerging diseases. Real-world testing and evaluation of these models are necessary to ensure their reliability and usefulness [21].
- Ethical and privacy concerns: The use of sensitive health data, particularly from socialmedia and other non-traditional sources, raises ethical and privacy concerns. Ensuring data protection, informed consent, and adherence to ethical guidelines is crucial when using AI techniques in outbreak prediction [76].
- Interdisciplinary collaboration: Effective outbreak prediction requires collaboration between various disciplines, including epidemiology, public health, computer science, and social sciences. Establishing strong interdisciplinary teams and fostering collaboration is essential for the successful application of AI in outbreak prediction [52].

Future Directions and Potential Applications

As AI technologies continue to advance, there are several potential applications and future directions in outbreak prediction:

- Integration of multiple data sources: Combining various data sources, such as electronic health records, social media, remote sensing, and mobile phone data, can enhance the accuracy and scope of outbreak prediction models [67].

- Real-time outbreak prediction: Developing AI-driven systems that can provide real-time predictions of disease outbreaks can enable rapid response measures and improve the overall effectiveness of public health interventions [52].
- Personalized outbreak prediction: AI algorithms can be used to develop personalized outbreak prediction models that consider individual risk factors, such as age, gender, comorbidities, and travel history. This can help identify vulnerable populations and inform targeted intervention strategies [77].
- Global health surveillance systems: AI-driven outbreak prediction models can be integrated into global health surveillance systems to improve the detection and monitoring of emerging infectious diseases worldwide [14].

By leveraging AI techniques in outbreak prediction, public health authorities can better anticipate the spread of diseases and allocate resources efficiently. Integrating multiple data sources, real-time outbreak prediction, personalized outbreak prediction, and global health surveillance systems are just a few of the future directions that can revolutionize the way we approach disease prevention and control. As technology continues to advance, so too will our ability to mitigate the impact of disease outbreaks on global health.

In conclusion, AI techniques have the potential to significantly improve outbreak prediction efforts, enabling more accurate, timely, and targeted public health interventions. By overcoming the current challenges and limitations, AI-driven outbreak prediction models can play a crucial role in preventing and mitigating the impact of disease outbreaks on global health.

8.3.3 Health Resource Allocation

Health resource allocation is a critical aspect of public health planning and decision-making, as it involves the distribution of scarce resources among various health services, programs, and populations in need. Effective health resource allocation can significantly improve the overall health outcomes of a population and reduce health disparities. The use of AI in health and disease mapping can provide valuable insights to inform better resource allocation strategies. This section will discuss the application of AI techniques in health resource allocation.

AI techniques can support health resource allocation in several ways, including identifying priority areas, predicting future health needs, and optimizing resource distribution. Some of the key methods used in AI for health resource allocation are machine learning algorithms, optimization techniques, simulation modeling, and decision support systems.

Identifying Priority Areas

One of the main challenges in health resource allocation is identifying the areas and populations that require the most attention and resources. AI can assist in this process by analyzing large datasets, such as health statistics, demographic data, and disease prevalence, to identify patterns and trends. By utilizing clustering algorithms, like K-means, and classification algorithms, like decision trees and support vector machines, AI can help identify priority areas based on factors such as disease prevalence, socioeconomic status, and accessibility to health services [39].

Predicting Future Health Needs

Another essential aspect of health resource allocation is predicting future health needs to ensure that adequate resources are available to meet the changing demands. AI can help forecast health needs by analyzing historical data and identifying trends that may indicate future changes in disease prevalence or health service utilization. Time-series analysis, regression models, and artificial neural networks are some of the techniques used in AI for predicting future health needs [43].

Optimizing Resource Distribution

Optimizing the distribution of health resources is critical to ensure that they are used efficiently and effectively. AI techniques can be used to develop decision support systems that help policymakers and health administrators make better resource allocation decisions. Linear programming, integer programming, and multi-objective optimization techniques can be used to determine the optimal allocation of resources, considering factors such as cost, accessibility, and effectiveness of interventions [1].

Simulation Modeling for Health Resource Allocation

Simulation modeling is another AI-based approach that can be employed to support health resource allocation. Agent-based models, system dynamics models, and discrete-event simulation models can help policymakers and health administrators understand the complex relationships between various factors affecting health resource allocation. These models can simulate the impact of different resource allocation strategies on health outcomes and costs, allowing decision-makers to test various scenarios and identify the most effective strategies [11].

Decision Support Systems for Health Resource Allocation

AI can also be used to develop decision support systems (DSS) that assist policymakers and health administrators in making more informed decisions about health resource allocation. These DSS can integrate data from various sources, such as disease prevalence, health service utilization, and cost-effectiveness analysis, to provide recommendations for resource allocation. Machine learning algorithms, optimization techniques, and simulation modeling can be combined in a DSS to offer more accurate and efficient resource allocation strategies [44].

The application of AI techniques in health and disease mapping allows for more efficient and effective resource allocation, leading to improved health outcomes and a better understanding of the factors affecting health resource distribution. As AI continues to advance, it is essential to address the challenges of data quality, privacy, and ethical considerations to ensure the responsible use of AI in health resource allocation. With proper implementation, AI has the potential to revolutionize the way we approach health and disease mapping and significantly improve public health planning and decision-making.

In conclusion, AI techniques have the potential to significantly improve health resource allocation by identifying priority areas, predicting future health needs, optimizing resource distribution, and providing decision support systems. The use of AI in health and disease mapping can lead to better health outcomes and more efficient use of resources. However, challenges remain in ensuring data quality, privacy, and ethical considerations in the application of AI in health resource allocation.

8.4 Challenges and Limitations of AI in Health and Disease Mapping

The application of artificial intelligence (AI) in health and disease mapping has shown great promise in improving public health outcomes, disease surveillance, and resource allocation. However, despite its potential benefits, there are several challenges and limitations (Table 8.2) associated with the use of AI in this field. This section discusses some of the key challenges and limitations, including data quality and availability, privacy and ethical considerations, interpretability and trust, and the need for interdisciplinary collaboration.

Table 8.2 Challenges in AI applications to health and disease mapping

Aspect	Challenge	Description
Data quality and availability	Ensuring access to high-quality and representative data for AI models is challenging, affecting the accuracy of health predictions and interventions	Data scarcity, particularly in LMICs, and biases in data collection can lead to ineffective AI applications, exacerbating health disparities
Privacy and ethical considerations	The use of sensitive health data raises concerns about privacy, consent, and potential biases in AI models	Addressing privacy concerns and ensuring fairness in AI models are paramount to maintaining trust in health systems and avoiding harm
Interpretability and trust	AI models, especially deep learning, can be complex and difficult to interpret, affecting their adoption and trustworthiness in health decision-making	Developing transparent and interpretable AI models is crucial for their acceptance and effective use in public health
Interdisciplinary collaboration	Effective AI applications require collaboration across disciplines, yet barriers exist due to differing methodologies, terminologies, and priorities	Fostering interdisciplinary collaboration is essential for developing innovative, accurate, and relevant AI solutions for complex health issues

Data Quality and Availability

One of the primary challenges in applying AI to health and disease mapping is the quality and availability of data. Accurate and reliable data is essential for the development and validation of AI models [13]. The lack of high-quality, representative data can lead to biased and inaccurate predictions, which may undermine the effectiveness of AI applications in public health [73].

Moreover, many low- and middle-income countries (LMICs) face significant challenges in collecting and maintaining comprehensive health data due to limited resources and infrastructure [26]. This data scarcity can hinder the development and application of AI models in these regions, exacerbating existing health disparities [20].

Privacy and Ethical Considerations

The use of AI in health and disease mapping raises several privacy and ethical concerns, particularly with regards to the handling of sensitive health data. Ensuring the privacy and security of patient data is crucial, as breaches can have severe consequences for individuals and healthcare systems [68].

Moreover, AI algorithms can potentially introduce or perpetuate biases in health and disease mapping, leading to unfair treatment of certain populations [62]. Ensuring that AI models are transparent and do not perpetuate existing health disparities is a critical ethical consideration in the application of AI in health and disease mapping [55].

Interpretability and Trust

The interpretability and trustworthiness of AI models are essential factors in their adoption and use in health and disease mapping. Complex AI models, such as deep learning algorithms, can often function as "black boxes," making it difficult for practitioners and policymakers to understand and trust their predictions [17].

Interpretability is particularly important in public health, where decisions can have significant consequences for individuals and communities [80]. Developing AI models that are both accurate and interpretable is an ongoing challenge in the field [35].

Interdisciplinary Collaboration

Successful AI applications in health and disease mapping require interdisciplinary collaboration between computer scientists, public health experts, geographers, and policymakers [56]. This collaboration is essential for addressing the complex challenges associated with AI in health and disease mapping, including data quality, privacy, and ethical considerations, as well as ensuring that AI models are appropriately designed and validated for their intended purposes [57].

Despite the potential benefits of interdisciplinary collaboration, there remain significant barriers to effective collaboration, including differences in terminologies, methodologies, and priorities across disciplines [13].

In conclusion, the application of AI in health and disease mapping has shown great potential for improving disease surveillance, outbreak prediction, and health resource allocation. However, it is essential to address the various challenges and limitations to ensure that AI techniques are implemented responsibly and ethically.

For instance, data privacy and security remain critical issues that need to be addressed when handling sensitive health information. Additionally, researchers should pay close attention to potential biases in AI models, as these biases can have serious consequences in health care settings. To mitigate these biases, fairness and transparency in AI systems should be prioritized, and the development of explainable AI should be encouraged to foster trust and understanding among end-users.

Moreover, the integration of AI technologies into health care systems should be performed cautiously, taking into consideration the need for an appropriate balance between automation and human expertise. This approach will ensure that AI tools are used effectively, while also preserving the essential human touch in health care.

Collaboration between AI researchers, health care professionals, and policy-makers is vital to address the challenges and limitations of AI in health and disease mapping. By working together, these stakeholders can develop strategies and guidelines that will harness the potential of AI to transform health care, while also ensuring the ethical, responsible, and equitable use of this technology.

8.5 Future Directions in AI Applications for Health and Disease Mapping

The application of artificial intelligence (AI) in health and disease mapping has shown significant potential for transforming the field of public health. As AI techniques continue to evolve and improve, new opportunities and challenges will emerge. This section discusses some of the future directions in AI applications for health and disease mapping, focusing on the potential advancements, collaborations, and ethical considerations that will shape the field in the coming years.

Advancements in AI Techniques and Technologies

One of the most promising future directions for AI in health and disease mapping involves the development and refinement of AI techniques and technologies. As AI algorithms become more sophisticated, they will be better equipped to handle complex, large-scale health data sets and derive meaningful insights from them. For example, deep learning techniques, such as convolutional neural networks (CNNs) and recurrent neural networks (RNNs), will likely continue to advance and become more accurate in detecting and predicting disease patterns [48].

Moreover, AI models will increasingly be integrated with other advanced technologies, such as the Internet of Things (IoT), to enable real-time health monitoring and disease surveillance [61]. This integration will facilitate the collection of more granular, diverse, and accurate health data, which can be used to improve health and disease mapping efforts.

Interdisciplinary Collaboration

The future of AI applications in health and disease mapping will also be characterized by increased interdisciplinary collaboration. Researchers from various fields, including computer science, public health, and social sciences, will need to work together to develop innovative AI solutions that address complex health issues [10].

Collaboration between AI researchers and health care professionals will be essential for ensuring that AI tools are developed with real-world health care needs in mind, and that these tools are effectively integrated into existing health care systems. Moreover, engaging with policymakers and other stakeholders will be crucial for creating supportive regulatory frameworks that enable the responsible and ethical use of AI in health and disease mapping.

Personalized and Precision Public Health

One of the most exciting potential applications of AI in health and disease mapping lies in the field of personalized and precision public health. By leveraging AI techniques to analyze large, diverse data sets, researchers can gain a deeper understanding of the underlying determinants of health and develop targeted interventions that address the unique needs of specific populations [45].

In the future, AI tools could be used to create highly detailed, individual-level health risk profiles, which can be used to inform the design of personalized health promotion strategies and resource allocation decisions. Furthermore, AI-driven disease mapping efforts could help to identify geographic hotspots of health disparities and inform the development of targeted public health interventions that address the specific needs of vulnerable communities.

Ethical considerations and responsible AI

As AI techniques become increasingly integrated into health and disease mapping efforts, it will be essential to address the various ethical considerations that arise from the use of these technologies. Researchers and practitioners will need to ensure that AI tools are developed and deployed in ways that are transparent, fair, and accountable, and that they do not exacerbate existing health disparities or create new ones [75].

To achieve this goal, future research in AI for health and disease mapping should prioritize the development of explainable AI techniques, which can help to build trust and understanding among end-users and stakeholders. Additionally, ongoing dialogue between AI researchers, health care professionals, policymakers, and community members will be essential for ensuring that AI tools are used responsibly and ethically.

Capacity Building and Education

As AI technologies become more widely adopted in health and disease mapping, there will be a growing need for capacity building and education initiatives aimed at training public health professionals, researchers, and policymakers in the use of these tools [47]. Developing and implementing AI-focused curricula within public health and medical education programs will be essential for ensuring that the next generation of health professionals is well-equipped to harness the power of AI in their work.

Additionally, the development of open-source AI tools and resources will help to democratize access to AI technologies and foster a more inclusive and diverse community of AI practitioners in the field of health and disease mapping.

Public Engagement and Citizen Science

Finally, the future of AI applications in health and disease mapping will likely involve greater public engagement and the integration of citizen science approaches. By involving community members in the collection, analysis, and interpretation of health data, AI-driven health and disease mapping efforts can become more responsive to local needs and priorities [31].

In the coming years, we can expect to see the development of new AI tools and platforms that enable members of the public to contribute to health and disease mapping efforts, as well as initiatives that promote greater transparency and public participation in the development of AI technologies for public health applications.

In conclusion, the future of AI applications in health and disease mapping is characterized by significant opportunities and challenges. As AI techniques continue to evolve and become more sophisticated, they hold the potential to transform the field of public health by enabling more accurate, timely, and targeted disease surveillance, outbreak prediction, and health resource allocation. However, it will be essential for researchers, practitioners, and policymakers to work together to address the various ethical, regulatory, and capacity-building challenges that arise from the use of AI in this context.

References

1. Ahmadi-Javid, A., Jalali, Z., & Klassen, K. J. (2017). Outpatient appointment systems in healthcare: A review of optimization studies. *European Journal of Operational Research, 258*(1), 3–34.
2. Alam, M. M., Benediktsson, J. A., & Shamsuddin, M. (2019). Deep learning in remote sensing applications: A comprehensive review and future directions. *IEEE Geoscience and Remote Sensing Magazine, 7*(4), 11–32.
3. Anderson, R. N., Miniño, A. M., Hoyert, D. L., & Rosenberg, H. M. (2012). Comparability of cause of death between ICD-9 and ICD-10: Preliminary estimates. *National Vital Statistics Reports, 49*(2), 1–32.
4. Anselin, L. (1995). Local Indicators of Spatial Association—LISA. *Geographical Analysis, 27*(2), 93–115.
5. Arrieta, A. B., Díaz-Rodríguez, N., Del Ser, J., Bennetot, A., Tabik, S., Barbado, A., García, S., Gil-López, S., Molina, D., Benjamins, R., Chatila, R., & Herrera, F. (2020). Explainable Artificial Intelligence (XAI): Concepts, taxonomies, opportunities, and challenges toward responsible AI. *Information Fusion, 58*, 82–115.

6. Bansal, S., Chowell, G., Simonsen, L., Vespignani, A., & Viboud, C. (2016). Big data for infectious disease surveillance and modeling. *The Journal of Infectious Diseases, 214*(Suppl 4), S375–S379.
7. Benjamin, R. (2019). Race after Technology: Abolitionist Tools for the New Jim Code. Polity.
8. Bhatt, S., Gething, P. W., Brady, O. J., Messina, J. P., Farlow, A. W., Moyes, C. L., Drake, J. M., Brownstein, J. S., Hoen, A. G., Sankoh, O., Myers, M. F., & Hay, S. I. (2013). The global distribution and burden of dengue. *Nature, 496*(7446), 504–507.
9. Birkhead, G. S., Klompas, M., & Shah, N. R. (2015). Uses of electronic health records for public health surveillance to advance public health. *Annual Review of Public Health, 36*, 345–359.
10. Boulos, M. N. K., Peng, G., & VoPham, T. (2019). An overview of GeoAI applications in health and healthcare. *International Journal of Health Geographics, 18*(1), 7.
11. Brailsford, S. C., Harper, P. R., Patel, B., & Pitt, M. (2009). An analysis of the academic literature on simulation and modelling in health care. *Journal of Simulation, 3*(3), 130–140.
12. Brauer, M., Freedman, G., Frostad, J., Van Donkelaar, A., Martin, R. V., Dentener, F., Dingenen, R. V., Estep, K., Amini, H., Apte, J. S., Balakrishnan, K., & Forouzanfar, M. H. (2016). Ambient air pollution exposure estimation for the Global Burden of Disease 2013. *Environmental Science & Technology, 50*(1), 79–88.
13. Brauer, M., Zhao, J. T., Bennitt, F. B., & Stanaway, J. D. (2021). Global access to high-resolution data on air pollution and public health. *Lancet Planetary Health, 5*(4), e164–e165.
14. Brownstein, J. S., Freifeld, C. C., & Madoff, L. C. (2008). Digital disease detection—Harnessing the web for public health surveillance. *New England Journal of Medicine, 360*(21), 2153–2157.
15. Brownstein, J. S., Holford, T. R., & Fish, D. (2005). Effect of climate change on Lyme disease risk in North America. *EcoHealth, 2*(1), 38–46.
16. Buczak, A. L., Koshute, P. T., Babin, S. M., Feighner, B. H., & Lewis, S. H. (2014). A data-driven epidemiological prediction method for dengue outbreaks using local and remote sensing data. *BMC Medical Informatics and Decision Making, 14*, 37.
17. Castelvecchi, D. (2016). Can we open the black box of AI? *Nature, 538*(7623), 20–23.
18. Chen, B., Khoshnevisan, B., Poggio, T., & Xiao, D. (2019). A Machine Learning Approach to Geophysical Inversion. arXiv:1905.02934
19. Chen, I. Y., Szolovits, P., & Ghassemi, M. (2018). Can AI help reduce disparities in general medical and mental health care? *AMA Journal of Ethics, 21*(2), 167–179.
20. Ching, T., Himmelstein, D. S., Beaulieu-Jones, B. K., Kalinin, A. A., Do, B. T., Way, G. P., Ferrero, E., Agapow, P. M., Zietz, M., Hoffman, M. M., Xie, W., & Greene, C. S. (2018). Opportunities and obstacles for deep learning in biology and medicine. *Journal of the Royal Society Interface, 15*(141), 20170387.
21. Chretien, J. P., Riley, S., & George, D. B. (2015). Mathematical modeling of the West Africa Ebola epidemic. *eLife, 4*, e09186.
22. Cromley, E. K., & McLafferty, S. L. (2011). *GIS and public health.* Guilford Press.
23. Culotta, A. (2010). Towards detecting influenza epidemics by analyzing Twitter messages. In *Proceedings of the First Workshop on Social Media Analytics* (pp. 115–122).
24. De Sherbinin, A. (2014). Remote sensing and human health: New sensors and new opportunities. *Emerging Infectious Diseases, 6*(3), 217.
25. Dombrowski, K., Khan, B., Wendel, T., McLean, K., Misshula, E., & Curtis, R. (2016). Estimating the size of the methamphetamine-using population in New York City using network sampling techniques. *Advances in Applied Sociology, 6*(11), 299.
26. Fang, H., Zhao, C., & Zhang, H. (2020). Artificial intelligence in health care: Bibliometric analysis. *Journal of Medical Internet Research, 22*(8), e19147.
27. Feldman, K., Chawla, N. V., & Meliker, J. R. (2015). Integrating artificial intelligence and GIS to improve disease surveillance. *Transactions in GIS, 19*(4), 510–529.
28. Fotheringham, A. S., & Rogerson, P. A. (2009). *The SAGE handbook of spatial analysis.* SAGE Publications Ltd.
29. Gliklich, R. E., Dreyer, N. A., & Leavy, M. B. (Eds.). (2014). Registries for Evaluating Patient Outcomes: A User's Guide. Agency for Healthcare Research and Quality (AHRQ).

30. Groves, R. M., Fowler, F. J., Couper, M. P., Lepkowski, J. M., Singer, E., & Tourangeau, R. (2009). *Survey methodology*. Wiley.
31. Haklay, M. (2013). Citizen science and volunteered geographic information: Overview and typology of participation. In *Crowdsourcing geographic knowledge* (pp. 105–122). Springer.
32. Hales, S., Weinstein, P., & Woodward, A. (2002). Dengue fever epidemics in the South Pacific: Driven by El Nino Southern Oscillation? *The Lancet, 360*(9348), 1849–1850.
33. Hay, S. I., George, D. B., Moyes, C. L., & Brownstein, J. S. (2013). Big data opportunities for global infectious disease surveillance. *PLoS Medicine, 10*(4), e1001413.
34. Hay, S. I., Snow, R. W., & Rogers, D. J. (2000). From predicting mosquito habitat to malaria seasons using remotely sensed data: Practice, problems and perspectives. *Parasitology Today, 16*(8), 306–313.
35. Holzinger, A., Biemann, C., Pattichis, C. S., & Kell, D. B. (2017). What do we need to build explainable AI systems for the medical domain? arXiv:1712.09923
36. Hosseini, P., Sokolow, S. H., Vandegrift, K. J., Kilpatrick, A. M., & Daszak, P. (2014). Predictive power of air travel and socio-economic data for early pandemic spread. *PLoS ONE, 9*(9), e107184.
37. Jerrett, M., Burnett, R. T., Ma, R., Pope, C. A., Krewski, D., Newbold, K. B., Thurston, G., Shi, Y., Finkelstein, N., Calle, E. E., & Thun, M. J. (2005). Spatial analysis of air pollution and mortality in Los Angeles. *Epidemiology, 16*(6), 727–736.
38. Jerrett, M., Turner, M. C., Beckerman, B. S., Pope, C. A., Van Donkelaar, A., Martin, R. V., Serre, M., Crouse, D., Gapstur, S. M., Krewski, D., Diver, W. R., & Burnett, R. T. (2013). Comparing the health effects of ambient particulate matter estimated using ground-based versus remote sensing exposure estimates. *Environmental Health Perspectives, 125*(4), 552–559.
39. Jia, P., Wang, F., Xierali, I. M., & Rosenfeld, R. (2016). Evaluating and re-demarcating the hospital service areas in Florida. *Applied Geography, 69*, 35–47.
40. Johnson, S. (2006). *The ghost map: The story of London's most terrifying epidemic—And How It changed science, cities, and the modern world*. Riverhead Books.
41. Kamel Boulos, M. N. (2004). Descriptive review of geographic mapping of severe acute respiratory syndrome (SARS) on the Internet. *International Journal of Health Geographics, 3*(1), 2.
42. Kamel Boulos, M. N., & Al-Shorbaji, N. M. (2014). On the Internet of Things, smart cities and the WHO Healthy Cities. *International Journal of Health Geographics, 13*(1), 10.
43. Kandwal, R., Garg, P. K., & Garg, R. D. (2009). Health GIS and HIV/AIDS studies: Perspective and retrospective. *Journal of Biomedical Informatics, 42*(4), 748–755.
44. Karnon, J., Stahl, J., Brennan, A., Caro, J. J., Mar, J., & Möller, J. (2011). Modeling using discrete event simulation: A report of the ISPOR-SMDM Modeling Good Research Practices Task Force-4. *Medical Decision Making, 32*(5), 701–711.
45. Khoury, M. J., Iademarco, M. F., & Riley, W. T. (2016). Precision public health for the era of precision medicine. *American Journal of Preventive Medicine, 50*(3), 398–401.
46. Klievink, B., Romijn, B. J., Cunningham, S., & De Bruijn, H. (2020). Big data in public: The tensions between personal privacy, government interests, and public benefits. In *Responsible data science* (pp. 49–66). Springer.
47. Kostkova, P., Brewer, H., & Smith, S. J. (2016). Who owns the Zika virus: A call for open science, open data, and open innovation. In *Digital health* (pp. 1–13). Springer.
48. LeCun, Y., Bengio, Y., & Hinton, G. (2015). Deep learning. *Nature, 521*(7553), 436–444.
49. Liang, F., & Gong, P. (2017). Integrating remote sensing, GIS, and dynamic modeling for landscape ecological health assessment. In *Integrating scale in remote sensing and GIS* (pp. 435–450). CRC Press.
50. McGough, S. F., Brownstein, J. S., Hawkins, J. B., & Santillana, M. (2017). Forecasting Zika incidence in the 2016 Latin America outbreak combining traditional disease surveillance with search, social media, and news report data. *PLoS Neglected Tropical Diseases, 11*(1), e0005295.
51. Milenković, A., Otasevic, P., & Vukmirovic, S. (2017). IoT wearable sensors and devices in elderly care: A literature review. *Sensors, 17*(6), 1191.

52. Milinovich, G. J., Williams, G. M., Clements, A. C., & Hu, W. (2014). Internet-based surveillance systems for monitoring emerging infectious diseases. *The Lancet infectious diseases, 14*(2), 160–168.
53. Miller, L., Lebina, L., & Lim, S. (2012). Remote sensing and GIS for mapping the risk of cholera: A case study in Mozambique. *Applied Geography, 33*, 9–16.
54. Mittelstadt, B., & Floridi, L. (2016). The ethics of big data: Current and foreseeable issues in biomedical contexts. *Science and Engineering Ethics, 22*(2), 303–341.
55. Mittelstadt, B., Allo, P., Taddeo, M., Wachter, S., & Floridi, L. (2016). The ethics of algorithms: Mapping the debate. *Big Data & Society, 3*(2), 2053951716679679.
56. Nuti, S. V., Wayda, B., Ranasinghe, I., Wang, S., Dreyer, R. P., Chen, S. I., & Murugiah, K. (2018). The use of Google Trends in health care research: A systematic review. *PLoS ONE, 13*(10), e0205759.
57. Obermeyer, Z., Powers, B., Vogeli, C., & Mullainathan, S. (2019). Dissecting racial bias in an algorithm used to manage the health of populations. *Science, 366*(6464), 447–453.
58. Openshaw, S. (1994). Developing GIS-relevant zone-based spatial analysis methods. In *Spatial analysis: Modelling in a GIS environment* (pp. 55–73). Wiley.
59. Patz, J. A., Olson, S. H., Uejio, C. K., & Gibbs, H. K. (2008). Disease emergence from global climate and land use change. *Medical Clinics, 92*(6), 1473–1491.
60. Pijanowski, B. C., Tayyebi, A., Del Castillo, C. R., & Sadowski, J. (2014). A fine-grained time series of land use and land cover change for the Indianapolis region. *Computers, Environment and Urban Systems, 46*, 31–42.
61. Raghupathi, W., & Raghupathi, V. (2014). Big data analytics in healthcare: Promise and potential. *Health Information Science and Systems, 2*(1), 3.
62. Rajkomar, A., Hardt, M., Howell, M. D., Corrado, G., & Chin, M. H. (2018). Ensuring fairness in machine learning to advance health equity. *Annals of Internal Medicine, 169*(12), 866–872.
63. Razavian, N., Blecker, S., Schmidt, A. M., Smith-McLallen, A., Nigam, S., & Sontag, D. (2015). Population-level prediction of type 2 diabetes from claims data and analysis of risk factors. *Big Data, 3*(4), 277–287.
64. Rocher, L., Hendrickx, J. M., & de Montjoye, Y. A. (2019). Estimating the success of re-identifications in incomplete datasets using generative models. *Nature Communications, 10*(1), 1–9.
65. Rogers, D. J., Olwoch, J. M., & Hay, S. I. (2014). Tsetse distribution in Africa: seeing the wood and the trees. In *Tsetse and trypanosomiasis information* (Vol. 25, pp. 55–69). Centre for Agriculture and Bioscience International.
66. Salathé, M., Bengtsson, L., & Merler, S. (2018). Digital epidemiology: What is it, and where is it going? *Life Sciences, Society and Policy, 14*(1), 1–5.
67. Salathé, M., Bengtsson, L., Bodnar, T. J., Brewer, D. D., Brownstein, J. S., Buckee, C., Campbell, E. M., Cattuto, C., Khandelwal, S., Mabry, P. L., Vespignani, A., & Freifeld, C. C. (2012). Digital epidemiology. *PLoS Computational Biology, 8*(7), e1002616.
68. Samuel, G., Derrick, G. E., & van Leeuwen, T. (2020). The ethics ecosystem: Personal ethics, network governance and regulating actors governing the use of social media research data. *Minerva, 58*(3), 317–343.
69. Stefanidis, A., Vraga, E., Lamprianidis, G., Radzikowski, J., Delamater, P. L., Jacobsen, K. H., Pfoser, D., & Croitoru, A. (2017). Zika in Twitter: Temporal variations of locations, actors, and concepts. *JMIR Public Health and Surveillance, 3*(2), e22.
70. Swan, M. (2012). Sensor mania! The internet of things, wearable computing, objective metrics, and the quantified self 2.0. *Journal of Sensor and Actuator Networks, 1*(3), 217–253.
71. Tatem, A. J., Hay, S. I., & Rogers, D. J. (2006). Global traffic and disease vector dispersal. *Proceedings of the National Academy of Sciences, 103*(16), 6242–6247.
72. Tatem, A. J., Huang, Z., Narib, C., Kumar, U., Kandula, D., Pindolia, D. K., Smith, D. L., Cohen, J. M., Graupe, B., Uusiku, P., Lourenço, C., & Gething, P. W. (2017). Integrating rapid risk mapping and mobile phone call record data for strategic malaria elimination planning. *Malaria Journal, 13*(1), 52.

73. Topol, E. J. (2019). High-performance medicine: The convergence of human and artificial intelligence. *Nature Medicine, 25*(1), 44–56.
74. Van Donkelaar, A., Martin, R. V., Brauer, M., Kahn, R., Levy, R., Verduzco, C., & Villeneuve, P. J. (2010). Global estimates of ambient fine particulate matter concentrations from satellite-based aerosol optical depth: Development and application. *Environmental Health Perspectives, 118*(6), 847–855.
75. Vayena, E., Blasimme, A., & Cohen, I. G. (2018). Machine learning in medicine: Addressing ethical challenges. *PLoS Medicine, 15*(11), e1002689.
76. Vayena, E., Salathé, M., Madoff, L. C., & Brownstein, J. S. (2015). Ethical challenges of big data in public health. *PLoS Computational Biology, 11*(2), e1003904.
77. Velasco-Hernandez, J. X., Peralta, A., & Hyman, J. M. (2021). Personalized outbreak predictions for COVID-19 with agent-based models. *PLoS ONE, 16*(3), e0248332.
78. Wang, L., & Alexander, C. A. (2015). Big data in health: opportunities and challenges. In *Big data-enabled nursing* (pp. 3–24). Springer.
79. Weiss, D. J., Nelson, A., Gibson, H. S., Temperley, W., Peedell, S., Lieber, A., Hancher, M., Poyart, E., Belchior, S., Fullman, N., Mappin, B., & Gething, P. W. (2018). A global map of travel time to cities to assess inequalities in accessibility in 2015. *Nature, 553*(7688), 333–336.
80. Weller, A. (2019). Transparency: Motivations and challenges. In *Explainable AI* (pp. 23–40). Springer.
81. Wesolowski, A., Eagle, N., Tatem, A. J., Smith, D. L., Noor, A. M., Snow, R. W., & Buckee, C. O. (2012). Quantifying the impact of human mobility on malaria. *Science, 338*(6104), 267–270.
82. Xu, B., Gong, P., Seto, E., Liang, F., Yang, C., & Zhuang, D. (2018). A review of remote sensing data and methods for urban health studies. *International Journal of Remote Sensing, 39*(11), 3581–3617.

Part II
AI Applications in Urban Planning

Chapter 9
Smart Cities and IoT Integration

9.1 Overview of Smart Cities and IoT Integration

The concept of smart cities has emerged as a response to the increasing urbanization and the need to create sustainable, efficient, and livable environments for the growing urban population [8]. Smart cities leverage advanced technologies, such as the Internet of Things (IoT), artificial intelligence (AI), big data, and geospatial analysis, to optimize urban planning, resource management, and service delivery (Table 9.1). IoT integration plays a crucial role in enabling smart cities by facilitating the collection, processing, and analysis of real-time data from various urban systems and infrastructures [101].

IoT refers to the interconnection of everyday objects and devices, enabling them to communicate and exchange data with each other and the internet [42]. IoT technologies, such as sensors, actuators, and communication networks, are integrated into urban infrastructures and services, collecting valuable data that can be analyzed to enhance decision-making, optimize resources, and improve the overall quality of life for city dwellers [80].

The integration of IoT and AI in smart cities provides numerous opportunities to address urban challenges, such as congestion, pollution, energy consumption, and public safety [15]. AI techniques, such as machine learning and deep learning, can be applied to analyze the vast amounts of data generated by IoT devices, enabling the development of predictive models, real-time monitoring, and intelligent decision support systems [74].

In this section, we will discuss the various aspects of smart cities and IoT integration, including data sources, AI techniques, applications, challenges, limitations, and future directions for AI applications in urban planning.

S. N. M. Moghaddam and H. Cao, *Artificial Intelligence-Driven Geographies*, City Development: Issues and Best Practices, https://doi.org/10.1007/978-981-97-5116-7_9

Table 9.1 Applications of AI in smart cities and IoT integration

Application area	AI integration
Urban infrastructure management	Utilizes AI for optimizing traffic signals, predicting energy consumption, automating waste sorting, and optimizing water distribution. Key implementations include AI-based traffic signal control systems, smart grid systems for energy, and IoT for waste management efficiency
Transportation and traffic management	AI enhances traffic flow predictions, supports Intelligent Transportation Systems (ITS), optimizes public transit, and enables autonomous vehicle navigation. Specific applications include machine learning models for traffic congestion prediction and optimization, and AI-powered public transit optimization for demand prediction
Public safety and security	Applies AI for surveillance, emergency response enhancements, crime prediction, and cyber-security. AI-powered video analytics for surveillance, machine learning for crime prediction, and AI algorithms for cyber threat detection are highlighted applications
Environmental monitoring and sustainability	AI contributes to air quality monitoring, energy management, waste management, water resource management, and climate change adaptation. Implementations include AI algorithms for air pollution level prediction, energy consumption optimization in buildings, and machine learning models for water demand forecasting
Citizen engagement and services	Leverages AI for improving communication between city authorities and citizens, enhancing public service delivery, and facilitating community-driven initiatives. AI-powered platforms for citizen feedback, sentiment analysis for gauging public sentiment, and AI in e-government services for streamlining processes are examples

9.2 Data Sources for Smart Cities and IoT Integration

Smart cities rely on various data sources to inform decision-making and optimize urban services. The proliferation of IoT devices, such as sensors, cameras, and smart meters, has facilitated the collection of a wide range of data types that can be harnessed to enhance urban planning and management [101]. These data sources include:

- IoT Sensors: Smart cities deploy a multitude of sensors to monitor environmental conditions, traffic flows, energy consumption, and other parameters that influence urban life. These sensors generate real-time data, which can be used to develop responsive and efficient urban systems [81].
- Mobile Devices: Smartphones and other mobile devices have become ubiquitous in modern society, serving as a valuable data source for smart cities. Location-based services, social media, and other mobile applications generate data that can be analyzed to understand human mobility patterns, social interactions, and preferences [15].

- Remote Sensing: Satellite and aerial imagery, along with other remote sensing data, provide critical information about land use, infrastructure, vegetation, and other urban features. These data sources can be used to monitor urban growth, assess environmental impacts, and inform urban planning and policy decisions [8].
- Open Data: Many cities and governments have embraced open data initiatives, making various datasets publicly available to citizens, researchers, and businesses. Open data sources can include census data, crime statistics, transportation data, and more, which can be used to identify trends, inform decision-making, and develop innovative urban solutions [74].

9.3 AI Techniques for Smart Cities and IoT Integration

The integration of AI techniques with IoT data can lead to significant advancements in smart city applications. Some key AI techniques employed in smart cities include:

- Machine Learning: Machine learning algorithms can be applied to analyze large-scale IoT datasets, enabling the identification of patterns, trends, and correlations. These algorithms can be used to develop predictive models, optimize resource allocation, and improve the overall efficiency of urban systems [15].
- Deep Learning: Deep learning, a subset of machine learning, leverages neural networks to process and analyze complex data, such as images and videos. Deep learning techniques can be employed in applications like traffic analysis, facial recognition, and environmental monitoring, which require the processing of high-dimensional data [74].
- Natural Language Processing (NLP): NLP techniques can be used to analyze and understand textual data, such as social media posts, news articles, and other unstructured data sources. NLP can help extract valuable insights from these data sources, informing urban planning and policy decisions [42].
- Reinforcement Learning: Reinforcement learning algorithms can be applied to optimize decision-making in smart city systems, such as traffic management, energy consumption, and public safety. These algorithms learn from data and improve over time, enabling the development of adaptive and responsive urban systems [81].

9.3.1 Machine Learning and IoT Data Analysis

The rapid growth of Internet of Things (IoT) technology has led to the generation of vast amounts of data, which can be utilized to create efficient, sustainable, and smart urban environments. Machine learning (ML) techniques play a crucial role in analyzing and making sense of the data collected from IoT devices, enabling smart cities to function more effectively. In this section, we will explore how machine learning and IoT data analysis can contribute to the development of smart cities.

Data Collection and Preprocessing

IoT devices, such as sensors, cameras, and smart meters, continuously collect data from various sources within a city. This data can include information on traffic, air quality, energy consumption, public transportation, and more. Before machine learning techniques can be applied to this data, it must be preprocessed to remove noise, fill in missing values, and normalize the data [74]. This step is essential for ensuring accurate and reliable results in subsequent analyses.

Feature Selection and Extraction

Due to the high dimensionality and complexity of IoT data, it is essential to identify the most relevant features for analysis. Feature selection and extraction techniques, such as principal component analysis (PCA) and mutual information, can be employed to reduce data dimensionality and identify the most significant variables for a particular task (Agrawal et al., 2021).

Machine Learning Models for IoT Data Analysis

Various machine learning models can be employed to analyze IoT data and provide actionable insights for smart city applications. Some of the most common models include:

- Decision Trees: Decision trees can be used for both classification and regression tasks. They are particularly useful for handling categorical data and can be easily visualized, making them an excellent choice for IoT data analysis [51].
- Random Forests: Random forests are an ensemble learning method that can be used for classification, regression, and feature selection. They have been widely used in IoT data analysis due to their robustness to noise and ability to handle large datasets [12].
- Support Vector Machines (SVM): SVM is a supervised learning algorithm that can be used for classification and regression tasks. It has been widely applied in IoT data analysis due to its ability to handle high-dimensional data and produce accurate results [5].
- Neural Networks: Neural networks are a type of deep learning model that can be used for various tasks, such as image recognition, natural language processing, and data classification. They have been increasingly applied to IoT data analysis due to their ability to model complex relationships and handle large amounts of data [104].

Applications of Machine Learning in Smart Cities

Machine learning and IoT data analysis have been utilized in various smart city applications, such as:

- Traffic Management: Machine learning models can be used to analyze traffic data collected from IoT sensors and cameras to optimize traffic flow, reduce congestion, and improve transportation efficiency [111].

- Energy Management: ML algorithms can analyze data from smart meters and other IoT devices to optimize energy consumption, predict equipment failures, and facilitate demand-response programs [11].
- Environmental Monitoring: Machine learning models can be employed to analyze data from IoT sensors to monitor air quality, predict pollution levels, and develop strategies to mitigate environmental issues [28].
- Public Safety: ML algorithms can be used to analyze data from IoT devices, such as security cameras and emergency response systems, to improve public safety and enhance disaster management [6].
- Smart Healthcare: IoT devices, such as wearables and medical sensors, generate large amounts of health data that can be analyzed using machine learning algorithms to monitor patient health, predict disease outbreaks, and optimize healthcare resource allocation [49].
- Waste Management: Machine learning can be employed to analyze data from IoT sensors installed in waste bins and waste collection vehicles to optimize waste collection routes and reduce operational costs [57].
- Urban Planning: ML algorithms can be utilized to analyze data from various sources, such as satellite imagery, social media, and IoT devices, to inform urban planning decisions and optimize land use [103].

Challenges and Future Directions

Despite the numerous advantages of employing machine learning and IoT data analysis in smart cities, several challenges need to be addressed:

- Data Privacy and Security: The massive amounts of data generated by IoT devices raise significant concerns regarding data privacy and security. It is essential to develop robust data protection mechanisms to ensure the privacy of citizens and protect sensitive information [76].
- Scalability: As the number of IoT devices and the volume of data generated continue to grow, it becomes increasingly important to develop scalable machine learning algorithms that can handle large-scale datasets and deliver real-time insights [4].
- Interoperability: IoT devices and systems often use different communication protocols and data formats, which can impede data integration and analysis. Standardization and interoperability among IoT systems are crucial for efficient data exchange and analysis [42].
- Model Interpretability: Many machine learning models, particularly deep learning models, can be difficult to interpret and explain. Developing explainable AI models is essential for fostering trust and understanding among stakeholders in smart city applications [9].

Future research in machine learning and IoT data analysis for smart cities should focus on addressing these challenges and exploring innovative approaches to enhance the efficiency, sustainability, and livability of urban environments.

9.3.2 Deep Learning for High-Dimensional Data Processing

Deep learning, a subfield of machine learning, has gained significant attention in recent years due to its ability to process and learn from high-dimensional and complex data. This powerful technique has shown great potential in various applications, including image and speech recognition, natural language processing, and autonomous vehicles. In the context of smart cities and IoT integration, deep learning can be employed to process the high-dimensional data generated by IoT devices and sensors, enabling efficient urban planning and management.

Deep Learning Techniques

Several deep learning techniques are commonly employed in smart city applications to process high-dimensional data:

Convolutional Neural Networks (CNNs): Convolutional Neural Networks (CNNs) are a popular deep learning technique for processing high-dimensional data, particularly image data. They consist of multiple layers, including convolutional, pooling, and fully connected layers, which help in automatically extracting relevant features from the data. In smart cities, CNNs can be employed for tasks such as traffic congestion prediction, urban growth modeling, and building recognition from satellite imagery [64, 107].

Recurrent Neural Networks (RNNs): Recurrent Neural Networks (RNNs) are designed to process sequential data, making them well-suited for time-series analysis. They possess feedback connections that allow them to maintain memory of past events and make predictions based on historical data. In the context of smart cities, RNNs can be applied for tasks like energy consumption forecasting, air quality prediction, and traffic flow prediction [56, 87].

Autoencoders: Autoencoders are unsupervised deep learning techniques that learn to encode and decode data in a way that minimizes the reconstruction error. They can be employed for dimensionality reduction and feature extraction, enabling the analysis of high-dimensional data in smart city applications. Autoencoders have been used for tasks such as anomaly detection in sensor data and building energy consumption prediction [94, 109].

Applications of Deep Learning in Smart Cities and IoT Integration

Deep learning techniques have been employed in various smart city applications to process and analyze high-dimensional data:

Traffic Management: Deep learning techniques can be utilized to predict traffic congestion and optimize traffic flow in smart cities. CNNs, for instance, have been used to predict traffic congestion levels using high-dimensional traffic data [60]. Similarly, RNNs, such as Long Short-Term Memory (LSTM) networks, can be employed to predict traffic flow patterns based on historical data [100].

Urban Growth Modeling: Deep learning techniques can be used to model urban growth and land-use change using high-dimensional satellite imagery. For example,

CNNs have been employed to classify land use and land cover from high-resolution satellite imagery, enabling the identification of urban growth patterns [23].

Environmental Monitoring: Deep learning techniques can be applied to monitor and predict air quality, noise levels, and other environmental factors in smart cities. RNNs, such as LSTMs, have been used to predict air quality based on historical data [56]. CNNs have also been utilized to classify and monitor urban noise levels using audio data collected by IoT sensors [89].

Public Safety: Deep learning techniques can be employed to enhance public safety in smart cities by detecting and preventing crime, monitoring crowd behavior, and analyzing emergency situations. For instance, CNNs can be used to detect and recognize suspicious activities in video surveillance data, allowing authorities to respond more quickly to potential threats [38]. RNNs, particularly LSTMs, can be applied to analyze social media data to identify and predict crime hotspots [93]. Deep learning techniques can also be employed to analyze crowd behavior during large events, such as detecting potential stampedes or monitoring crowd density for public safety purposes [37].

Energy Management: Deep learning techniques can contribute to energy management in smart cities by forecasting energy consumption, optimizing energy distribution, and detecting energy theft. Autoencoders, for example, have been employed for building energy consumption prediction, enabling more efficient energy management [109]. RNNs, particularly LSTMs, have also been utilized for short-term load forecasting in smart grids, which can help optimize energy distribution and reduce energy wastage [29].

Challenges and Limitations

Despite the promising applications of deep learning techniques in smart cities and IoT integration, several challenges and limitations need to be addressed:

The performance of deep learning models heavily depends on the quality and quantity of available data. In the context of smart cities, obtaining high-quality data can be challenging due to issues such as missing data, sensor malfunctions, and data privacy concerns.

Deep learning techniques often require significant computational resources, which can be a challenge when processing large volumes of high-dimensional data generated by IoT devices. Moreover, scaling deep learning models to handle growing data volumes and new IoT devices can be difficult.

Deep learning models are often considered "black boxes" due to their complex structure and lack of interpretability. In the context of smart cities, where decision-making may have significant consequences for citizens, understanding and explaining the underlying decision-making processes of deep learning models is crucial.

Deep learning techniques have shown great potential for processing and analyzing high-dimensional data in smart cities and IoT integration. Applications range from traffic management and urban growth modeling to environmental monitoring, public safety, and energy management. However, challenges such as data quality, scalability, and interpretability must be addressed to fully realize the potential of deep learning in smart city applications.

9.3.3 Natural Language Processing for Textual Data Analysis

Natural Language Processing (NLP) plays a vital role in the analysis of textual data generated within smart cities and IoT integrations. As urban planning becomes more reliant on data-driven decision-making, the importance of extracting valuable insights from a wide range of textual data sources, such as social media, online forums, news articles, and IoT sensor descriptions, cannot be understated. This section provides an overview of NLP techniques and their applications in smart cities and IoT integration.

NLP is a subfield of artificial intelligence that focuses on the interaction between computers and human languages. It deals with the processing, understanding, and generation of human languages by machines [50]. NLP techniques can be broadly categorized into two types: traditional NLP techniques and deep learning-based NLP techniques.

Traditional NLP techniques involve rule-based methods, statistical methods, and machine learning methods, such as decision trees, support vector machines, and Naïve Bayes classifiers. These techniques have been widely used for various text analysis tasks in smart cities, including sentiment analysis, topic modeling, and information extraction [2].

In recent years, deep learning-based NLP techniques have gained popularity due to their ability to capture complex patterns in textual data. These techniques include recurrent neural networks (RNNs), long short-term memory networks (LSTMs), convolutional neural networks (CNNs), and transformer models [88]. Deep learning-based NLP methods have been shown to outperform traditional techniques in various text analysis tasks, such as text classification, sentiment analysis, and named entity recognition [99].

In the context of smart cities and IoT integration, NLP techniques have been applied to a wide range of problems, including:

1. Sentiment analysis: Analyzing public opinions and sentiments toward urban issues and policies is essential for effective urban planning. NLP techniques, such as sentiment analysis, have been used to extract valuable insights from social media platforms like Twitter, Facebook, and online forums [21]. These insights can help policymakers understand citizens' concerns and preferences, leading to better decision-making.
2. Topic modeling: Understanding the key topics discussed in textual data related to urban issues can help urban planners identify emerging trends and patterns. Topic modeling techniques, such as Latent Dirichlet Allocation (LDA), have been used to extract topics from news articles, research papers, and social media data [16]. This information can help planners prioritize areas of concern and allocate resources more effectively.
3. Information extraction: Extracting structured information from unstructured textual data is crucial for smart city applications. NLP techniques, such as named entity recognition (NER), can be used to identify and extract key entities, such

as locations, organizations, and dates, from various data sources [69]. This structured information can then be used for spatial analysis, event detection, and decision support systems.

4. Natural language generation: Generating human-like text based on structured data can be useful for creating automated reports, summaries, and recommendations for urban planners. Techniques such as sequence-to-sequence models and transformer models have been used to generate coherent and relevant text based on input data [83, 88].

Despite the promising applications of NLP techniques in smart cities and IoT integration, several challenges remain. These challenges include:

1. Data quality and preprocessing: Textual data from various sources can be noisy, incomplete, or inconsistent. Preprocessing steps, such as data cleaning, normalization, and tokenization, are crucial for improving data quality and ensuring the effectiveness ofNLP techniques. However, these preprocessing steps can be time-consuming and may require domain-specific knowledge [44].
2. Language diversity and multilingualism: The textual data generated in smart cities often contains multiple languages, dialects, and informal expressions. Developing NLP techniques that can handle this language diversity and adapt to different linguistic contexts is challenging [77].
3. Domain-specific language: Urban planning and IoT integration involve various domain-specific terms, jargon, and abbreviations that may not be well-represented in general-purpose NLP models. Developing domain-specific NLP models or fine-tuning existing models for these specific contexts can improve performance but may require substantial labeled data and computational resources [32].
4. Interpretable and explainable AI: For urban planners and decision-makers to trust and adopt AI-driven recommendations, the outputs of NLP techniques need to be interpretable and explainable. However, many deep learning-based NLP models are often considered black boxes, making it difficult to understand their inner workings and decision-making processes [9].
5. Ethical and privacy concerns: Using NLP techniques to analyze social media data and other textual data sources raises ethical and privacy concerns, such as data privacy, data ownership, and potential biases in the data or models. Addressing these concerns requires careful consideration of data collection and analysis practices, as well as the development of privacy-preserving and fair AI techniques [47].

In conclusion, NLP techniques have shown great promise in addressing various challenges associated with smart cities and IoT integration. By leveraging the rich textual data generated in urban environments, NLP techniques can help urban planners and policymakers make better-informed decisions, improve resource allocation, and enhance the overall quality of life for citizens. Despite the challenges, continued research and development in NLP techniques, along with the integration of other AI technologies, are expected to drive significant advancements in the field of smart cities and IoT integration.

9.3.4 Reinforcement Learning for Decision Optimization

Reinforcement Learning (RL) is an area of machine learning that focuses on training agents to make decisions by interacting with an environment. RL algorithms are designed to learn an optimal policy, which is a mapping from states to actions, such that the agent can maximize its cumulative reward over time. In the context of smart cities and IoT integration, reinforcement learning can be applied to optimize various decision-making processes, ranging from traffic management and energy consumption to public transportation and emergency response systems.

Reinforcement Learning Basics

Reinforcement learning is founded on the principles of trial-and-error learning and delayed rewards. It involves an agent that takes actions within an environment to achieve a goal, and the environment responds to those actions by providing feedback in the form of rewards or penalties [84]. The objective of the agent is to learn a policy that maximizes the expected cumulative reward over time.

Reinforcement learning is typically modeled as a Markov Decision Process (MDP), which consists of a set of states (S), a set of actions (A), a reward function (R), and a state transition probability function (P). The agent's goal is to learn a policy (π) that maps states to actions in a way that maximizes the expected cumulative reward.

Key Reinforcement Learning Algorithms

Several reinforcement learning algorithms have been developed to tackle different aspects of decision optimization. Some of the most widely used algorithms are:

1. Q-learning: Q-learning is a model-free, value-based RL algorithm that learns an action-value function (Q) to estimate the expected future reward for taking a particular action in a given state [95]. The agent updates its Q-values based on the observed rewards and the maximum Q-value of the next state-action pair, following a greedy exploration strategy.
2. Deep Q-Networks (DQN): DQN extends the Q-learning algorithm by using deep neural networks as function approximators to estimate the Q-values [65]. This allows the algorithm to handle high-dimensional state spaces and complex decision-making problems, such as those found in smart cities and IoT applications.
3. Policy Gradient Methods: Policy gradient methods are model-free, policy-based RL algorithms that directly optimize the policy function rather than the value function [85]. The objective is to maximize the expected cumulative reward by updating the policy's parameters using gradient ascent.
4. Actor-Critic Methods: Actor-critic methods combine value-based and policy-based approaches to reinforcement learning [53]. The actor component is responsible for selecting actions based on the current policy, while the critic component estimates the value function and provides feedback to the actor for policy improvement.

Applications of Reinforcement Learning in Smart Cities and IoT Integration

Reinforcement learning techniques have been applied to various aspects of smart cities and IoT integration to optimize decision-making processes:

1. Traffic management: RL algorithms have been used to optimize traffic signal control, route planning, and congestion management [105]. Agents can learn to adapt to changing traffic conditions in real-time, leading to reduced travel times, improved road safety, and lower emissions.
2. Energy management: Reinforcement learning can help optimize energy consumption in smart buildings and city-wide energy systems [71]. Agents can learn to balance energy production, storage, and consumption, taking into account factors such as weather conditions, energy prices, and user behavior.
3. Public transportation: RL algorithms can be used to optimize public transportation schedules, routing, and passenger demand forecasting [30]. By learning from historical data and real-time updates, agents can improve service efficiency, reduce waiting times, and minimize operational costs.
4. Emergency response systems: Reinforcement learning can be employed to optimize emergency response strategies, such as ambulance routing, disaster management, and resource allocation [36]. By learning from past experiences and adapting to dynamic environments, RL algorithms can help reduce response times and save lives.
5. Waste management: Reinforcement learning can be applied to optimize waste collection and recycling processes in smart cities [33]. Agents can learn to balance resource usage, environmental impact, and cost-effectiveness while taking into account the dynamic nature of waste generation and disposal.
6. Urban planning: RL algorithms can be used to support urban planning decisions, such as land use allocation and infrastructure development [108]. By simulating different scenarios and learning from past experiences, agents can recommend optimal strategies that promote sustainable urban development.

Challenges and Future Directions

Despite the successful application of reinforcement learning in smart cities and IoT integration, several challenges remain:

1. Scalability: Many smart city applications involve large state and action spaces, making it difficult for RL algorithms to scale effectively [46]. Future research should focus on developing more efficient algorithms and leveraging parallel computing resources to tackle large-scale problems.
2. Data privacy and security: The use of IoT devices and big data in smart city applications raises concerns about data privacy and security [48]. Reinforcement learning algorithms should be designed to ensure the protection of sensitive information and to comply with data protection regulations.
3. Explainability and interpretability: Reinforcement learning algorithms, especially those based on deep learning, can be difficult to interpret and explain [10]. Developing more explainable and interpretable RL models will be crucial

for gaining the trust of stakeholders and promoting the widespread adoption of AI in smart city applications.

4. Real-world deployment: Transferring reinforcement learning algorithms from simulation environments to real-world applications remains a significant challenge [34]. Future work should focus on developing algorithms that are robust to model inaccuracies, partial observability, and other real-world constraints.

In conclusion, reinforcement learning has demonstrated significant potential for optimizing decision-making processes in smart cities and IoT integration. By addressing the challenges and building on the successes of existing applications, reinforcement learning can play a pivotal role in shaping the future of urban planning and development.

9.4 Applications of AI in Smart Cities and IoT Integration

9.4.1 Urban Infrastructure Management

Urban Infrastructure Management (UIM) is an essential aspect of smart cities and IoT integration. By leveraging AI techniques, city planners and administrators can improve the efficiency, sustainability, and resilience of urban infrastructure systems. This section will discuss the role of AI in urban infrastructure management, covering topics such as traffic management, energy and water resource management, waste management, and public transportation systems.

Traffic Management

Traffic congestion is a major challenge in urban areas, causing increased air pollution, wasted time, and increased fuel consumption. AI techniques can be applied to improve traffic management by optimizing traffic signal timings, predicting traffic congestion, and providing real-time routing suggestions for vehicles.

For instance, AI-based traffic signal control systems can process data from multiple sources, such as traffic cameras, sensors, and GPS data, to optimize traffic signal timings and reduce traffic congestion [102]. Machine learning algorithms can also be employed to predict traffic congestion based on historical data and real-time traffic conditions, allowing authorities to implement proactive measures to alleviate congestion [92]. Furthermore, AI-powered navigation systems can provide drivers with real-time routing suggestions to avoid congested areas and minimize travel times (Alemi, Circella, Handy, & Mokhtarian, 2018).

Energy and Water Resource Management

AI can be instrumental in optimizing energy and water resource management in smart cities. AI techniques can be applied to analyze consumption patterns, identify inefficiencies, and optimize energy and water use in various urban sectors.

For example, AI-based smart grid systems can help balance energy supply and demand by predicting energy consumption, managing distributed energy resources, and optimizing energy storage systems [71]. In addition, AI can be used to optimize water distribution systems by predicting water demand, detecting leaks, and implementing demand-responsive pricing [43].

AI can also contribute to the efficient management of renewable energy resources. Machine learning algorithms can be employed to predict solar and wind energy generation, allowing grid operators to integrate renewable energy resources more effectively [75].

Waste Management

Waste management is another critical aspect of urban infrastructure management. AI can help improve waste management systems by optimizing waste collection routes, predicting waste generation, and automating waste sorting and recycling processes.

For example, AI-based route optimization algorithms can minimize fuel consumption and transportation costs associated with waste collection by suggesting the most efficient routes for waste collection vehicles (Ghoseiri, Hamedi, & Ziarati, 2017). Additionally, machine learning models can be developed to predict waste generation based on socioeconomic factors, allowing waste management authorities to allocate resources more effectively [72]. Lastly, AI-powered robots and computer vision systems can be utilized to automate waste sorting and recycling processes, increasing the efficiency of waste management systems and reducing the need for manual labor (Matsumoto, Szabo, & Hashimoto, 2019).

Public Transportation Systems

AI can play a vital role in improving the efficiency, accessibility, and sustainability of public transportation systems in smart cities. AI techniques can be applied to optimize transportation networks, predict demand, and enhance the passenger experience.

AI-based algorithms can optimize public transportation networks by analyzing historical and real-time data, such as passenger flows, travel times, and network capacity, to identify bottlenecks and suggest improvements [25]. Demand prediction models can be employed to forecast public transportation usage, allowing authorities to allocate resources more effectively and reduce overcrowding [68]. Furthermore, AI-powered applications can provide personalized travel recommendations to passengers based on their preferences, real-time traffic conditions, and public transportation schedules [110].

AI can also be used to enhance the safety and security of public transportation systems. For instance, computer vision and machine learning algorithms can be employed to analyze video feeds from surveillance cameras, detect suspicious activities or unattended objects, and alert security personnel [112]. In addition, AI-based predictive maintenance systems can help minimize disruptions and ensure the reliability of public transportation infrastructure by analyzing sensor data, identifying potential failures, and scheduling maintenance activities [66].

Challenges and Limitations

Despite the promising applications of AI in urban infrastructure management, several challenges and limitations need to be considered. These include data privacy and security concerns, the need for reliable and accurate data, the integration of AI systems with existing infrastructure, and the potential for biased decision-making.

Data privacy and security are significant concerns when implementing AI systems in smart cities. The collection, storage, and processing of vast amounts of data from various sources, such as IoT devices, sensors, and surveillance cameras, can raise privacy concerns and increase the risk of data breaches [7]. Ensuring data privacy and security will be crucial to gain public trust and ensure the successful implementation of AI in urban infrastructure management.

The reliability and accuracy of data are essential for the effective application of AI techniques. Inadequate or incomplete data can lead to inaccurate predictions and suboptimal decision-making [26]. Therefore, it is crucial to invest in high-quality data collection and processing systems to ensure the effectiveness of AI applications in urban infrastructure management.

Integrating AI systems with existing urban infrastructure can be a complex and resource-intensive process. This may require significant investments in hardware, software, and human resources, as well as the development of new regulatory frameworks and standards [22]. Addressing these challenges will be critical to ensure the successful adoption of AI in urban infrastructure management.

Finally, biased decision-making is a potential concern when applying AI techniques. AI algorithms can inadvertently perpetuate existing biases in data, leading to unfair or discriminatory outcomes [73]. It is crucial to develop transparent and accountable AI systems and to incorporate diverse perspectives and stakeholders in the decision-making process to minimize the risk of biased outcomes.

9.4.2 Transportation and Traffic Management

Transportation and traffic management are critical aspects of smart cities, and the integration of AI and IoT technologies is revolutionizing the way cities manage their transportation systems. By utilizing AI and IoT, cities can optimize transportation networks, improve traffic flow, reduce congestion, and enhance public transit services. This section will explore how AI techniques can be applied in transportation and traffic management.

Traffic Flow Prediction and Optimization

One of the primary applications of AI in transportation and traffic management is predicting traffic flow and optimizing its patterns. Machine learning algorithms, particularly deep learning models, can analyze large volumes of historical and real-time traffic data to predict traffic conditions and identify potential bottlenecks [60]. This information can be used to optimize traffic signal timings, reroute traffic, and implement dynamic traffic management strategies that adapt to changing traffic conditions.

Intelligent Transportation Systems (ITS)

Intelligent Transportation Systems (ITS) use AI and IoT technologies to optimize the operation and management of transportation networks. ITS applications include traffic signal control, incident detection and response, traveler information systems, and public transit management [82]. AI techniques such as reinforcement learning can be applied to optimize traffic signal timings and adapt to changing traffic patterns [102].

Public Transit Optimization and Demand Prediction

AI can be used to optimize public transit schedules, routes, and capacities by analyzing historical and real-time data, such as passenger volumes, travel times, and service disruptions. Machine learning techniques can predict transit demand, enabling transit agencies to adjust service levels accordingly [68]. Furthermore, AI-powered trip planning tools can provide personalized, multimodal transportation options for individual users [110].

Autonomous Vehicles and Connected Cars

AI is playing a crucial role in the development of autonomous vehicles and connected cars, which promise to transform urban transportation. Advanced AI algorithms enable vehicles to perceive their environment, make decisions, and navigate complex urban environments [79]. Additionally, Vehicle-to-Everything (V2X) communication allows vehicles to share information with other vehicles, infrastructure, and pedestrians, enhancing safety and efficiency [59].

Mobility-as-a-Service (MaaS) Platforms

AI plays a critical role in the development of Mobility-as-a-Service (MaaS) platforms, which integrate various transportation services, such as public transit, ride-hailing, bike-sharing, and car-sharing, into a single, user-friendly platform. AI algorithms can analyze user preferences, real-time traffic data, and available transportation options to provide personalized, efficient, and sustainable travel solutions [40].

Traffic Incident Detection and Response

AI techniques can be employed to automatically detect traffic incidents, such as accidents or road closures, by analyzing data from sensors, cameras, and social media feeds. Once detected, AI-powered systems can respond to incidents by rerouting traffic, adjusting traffic signals, and dispatching emergency services as needed [91].

Environmental Impact and Sustainable Transportation

AI can help cities reduce the environmental impact of transportation by optimizing traffic flow, promoting public transit use, and encouraging the adoption of electric vehicles (EVs). Machine learning models can predict traffic-related emissions, enabling cities to implement targeted interventions to reduce air pollution [106].

AI algorithms can also be used to optimize EV charging infrastructure and manage energy demand [96].

In conclusion, AI and IoT technologies offer significant potential for improving transportation and traffic management in smart cities. Applications such as traffic flow optimization, public transit management, and autonomous vehicles can contribute to more efficient, sustainable, and user-friendly urban transportation systems. However, it is crucial to address the challenges and limitations of AI in transportation, such as data privacy, security, and ethical concerns, to ensure the successful implementation of these technologies.

9.4.3 Public Safety and Security in Smart Cities and IoT Integration

The integration of AI and IoT in smart cities has the potential to significantly improve public safety and security. By leveraging advanced technologies such as video analytics, natural language processing, and machine learning, smart cities can enhance emergency response, monitor public spaces, and detect potential threats more effectively. In this section, we discuss various AI applications in public safety and security, including surveillance, emergency response, crime prediction, and cyber-security.

AI-Powered Surveillance

One of the most significant applications of AI in public safety is the use of video analytics for surveillance purposes. AI algorithms can analyze video feeds from cameras installed throughout a city to detect and recognize faces, objects, and behaviors in real-time. These systems can automatically identify suspicious activities, such as unattended bags, vandalism, or trespassing, and alert law enforcement or security personnel [55].

Emergency Response and Disaster Management

AI can enhance emergency response by analyzing real-time data from various sources, such as social media, weather forecasts, and IoT sensors, to predict and respond to natural disasters or other emergencies more effectively. For example, machine learning algorithms can predict the occurrence and severity of floods, earthquakes, or wildfires, enabling authorities to take preventive measures and allocate resources more efficiently [70].

Crime Prediction and Prevention

AI can help law enforcement agencies predict and prevent crime by analyzing large volumes of data from various sources, such as police reports, demographic information, and social media. Machine learning models can identify patterns and correlations in this data to predict the likelihood of criminal activity in specific locations and times, enabling police to deploy resources more effectively and prevent crime proactively [67].

Cyber-Security

The increasing reliance on digital infrastructure in smart cities also raises concerns about cyber-security. AI can help detect and respond to cyber threats by analyzing large volumes of network traffic, identifying anomalies, and automatically deploying countermeasures [20]. Machine learning algorithms can also be used to detect and prevent social engineering attacks, such as phishing, by analyzing email content and sender behavior [61].

Ethical Considerations and Privacy Concerns

While AI applications in public safety and security offer significant benefits, they also raise ethical and privacy concerns. The widespread use of surveillance systems, for instance, can potentially infringe on individual privacy rights and lead to discrimination against certain communities. Moreover, AI systems may exhibit biases based on the data used to train them, which could result in unfair treatment of specific groups [62]. Therefore, it is crucial to strike a balance between public safety and individual privacy and ensure that AI applications are transparent, accountable, and respect human rights.

AI applications in public safety and security within the context of smart cities and IoT integration offer numerous opportunities for enhancing the well-being and safety of urban residents. By utilizing advanced technologies such as machine learning, deep learning, and natural language processing, cities can improve surveillance, emergency response, crime prediction, and cyber-security measures. However, it is essential to address the ethical and privacy concerns associated with these technologies to ensure a more secure and equitable urban environment.

The future of AI applications in public safety and security will likely involve further advancements in predictive analytics, real-time decision-making, and automated response systems. Additionally, the integration of AI with other emerging technologies, such as drones, autonomous vehicles, and wearable devices, could lead to novel applications in public safety and security. Finally, addressing the ethical and privacy concerns associated with AI applications will remain a critical challenge in the development and implementation of these technologies in smart cities.

As AI applications in public safety and security continue to advance, it is crucial to promote cross-sector collaboration between government agencies, technology companies, and research institutions. This collaboration can help ensure that the development and implementation of AI technologies are aligned with the broader goals of urban planning and sustainability, as well as the specific needs of individual communities.

To address the ethical and privacy concerns associated with AI applications in public safety and security, governments and relevant stakeholders should work together to develop clear guidelines and regulations. These guidelines should focus on issues such as data protection, algorithmic transparency, and fairness, while also promoting the responsible use of AI technologies for public safety purposes.

Public engagement and education are essential components of the successful implementation of AI applications in public safety and security. By involving citizens in the development and deployment of AI technologies, cities can help ensure that these applications are both effective and socially acceptable. Moreover, raising public awareness about the potential benefits and challenges of AI can help foster a more informed and balanced discussion about the role of these technologies in urban environments.

The successful implementation of AI applications in public safety and security also depends on addressing the underlying infrastructure and resource constraints that many cities face. This may involve investing in the development and maintenance of robust data networks, IoT devices, and computing resources, as well as ensuring that public safety agencies have the necessary skills and expertise to effectively leverage AI technologies.

In conclusion, AI applications in public safety and security have the potential to significantly improve the well-being and safety of urban residents. However, to fully realize the benefits of these technologies, it is crucial to address the various ethical, privacy, and resource challenges that they present. By fostering cross-sector collaboration, developing clear ethical guidelines, engaging the public, and addressing infrastructure constraints, cities can harness the power of AI to create safer and more equitable urban environments.

AI applications in public safety and security can also contribute to enhancing the resilience of urban environments and improving emergency response capabilities. By integrating AI technologies with existing emergency management systems, cities can improve their ability to predict, prepare for, and respond to natural disasters, terrorist attacks, and other emergencies. For example, AI-based early warning systems can help detect potential threats and trigger appropriate response measures, while AI-assisted emergency operations centers can optimize resource allocation and coordination during crisis situations.

The development and deployment of AI applications in public safety and security can also stimulate innovation and drive economic growth in urban areas. As cities invest in AI technologies and related infrastructure, they create new business opportunities and attract investment from technology companies, startups, and research institutions. This, in turn, can help catalyze the growth of a thriving innovation ecosystem centered around AI and other emerging technologies, contributing to job creation and economic development.

Given the global nature of many public safety and security challenges, it is crucial to strengthen international collaboration in the development and deployment of AI technologies. Cities and countries can benefit from sharing best practices, data, and technological resources, as well as collaborating on joint research and development initiatives. This can help accelerate the adoption of AI in public safety and security and promote a more unified and coordinated approach to addressing common challenges.

As AI applications in public safety and security continue to evolve, it is essential to establish robust monitoring and evaluation mechanisms to assess their impact and effectiveness. This can help cities identify areas where AI technologies are delivering the most significant benefits, as well as areas where improvements or adjustments may be needed. By continuously monitoring the performance of AI applications and learning from successes and failures, cities can ensure that their public safety and security strategies remain dynamic, effective, and aligned with the changing needs of their communities.

In summary, AI applications in public safety and security hold enormous potential for improving the lives of urban residents and addressing many of the complex challenges that cities face. By embracing a comprehensive and forward-thinking approach that addresses ethical, privacy, infrastructure, and collaboration issues, cities can fully harness the power of AI to create safer, more resilient, and more prosperous urban environments.

9.4.4 Environmental Monitoring and Sustainability

Environmental monitoring and sustainability are essential aspects of smart cities and IoT integration. The use of AI in this context can contribute to the efficient management of resources, reduction of environmental impact, and the achievement of sustainability goals. In this section, we will discuss the various ways AI can be applied to environmental monitoring and sustainability in smart cities and IoT integration.

Air Quality Monitoring and Management

Air pollution is a major concern in urban environments, and AI can play a significant role in monitoring and managing air quality. By analyzing data from IoT sensors and other sources, AI algorithms can identify patterns and trends in air pollution levels, allowing for better understanding of the factors contributing to poor air quality. This information can then be used to develop targeted interventions, such as traffic restrictions or industrial emissions controls, to improve air quality. Moreover, AI can help forecast air pollution levels, enabling city authorities to issue timely warnings and take preventive measures.

Energy Management and Optimization

AI can be used to optimize energy consumption in urban environments by analyzing data from various sources, such as smart meters, weather forecasts, and building management systems. AI algorithms can learn from historical data and real-time information to predict energy demand patterns and optimize energy generation, distribution, and storage. This can help reduce energy costs, minimize greenhouse gas emissions, and increase the use of renewable energy sources. Additionally, AI can help identify energy inefficiencies in buildings, enabling targeted retrofitting and the development of energy-efficient infrastructure.

Waste Management and Recycling

Effective waste management is essential for sustainable urban environments. AI can be employed to optimize waste collection routes and schedules, leading to reduced fuel consumption and lower emissions. By analyzing data on waste generation and recycling rates, AI algorithms can identify areas that require targeted interventions to improve recycling rates and reduce waste. Furthermore, AI can be employed in sorting facilities to automatically identify and separate recyclable materials, increasing the efficiency of recycling processes and reducing the amount of waste sent to landfills.

Water Resource Management

AI can play a crucial role in water resource management in smart cities. By analyzing data from IoT sensors, AI algorithms can monitor water consumption patterns, detect leaks and inefficiencies in water distribution systems, and optimize water usage. This can help conserve water resources and reduce the environmental impact of water consumption. AI can also be used to monitor water quality in real-time, enabling early detection of contamination and ensuring the safety of drinking water supplies.

Climate Change Adaptation and Mitigation

As cities face the challenges of climate change, AI can support adaptation and mitigation efforts. AI-based climate models can help predict the impacts of climate change on urban environments, enabling city planners to design infrastructure and policies that are resilient to future climatic conditions. AI can also support the development of low-carbon transportation systems, such as autonomous electric vehicles, and optimize urban green spaces to enhance carbon sequestration and biodiversity.

Biodiversity Monitoring and Conservation

AI can help monitor and conserve urban biodiversity by processing data from IoT sensors, remote sensing, and citizen science initiatives. AI algorithms can identify patterns and trends in species distribution and abundance, allowing city authorities to develop targeted conservation strategies and monitor the effectiveness of these measures. This can contribute to the creation of healthier and more biodiverse urban ecosystems.

Noise Pollution Management

Noise pollution is a significant concern in urban environments, affecting human health and well-being. AI can be employed to monitor noise levels in real-time, using data from IoT sensors and other sources. By analyzing this data, AI algorithms can identify patterns and trends in noise pollution, enabling targeted interventions to reduce noise levels, such as traffic restrictions, noise barriers, or urban planning initiatives.

Citizen Engagement and Behavior Change

AI can support citizen engagement and behavior change efforts by analyzing data on individual and community behaviors related to environmental sustainability. AI algorithms can identify trends, preferences, and areas of interest, enabling the development of targeted campaigns and initiatives that encourage sustainable behaviors. For example, AI can be used to develop personalized recommendations for energy-saving measures or to provide real-time feedback on individual carbon footprints. By leveraging social media and other digital platforms, AI can help engage citizens in sustainability initiatives and foster a sense of collective responsibility for environmental stewardship.

Urban Planning and Design

AI can play a crucial role in sustainable urban planning and design by analyzing data on land use, infrastructure, transportation, and environmental factors. AI algorithms can support planners in identifying the most sustainable and efficient configurations for urban development, taking into account factors such as energy consumption, emissions, green spaces, and accessibility to services. This can lead to the creation of more sustainable, resilient, and livable urban environments.

Disaster Risk Management and Response

AI can be employed to improve disaster risk management and response in smart cities. By analyzing data from IoT sensors, remote sensing, and other sources, AI algorithms can help identify areas at risk of natural disasters, such as floods, storms, or earthquakes. This information can be used to develop targeted risk reduction measures, such as early warning systems or infrastructure improvements. Moreover, AI can support disaster response efforts by analyzing real-time data on affected areas and providing decision-makers with timely and accurate information on the impacts of a disaster and the most effective response strategies.

Policy Development and Evaluation

AI can contribute to the development and evaluation of policies related to environmental sustainability in smart cities. By analyzing data on the impacts of existing policies and identifying the factors that contribute to their success or failure, AI algorithms can help inform the design of more effective policies and interventions. AI can also be employed to monitor and evaluate the effectiveness of policies in real-time, allowing for continuous improvement and adaptation to changing conditions.

AI has significant potential to support environmental monitoring and sustainability in smart cities and IoT integration. Through the application of advanced algorithms and data analysis techniques, AI can contribute to the optimization of resource management, the reduction of environmental impact, and the achievement of sustainability goals. By harnessing the power of AI, cities can become more resilient, sustainable, and livable environments for their citizens. However, it is crucial to address the challenges and limitations associated with AI deployment, such as data privacy, security, and ethical considerations, to ensure that the benefits of AI are realized in a responsible and equitable manner.

9.4.5 Citizen Engagement and Services

Citizen engagement and services are essential components of smart cities, as they contribute to a more inclusive, democratic, and participatory urban environment. With the integration of artificial intelligence (AI) and Internet of Things (IoT) technologies, smart cities can enhance the delivery of public services, improve communication between local governments and citizens, and facilitate community-driven decision-making processes. This section will discuss how AI and IoT technologies can be applied to citizen engagement and services in smart cities.

AI and IoT in Citizen Engagement

Citizen engagement refers to the involvement of citizens in various aspects of urban planning, decision-making, and service delivery. AI and IoT technologies can play a crucial role in facilitating meaningful engagement and empowering citizens to actively participate in shaping their cities. Some potential applications include:

(a) Online platforms and mobile applications: AI-powered platforms and mobile applications can be used to gather citizen feedback, suggestions, and ideas on various urban issues. Natural language processing (NLP) techniques can be used to analyze large volumes of textual data, identify patterns and trends, and prioritize citizen concerns. IoT devices, such as sensors and smart meters, can also provide real-time data that can be used to inform policy decisions and improve public services.
(b) Sentiment analysis and opinion mining: AI algorithms can be employed to analyze social media data, online forums, and other digital communication channels to gauge public sentiment on specific topics or issues. This can help local governments to identify areas of concern, monitor the effectiveness of their policies, and adapt their strategies based on citizen feedback.
(c) Participatory budgeting and decision-making: AI and IoT technologies can support participatory budgeting and decision-making processes by providing data-driven insights and facilitating communication between citizens and local governments. For example, AI algorithms can be used to analyze budget proposals, identify potential cost savings, and prioritize projects based on community needs and preferences.

AI and IoT in Public Services

AI and IoT technologies can significantly improve the delivery of public services in smart cities, leading to increased efficiency, cost savings, and enhanced citizen satisfaction. Some examples include:

(a) Smart waste management: IoT sensors can be installed in waste bins and collection vehicles to monitor waste levels and optimize collection routes. AI algorithms can then be used to analyze this data and predict future waste generation patterns, enabling more efficient resource allocation and reducing the environmental impact of waste management operations.
(b) Intelligent street lighting: IoT-enabled streetlights can automatically adjust their brightness based on ambient light conditions, pedestrian and vehicle traffic, and other factors. AI algorithms can be used to analyze data from these sensors and optimize energy consumption, leading to significant cost savings and reduced carbon emissions.
(c) E-government services: AI can be used to streamline and automate various e-government services, such as permit applications, tax filings, and benefits claims. For example, AI-powered chatbots can be deployed to answer citizen queries, provide personalized recommendations, and guide users through complex processes, improving the overall user experience and reducing the burden on government staff.
(d) Emergency response and disaster management: AI and IoT technologies can be leveraged to enhance emergency response and disaster management efforts in smart cities. For example, IoT sensors can be used to monitor critical infrastructure and detect potential hazards, such as gas leaks, structural failures, or flooding. AI algorithms can then analyze this data and predict the likelihood of future incidents, enabling local authorities to take preventive measures and allocate resources more effectively.

Challenges and Limitations

While AI and IoT technologies have the potential to revolutionize citizen engagement and public services in smart cities, several challenges and limitations must be addressed to ensure their successful implementation:

(a) Data privacy and security: The widespread deployment of IoT devices and the collection of large volumes of data raise significant privacy and security concerns. Smart cities must implement robust data protection measures and privacy policies to ensure that citizens' personal information is safeguarded and that sensitive data is not misused or accessed by unauthorized parties.
(b) Digital divide: The integration of AI and IoT technologies in citizen engagement and public services may exacerbate existing digital divides, particularly among marginalized or underserved communities. To ensure equitable access to these services, smart cities must invest in digital infrastructure and develop targeted initiatives to bridge the digital gap and promote digital literacy.
(c) Ethical considerations: The use of AI algorithms in decision-making processes raises several ethical concerns, such as algorithmic bias, fairness, and transparency. It is crucial for smart cities to adopt ethical guidelines and implement mechanisms to monitor and evaluate the performance of AI systems to ensure that they do not perpetuate existing inequalities or unfairly disadvantage certain groups.

(d) Interoperability and integration: The successful implementation of AI and IoT technologies in smart cities requires seamless integration between various systems, devices, and data sources. This can be challenging due to the heterogeneous nature of these technologies and the lack of standardized protocols. Smart cities must invest in developing interoperable systems and fostering collaboration between stakeholders to overcome these challenges.

In conclusion, AI and IoT technologies have the potential to significantly enhance citizen engagement and public services in smart cities, leading to more inclusive, efficient, and sustainable urban environments. However, to realize these benefits, it is essential to address the various challenges and limitations associated with these technologies, including data privacy, the digital divide, ethical considerations, and interoperability issues. By addressing these challenges and prioritizing the needs and preferences of citizens, smart cities can harness the power of AI and IoT to improve the quality of life for all residents and foster more equitable and resilient urban communities.

9.5 Challenges and Limitations of AI in Smart Cities and IoT Integration

As AI and IoT technologies are increasingly being adopted to support and improve urban management and services in smart cities, several challenges and limitations have emerged. Some of the main concerns include data privacy and security, data quality and integration, scalability, energy efficiency, and ethical considerations. This section explores these challenges and limitations in detail and suggests possible ways to address them (Table 9.2).

One of the most significant challenges in implementing AI and IoT solutions in smart cities is ensuring data privacy and security [101]. The massive amounts of data collected from various sources, such as sensors, cameras, and social media, can potentially reveal sensitive information about individuals and organizations. Unauthorized access to this data may lead to privacy breaches and identity theft, which can have severe consequences for the individuals and the city as a whole [76].

To address this issue, effective data encryption techniques and secure communication protocols should be employed to protect the data transmitted between IoT devices and the cloud [97]. Moreover, privacy-preserving data mining and machine learning algorithms can be developed to analyze the data without revealing sensitive information [35]. Data anonymization techniques can also be employed to de-identify personal information before data processing [86].

The accuracy and reliability of AI and IoT solutions in smart cities depend heavily on the quality of the data collected [3]. However, data from various sources and devices may contain inconsistencies, errors, and missing values, which can negatively impact the performance of AI models and the overall effectiveness of the system [15].

Table 9.2 Challenges and future directions in AI for smart cities

Aspect	Challenges	Future directions
Data privacy and security	Ensuring the protection of sensitive information amid the vast data collection from IoT devices is a major concern	Develop robust encryption techniques, privacy-preserving algorithms, and secure communication protocols. Focus on transparent and accountable AI systems that respect privacy and data protection regulations
Data quality and integration	The heterogeneity and inconsistency in data from diverse sources can impact AI model performance	Employ data preprocessing, data fusion techniques, and develop interoperability standards to enhance data quality and facilitate seamless data integration
Scalability	Handling the increasing volume of data and the growing number of IoT devices poses scalability challenges	Explore edge computing solutions and develop scalable machine learning algorithms that can efficiently manage large datasets and support a growing number of devices
Energy efficiency	The energy consumption of AI algorithms and IoT devices, especially in large-scale deployments, is a concern	Design energy-efficient IoT devices and communication protocols, develop energy-aware AI algorithms, and integrate renewable energy sources to mitigate energy consumption issues
Ethical considerations	AI-driven decision-making raises issues related to surveillance, potential bias, fairness, and transparency	Adopt ethical guidelines, ensure algorithmic fairness, and promote transparent AI systems to address ethical concerns. Involve diverse stakeholders in the development and deployment of AI solutions to ensure equity and fairness
Interoperability and integration	Integrating AI systems with existing urban infrastructure and ensuring interoperability among different IoT platforms and AI models is challenging	Invest in developing interoperable systems and fostering collaboration between stakeholders to overcome integration challenges. Standardize protocols and data formats to facilitate integration and collaboration between various smart city applications

(continued)

Table 9.2 (continued)

Aspect	Challenges	Future directions
Future technological advancements	Continuous innovation is needed to enhance AI capabilities and IoT integration for addressing complex urban challenges	Focus on advancing AI and IoT technologies, exploring new models and architectures, and innovating in areas such as edge AI, sustainable AI applications, and AI for climate resilience. Emphasize the development of next-generation IoT devices that are more efficient, secure, and capable of supporting advanced AI functionalities
Public engagement and policy development	Ensuring public acceptance and addressing the social implications of AI and IoT technologies in urban management require active engagement and thoughtful policy development	Enhance citizen participation through AI-driven platforms, develop policies that promote ethical use of AI and IoT, and ensure that smart city initiatives are inclusive and beneficial to all segments of the population. Work towards establishing a regulatory framework that balances innovation with ethical considerations and privacy concerns

Furthermore, integrating data from different sources with varying formats, spatial resolutions, and temporal frequencies can be challenging [45].

To address these challenges, data preprocessing techniques, such as data cleansing, imputation, and normalization, should be employed to improve data quality [27]. Additionally, data fusion techniques can be used to combine data from various sources and devices to create a more comprehensive and accurate representation of the city's environment [42]. Data interoperability standards and ontologies can also be developed to facilitate seamless data integration and exchange between different systems [13].

As smart cities continue to grow and evolve, the number of connected devices and the volume of data generated will also increase [101]. This growth presents challenges in terms of system scalability and the ability to handle large amounts of data efficiently. Traditional centralized cloud computing solutions may not be able to provide the necessary computational resources, response time, and energy efficiency required for real-time processing and analysis of massive datasets [18].

To overcome these scalability challenges, edge computing and fog computing paradigms can be employed to distribute computation and data storage closer to the source of data generation [17]. This approach reduces the amount of data transmitted to the cloud and allows for more efficient processing and real-time decision-making [98]. Moreover, scalable machine learning and data processing algorithms should be developed to handle the increasing volume of data and support the growing number of IoT devices [27].

Energy efficiency is another critical challenge in implementing AI and IoT solutions in smart cities. The large number of sensors, devices, and communication networks involved in smart city applications consume a significant amount of energy

[1]. Moreover, the energy consumption of AI algorithms for data processing and analysis can also be considerable, especially in large-scale deployments [58].

To address this issue, energy-efficient IoT devices and communication protocols should be designed to minimize energy consumption while maintaining adequate performance levels [4]. Furthermore, energy-aware AI algorithms and optimization techniques can be developed to reduce the computational complexity and energy requirements of data processing and decision-making tasks [58]. Renewable energy sources, such as solar and wind power, can also be integrated into smart city infrastructure to provide more sustainable and environmentally friendly energy solutions [39].

The deployment of AI and IoT technologies in smart cities raises several ethical concerns. These concerns include surveillance and the potential loss of privacy, potential bias in AI algorithms, and the impact of automation on employment and social inequality [63]. Ensuring that AI applications in smart cities are designed and implemented ethically is crucial to maintain public trust and ensure the long-term success of these initiatives [24].

To address these ethical concerns, transparent and accountable AI systems should be developed, which provide clear explanations of their decision-making processes and allow for human oversight [90]. Additionally, efforts should be made to ensure that AI algorithms are unbiased and do not discriminate against specific groups or individuals [31]. Finally, policymakers and urban planners should carefully consider the potential social implications of AI and IoT technologies and work to mitigate any negative consequences, such as job displacement or increased social inequality [19].

While AI and IoT technologies offer significant potential for improving urban management and services in smart cities, several challenges and limitations must be addressed. By focusing on data privacy and security, data quality and integration, scalability, energy efficiency, and ethical considerations, researchers, urban planners, and policymakers can work together to develop and deploy AI and IoT solutions that are both effective and responsible.

9.6 Future Directions in AI Applications for Smart Cities and IoT Integration

As urbanization continues to expand globally, cities face mounting challenges in delivering efficient services, managing resources, and ensuring a high quality of life for their inhabitants. Smart cities and IoT integration have emerged as promising solutions to address these challenges by harnessing the power of artificial intelligence (AI), big data, and advanced analytics. In this section, we discuss the future directions of AI applications in smart cities and IoT integration, highlighting potential research avenues, technological advancements, and policy considerations.

To accommodate the increasing complexity and scale of smart city and IoT deployments, future research should focus on developing scalable AI and IoT frameworks that can efficiently manage vast amounts of data and devices. These frameworks must be capable of dynamically adapting to changing environmental conditions and evolving requirements [42]. Moreover, interoperability among different IoT platforms and AI models will be crucial for seamless integration and collaboration between various smart city applications [101].

With the vast amount of data generated by IoT devices, processing and analyzing the information centrally in the cloud can lead to latency issues and increased network congestion. Therefore, incorporating edge computing and AI integration into smart cities can help to overcome these challenges by processing data closer to the source [78]. Future research should explore efficient AI algorithms and architectures tailored for edge computing, enabling real-time analytics and decision-making in smart city applications.

As climate change and environmental issues become increasingly pressing concerns, AI can play a crucial role in promoting sustainable urban development. Future research should focus on AI applications that optimize energy consumption, enhance waste management, and support urban agriculture [14]. Moreover, AI can be employed to identify and mitigate environmental risks and vulnerabilities, fostering climate change resilience in urban areas [15].

Future research should also investigate the development of AI-driven decision support systems for urban planning and policy-making. These systems can analyze massive amounts of data from various sources and provide actionable insights for city officials, helping them make informed decisions on infrastructure investments, resource allocation, and public policies [41]. Additionally, AI can be utilized for predictive analytics, enabling city planners to anticipate and address potential issues before they escalate.

Engaging citizens in the decision-making process is vital for the successful implementation of smart city initiatives. Future research should explore AI applications that facilitate citizen participation, such as sentiment analysis, social media mining, and virtual town halls [52]. Furthermore, AI can be employed to deliver personalized services tailored to individual needs and preferences, enhancing the overall quality of life for urban dwellers [54].

As AI becomes increasingly integrated into smart cities and IoT applications, it is essential to address the ethical, legal, and social implications associated with these technologies. Future research should examine issues such as data privacy, security, and ownership, as well as the potential for AI-driven decision-making to perpetuate existing inequalities and biases [90]. Additionally, researchers should explore frameworks for AI governance and accountability to ensure that these technologies are employed responsibly and transparently.

In conclusion, the future of AI applications in smart cities and IoT integration promises a wide range of opportunities for enhancing urban life and addressing pressing global challenges. By exploring scalable AI and IoT frameworks, edge computing, sustainable urban development, AI-driven decision support systems, citizen engagement, and the ethical, legal, and social implications of these technologies, researchers and practitioners can contribute to the realization of more efficient, resilient, and sustainable urban environments. As AI continues to advance and become more integrated into the fabric of our cities, it is essential to ensure that these technologies are employed responsibly and inclusively, harnessing their full potential for the betterment of all urban dwellers.

References

1. Aazam, M., Huh, E. N., & St-Hilaire, M. (2016). Cloudlet-based smart gateway for heterogeneous IoT environments. *Computer Communications, 89–90*, 99–106.
2. Aggarwal, C. C., & Zhai, C. (Eds.). (2012). *Mining text data*. Springer Science & Business Media.
3. Al Nuaimi, E., Al Neyadi, H., Mohamed, N., & Al-Jaroodi, J. (2015). Applications of big data to smart cities. *Journal of Internet Services and Applications, 6*(1), 1–15.
4. Al-Fuqaha, A., Guizani, M., Mohammadi, M., Aledhari, M., & Ayyash, M. (2015). Internet of things: A survey on enabling technologies, protocols, and applications. *IEEE Communications Surveys & Tutorials, 17*(4), 2347–2376.
5. Aljawarneh, S., Yassein, M. B., Al-Rahayfeh, A., & Bani Yassein, M. (2018). A survey on the role of IoT in agriculture for the implementation of smart farming. *IEEE Access, 6*, 32911–32927.
6. Alomari, E., Alkasassbeh, M., Almomani, A., & Alnsour, Y. (2020). Internet of things-based smart cities: Recent advances and challenges. *Computers, Materials & Continua, 62*(3), 899–924.
7. Alrawais, A., Alhothaily, A., Hu, C., & Cheng, X. (2017). Fog computing for the internet of things: Security and privacy issues. *IEEE Internet of Things Journal, 4*(5), 1189–1200.
8. Angelidou, M. (2017). The role of smart city characteristics in the plans of fifteen cities. *Journal of Urban Technology, 24*(4), 3–28.
9. Arrieta, A. B., Díaz-Rodríguez, N., Del Ser, J., Bennetot, A., Tabik, S., Barbado, A., García, S., Gil-López, S., Molina, D., Benjamins, R., Chatila, R., & Herrera, F. (2020). Explainable Artificial Intelligence (XAI): Concepts, taxonomies, opportunities and challenges toward responsible AI. *Information Fusion, 58*, 82–115.
10. Arulkumaran, K., Deisenroth, M. P., Brundage, M., & Bharath, A. A. (2017). A brief survey of deep reinforcement learning. arXiv:1708.05866
11. Aydin, M., Hariri, S., & Tekinerdogan, B. (2019). IoT-based smart cities: A survey. In *Environment and Electrical Engineering and 2019 IEEE Industrial and Commercial Power Systems Europe (EEEIC/I&CPS Europe)* (pp. 1–6). IEEE.
12. Bandyopadhyay, S., & Sengupta, M. (2019). Machine learning applications for smart city: A survey. *Journal of Ambient Intelligence and Humanized Computing, 10*(12), 4725–4743.
13. Barnaghi, P., Bermudez-Edo, M., & Tönjes, R. (2015). Challenges for quality of data in smart cities. *Journal of Data and Information Quality, 6*(2–3), 1–4.
14. Bibri, S. E. (2018). The IoT for smart sustainable cities of the future: A systematic review of the literature. *Energy Procedia, 156*, 759–765.
15. Bibri, S. E., & Krogstie, J. (2017). Smart sustainable cities of the future: An extensive interdisciplinary literature review. *Sustainable Cities and Society, 31*, 183–212.
16. Blei, D. M., Ng, A. Y., & Jordan, M. I. (2003). Latent dirichlet allocation. *Journal of Machine Learning Research, 3*(Jan), 993–1022.
17. Bonomi, F., Milito, R., Zhu, J., & Addepalli, S. (2012). Fog computing and its role in the internet of things. In *Proceedings of the First Edition of the MCC Workshop on Mobile Cloud Computing* (pp. 13–16).
18. Botta, A., De Donato, W., Persico, V., & Pescapé, A. (2016). Integration of cloud computing and internet of things: A survey. *Future Generation Computer Systems, 56*, 684–700.
19. Brynjolfsson, E., & McAfee, A. (2014). *The second machine age: Work, progress, and prosperity in a time of brilliant technologies*. WW Norton & Company.

20. Buczak, A. L., & Guven, E. (2016). A survey of data mining and machine learning methods for cyber security intrusion detection. *IEEE Communications Surveys & Tutorials, 18*(2), 1153–1176.
21. Cambria, E., Schuller, B., Xia, Y., & Havasi, C. (2013). New avenues in opinion mining and sentiment analysis. *IEEE Intelligent Systems, 28*(2), 15–21.
22. Caragliu, A., Del Bo, C., & Nijkamp, P. (2011). Smart cities in Europe. *Journal of Urban Technology, 18*(2), 65–82. https://doi.org/10.1080/10630732.2011.601117
23. Castelluccio, M., Poggi, G., Sansone, C., & Verdoliva, L. (2015). Land use classification in remote sensing images by convolutional neural networks. arXiv:1508.00092
24. Cath, C., Wachter, S., Mittelstadt, B., Taddeo, M., & Floridi, L. (2018). Artificial intelligence and the 'good society': The US, EU, and UK approach. *Science and Engineering Ethics, 24*(2), 505–528.
25. Cats, O., Yap, M., & van Oort, N. (2016). Exposing the role of exposure: Public transport network risk analysis. *Transportation Research Part A: Policy and Practice, 88*, 1–14.
26. Chen, M. (2018). Challenges and opportunities of big data for urban infrastructure management. *Journal of Infrastructure Systems, 24*(4), 04018032.
27. Chen, M., Mao, S., & Liu, Y. (2014). Big data: A survey. *Mobile Networks and Applications, 19*(2), 171–209.
28. Chen, M., Zhou, Y., & Wang, Y. (2018). A machine learning method for the large-scale evaluation of the qualities of the urban environment. *Computers, Environment and Urban Systems, 70*, 104–114.
29. Chen, T., He, Y., Benesty, M., Khotilovich, V., & Tang, Y. (2018). Xgboost: extreme gradient boosting. R package version 0.4-2, 1–4.
30. Chen, X., Wei, Y., Wang, P., & Liu, Y. (2018). A review of machine learning for IoT-data-driven urban computing. *IEEE Internet of Things Journal, 5*(2), 819–839.
31. Crawford, K., & Calo, R. (2016). There is a blind spot in AI research. *Nature, 538*(7625), 311–313.
32. Devlin, J., Chang, M. W., Lee, K., & Toutanova, K. (2018). BERT: Pre-training of deep bidirectional transformers for language understanding. arXiv:1810.04805
33. Di Felice, M., Paolanti, M., & Ravanelli, R. (2019). Waste management in smart cities: Reinforcement learning for waste collection. In *2019 IEEE International Conference on Smart Computing (SMARTCOMP)* (pp. 244–251). IEEE.
34. Dulac-Arnold, G., Mankowitz, D., & Hester, T. (2019). Challenges of real-world reinforcement learning. arXiv:1904.12901
35. Dwork, C. (2006). Differential privacy. In *International Colloquium on Automata, Languages, and Programming* (pp. 1–12).
36. Ferreira, G., Pinto, T., Rosseti, R. J., & Rodrigues, F. (2019). A reinforcement learning approach for dynamic dispatching of emergency medical services. In *2019 IEEE Symposium Series on Computational Intelligence (SSCI)* (pp. 2786–2793). IEEE.
37. Gallagher, J., Puri, A., & Ramamoorthy, S. (2019). Crowdsourcing for large-scale monitoring and intervention in smart cities. In *2019 IEEE International Conference on Pervasive Computing and Communications Workshops (PerCom Workshops)* (pp. 607–612). IEEE.
38. Gao, Z., Xie, H., & Zhang, X. (2016). A deep learning-based method for detecting suspicious behavior in video surveillance. In *2016 9th International Congress on Image and Signal Processing, BioMedical Engineering and Informatics (CISP-BMEI)* (pp. 1389–1394). IEEE.
39. García, N. M., Ortego, M. I., & Alaña, J. G. (2017). Renewable energy for smart cities. *Smart Cities, 1*(1), 29–46.
40. Giesecke, R., Surakka, T., & Sanchez, A. (2018). *Mobility as a service: Implications for urban and regional transport*. Routledge.
41. Goodchild, M. F. (2018). The future of digital earth. *Annals of GIS, 24*(2), 87–98.
42. Gubbi, J., Buyya, R., Marusic, S., & Palaniswami, M. (2013). Internet of Things (IoT): A vision, architectural elements, and future directions. *Future Generation Computer Systems, 29*(7), 1645–1660.

43. Gupta, R. K. (2019). IoT and AI-driven water management for smart cities. In P. Sharma (Ed.), *Sustainable smart cities in India* (pp. 215–226). Springer.
44. Haddi, E., Liu, X., & Shi, Y. (2013). The role of text preprocessing in sentiment analysis. *Procedia Computer Science, 17*, 26–32.
45. Hashem, I. A. T., Chang, V., Anuar, N. B., Adewole, K., Yaqoob, I., & Gani, A. (2016). The role of big data in smart city. *International Journal of Information Management, 36*(5), 748–758.
46. Hernández-Ramos, J. L., Oyedele, L. O., Ajayi, A. O., Akinade, O. O., Bilal, M., & Jaiyesimi, B. (2021). Reinforcement learning for intelligent building energy management: Challenges, trends, and research agenda. *Sustainable Cities and Society, 68*, 102767.
47. Hovy, D., & Spruit, S. L. (2016). The social impact of natural language processing. In *Proceedings of the 54th Annual Meeting of the Associationfor Computational Linguistics* (Vol. 2, Short Papers, pp. 591–598).
48. Hussain, R., Iqbal, W., Abbas, H., & Kim, D. (2021). Security and privacy for IoT and fog computing paradigm: A comprehensive analysis. *IEEE Communications Surveys & Tutorials, 23*(1), 439–473.
49. Islam, S. M. R., Kwak, D., Kabir, M. H., Hossain, M., & Kwak, K. S. (2015). The internet of things for health care: A comprehensive survey. *IEEE Access, 3*, 678–708.
50. Jurafsky, D., & Martin, J. H. (2018). *Speech and language processing*. Prentice Hall.
51. Karnatak, R., Rathore, H., & Sharma, D. (2017). Multivariate decision tree analysis for smart cities. *Journal of Urban Management, 6*(2), 61–69.
52. Kitchin, R. (2014). The real-time city? *Big Data and Smart Urbanism. GeoJournal, 79*(1), 1–14. https://doi.org/10.1007/s10708-013-9516-8
53. Konda, V. R., & Tsitsiklis, J. N. (2000). Actor-critic algorithms. In *Advances in neural information processing systems* (pp. 1008–1014).
54. Kummitha, R. K. R., & Crutzen, N. (2017). How do we understand smart cities? An evolutionary perspective. *Cities, 67*, 43–52.
55. Li, H., Li, Y., Porikli, F., & Zhang, X. (2018). Deep learning in video surveillance: Object detection, tracking, and recognition. arXiv:1803.04558
56. Li, X., Peng, L., Yao, X., Cui, S., Lei, M., & Deng, W. (2015). Long short-term memory neural network for air pollutant concentration predictions: Method development and evaluation. *Environmental Pollution, 206*, 93–101.
57. Li, X., Zhang, C., Li, W., & Ricard, R. (2019). Urban intelligent waste collection and transportation system based on internet of things. *IEEE Access, 7*, 179697–179709.
58. Liu, Y., Yuen, C., Huang, X., Hassan, M. M., Alam, M. M., & Alelaiwi, A. (2015). Energy-efficient dynamic offloading and resource scheduling in mobile cloud computing. *ACM Transactions on Embedded Computing Systems, 14*(3), 1–23.
59. Lu, N., Cheng, N., Zhang, N., Shen, X., & Mark, J. W. (2014). Connected vehicles: Solutions and challenges. *IEEE Internet of Things Journal, 1*(4), 289–299.
60. Lv, Y., Duan, Y., Kang, W., Li, Z., & Wang, F. Y. (2015). Traffic flow prediction with big data: A deep learning approach. *IEEE Transactions on Intelligent Transportation Systems, 16*(2), 865–873.
61. Mei, J., & Zhu, H. (2018). Robust and interpretable phishing email detection using deep learning. arXiv:1805.02987
62. Mittal, S., Vaithianathan, R., Zayyad, M., & Selvarajah, S. (2019). Using machine learning to assess the risk of and prevent reoffending in youth. *Nature Human Behaviour, 3*(10), 1094–1101.
63. Mittelstadt, B. D., Allo, P., Taddeo, M., Wachter, S., & Floridi, L. (2016). The ethics of algorithms: Mapping the debate. *Big Data & Society, 3*(2), 1–21.
64. Mnih, V., & Hinton, G. E. (2012). Learning to detect roads in high-resolution aerial images. In *European Conference on Computer Vision* (pp. 210–223). Springer.
65. Mnih, V., Kavukcuoglu, K., Silver, D., Rusu, A. A., Veness, J., Bellemare, M. G., Graves, A., Riedmiller, M., Fidjeland, A. K., Ostrovski, G., & Petersen, S. (2015). Human-level control through deep reinforcement learning. *Nature, 518*(7540), 529–533.

66. Mobley, R. K. (2019). *An introduction to predictive maintenance*. Wiley.
67. Mohler, G., Short, M. B., Malinowski, S., Johnson, M., Tita, G. E., Bertozzi, A. L., & Brantingham, P. J. (2015). Randomized controlled field trials of predictive policing. *Journal of the American Statistical Association, 110*(512), 1399–1411.
68. Moreira-Matias, L., Gama, J., Ferreira, M., Mendes-Moreira, J., & Damas, L. (2017). Predicting taxi–passenger demand using streaming data. *IEEE Transactions on Intelligent Transportation Systems, 18*(6), 1632–1641.
69. Nadeau, D., & Sekine, S. (2007). A survey of named entity recognition and classification. *Lingvisticae Investigationes, 30*(1), 3–26.
70. Nguyen, K. A., Armitage, G. H., & Islam, S. (2019). A survey of techniques for internet traffic classification using machine learning. *IEEE Communications Surveys & Tutorials, 21*(1), 349–392.
71. Nguyen, P., Abbass, H. A., & Petraki, E. (2021). A review on AI techniques for the smart grid: A reinforcement learning perspective. *Artificial Intelligence Review, 54*(5), 3475–3511.
72. Niska, H., Serkkola, A., & Hiltunen, V. (2018). Waste generation in Finnish households: Data from a waste sorting and weighing campaign. *Waste Management, 79*, 586–596.
73. O'Neil, C. (2016). *Weapons of math destruction: How big data increases inequality and threatens democracy*. Broadway Books.
74. Rathore, M. M., Ahmad, A., Paul, A., & Rho, S. (2016). Urban planning and building smart cities based on the Internet of Things using Big Data analytics. *Computer Networks, 101*, 63–80.
75. Raza, M. Q., & Khosravi, A. (2015). A review on artificial intelligence based load demand forecasting techniques for smart grid and buildings. *Renewable and Sustainable Energy Reviews, 50*, 1352–1372.
76. Roman, R., Zhou, J., & Lopez, J. (2013). On the features and challenges of security and privacy in distributed internet of things. *Computer Networks, 57*(10), 2266–2279.
77. Ruder, S., Vulic, I., & Søgaard, A. (2019). A survey of cross-lingual word embedding models. *Journal of Artificial Intelligence Research, 65*, 569–631.
78. Shi, W., Cao, J., Zhang, Q., Li, Y., & Xu, L. (2016). Edge computing: Vision and challenges. *IEEE Internet of Things Journal, 3*(5), 637–646.
79. Siciliano, B., & Khatib, O. (Eds.). (2016). *Springer handbook of robotics*. Springer.
80. Silva, B. N., Khan, M., & Han, K. (2018). Integration of Internet of Things (IoT) and cloud computing for health monitoring using wearable devices. In *2018 IEEE 4th International Conference on Collaboration and Internet Computing (CIC)* (pp. 407–415). IEEE.
81. Silva, B. N., Khan, M., & Han, K. (2018). Towards sustainable smart cities: A review of trends, architectures, components, and open challenges in smart cities. *Sustainable Cities and Society, 38*, 697–713.
82. Sussman, J. M. (2005). *Perspectives on intelligent transportation systems (ITS)*. Springer Science & Business Media.
83. Sutskever, I., Vinyals, O., & Le, Q. V. (2014). Sequence to sequence learning with neural networks. In *Advances in neural information processing systems* (pp. 3104–3112).
84. Sutton, R. S., & Barto, A. G. (2018). *Reinforcement learning: An introduction*. MIT press.
85. Sutton, R. S., McAllester, D. A., Singh, S. P., & Mansour, Y. (2000). Policy gradient methods for reinforcement learning with function approximation. In *Advances in neural information processing systems* (pp. 1057–1063).
86. Sweeney, L. (2002). K-Anonymity: A model for protecting privacy. *International Journal of Uncertainty, Fuzziness and Knowledge-Based Systems, 10*(05), 557–570.
87. Tieleman, T., & Hinton, G. (2012). Lecture 6.5-RMSprop: Divide the gradient by a running average of its recent magnitude. *COURSERA: Neural Networks for Machine Learning, 4*(2), 26–31.
88. Vaswani, A., Shazeer, N., Parmar, N., Uszkoreit, J., Jones, L., Gomez, A. N., Kaiser, L., & Polosukhin, I. (2017). Attention is all you need. In *Advances in neural information processing systems* (pp. 5998–6008).

89. Vesperini, F., Schuller, B., & Maier, A. (2016). Urban noise monitoring with noise sensors and social media data. In *Proceedings of the 3rd ACM International Workshop on Social Sensing* (pp. 37–42).
90. Wachter, S., Mittelstadt, B., & Floridi, L. (2017). Transparent, explainable, and accountable AI for robotics. *Science Robotics, 2*(6), 1–8.
91. Wang, H., Zheng, V. W., Yuan, N. J., Xie, X., & Zheng, K. (2018). Automatic traffic incident detection and response framework based on spatiotemporal data. *IEEE Transactions on Intelligent Transportation Systems, 19*(12), 3915–3925.
92. Wang, S., Zheng, L., & Ye, X. (2020). Real-time urban traffic congestion prediction based on deep learning. *Computers, Environment and Urban Systems, 81*, 101452.
93. Wang, X., & Ye, X. (2018). Spatial-temporal recurrent neural network for crime prediction. In *Proceedings of the 26th ACM SIGSPATIAL International Conference on Advances in Geographic Information Systems* (pp. 521–524).
94. Wang, Y., Yao, Q., Kwok, J. T., & Ni, L. M. (2016). Generalized autoencoder: A neural network framework for dimensionality reduction. In *2016 IEEE Conference on Computer Vision and Pattern Recognition (CVPR)* (pp. 4906–4914). IEEE.
95. Watkins, C. J., & Dayan, P. (1992). Q-learning. *Machine Learning, 8*(3–4), 279–292.
96. Xu, Y., Zhang, C., Li, W., Zhu, Y., Liu, J., & Wang, Q. (2018). A big data enabled method for energy-efficient scheduling of electric vehicle charging. *Applied Energy, 230*, 1645–1658.
97. Yang, L., Wu, M., Valli, C., & Rabadia, P. (2017). A survey of IoT protocol security vulnerabilities and mitigation techniques. *Journal of Cyber Security Technology, 1*(3–4), 161–183.
98. Yi, S., Qin, Z., & Li, Q. (2015). Security and privacy issues of fog computing: A survey. In *International Conference on Wireless Algorithms, Systems, and Applications* (pp. 685–695).
99. Young, T., Hazarika, D., Poria, S., & Cambria, E. (2018). Recent trends in deep learning-based natural language processing. *IEEE Computational Intelligence Magazine, 13*(3), 55–75.
100. Yu, R., Li, Y., Shahabi, C., Demiryurek, U., & Liu, Y. (2017). Deep learning: A generic approach for extreme condition traffic forecasting. In *Proceedings of the 2017 SIAM International Conference on Data Mining* (pp. 777–785). Society for Industrial and Applied Mathematics.
101. Zanella, A., Bui, N., Castellani, A., Vangelista, L., & Zorzi, M. (2014). Internet of Things for smart cities. *IEEE Internet of Things Journal, 1*(1), 22–32. https://doi.org/10.1109/JIOT.2014.2306328
102. Zhang, L., Ghaderi, J., & Lagoa, C. (2020). Reinforcement learning for traffic signal control with adaptive state space construction. *IEEE Transactions on Intelligent Transportation Systems, 21*(8), 3251–3262.
103. Zhang, L., Wen, J., & Zhang, X. (2017). Land use change detection based on multi-scale fusion and optimized deep learning model. *Remote Sensing, 9*(12), 1283.
104. Zhang, Q., Yang, L. T., Chen, Z., & Li, P. (2020). A survey on deep learning for big data. *Information Fusion, 42*, 146–157.
105. Zhang, S., Ghaderi, A., & Lagoa, C. (2020). Adaptive traffic signal control using reinforcement learning: A survey. *ACM Computing Surveys (CSUR), 53*(1), 1–37.
106. Zhang, S., Yang, Y., Grunder, O., & Wong, K. I. (2017). Air quality prediction using spatiotemporal correlation of traffic data: A deep learning approach. In *2017 IEEE International Conference on Industrial Technology (ICIT)* (pp. 1278–1283). IEEE.
107. Zhang, Y., Du, B., Zhang, L., & Wang, S. (2017). A CNN-RNN framework for remote sensing image scene classification. *IEEE Geoscience and Remote Sensing Letters, 14*(11), 2075–2079.
108. Zhang, Y., Liu, H., Yang, S., & Liu, Y. (2019). Urban land use and cover change prediction and uncertainty analysis using ensemble model combination: A case study of Mainland China. *Computers, Environment and Urban Systems, 75*, 21–35.
109. Zhao, H., Chen, Q., & Dai, W. (2017). Building energy consumption prediction: An extreme deep learning approach. In *Proceedings of the 2nd International Conference on Building Energy and Environment* (pp. 1–6).

110. Zheng, Y., Chen, X., & Yuan, N. J. (2018). Joint optimization of personalized trip recommendation and user satisfaction. *Transportation Research Part C: Emerging Technologies, 93*, 372–385.
111. Zheng, Y., Liu, F., & Hsieh, H. P. (2019). U-Air: When urban air quality inference meets big data. In *Proceedings of the 19th ACM SIGKDD International Conference on Knowledge Discovery and Data Mining* (pp. 1436–1444).
112. Zhu, X., Zhang, S., Zeng, D. D., & Yan, J. (2018). Public transport security under terrorist threat: A literature review. *Transportation Research Part A: Policy and Practice, 110*, 198–217.

Chapter 10
Transportation and Traffic Management

10.1 Overview of Transportation and Traffic Management

The field of transportation and traffic management focuses on the planning, design, operation, and maintenance of transportation systems to ensure safe, efficient, and accessible movement of people and goods (Table 10.1). As urban populations continue to grow, the demand for efficient transportation systems has increased, leading to an expansion in transportation infrastructure, modes of transportation, and the complexity of managing transportation systems [107]. With the advent of artificial intelligence (AI) and related technologies, there are now new opportunities to enhance transportation and traffic management, making it more efficient, safer, and sustainable.

The traditional transportation planning process is often based on manual data collection, statistical analysis, and simulation models. Although these methods have been successful in addressing transportation problems to some extent, they are often limited by their reliance on historical data, their inability to adapt to real-time changes in transportation systems, and the complexity of the models used [109]. AI has the potential to overcome these limitations by using advanced computational algorithms, real-time data, and predictive analytics to optimize the performance of transportation systems.

AI technologies have been widely used in transportation and traffic management to address various challenges, such as traffic congestion, transportation network optimization, public transportation planning, and autonomous vehicles [100]. AI techniques, such as machine learning, deep learning, reinforcement learning, and natural language processing, have been applied to transportation data to improve traffic prediction, route optimization, and traffic signal control. These applications have resulted in significant improvements in transportation efficiency, safety, and environmental sustainability.

In addition to AI techniques, transportation and traffic management have also benefited from advances in data collection, storage, and processing technologies. The

S. N. M. Moghaddam and H. Cao, *Artificial Intelligence-Driven Geographies*, City Development: Issues and Best Practices, https://doi.org/10.1007/978-981-97-5116-7_10

Table 10.1 Application Areas of AI in Transportation and Traffic Management

Application Area	Description	Applications	Challenges
Traffic Flow Prediction and Optimization	AI technologies predict future traffic conditions and optimize traffic flow, contributing to safer, more efficient, and sustainable transportation systems.	–Machine learning algorithms for predicting traffic volumes and travel times –Reinforcement learning for optimizing traffic signal timings	–Ensuring the quality and comprehensiveness of traffic data –Addressing the computational complexity of AI algorithms
Intelligent Transportation Systems (ITS)	AI integrates with ITS to enhance transportation safety and efficiency through advanced monitoring and management of traffic flow.	–Incident detection and management using AI algorithms –Real-time traffic condition monitoring and control	–Achieving interoperability among diverse ITS components –Managing privacy and security concerns related to data collection and analysis
Public Transportation Planning and Management	AI aids in efficiently organizing and operating public transportation systems by optimizing routes, schedules, and resource allocation.	–AI-driven route planning and schedule optimization –Demand forecasting for resource allocation	–Designing for inclusivity and accessibility –Balancing efficiency with user satisfaction
Autonomous Vehicles and Connected Mobility	AI powers autonomous vehicles and connected mobility solutions, allowing for real-time navigation and decision-making without human input.	–Development and operation of autonomous taxis and ride-hailing services –Vehicle-to-everything (V2X) communication systems	–Ensuring safety and reliability of autonomous systems –Developing necessary infrastructure and connectivity
Multimodal Transportation Integration	AI enhances the integration of various transportation modes to provide seamless and efficient transportation solutions across public and private transport options.	–AI for dynamic and real-time multimodal route planning –Integration of shared mobility services with traditional public transit systems	–Navigating complex legal and regulatory landscapes –Encouraging public adoption and trust in integrated transportation solutions

growing availability of real-time, high-resolution data from various sources, such as traffic sensors, connected vehicles, and social media, has enabled the development of data-driven transportation models and applications [105]. Furthermore, advances in geospatial analysis and remote sensing technologies have facilitated the integration of transportation data with other urban data sources, enabling a more comprehensive understanding of transportation systems and their impacts on urban environments [109].

However, the implementation of AI in transportation and traffic management is not without challenges. Some of the key issues include data privacy and security, algorithmic bias, and the need for human expertise in interpreting and validating AI-generated insights [102]. Moreover, the integration of AI technologies into existing transportation infrastructure and systems requires significant investments in hardware, software, and workforce training.

Despite these challenges, the potential benefits of AI in transportation and traffic management are substantial. As AI technologies continue to advance and become more accessible, they will play a critical role in addressing the growing transportation demands of urban populations and contribute to the development of more sustainable and livable cities.

10.2 Data Sources for Transportation and Traffic Management

The effectiveness of artificial intelligence (AI) applications in transportation and traffic management depends largely on the availability and quality of data. In recent years, the emergence of various data sources has facilitated the development of AI-driven transportation models and applications, enabling a more comprehensive understanding of transportation systems and their impacts on urban environments. This section discusses the main data sources for transportation and traffic management and their respective roles in AI-driven applications.

Traditional Traffic Data Collection Methods

Traditional data collection methods, such as manual traffic counts, automated traffic recorders, and travel surveys, have long been used to gather information on vehicle volumes, speeds, and occupancies, as well as travel behavior and preferences [102]. Although these methods can provide accurate and detailed data, they are typically resource-intensive, time-consuming, and limited in their spatial and temporal coverage.

Roadside Sensors and Traffic Cameras

Roadside sensors, such as inductive loop detectors, infrared sensors, and radar detectors, are commonly used to monitor traffic conditions and collect real-time data on vehicle volumes, speeds, and occupancies [110]. Similarly, traffic cameras capture visual data that can be used to monitor traffic flow, detect incidents, and analyze road user behavior. Data from roadside sensors and traffic cameras can be used to develop AI-driven traffic models and applications, such as traffic prediction, congestion management, and incident detection.

Connected and Autonomous Vehicles

Connected vehicles, which use wireless communication technologies to exchange information with other vehicles, infrastructure, and devices, generate a wealth of data on vehicle location, speed, and direction, as well as environmental and traffic conditions [109]. Autonomous vehicles, equipped with various sensors and cameras, also collect data on their surroundings, including other road users and obstacles. Data from connected and autonomous vehicles can be used in AI-driven transportation applications, such as dynamic routing, traffic signal optimization, and cooperative driving strategies.

Mobile Devices and Social Media

With the widespread use of mobile devices, such as smartphones and tablets, location-based data generated by GPS-enabled applications have become an important source of information on travel behavior and traffic conditions [53]. Additionally, social media platforms, such as Twitter and Facebook, provide user-generated data on traffic incidents, road closures, and travel experiences. AI techniques, such as natural language processing and machine learning, can be applied to these data sources to extract valuable insights for transportation planning and management.

Public Transportation Data

Public transportation agencies collect a variety of data on transit operations, ridership, and infrastructure, such as bus and train schedules, real-time vehicle locations, and passenger counts [109]. These data sources can be used in AI-driven applications to optimize public transportation planning and management and management, such as demand forecasting, service reliability, and dynamic routing.

Geospatial Data and Remote Sensing

Geospatial data, including geographic information systems (GIS) and remote sensing data, play a crucial role in transportation and traffic management. GIS data provide spatial information on transportation networks, land use patterns, and population distribution, which can be used to analyze the accessibility, connectivity, and efficiency of transportation systems [53]. Remote sensing data, obtained from satellites or aerial platforms, offer high-resolution images and measurements of land use, vegetation, and urban heat islands, which can be used to assess the environmental impacts of transportation systems and inform sustainable transportation planning [102].

Open Data and Big Data Platforms

Open data initiatives by governments and organizations have led to the increased availability of transportation-related data, such as traffic incidents, road closures, and public transportation schedules [102]. These data sources can be combined with other urban data sets to develop a comprehensive understanding of transportation systems and their impacts on urban environments. Big data platforms, such as Hadoop and Spark, facilitate the storage, processing, and analysis of large-scale transportation data, enabling the development of data-driven transportation models and applications [107].

In summary, a variety of data sources are available for transportation and traffic management, each with its unique characteristics and potential applications in AI-driven models and solutions. By leveraging these diverse data sources, AI algorithms can provide valuable insights and support decision-making processes in transportation planning, operation, and management. However, it is essential to consider data quality, privacy, and security issues when using these data sources to ensure the reliability and integrity of AI-driven transportation applications.

10.3 AI Techniques for Transportation and Traffic Management

10.3.1 Machine Learning for Traffic Prediction and Optimization

Machine learning (ML), a subset of artificial intelligence (AI), has emerged as a powerful tool for transportation and traffic management. ML algorithms learn from data, identify patterns, and make predictions or decisions without being explicitly programmed. In the context of traffic prediction and optimization, ML techniques can be applied to analyze historical and real-time transportation data, predict future traffic conditions, and optimize transportation systems, contributing to improved efficiency, safety, and sustainability. This section discusses the applications of ML in traffic prediction and optimization, covering various techniques, challenges, and future directions.

Traffic Prediction

Traffic prediction is a critical component of transportation planning and management, as accurate forecasts of traffic conditions facilitate informed decision-making and resource allocation. ML techniques, such as regression, neural networks, and time series analysis, have been widely applied to predict traffic volumes, travel times, and congestion levels [100]. These techniques can handle complex, non-linear relationships between input variables, such as weather conditions, time of day, and road network characteristics, and output variables, such as traffic flow and travel times.

One popular approach for traffic prediction is deep learning, a subfield of ML that focuses on artificial neural networks with multiple layers. Deep learning models, such as Convolutional Neural Networks (CNNs) and Recurrent Neural Networks (RNNs), can automatically learn hierarchical representations of input data, making them well-suited for traffic prediction tasks that involve spatiotemporal data . These models have shown promising results in predicting short-term and long-term traffic conditions, as well as detecting anomalies, such as accidents and congestion.

Traffic Optimization

Traffic optimization aims to improve the efficiency and performance of transportation systems by allocating resources and adjusting system parameters based on predicted or observed traffic conditions. ML techniques can be applied to various traffic optimization tasks, such as route planning, traffic signal control, and transportation network design.

Route planning involves determining the optimal path between an origin and destination, considering factors such as travel time, distance, and traffic conditions. ML algorithms, such as reinforcement learning and genetic algorithms, have been used to optimize route planning by learning from historical and real-time traffic data, adapting to changing traffic conditions, and considering multiple objectives, such as minimizing travel time and fuel consumption [107].

Traffic signal control involves adjusting the timing and coordination of traffic signals to facilitate smooth traffic flow and minimize delays. ML techniques, such as reinforcement learning and fuzzy logic, have been applied to optimize traffic signal control by considering dynamic traffic conditions, vehicle priorities, and pedestrian demands [53]. These approaches have shown significant improvements in reducing vehicle delays, queue lengths, and emissionscompared to traditional fixed-time or actuated signal control strategies.

Transportation network design involves the strategic planning and optimization of transportation infrastructure, such as roads, public transit networks, and bike lanes. ML techniques, such as genetic algorithms and swarm intelligence, have been applied to optimize transportation network design by considering multiple objectives, such as minimizing travel times, costs, and environmental impacts, while maximizing accessibility and connectivity [53]. These approaches can help transportation planners and decision-makers evaluate alternative network configurations and identify optimal solutions that balance competing objectives.

Challenges and Future Directions

Despite the significant progress and promising results of ML applications in traffic prediction and optimization, several challenges remain to be addressed. Some of these challenges include:

1. Data quality and availability: The accuracy and reliability of ML-based traffic prediction and optimization models depend on the quality and availability of input data. Incomplete, noisy, or biased data can lead to poor model performance and misleading predictions. Future research should focus on developing robust ML techniques that can handle data uncertainties and improve data quality through preprocessing, imputation, and fusion techniques.
2. Model interpretability and transparency: Many ML models, particularly deep learning models, are often considered "black boxes" due to their complex structures and lack of interpretability. This can limit their adoption in transportation planning and management, where stakeholders require transparent and understandable decision-making processes. Developing more interpretable and

explainable ML models is crucial for facilitating trust and adoption in the transportation domain.

3. Scalability and real-time applicability: Traffic prediction and optimization tasks often require real-time processing and analysis of large-scale, high-dimensional data. This poses computational challenges for many ML techniques, particularly those with high complexity and memory requirements. Developing scalable and efficient ML algorithms that can handle large-scale transportation data and provide real-time predictions and optimizations is an important area of future research.
4. Integration of multi-modal and multi-objective approaches: Transportation systems are complex and interconnected, involving multiple modes of transportation and competing objectives. Developing ML techniques that can integrate and optimize multi-modal transportation systems, considering various objectives and constraints, is essential for achieving sustainable and efficient urban mobility.

In conclusion, ML techniques have shown great potential for improving traffic prediction and optimization in transportation and traffic management. By addressing the challenges and exploring new directions in ML research, transportation planners and decision-makers can leverage these advanced techniques to develop more efficient, sustainable, and intelligent transportation systems.

10.3.2 Deep Learning for Traffic Analysis and Control

Deep learning, a subset of machine learning and artificial intelligence, has gained significant attention in recent years due to its ability to learn complex hierarchical representations from raw data. This powerful learning technique has shown remarkable success in a wide range of applications, including image recognition, natural language processing, and speech recognition. In the context of transportation and traffic management, deep learning techniques have been applied to traffic analysis and control tasks, such as traffic prediction, traffic signal control, vehicle detection, and incident detection. This section provides an overview of deep learning techniques used for traffic analysis and control, discussing their strengths, challenges, and future directions.

Traffic Prediction

Accurate traffic prediction is crucial for efficient transportation planning and management. Deep learning techniques, such as Convolutional Neural Networks (CNNs), Recurrent Neural Networks (RNNs), and Long Short-Term Memory (LSTM) networks, have shown great potential in predicting short-term and long-term traffic conditions [109]. These techniques can effectively capture the spatiotemporal dependencies in traffic data, which traditional machine learning techniques often struggle with.

For example, CNNs have been applied to capture spatial patterns in traffic data, such as the distribution of vehicle speeds or densities across different road segments [108]. RNNs, especially LSTM networks, have been used to model the temporal dependencies in traffic data, such as the influence of historical traffic conditions on future traffic patterns [109]. By combining these deep learning techniques, researchers can develop more accurate and robust traffic prediction models that account for both spatial and temporal dependencies in traffic data.

Traffic Signal Control

Deep learning techniques have also been applied to optimize traffic signal control strategies, with the goal of minimizing delays, reducing queue lengths, and improving overall traffic flow. Reinforcement learning, a type of deep learning that learns to make decisions by interacting with an environment, has been used to develop adaptive traffic signal control strategies that consider dynamic traffic conditions, vehicle priorities, and pedestrian demands [110].

For example, Deep Q-Networks (DQNs), a popular deep reinforcement learning technique, have been applied to learn optimal traffic signal timings that minimize vehicle delays and improve traffic flow . These techniques have shown significant improvements in traffic signal control performance compared to traditional fixed-time or actuated signal control strategies.

Vehicle Detection and Classification

Deep learning techniques, particularly CNNs, have demonstrated exceptional performance in vehicle detection and classification tasks, which are essential for traffic analysis and control applications such as traffic monitoring, incident detection, and automated toll collection. CNNs can automatically learn features from raw images, such as vehicle shapes, colors, and textures, and use these features to accurately detect and classify vehicles in real-time [109].

For example, the Single Shot MultiBox Detector (SSD) and You Only Look Once (YOLO) are two popular deep learning architectures that have been applied to vehicle detection and classification tasks, achieving high detection accuracy and processing speed [109, 103]. These techniques can be integrated with other traffic analysis and control applications, such as traffic signal control and incident detection, to provide more accurate and responsive traffic management solutions.

Incident Detection

Incident detection, such as identifying traffic accidents or congestion, is a critical task for traffic management systems to ensure timely response and minimize the impact on traffic flow. Deep learning techniques, especially CNNs and LSTMs, have been applied to detect traffic incidents from various data sources, such as traffic flow data, social media data, and traffic camera images [107].

For instance, researchers have developed hybrid CNN-LSTM models to analyze spatiotemporal traffic data and detect anomalies, such as sudden changes in vehicle speeds or densities, which may indicate incidents or congestion [53]. These models can provide real-time incident detection and facilitate proactive traffic management strategies, such as congestion mitigation and incident response.

Challenges and Future Directions

Despite the promising results of deep learning techniques in traffic analysis and control, several challenges remain to be addressed:

1. Data quality and availability: The performance of deep learning models relies heavily on the quality and availability of input data. Incomplete, noisy, or biased data can lead to poor model performance and misleading predictions. Future research should focus on developing robust deep learning techniques that can handle data uncertainties and improve data quality through preprocessing, imputation, and fusion techniques.
2. Model interpretability and transparency: Many deep learning models, particularly CNNs and LSTMs, are often considered "black boxes" due to their complex structures and lack of interpretability. This can limit their adoption in transportation planning and management, where stakeholders require transparent and understandable decision-making processes. Developing more interpretable and explainable deep learning models is crucial for facilitating trust and adoption in the transportation domain.
3. Scalability and real-time applicability: Traffic analysis and control tasks often require real-time processing and analysis of large-scale, high-dimensional data. This poses computational challenges for many deep learning techniques, particularly those with high complexity and memory requirements. Developing scalable and efficient deep learning algorithms that can handle large-scale transportation data and provide real-time predictions and optimizations is an important area of future research.
4. Integration of multi-modal and multi-objective approaches: Transportation systems are complex and interconnected, involving multiple modes of transportation and competing objectives. Developing deep learning techniques that can integrate and optimize multi-modal transportation systems, considering various objectives and constraints, is essential for achieving sustainable and efficient urban mobility.

In conclusion, deep learning techniques have shown great potential for improving traffic analysis and control in transportation and traffic management. By addressing the challenges and exploring new directions in deep learning research, transportation planners and decision-makers can leverage these advanced techniques to develop more efficient, sustainable, and intelligent transportation systems.

10.3.3 Reinforcement Learning for Traffic Signal Optimization

Reinforcement Learning (RL) is a subset of machine learning that focuses on decision-making processes by learning from interactions with an environment. In the context of transportation and traffic management, RL has been widely used for traffic signal optimization, aiming to minimize delays, reduce queue lengths, and enhance overall traffic flow. This section will provide an overview of reinforcement learning techniques for traffic signal optimization, discussing their advantages, challenges, and future directions.

Introduction to Reinforcement Learning

In reinforcement learning, an agent learns to make decisions by interacting with an environment to achieve a specific goal. The agent chooses actions based on its current state, and the environment responds to the action, providing a reward signal that indicates how well the action aligns with the goal [98]. The agent's objective is to learn a policy that maximizes the cumulative reward over time. In traffic signal optimization, the agent represents the traffic signal controller, the environment is the traffic network, and the goal is to optimize traffic flow.

Q-Learning for Traffic Signal Optimization

Q-Learning is a popular model-free reinforcement learning algorithm that has been applied to traffic signal optimization [1]. In Q-Learning, the agent learns an action-value function (Q-function), which estimates the expected cumulative reward of taking an action in a given state. The Q-function is updated iteratively using the Temporal Difference (TD) learning rule, which incorporates the difference between successive state-action value estimates. This update rule allows the agent to learn the optimal policy without requiring a model of the environment.

Q-Learning has been used to optimize fixed-time, actuated, and adaptive traffic signal control strategies [53]. Researchers have demonstrated that Q-Learning-based traffic signal controllers can effectively reduce vehicle delays, queue lengths, and travel times compared to traditional traffic signal control strategies.

Deep Reinforcement Learning for Traffic Signal Optimization

Deep Reinforcement Learning (DRL) combines deep learning techniques, such as Convolutional Neural Networks (CNNs) or Recurrent Neural Networks (RNNs), with reinforcement learning algorithms to learn more complex policies and handle high-dimensional state spaces. In traffic signal optimization, DRL techniques have been used to tackle the challenges of scalability and adaptability, as they can handle large-scale traffic networks and adapt to dynamic traffic conditions .

Deep Q-Networks (DQNs) are a popular DRL technique that employs deep learning to approximate the Q-function in Q-Learning [108]. In traffic signal optimization, DQNs have been used to learn traffic signal policies that adapt to real-time traffic conditions and improve traffic efficiency [53]. Studies have shown that DQN-based traffic signal controllers can outperform traditional traffic signal control strategies in terms of reducing vehicle delays, travel times, and emissions.

Another DRL technique, Proximal Policy Optimization (PPO), has been applied to traffic signal control, demonstrating improved performance and stability compared to other RL algorithms [107]. PPO is a policy gradient method that optimizes the policy directly by estimating the gradient of the objective function and updating the policy parameters iteratively. PPO-based traffic signal controllers have been shown to reduce delays and travel times while maintaining stable performance across different traffic conditions.

Challenges and Future Directions

Despite the promising results of reinforcement learning techniques in traffic signal optimization, several challenges remain to be addressed:

1. Exploration versus exploitation: In reinforcement learning, the agent needs to balance exploration (trying new actions to discover better policies) with exploitation (using the best-known policy to maximize rewards). This trade-off is particularly challenging in traffic signal optimization due to the dynamic nature of traffic conditions and the need for real-time decision-making. Developing more efficient exploration strategies and adaptive learning rates is essential for improving the performance of RL-based traffic signal controllers.
2. Multi-agent coordination: In large-scale traffic networks, multiple traffic signals need to coordinate their actions to optimize global traffic flow. This introduces additional complexities, such as the need for multi-agent reinforcement learning and the possibility of non-stationary environments. Developing techniques for coordinated multi-agent RL and communication strategies among traffic signal controllers is crucial for achieving large-scale traffic optimization.
3. Robustness and adaptability: Traffic conditions can change rapidly due to various factors, such as incidents, weather, and special events. Reinforcement learning algorithms need to be robust and adaptable to handle these uncertainties and provide reliable traffic signal control. Future research should focus on developing RL techniques that can learn from and adapt to diverse traffic scenarios, as well as incorporating uncertainty estimation into the learning process.
4. Transfer learning and generalization: In practice, traffic signal controllers need to be deployed in different traffic networks with varying characteristics. Developing reinforcement learning techniques that can generalize across different traffic scenarios and leverage transfer learning to speed up the learning process is an important area of future research.

In conclusion, reinforcement learning techniques have shown great potential for traffic signal optimization, providing adaptive and efficient traffic signal control strategies that can improve traffic flow and reduce congestion. By addressing the challenges and exploring new directions in reinforcement learning research, transportation planners and decision-makers can leverage these advanced techniques to develop more efficient, sustainable, and intelligent transportation systems.

10.3.4 Natural Language Processing for Public Transportation Feedback Analysis

Natural Language Processing (NLP) is a subfield of artificial intelligence that focuses on the interaction between computers and human languages. NLP techniques have been increasingly applied to transportation and traffic management, particularly in the analysis of public transportation feedback. This section will provide an overview of NLP applications in public transportation feedback analysis, discussing the benefits, challenges, and future directions in this area.

NLP techniques encompass various tasks, such as text classification, sentiment analysis, topic modeling, and information extraction. These tasks aim to derive meaningful information from unstructured text data, enabling computers to understand and process human language . In the context of public transportation feedback analysis, NLP techniques can be used to automatically process and analyze large volumes of user-generated feedback data, such as social media posts, online reviews, and survey responses.

Sentiment Analysis for Public Transportation Feedback

Sentiment analysis, also known as opinion mining, is an NLP technique used to determine the sentiment or emotion expressed in a text. In public transportation feedback analysis, sentiment analysis can be used to gauge customer satisfaction, identify areas of concern, and monitor the performance of transportation services over time [109].

For example, researchers have used sentiment analysis to analyze Twitter data related to public transportation, identifying the most common complaints and positive aspects of the services [2]. This information can be used by transportation planners and decision-makers to prioritize improvements and enhance customer satisfaction.

Topic Modeling for Public Transportation Feedback

Topic modeling is an NLP technique used to discover hidden semantic structures in large collections of documents. In public transportation feedback analysis, topic modeling can be employed to uncover common themes and trends within user-generated feedback data [109].

For instance, researchers have applied Latent Dirichlet Allocation (LDA), a popular topic modeling technique, to analyze user reviews of public transportation services [102]. The identified topics can provide valuable insights into the most pressing issues and areas of interest for customers, helping transportation authorities prioritize their efforts and resources.

Information Extraction for Public Transportation Feedback

Information extraction is an NLP task that involves automatically identifying and extracting specific information from unstructured text. In public transportation feedback analysis, information extraction can be used to identify and extract relevant entities, such as transportation modes, routes, stations, and times, as well as events, such as delays, accidents, or service disruptions [53]. This extracted information can be leveraged by transportation authorities to gain a more comprehensive understanding of customer experiences and to address specific issues in a timely manner.

For example, researchers have applied Named Entity Recognition (NER), an information extraction technique, to identify and categorize entities mentioned in public transportation feedback [102]. By extracting and analyzing this information, transportation authorities can better understand the context of customer feedback and identify patterns or trends related to specific aspects of the transportation system.

Challenges and Future Directions

While NLP techniques have shown promising results in public transportation feedback analysis, several challenges remain to be addressed:

1. Ambiguity and noise in user-generated content: User-generated feedback data, such as social media posts and online reviews, often contain informal language, abbreviations, misspellings, and slang. This can pose challenges for NLP techniques in accurately processing and analyzing the text. Developing more robust and adaptable NLP algorithms that can handle the inherent noise and ambiguity in user-generated content is essential for improving the quality and reliability of feedback analysis.
2. Multilingual and multicultural aspects: Public transportation feedback data may be written in various languages and originate from diverse cultural contexts. Developing NLP techniques that can handle multilingual and multicultural data is necessary for providing a comprehensive analysis of public transportation feedback, especially in regions with diverse populations.
3. Integration with other data sources: Public transportation feedback data can be enriched by integrating it with other sources of transportation data, such as GPS traces, ticketing data, or sensor data. This integration can provide more complete and accurate insights into customer experiences and transportation system performance. Developing techniques for data integration and fusion in the context of public transportation feedback analysis is an important area of future research.
4. Real-time analysis and decision-making: To effectively address customer concerns and improve transportation system performance, it is crucial to analyze public transportation feedback data in real-time and incorporate the derived insights into decision-making processes. Developing NLP techniques that can process and analyze feedback data in real-time, as well as designing decision support systems that leverage these insights, is a key direction for future research.

In conclusion, NLP techniques have demonstrated great potential in public transportation feedback analysis, providing valuable insights into customer experiences, satisfaction, and concerns. By addressing the challenges and exploring new directions in NLP research, transportation planners and decision-makers can leverage these advanced techniques to develop more responsive, efficient, and customer-centric transportation systems.

10.4 Applications of AI in Transportation and Traffic Management

10.4.1 Traffic Flow Prediction and Optimization

The application of Artificial Intelligence (AI) techniques in transportation and traffic management has the potential to revolutionize how traffic flow is predicted and optimized. This section will discuss various AI techniques used for traffic flow prediction and optimization, outlining their benefits, challenges, and real-world applications.

Traffic flow prediction is the process of estimating future traffic conditions based on historical and real-time data. Accurate traffic flow prediction is essential for effective traffic management, as it allows transportation planners and traffic managers to make informed decisions about traffic control, route guidance, and infrastructure planning. Traffic optimization, on the other hand, focuses on improving traffic flow and reducing congestion by adjusting traffic signal timings, coordinating traffic signals, and providing real-time traffic information to drivers [100].

AI Techniques for Traffic Flow Prediction

Various AI techniques have been employed for traffic flow prediction, including machine learning, deep learning, and hybrid approaches. Machine learning techniques, such as linear regression, decision trees, and support vector machines, have been widely used to model and predict traffic flow based on historical data [109].

Deep learning techniques, particularly recurrent neural networks (RNNs) and long short-term memory (LSTM) networks, have gained popularity in traffic flow prediction due to their ability to model complex temporal dependencies and handle large-scale traffic data [109].

Hybrid approaches, which combine traditional time series models with machine learning or deep learning techniques, have also been proposed to improve the accuracy and robustness of traffic flow prediction [106].

AI Techniques for Traffic Optimization

Traffic optimization techniques aim to improve traffic flow and reduce congestion by adjusting traffic signal timings, coordinating traffic signals, and providing real-time traffic information to drivers. AI techniques, such as reinforcement learning, have been employed for traffic signal optimization, demonstrating significant improvements in reducing delays, travel times, and emissions [109].

Real-time traffic information systems, which provide drivers with up-to-date traffic conditions and route guidance, can also benefit from AI techniques. For example, AI algorithms can be used to analyzelarge-scale traffic data and predict congestion patterns, allowing the system to provide more accurate and timely route recommendations [104].

Real-world Applications of AI in Traffic Flow Prediction and Optimization

Several real-world applications of AI techniques for traffic flow prediction and optimization have been implemented, demonstrating their potential for improving transportation systems. Some examples include:

a. Traffic management centers: AI algorithms have been deployed in traffic management centers to predict traffic flow and optimize traffic signal timings, resulting in reduced congestion and improved traffic flow [109].
b. Intelligent transportation systems: AI techniques have been integrated into intelligent transportation systems to provide real-time traffic information and route guidance to drivers, improving the overall efficiency of transportation networks [109].
c. Public transportation: AI algorithms have been employed to predict and optimize public transportation schedules and routes, enhancing the reliability and efficiency of public transportation systems [106].

Challenges and Future Directions

While AI techniques have shown promise in traffic flow prediction and optimization, several challenges remain to be addressed:

1. Data quality and availability: The accuracy and effectiveness of AI techniques depend on the quality and availability of traffic data. Ensuring the collection of accurate, reliable, and timely traffic data is essential for improving the performance of AI algorithms in traffic flow prediction and optimization.
2. Scalability: As traffic networks continue to grow and become more complex, AI algorithms need to be scalable and adaptable to handle the increasing volume and complexity of traffic data.
3. Integration with existing traffic management systems: AI techniques must be effectively integrated with existing traffic management systems to fully realize their potential in improving traffic flow and reducing congestion.
4. Evaluation and validation: Developing rigorous evaluation and validation methods for AI techniques in traffic flow prediction and optimization is necessary to ensure their reliability and effectiveness in real-world applications.
5. Ethical and privacy concerns: The use of AI techniques in traffic management raises ethical and privacy concerns related to data collection, sharing, and storage. Addressing these concerns while maintaining the benefits of AI applications in traffic management is an important area of future research.

In conclusion, AI techniques have demonstrated significant potential for improving traffic flow prediction and optimization in transportation and traffic management. By addressing the challenges and exploring new directions in AI research, transportation planners and traffic managers can leverage these advanced techniques to develop more efficient, responsive, and sustainable transportation systems.

10.4.2 Intelligent Transportation Systems

Intelligent Transportation Systems (ITS) are advanced applications that aim to provide innovative services related to different modes of transportation and traffic management. This section will discuss various AI techniques used in ITS, outlining their benefits, challenges, and real-world applications.

ITS utilize advanced information and communication technologies to improve transportation safety, mobility, and efficiency while reducing negative environmental impacts. These systems encompass a wide range of applications, including traffic management, public transportation, road user charging, logistics, traveler information, and vehicle control systems [102].

AI Techniques in Intelligent Transportation Systems

AI techniques have been increasingly employed in ITS to optimize various aspects of transportation, including traffic flow prediction, incident detection, route planning, and vehicle control. Some of the key AI techniques used in ITS are:

a. Machine learning: Machine learning techniques, such as decision trees, support vector machines, and clustering algorithms, have been used for traffic flow prediction, incident detection, and route planning in ITS [107].
b. Deep learning: Deep learning techniques, such as convolutional neural networks (CNNs) and recurrent neural networks (RNNs), have been employed for image and video processing in vehicle detection, vehicle classification, and traffic sign recognition in ITS [53].
c. Reinforcement learning: Reinforcement learning techniques have been used for traffic signal optimization, route guidance, and adaptive cruise control in ITS [108].
d. Natural language processing: Natural language processing techniques have been employed for information extraction and sentiment analysis of user-generated content, such as social media posts and online reviews, to improve public transportation services and user satisfaction [2].

Real-world Applications of AI in Intelligent Transportation Systems

Several real-world applications of AI techniques in ITS have been implemented, demonstrating their potential for improving transportation systems. Some examples include:

a. Adaptive traffic signal control: AI techniques, such as reinforcement learning, have been used to optimize traffic signal timings in response to real-time traffic conditions, resulting in reduced delays, shorter travel times, and lower emissions .
b. Incident detection and management: AI techniques, such as machine learning and deep learning, have been employed to detect and predict traffic incidents, allowing for more efficient management and faster response times [53].

c. Public transportation optimization: AI techniques have been used to optimize public transportation schedules, routes, and user satisfaction by analyzing user-generated content and predicting passenger demand .
d. Autonomous vehicles: AI techniques, such as deep learning and reinforcement learning, have been employed in the development and operation of autonomous vehicles, enabling safer and more efficient transportation [109].

Challenges and Future Directions

While AI techniques have shown promise in ITS, several challenges remain to be addressed:

1. Data quality and availability: The accuracy and effectiveness of AI techniques in ITS depend on the quality and availability of transportation data. Ensuring the collection of accurate, reliable, and timely data is essential for improving the performance of AI algorithms in ITS.
2. Interoperability and standardization: As ITS encompass a wide range of applications and technologies, interoperability and standardization are crucial for the successful implementation and integration of AI techniques in ITS.
3. Security and privacy concerns: The use of AI techniques in ITS raises security and privacy concerns related to data collection, sharing, and storage. Addressing these concerns while maintaining the benefits of AI applications in ITS is an important area of future research.
4. Legal and regulatory frameworks: The deployment of AI techniques in ITS requires the development of appropriate legal and regulatory frameworks to ensure their safe and responsible use in transportation systems.
5. Human factors: The adoption of AI techniques in ITS should consider human factors, such as user acceptance, trust, and behavior, to ensure the successful implementation and adoption of AI-based ITS solutions.

In conclusion, AI techniques have demonstrated significant potential for improving ITS in transportation and traffic management. By addressing the challenges and exploring new directions in AI research, transportation planners and traffic managers can leverage these advanced techniques to develop more efficient, responsive, and sustainable transportation systems.

10.4.3 Public Transportation Planning and Management

Public transportation planning and management involve the organization, operation, and optimization of public transportation services, such as buses, trains, and subways. This section will discuss various AI techniques used in public transportation planning and management, outlining their benefits, challenges, and real-world applications.

Public transportation planning and management aim to provide efficient, accessible, and sustainable transportation services for urban and rural populations.

Effective public transportation planning and management involve several aspects, including route planning, schedule optimization, demand forecasting, and resource allocation [53].

AI Techniques in Public Transportation Planning and Management

AI techniques have been increasingly employed in public transportation planning and management to optimize various aspects of public transportation services, including:

a. Route planning: AI techniques, such as genetic algorithms and ant colony optimization, have been used to optimize public transportation routes, considering factors such as travel time, distance, and passenger demand [109].
b. Schedule optimization: AI techniques, such as simulated annealing and particle swarm optimization, have been employed to optimize public transportation schedules, considering factors such as vehicle capacity, passenger demand, and service frequency [98].
c. Demand forecasting: AI techniques, such as machine learning and deep learning, have been used to predict passenger demand for public transportation services, enabling more efficient resource allocation and service planning [109].
d. Resource allocation: AI techniques, such as reinforcement learning and multi-agent systems, have been employed to optimize resource allocation in public transportation systems, considering factors such as vehicle capacity, maintenance requirements, and workforce management .

Real-world Applications of AI in Public Transportation Planning and Management

Several real-world applications of AI techniques in public transportation planning and management have been implemented, demonstrating their potential for improving public transportation services. Some examples include:

a. Bus rapid transit systems: AI techniques, such as genetic algorithms and simulated annealing, have been employed to optimize bus rapid transit (BRT) routes and schedules, resulting in reduced travel times, increased service reliability, and improved passenger satisfaction [110].
b. Metro systems: AI techniques, such as deep learning and reinforcement learning, have been used to optimize metro system operations, including train scheduling, headway control, and energy consumption optimization [109].
c. Demand-responsive transit systems: AI techniques, such as machine learning and multi-agent systems, have been employed to optimize demand-responsive transit systems, which adjust service routes and schedules based on real-time passenger demand .
d. Integrated multi-modal transportation systems: AI techniques, such as genetic algorithms and ant colony optimization, have been used to optimize integrated multi-modal transportation systems, which combine various transportation modes, such as buses, trains, and bicycles, to provide seamless and efficient transportation services [53].

Challenges and Future Directions

While AI techniques have shown promise in public transportation planning and management, several challenges remain to be addressed:

1. Data quality and availability: The accuracy and effectiveness of AI techniques in public transportation planning and management depend on the quality and availability of transportation data. Ensuring the collection of accurate, reliable, and timely data is essential for improving the performance of AI algorithms in public transportation systems.
2. Interoperability and integration: As public transportation systems encompass various modes and services, interoperability and integration are crucial for the successful implementation and integration of AI techniques in public transportation planning and management.
3. User-centered design: The adoption of AI techniques in public transportation planning and management should consider user needs, preferences, and behaviors to ensure the successful implementation and adoption of AI-based solutions.
4. Equity and accessibility: The deployment of AI techniques in public transportation planning and management should ensure equitable and accessible transportation services for all users, including vulnerable populations, such as low-income individuals, people with disabilities, and older adults.
5. Collaboration and coordination: The successful implementation of AI techniques in public transportation planning and management requires collaboration and coordination among various stakeholders, including transportation agencies, local governments, and private sector partners.

In conclusion, AI techniques have demonstrated significant potential for improving public transportation planning and management. By addressing the challenges and exploring new directions in AI research, transportation planners and managers can leverage these advanced techniques to develop more efficient, accessible, and sustainable public transportation systems.

10.4.4 Autonomous Vehicles and Connected Mobility

Autonomous vehicles (AVs) and connected mobility have the potential to revolutionize transportation and traffic management. This section will discuss various AI techniques used in the development and operation of autonomous vehicles and connected mobility systems, outlining their benefits, challenges, and real-world applications.

Autonomous vehicles are capable of sensing their environment and navigating without human input. They rely on advanced AI techniques, such as machine learning, deep learning, and reinforcement learning, to perceive their surroundings, make decisions, and control their movements. Connected mobility refers to the integration of various transportation modes, services, and technologies, enabling seamless and efficient transportation systems .

AI Techniques in Autonomous Vehicles and Connected Mobility

AI techniques play a critical role in the development and operation of autonomous vehicles and connected mobility systems. Some of the key AI techniques employed in this domain include:

a. Perception and sensing: AI techniques, such as deep learning and computer vision, are used to process and analyze data from various sensors, such as cameras, lidar, radar, and ultrasonic sensors, enabling autonomous vehicles to perceive their environment and detect objects, pedestrians, and other vehicles [109].
b. Decision-making and control: AI techniques, such as reinforcement learning and Bayesian networks, are employed to enable autonomous vehicles to make decisions and control their movements, considering factors such as traffic conditions, road infrastructure, and safety constraints [109].
c. Localization and mapping: AI techniques, such as simultaneous localization and mapping (SLAM) and graph-based optimization, are used to enable autonomous vehicles to localize themselves within their environment and build accurate maps of their surroundings [53].
d. Vehicle-to-everything (V2X) communication: AI techniques, such as machine learning and deep learning, are employed to enable vehicle-to-everything communication, which refers to the exchange of information between vehicles, infrastructure, and other road users, improving traffic efficiency and safety [108].

Real-world Applications of AI in Autonomous Vehicles and Connected Mobility

Several real-world applications of AI techniques in autonomous vehicles and connected mobility have been implemented, demonstrating their potential for improving transportation and traffic management. Some examples include:

a. Autonomous taxis and ride-hailing services: AI techniques have been employed to develop and operate autonomous taxis and ride-h ailing services, such as Waymo and Cruise, which aim to provide efficient, safe, and sustainable transportation solutions .
b. Autonomous buses and shuttles: AI techniques have been employed to develop and operate autonomous buses and shuttles, such as the Navya Arma and EasyMile EZ10, providing efficient and flexible public transportation services in urban and suburban areas [109].
c. Autonomous trucking and freight transportation: AI techniques have been employed to develop and operate autonomous trucks and freight transportation systems, such as the Otto self-driving truck and Volvo's Vera, aiming to improve transportation efficiency, reduce fuel consumption, and enhance safety [102].
d. Connected traffic management systems: AI techniques, such as machine learning and deep learning, have been employed to develop connected traffic management systems, which integrate real-time data from various sources, such as vehicles, infrastructure, and road users, to optimize traffic flow and reduce congestion .

Challenges and Future Directions

While AI techniques have shown promise in autonomous vehicles and connected mobility, several challenges remain to be addressed:

1. Safety and security: Ensuring the safety and security of autonomous vehicles and connected mobility systems is paramount, requiring rigorous testing, validation, and certification processes, as well as the development of robust AI algorithms that can handle various driving conditions, scenarios, and potential cyberattacks [109].
2. Infrastructure and connectivity: The successful implementation and operation of autonomous vehicles and connected mobility systems depend on the availability of adequate infrastructure and connectivity, such as high-speed communication networks, dedicated lanes, and smart traffic signals.
3. Regulatory and policy frameworks: The development and deployment of autonomous vehicles and connected mobility systems require clear regulatory and policy frameworks that address various issues, such as liability, data privacy, and ethical considerations [109].
4. Public acceptance and adoption: Gaining public acceptance and trust in autonomous vehicles and connected mobility systems is crucial for their successful implementation and adoption, requiring effective communication, education, and engagement strategies.
5. Integration with existing transportation systems: The successful deployment of autonomous vehicles and connected mobility systems requires their seamless integration with existing transportation systems, including public transportation, traffic management systems, and urban planning initiatives.

In conclusion, AI techniques have demonstrated significant potential for improving transportation and traffic management through the development and operation of autonomous vehicles and connected mobility systems. By addressing the challenges and exploring new directions in AI research, transportation planners, and managers can leverage these advanced techniques to develop more efficient, safe, and sustainable transportation systems.

10.4.5 Multimodal Transportation Integration

The growing urbanization and the increasing demand for efficient transportation systems have led to the development of multimodal transportation networks. Multimodal transportation refers to the integration of various modes of transportation, such as private vehicles, public transit, cycling, walking, and shared mobility services, to provide seamless and efficient transportation solutions to the users. Artificial intelligence (AI) has the potential to revolutionize the way multimodal transportation systems are designed, managed, and optimized. This section discusses the various applications of AI in multimodal transportation integration.

Real-time Route Planning and Optimization

AI-powered algorithms can leverage real-time data from different transportation modes to provide users with optimal routes based on their preferences, such as travel time, cost, and mode of transportation. By considering factors like traffic congestion, public transit schedules, and walking or cycling distances, AI algorithms can offer personalized recommendations and automatically update routes in response to changing conditions (Furuhata et al., 2013). These algorithms can also help transportation agencies optimize their transit routes and schedules to maximize efficiency and reduce operational costs.

Demand Forecasting and Resource Allocation

AI techniques, such as machine learning and deep learning, can be used to analyze historical and real-time data to predict future transportation demand. This information can help transportation planners allocate resources more effectively, such as adjusting public transit schedules, identifying areas where additional shared mobility services are needed, and planning for infrastructure improvements (Zhang, 2017). Demand forecasting can also inform dynamic pricing strategies for transportation services, incentivizing users to choose less congested routes or travel at off-peak times.

Traveler Information Systems

AI-powered traveler information systems can provide real-time information on transportation conditions, such as traffic congestion, public transit schedules, and shared mobility service availability. These systems can also offer personalized notifications and alerts based on users' preferences and travel patterns, helping them make informed decisions about their transportation options (Kaplan & Haenlein, 2019). For instance, if a bus is delayed or a bike-sharing station is out of bikes, the system can suggest alternative routes or modes of transportation.

AI-enhanced Shared Mobility Services

AI has the potential to improve the efficiency and user experience of shared mobility services, such as bike-sharing, car-sharing, and ride-hailing platforms. Machine learning algorithms can predict demand and optimize the allocation and rebalancing of shared resources, such as bikes and cars, to ensure their availability at the right locations and times [109]. AI can also facilitate dynamic pricing and matching of riders and drivers in ride-hailing services, improving service quality and reducing wait times [109].

Smart Infrastructure for Multimodal Transportation

Intelligent transportation infrastructure, such as connected traffic signals and sensors, can enable AI-driven optimization of traffic flow across different transportation modes. AI algorithms can analyze real-time data from connected infrastructure to optimize traffic signal timings, prioritize public transit vehicles, and manage traffic flow at intersections . This can help improve traffic efficiency, reduce congestion, and minimize the environmental impact of transportation systems.

Enhancing Accessibility and Equity in Multimodal Transportation

AI can help address accessibility and equity concerns in multimodal transportation systems by identifying underserved areas and populations, analyzing barriers to access, and suggesting targeted interventions . For example, AI algorithms can identify gaps in public transit coverage or analyze the availability of shared mobility services in low-income neighborhoods. This information can guide transportation planners in making data-driven decisions to improve accessibility and equity in transportation systems.

Challenges and Future Directions

While AI offers numerous opportunities for enhancing multimodal transportation integration, there are several challenges that need to be addressed to fully realize its potential. Some of the key challenges include:

The collection, storage, and analysis of large-scale data from various transportation modes and users raise concerns about data privacy and security. Ensuring the protection of personal information and preventing unauthorized access to sensitive data are crucial for maintaining user trust and promoting the adoption of AI-driven transportation services [102].

The integration of various transportation modes and data sources requires effective interoperability and standardization. This includes the development of common data formats, protocols, and APIs for seamless communication between different systems and services [101]. The lack of interoperability and standardization can hinder the implementation of AI-driven solutions in multimodal transportation systems.

The implementation of AI solutions in multimodal transportation systems requires significant investments in infrastructure, such as connected sensors, communication networks, and computing resources. Additionally, ongoing maintenance and support for these systems can be resource-intensive. Securing the necessary funding and resources for these investments can be a challenge, particularly for cities and regions with limited budgets [108].

AI-driven transportation services often operate in a complex regulatory environment, with various laws and policies governing data collection, privacy, and service provision. Navigating these regulations and developing policies that support innovation while ensuring public safety and equity can be a challenge for transportation planners and policymakers [101].

Future Directions

Despite these challenges, the potential of AI in multimodal transportation integration is immense, and there are several promising directions for future research and development:

The development of more advanced AI algorithms and models can further improve the accuracy and efficiency of transportation prediction, optimization, and decision-making. This includes the incorporation of additional data sources, such as social media and crowdsourced data, and the development of novel machine learning and deep learning techniques tailored to the specific needs of multimodal transportation systems .

The integration of AI-driven route planning, demand forecasting, and dynamic pricing into comprehensive Mobility-as-a-Service (MaaS) platforms can provide users with a seamless, personalized, and efficient transportation experience across multiple modes [107]. These platforms can enable better decision-making and resource allocation for transportation providers, leading to improved service quality and sustainability.

AI can play a crucial role in promoting the sustainability and resilience of multimodal transportation systems by optimizing energy consumption, reducing emissions, and enabling better adaptation to changing environmental conditions and extreme events . Future research can focus on the development of AI-driven solutions for green transportation, such as electric vehicle charging infrastructure management and the optimization of public transit for reduced emissions.

As AI becomes more integrated into transportation systems, understanding and enhancing the collaboration between humans and AI will be essential for ensuring safety, efficiency, and user satisfaction. This includes research on human factors, such as trust, perception, and decision-making, in the context of AI-driven transportation services .

In conclusion, AI has the potential to transform multimodal transportation systems, making them more efficient, sustainable, and user-friendly. Despite the challenges, ongoing research and development in this area promise to further enhance the capabilities of AI in addressing the complex needs of urban transportation systems.

Future research and development efforts should focus on leveraging AI to create more inclusive transportation systems that cater to the diverse needs of different user groups, including people with disabilities, the elderly, and those from low-income communities (Ma et al., 2018). By incorporating accessibility and equity considerations into AI-driven decision-making, planners and policymakers can ensure that the benefits of multimodal transportation integration are equitably distributed among all users.

The integration of AI in transportation infrastructure maintenance and asset management can lead to more efficient and cost-effective strategies for preserving and optimizing the performance of transportation assets [102]. By leveraging AI for predictive maintenance, transportation agencies can better allocate resources, prioritize investments, and minimize disruptions to the transportation network.

AI can also be applied to inform transportation policy and planning decisions, providing data-driven insights to support the development of more effective strategies for multimodal transportation integration [111]. This includes the use of AI for scenario analysis, impact assessment, and policy evaluation, enabling more informed and evidence-based decision-making by planners and policymakers.

As AI becomes increasingly integrated into multimodal transportation systems, ethical considerations related to fairness, accountability, transparency, and privacy will become more critical [102]. Future research should address these ethical concerns, developing guidelines and best practices for the responsible use of AI in transportation applications to ensure that the technology serves the public interest while respecting individual rights and values.

In summary, AI offers tremendous opportunities to revolutionize multimodal transportation systems by enhancing their efficiency, sustainability, and user experience. While challenges remain, continued research and development in this area hold the promise of overcoming these obstacles and unlocking the full potential of AI for transforming urban transportation systems. As cities around the world continue to grow and evolve, AI-driven multimodal transportation integration will play an increasingly important role in shaping the future of urban mobility, contributing to more livable, equitable, and sustainable urban environments.

As climate change and other global challenges present new and evolving threats to transportation systems, incorporating AI-driven resilience and adaptability strategies becomes crucial (Hanna et al., 2020). Future research should explore how AI can be employed to anticipate, respond to, and recover from disruptions caused by extreme weather events, natural disasters, or other unforeseen circumstances. This may include the development of AI-powered early warning systems, real-time response coordination, and adaptive infrastructure design.

The future of multimodal transportation will likely involve increased collaboration and coordination between various transportation modes, agencies, and stakeholders [101]. AI can facilitate such collaboration by streamlining communication, sharing data, and integrating decision-making processes. As a result, research should focus on developing collaborative AI systems and platforms that enhance the cooperation between different transportation actors and help achieve a more integrated and seamless multimodal transportation experience.

As the demand for personalized and on-demand transportation services continues to grow, user-centric AI systems can play a significant role in meeting these needs [101]. Future research should explore how AI can be used to deliver personalized transportation recommendations, real-time route planning, and tailored mobility services that cater to individual preferences and needs. This may involve the development of AI-powered recommendation engines, intelligent routing algorithms, and dynamic pricing models that consider user preferences and external factors such as traffic conditions, weather, and local events.

Sustainability and environmental considerations are becoming increasingly important in urban transportation planning and policy [110]. AI can contribute to more sustainable and green multimodal transportation systems by optimizing energy use, reducing emissions, and promoting environmentally friendly transportation options. For example, AI can be employed to optimize electric vehicle charging infrastructure, improve the efficiency of public transportation systems, and incentivize the use of active transportation modes such as walking and cycling.

In conclusion, the potential of AI to transform multimodal transportation systems is immense. By addressing the challenges and seizing the opportunities presented by AI, urban planners, policymakers, and transportation stakeholders can build more efficient, sustainable, inclusive, and resilient transportation systems that cater to the diverse needs of urban dwellers. As cities continue to grow and evolve, harnessing the power of AI for multimodal transportation integration will be crucial for creating livable, equitable, and sustainable urban environments that are prepared to face the challenges of the 21st century and beyond.

10.5 Challenges and Limitations of AI in Transportation and Traffic Management

One of the significant challenges in applying AI to transportation and traffic management is the quality and availability of data. AI algorithms require a large amount of high-quality, accurate, and representative data to produce reliable and generalizable results [103]. However, obtaining such data can be difficult due to the complexity of transportation systems, the dynamic nature of traffic patterns, and the heterogeneity of data sources (e.g., traffic sensors, GPS devices, social media, etc.). In many cases, data may be incomplete, outdated, or biased, leading to suboptimal or inaccurate AI-driven decision-making [108]. Future research should focus on developing techniques for data preprocessing, imputation, and fusion to overcome these challenges and enhance the effectiveness of AI in transportation and traffic management (Table 10.2).

Another challenge is the scalability and computational complexity of AI algorithms, particularly for large-scale transportation systems . Many AI techniques, such as deep learning and reinforcement learning, involve complex computations and large-scale optimization problems, which can be resource-intensive and time-consuming. This can be a significant limitation for real-time traffic management applications where timely decision-making is crucial [99]. Future research should explore more efficient and scalable AI algorithms and techniques, as well as leverage advances in parallel and distributed computing, edge computing, and hardware acceleration to overcome these challenges.

Transportation systems are characterized by inherent uncertainty and unpredictability due to various factors, such as weather conditions, accidents, and human behavior [108]. This uncertainty poses challenges for AI algorithms, which often rely on deterministic models and assumptions. Developing AI techniques that can effectively handle uncertainty and make robust predictions in the presence of unpredictable events is essential for transportation and traffic management applications [109]. Probabilistic models, such as Bayesian networks and stochastic optimization techniques, can be employed to address these challenges and improve the reliability of AI-driven decision-making in transportation.

The integration and interoperability of AI systems with existing transportation infrastructure and systems pose significant challenges [100]. Transportation systems often involve various stakeholders, such as government agencies, private companies, and users, each with their own data formats, protocols, and standards. Integrating AI-driven solutions with these diverse systems requires addressing issues related to data exchange, communication, and standardization. Future research should focus on developing open and flexible architectures, standardized data formats, and

Table 10.2 Challenges and Limitations of AI in Transportation and Traffic Management

Challenge/limitation	Description	Potential Solutions
Data quality and availability	AI algorithms require high-quality, comprehensive data, which can be difficult to obtain due to various challenges in transportation systems.	–Development of techniques for data preprocessing, imputation, and fusion
Scalability and computational complexity	Many AI techniques involve complex computations, posing challenges for large-scale transportation systems.	–Exploration of more efficient AI algorithms and leveraging advances in computing technologies
Interoperability and integration	Integrating AI systems with existing infrastructure requires overcoming data exchange and standardization issues.	–Development of standardized data formats and interoperable communication protocols
Security and privacy concerns	The sensitive nature of location data and potential for misuse raise privacy and security concerns in AI-driven transportation systems.	–Implementation of privacy-preserving techniques and robust security measures
Ethical and social implications	AI-driven decision-making processes in transportation must be transparent, fair, and accountable to prevent biases and unintended consequences.	–Development of ethical guidelines and frameworks for AI applications in transportation
Legal and regulatory frameworks	Developing appropriate policies and regulations to govern the use of AI in transportation is challenging due to rapidly advancing technologies.	–Engaging in dialogues with stakeholders to develop comprehensive policies
Public acceptance and trust	Building trust in AI-driven transportation systems is crucial for their adoption, requiring attention to usability, ergonomics, and addressing user concerns.	–Designing user-centric AI systems and developing strategies for promoting user acceptance
Infrastructure and connectivity	Successful operation of autonomous vehicles and connected mobility systems depends on adequate infrastructure and connectivity.	–Securing investments in necessary infrastructure and connectivity

interoperable communication protocols to facilitate seamless integration of AI in transportation and traffic management.

The widespread use of AI in transportation and traffic management raises privacy and security concerns due to the sensitive nature of location data and the potential for misuse [108]. Ensuring that AI-driven transportation systems respect user privacy and protect sensitive data is crucial for maintaining public trust and promoting the adoption of AI technologies. Future research should explore privacy-preserving techniques, such as differential privacy and federated learning, as well as develop robust security measures to protect AI systems from cyberattacks and other threats.

The ethical and social implications of AI in transportation and traffic management are also critical challenges that need to be addressed . AI-driven decision-making processes should be transparent, fair, and accountable to prevent unintended consequences and biases in transportation planning and management. Future research should focus on developing ethical guidelines and frameworks for AI applications in transportation, as well as exploring techniques for ensuring fairness, transparency, and explainability in AI-driven decision-making processes [107]. In addition, AI should be used to promote social equity and inclusiveness in transportation systems, ensuring that all users, including vulnerable populations, have access to safe, affordable, and efficient transportation options [100].

As AI technologies are increasingly adopted in transportation and traffic management, policymakers and regulators face challenges in developing appropriate policies and regulations to govern their use [53]. These challenges include addressing issues related to liability, safety, and certification of AI-driven transportation systems, as well as ensuring that AI technologies are deployed in a manner that is consistent with public interests and societal values. Policymakers and regulators should engage in proactive dialogues with AI researchers, industry stakeholders, and the public to develop comprehensive and forward-looking policies and regulations that foster innovation while protecting public safety and welfare [109].

User acceptance and trust in AI-driven transportation systems are crucial for their widespread adoption and success [102]. Addressing human factors, such as usability, ergonomics, and user experience, is essential to ensure that AI technologies are user-friendly, accessible, and reliable. Moreover, understanding and addressing potential user concerns, such as job displacement, privacy, and safety, are critical for building public trust in AI-driven transportation systems . Future research should focus on exploring user needs and preferences, designing user-centric AI systems, and developing strategies for promoting user acceptance and trust in AI technologies for transportation and traffic management.

10.6 Future Directions in AI Applications for Transportation and Traffic Management

The rapid advancements in artificial intelligence (AI) and machine learning technologies have revolutionized transportation and traffic management, leading to innovative solutions and improvements in safety, efficiency, and sustainability. As urban populations continue to grow, the demand for more advanced, intelligent, and integrated transportation systems will increase. This section highlights the future directions in AI applications for transportation and traffic management, which could potentially transform the way we travel and interact with transportation systems.

One of the key future directions in AI applications for transportation and traffic management is the integration of AI technologies with Intelligent Transportation Systems (ITS). ITS use advanced sensor technologies, communication networks, and data analysis techniques to monitor and manage traffic flow, reduce congestion, and enhance transportation safety and efficiency [101]. Future AI applications could help develop more advanced ITS, capable of predicting traffic patterns, optimizing traffic signal timings, and implementing adaptive traffic management strategies to address dynamic traffic conditions [104].

The development of autonomous vehicles (AVs) and connected mobility is another promising future direction for AI applications in transportation and traffic management. AVs are expected to play a significant role in shaping the future of transportation, reducing human error, and enhancing overall traffic safety [109]. AI technologies, such as machine learning and deep learning algorithms, can enable AVs to make real-time decisions, navigate complex environments, and interact with other vehicles and road users. Furthermore, the integration of connected vehicle technologies with AI systems could enable cooperative driving strategies, enhancing traffic flow efficiency and reducing emissions [53].

As urban populations continue to grow, there is an increasing need for efficient and sustainable transportation systems that integrate various modes of transport, such as public transit, private vehicles, and non-motorized modes (e.g., walking and cycling). AI technologies can play a crucial role in optimizing and managing multimodal transportation systems, enabling real-time route planning, demand-responsive transit services, and dynamic pricing schemes [109]. Moreover, AI can support the development of Mobility-as-a-Service (MaaS) platforms, which provide users with seamless access to various transportation modes through a single interface (MaaS [108]).

The transportation sector is a significant contributor to global greenhouse gas emissions, and there is a growing need to develop sustainable and environmentally friendly transportation solutions. AI technologies can help address environmental and sustainability challenges by optimizing traffic flow, enabling energy-efficient driving strategies, and supporting the deployment of electric vehicles and charging infrastructure [98]. Moreover, AI can contribute to the development of advanced traffic management strategies that consider environmental impacts, such as traffic signal optimization algorithms that minimize vehicle emissions [108].

The widespread adoption of AI technologies in transportation and traffic management raises several ethical, legal, and social implications that need to be addressed. For example, the deployment of autonomous vehicles raises questions about liability, privacy, and security, as well as potential job displacement [53]. Moreover, the use of AI algorithms in transportation decision-making processes can introduce biases and unfairness, which could exacerbate existing social inequalities [102]. Future research should explore the ethical, legal, and social implications of AI applications in transportation and traffic management and develop guidelines and best practices to ensure the responsible and equitable adoption of these technologies.

In conclusion, the future of AI applications in transportation and traffic management holds immense potential for transforming the way we travel and interact with transportation systems. The integration of AI technologies with Intelligent Transportation Systems, the development of autonomous vehicles and connected mobility, the optimization of multimodal transportation systems, and the consideration of environmental and sustainability aspects will all contribute to safer, more efficient, and sustainable transportation solutions. However, addressing the ethical, legal, and social implications of these technologies is crucial to ensure their responsible and equitable adoption.

As we move forward, continued research, collaboration, and investment in AI technologies for transportation and traffic management will be necessary to overcome the challenges and harness the potential benefits. Policymakers, industry stakeholders, and researchers must work together to develop and implement innovative AI solutions that address the complex and evolving transportation needs of our urban environments.

References

1. Abdoos, M., Mozayani, N., & Bazzan, A. L. (2011). Holonic multi-agent learning for traffic signals control. In *IJCAI Proceedings-International Joint Conference on Artificial Intelligence* (Vol. 22, No. 1, pp. 1861–1866).
2. Abdullah, N., Raj, R. G., & Ward, R. (2016). A Twitter sentiment analysis of public transportation user satisfaction. In *2016 International Conference on Informatics and Applications (ICIA)* (pp. 69-74). IEEE.
3. Agatz, N., Erera, A., Savelsbergh, M., & Wang, X. (2011). Optimization for dynamic ride-sharing: A review. *European Journal of Operational Research, 223*(2), 295–303.
4. Aggarwal, C. C., & Zhai, C. (2012). A survey of text classification algorithms. In *Mining text data* (pp. 163-222). Springer.
5. Alessandrini, A., Cattivera, A., Holguin, C., Stam, D., & Alonso, L. (2020). Integration of public transport and shared mobility services. In C. Macharis, L. A. Tavasszy, & J. A. A. Witlox (Eds.), *City distribution and urban freight transport: Multiple perspectives* (pp. 58–76). Edward Elgar Publishing.
6. Alonso-Mora, J., Samaranayake, S., Wallar, A., Frazzoli, E., & Rus, D. (2017). On-demand high-capacity ride-sharing via dynamic trip-vehicle assignment. *Proceedings of the National Academy of Sciences, 114*(3), 462–467.

7. Amin, S. U., Alhaisoni, M., Badr, E., & Mostafa, S. A. (2018). A deep learning-based approach for the detection and localization of vehicles in traffic surveillance systems. *Soft Computing, 22*(13), 4377–4390.
8. Anderson, J. M., Kalra, N., Stanley, K. D., Sorensen, P., Samaras, C., & Oluwatola, O. A. (2014). *Autonomous vehicle technology: A guide for policymakers*. Rand Corporation.
9. Anderson, J. M., Nidhi, K., Stanley, K. D., Sorensen, P., Samaras, C., & Oluwatola, O. A. (2016). *Autonomous Vehicle Technology: A Guide for Policymakers*. Rand Corporation.
10. Ban, Y., Gamba, P., & Jakobsson, M. (2017). A new backscatter-radiation model for the built environment: Application to urban exposure assessment. *IEEE Transactions on Geoscience and Remote Sensing, 55*(5), 2810–2822.
11. Bar-Gera, H. (2013). Traffic data from mobile devices. In *Proceedings of the 16th International IEEE Conference on Intelligent Transportation Systems (ITSC 2013)* (pp. 2227–2232). IEEE.
12. Bielli, M., Boulmakoul, A., & Mouncif, H. (2012). Object-oriented model for public transport network optimisation. *Journal of Artificial Intelligence, 5*(2), 96–107.
13. Bishop, R. (2020). Autonomous vehicle technology: How to best realize its social benefits. *Philosophy & Technology, 33*(1), 125–137.
14. Blei, D. M., Ng, A. Y., & Jordan, M. I. (2003). Latent Dirichlet allocation. *Journal of Machine Learning Research, 3*(Jan), 993–1022.
15. Blyth, P. L., & Laperrière-Robillard, T. (2020). Policy implications of the automation of mobility services. *Transport Policy, 99*, 15–26.
16. Bojarski, M., Del Testa, D., Dworakowski, D., Firner, B., Flepp, B., Goyal, P., Jackel, L. D., Monfort, M., Muller, U., Zhang, J., & Zhang, X. (2016). End to end learning for self-driving cars. arXiv:1604.07316
17. Cadena, C., Carlone, L., Carrillo, H., Latif, Y., Scaramuzza, D., Neira, J., Reid, I., & Leonard, J. J. (2016). Past, present, and future of simultaneous localization and mapping: Toward the robust-perception age. *IEEE Transactions on Robotics, 32*(6), 1309–1332.
18. Castro-Neto, M., Jeong, Y. S., Jeong, M. K., & Han, L. D. (2009). Online-SVR for short-term traffic flow prediction under typical and atypical traffic conditions. *Expert Systems with Applications, 36*(3), 6164–6176.
19. Cats, O. (2016). The future of public transport planning. *Public Transport, 8*(3), 161–164.
20. Chakirov, A., & Erath, A. (2012). Processing and analysis of multiday electronic smart card data for public transportation. In *Proceedings of the 12th Swiss Transport Research Conference*.
21. Chen, C., Wang, K. C., Li, Y., & Hu, Z. (2017). A deep learning approach for real-time detection of traffic incidents using social media data. In *2017 IEEE 20th International Conference on Intelligent Transportation Systems (ITSC)* (pp. 1–7). IEEE.
22. Chen, M. Y., Wu, Y. J., & Guo, J. H. (2012). A traffic flow forecasting model for an intelligent transportation system based on a deep belief network. In *Proceedings of the 11th International Conference on Machine Learning and Applications (ICMLA)* (pp. 291–296). IEEE.
23. Chen, S., Wang, H., & Yu, F. (2017). Deep learning for intelligent transportation systems: A survey. In *2017 IEEE 20th International Conference on Intelligent Transportation Systems (ITSC)* (pp. 1–7). IEEE.
24. Chen, X., Wang, S., & Jin, Z. (2020). AI-driven predictive maintenance for smart cities' transportation systems. *IEEE Access, 8*, 103801–103811.
25. Choy, M. C., Srinivasan, D., & Cheu, R. L. (2003). Neural networks for continuous online learning and control. *IEEE Transactions on Neural Networks, 14*(6), 1511–1519.
26. Cools, M., Moons, E., & Wets, G. (2010). Assessing the impact of weather on traffic intensity. *Weather, Climate, and Society, 2*(1), 60–68.
27. Crawford, K., & Calo, R. (2016). There is a blind spot in AI research. *Nature, 538*(7625), 311–313.
28. Demir, E., Huang, Y., Scholts, S., & Van Woensel, T. (2019). A selected review on the negative externalities of the freight transportation: Modeling and pricing. *Transportation Research Part E: Logistics and Transportation Review, 77*, 95–114.

29. Dimitrakopoulos, G., & Demestichas, P. (2010). Intelligent transportation systems. *IEEE Vehicular Technology Magazine, 5*(1), 77–84.
30. El-Tantawy, S., Abdulhai, B., & Abdelgawad, H. (2013). Multiagent reinforcement learning for integrated network of adaptive traffic signal controllers (MARLIN-ATSC): Methodology and large-scale application on downtown Toronto. *IEEE Transactions on Intelligent Transportation Systems, 14*(3), 1140–1150.
31. Erdogan, S., Argin, G., & Tatli, O. (2017). The dark side of the force: The side effects of intelligent transportation systems. In *2017 5th International Istanbul Smart Grids and Cities Congress and Fair (ISG&CC)* (pp. 48–52). IEEE.
32. Fagnant, D. J., & Kockelman, K. (2018). The impacts of autonomous vehicles and e-commerce on local government budgeting and finance. National League of Cities.
33. Fu, R., Zhang, Z., & Li, L. (2016). Using LSTM and GRU neural network methods for traffic flow prediction. In *2016 31st Youth Academic Annual Conference of Chinese Association of Automation (YAC)* (pp. 324–328). IEEE.
34. Gakis, K., Piliouras, G., Rontogiannis, A., Yannacopoulos, A., & Kaliampakos, D. (2018). A multi-objective decision-making approach for the planning of multi-modal transportation systems. *Transportation Research Part E: Logistics and Transportation Review, 110*, 1–20.
35. Gama, K., Fonseca, R., & Santos, M. Y. (2017). Open data for smart cities: A case of study for urban mobility. In *Proceedings of the 19th International Conference on Enterprise Information Systems-Volume 1: ICEIS* (pp. 611–618).
36. Gao, S., Rao, J., Kang, Y., Liang, Y., & Krizek, K. J. (2016). Data-enabled evidence-based urban planning: The case of transit-induced gentrification. *Computers, Environment and Urban Systems, 57*, 101–110.
37. Gasser, T. M., & Westhoff, D. (2012). BASt-study: Definitions of automation and legal issues in Germany. In *Proceedings of the 2012 Road Vehicle Automation Workshop.*
38. Genders, W., & Razavi, S. (2016). Using a deep reinforcement learning agent for traffic signal control. arXiv:1611.01142
39. Giannopoulos, G. A. (2004). The application of information and communication technologies in transport. *European Journal of Operational Research, 152*(2), 302–320.
40. Gkiotsalitis, K., & Cats, O. (2019). Real-time vehicle and crew scheduling for public transport services with rolling horizons. *Transportation Research Part C: Emerging Technologies, 103*, 299–316.
41. Goulias, K. G., & Shiftan, Y. (2017). Activity-based models of travel demand: Promises, progress and prospects. *International Journal of Urban Sciences, 21*(sup1), S49–S63.
42. Grishman, R. (2015). Information extraction. In *The Oxford Handbook of Computational Linguistics 2.0* (Vol. 2).
43. Hallenbeck, M. E. (2013). *Traffic data collection and its standardization*. Springer Science & Business Media.
44. Horni, A., Nagel, K., & Axhausen, K. W. (Eds.). (2016). *The multi-agent transport simulation MATSim*. Ubiquity Press.
45. Hossain, M. (2018). Framework for AI-driven intelligent transportation systems: Realizing smart cities. *IEEE Transactions on Industrial Informatics, 14*(4), 1594–1601.
46. Huang, A. Q., Zheng, Y., Wang, L., & Luo, J. (2021). A survey on traffic prediction for smart mobility. *ACM Computing Surveys, 54*(2), 1–32.
47. Huang, B. (2018). GIS-supported research on urban and transportation planning. In *GIScience for Intelligent Services* (pp. 123142). Springer.
48. Jiang, Y., Zhang, X., & Ma, L. (2018). An empirical study on public transit travel behavior based on smart card data and online reviews. In *Proceedings of the 21st International Conference on Intelligent Transportation Systems (ITSC)* (pp. 3791–3796). IEEE.
49. Jurafsky, D., & Martin, J. H. (2019). *Speech and language processing*. Prentice Hall.
50. Kamargianni, M., & Matyas, M. (2017). The business ecosystem of Mobility-as-a-Service. *Transportation Research Part A: Policy and Practice, 131*, 283–295.
51. Kumar, N., & Dave, M. (2019). A review on the applications of deep learning in connected and autonomous vehicles. *Vehicular Communications, 19*, 100178.

52. Kumar, S., & Nandagopal, S. (2020). Connected vehicles: Applications, challenges, and performance. In *Wireless Networks* (pp. 453–472). Springer.
53. Kyriakidis, M., Happee, R., & De Winter, J. C. F. (2021). Public opinion on automated driving: Results of an international questionnaire among 5000 respondents. *Transportation Research Part F: Traffic Psychology and Behaviour, 32*, 127–140.
54. Lavasani, M., Jin, X., & Du, Y. (2016). *Autonomous trucking: A supply chain game changer*. IHS Markit.
55. Li, L., Lv, Y., & Wang, F. (2016). Traffic signal timing optimization based on deep reinforcement learning. In *2016 IEEE 19th International Conference on Intelligent Transportation Systems (ITSC)* (pp. 278–283). IEEE.
56. Li, S., Wen, D., & Yao, Y. (2018). A survey of traffic control with vehicular cloud computing. *IET Intelligent Transport Systems, 12*(2), 99–106.
57. Li, X., Lv, Y., Wang, W., & He, Q. (2019). Traffic signal timing optimization based on vehicle emissions: A review. *Journal of Advanced Transportation, 2019*, 1–18.
58. Li, X., Pan, G., & Wang, Z. (2018). Traffic flow prediction with big data: A deep learning approach. *IEEE Transactions on Intelligent Transportation Systems, 19*(10), 3320–3330.
59. Litman, T. (2017). *Autonomous vehicle implementation predictions*. Victoria Transport Policy Institute.
60. Liu, H. X., Zhang, J., & Zheng, W. (2017). Artificial intelligence in transportation: status quo and future directions. *Journal of Intelligent Transportation Systems, 21*(4), 384–399.
61. Liu, J., Li, Z., Li, W., & Li, J. (2021). A survey on security and privacy issues in intelligent transportation systems: From the perspective of artificial intelligence. *IEEE Transactions on Intelligent Transportation Systems, 22*(3), 1403–1419.
62. Liu, W., Anguelov, D., Erhan, D., Szegedy, C., & Reed, S. (2016). SSD: Single shot multibox detector. In *European Conference on Computer Vision* (pp. 21–37). Springer.
63. Lu, D., & Weng, Q. (2014). A survey of image classification methods and techniques for improving classification performance. *International Journal of Remote Sensing, 28*(5), 823–870.
64. Lv, Y., Duan, Y., Kang, W., Li, Z., & Wang, F. (2014). Traffic flow prediction with big data: A deep learning approach. *IEEE Transactions on Intelligent Transportation Systems, 16*(2), 865–873.
65. Ma, X., Ma, Z., & Li, L. (2019). Learning traffic as images: A deep convolutional neural network for large-scale transportation network speed prediction. *Sensors, 17*(4), 818.
66. MaaS Alliance. (2021). Mobility as a Service (MaaS): A global market perspective. Retrieved from https://maas-alliance.eu/wp-content/uploads/sites/7/2021/02/Global-Market-Perspective-on-MaaS.pdf
67. Mannion, P., Duggan, J., & Howley, E. (2016). An experimental review of reinforcement learning algorithms for adaptive traffic signal control. *Autonomic Road Transport Support Systems, 31*, 45–66.
68. Mersky, A. C., & Samaras, C. (2020). Fuel economy and greenhouse gas emissions testing of connected and automated vehicles. *Transportation Research Part D: Transport and Environment, 85*, 102380.
69. Milakis, D., Snelder, M., Van Arem, B., Van Wee, B., & Correia, G. (2017). Development and transport implications of automated vehicles in the Netherlands: Scenarios for 2030 and 2050. *European Journal of Transport and Infrastructure Research, 17*(1), 63–85.
70. Mladenovic, M. N., & McPherson, T. (2016). Engineering social justice into traffic control for self-driving vehicles? *Science and Engineering Ethics, 22*, 1131–1149.
71. Mladenović, M. N., & McPherson, T. (2021). Artificial intelligence and urban transportation systems. *Transport Reviews, 41*(1), 97–114.
72. Mnih, V., Kavukcuoglu, K., Silver, D., & Graves, A. (2013). Playing Atari with deep reinforcement learning. arXiv:1312.5602
73. Mnih, V., Kavukcuoglu, K., Silver, D., Rusu, A. A., Veness, J., Bellemare, M. G., Graves, A., Riedmiller, M., Fidjeland, A. K., Ostrovski, G., & Petersen, S. (2015). Human-level control through deep reinforcement learning. *Nature, 518*(7540), 529–533.

74. Moreira-Matias, L., Gama, J., Ferreira, M., Mendes-Moreira, J., & Damas, L. (2013). Predicting taxi-passenger demand using streaming data. *IEEE Transactions on Intelligent Transportation Systems, 14*(3), 1393–1402.
75. Mousavi, S. S., Schukat, M., & Howley, E. (2017). Traffic light control using deep policy-gradient and value-function-based reinforcement learning. *IET Intelligent Transport Systems, 11*(7), 417–423.
76. Naphade, M., Banavar, G., Harrison, C., Paraszczak, J., & Morris, R. (2011). Smarter cities and their innovation challenges. *Computer, 44*(6), 32–39.
77. Pereira, R. H. M., Schwanen, T., & Banister, D. (2021). Distributive justice and equity in transportation. *Transport Reviews, 41*(2), 235–252.
78. Rasouli, S., & Timmermans, H. (2018). Mobility as a service and sustainable travel behavior: A research agenda. *Journal of Transportation Research Part D: Transport and Environment, 64*, 72–91.
79. Rasouli, S., & Timmermans, H. J. P. (2020). Autonomous vehicles, artificial intelligence, and public transport: A review and prospects. *Transport Reviews, 40*(6), 759–776.
80. Redmon, J., Divvala, S., Girshick, R., & Farhadi, A. (2016). You only look once: Unified, real-time object detection. In *Proceedings of the IEEE Conference on Computer Vision and Pattern Recognition* (pp. 779–788).
81. Rios-Torres, J., & Malikopoulos, A. A. (2017). A survey on the coordination of connected and automated vehicles at intersections and merging at highway on-ramps. *IEEE Transactions on Intelligent Transportation Systems, 18*(5), 1066–1077.
82. Rodrigue, J. P., Comtois, C., & Slack, B. (2016). *The geography of transport systems.* Routledge.
83. Seshadri, R., Kumar, A., & Chatterjee, K. (2021). Handling uncertainty in urban traffic management: A review. *IEEE Transactions on Intelligent Transportation Systems, 22*(2), 1265–1281.
84. Shladover, S. E. (2020). Connected and automated vehicle systems: Introduction and overview. *Annual Review of Control, Robotics, and Autonomous Systems, 3*, 1–28.
85. Sun, S., Zhang, C., & Wu, Y. (2016). Roadside sensing for intelligent transportation systems. In *Proceedings of the 2016 ACM International Joint Conference on Pervasive and Ubiquitous Computing: Adjunct* (pp. 1039–1046).
86. Sutton, R. S., & Barto, A. G. (2018). *Reinforcement learning: An introduction.* MIT Press.
87. Vlahogianni, E. I., Karlaftis, M. G., & Golias, J. C. (2014). Short-term traffic forecasting: Where we are and where we're going. *Transportation Research Part C: Emerging Technologies, 43*, 3–19.
88. Wang, D., Zhang, Q., Zhang, W., & Liu, L. (2021). Intelligent Transportation Systems (ITS) for sustainable cities: A review of research trends and future directions. *Sustainable Cities and Society, 66*, 102607.
89. Wang, H., Gerber, M. S., & Brown, D. (2012). Automatic crime prediction using events extracted from Twitter posts. In *International Conference on Social Computing, Behavioral-Cultural Modeling, and Prediction* (pp. 231–238). Springer.
90. Wang, L., Zhang, X., Yao, Y., Wang, H., & Xie, K. (2021). A survey on transportation data: Acquisition, preprocessing, and application. *IEEE Transactions on Intelligent Transportation Systems, 22*(8), 5183–5202.
91. Wang, Y., Zheng, Y., & Xu, X. (2016). ST-MVL: Filling missing values in geo-sensory time series data. In *Proceedings of the 25th International Conference on World Wide Web* (pp. 511–521).
92. Wang, K., & Kockelman, K. M. (2019). Market penetration of autonomous vehicles on household vehicle ownership and trip generation: Results from a general household survey. *Transportation Research Part C: Emerging Technologies, 99*, 1–20.
93. Wei, H., Chen, L., & Liu, Y. (2018). A deep reinforcement learning approach for traffic signal control at an isolated intersection. In *2018 21st International Conference on Intelligent Transportation Systems (ITSC)* (pp. 2170–2175). IEEE.

94. Xu, G., Zhang, W., & Wang, J. (2020). A review of data-driven approaches for prediction and classification of building energy consumption. *Renewable and Sustainable Energy Reviews, 127*, 109837.
95. Yang, L., Li, K., & Gao, Z. (2018). Energy-efficient metro train rescheduling with uncertain time-variant passenger demands: An approximate dynamic programming approach. *Transportation Research Part B: Methodological, 108*, 55–80.
96. Yin, J., Ma, J., Wang, Y., Wu, J., & Wang, Y. (2018). A multi-objective optimization model for bus rapid transit network design under uncertainty. *Transportation Research Part C: Emerging Technologies, 86*, 585–612.
97. Yin, Y., Wong, S. C., Xu, J., & Yang, H. (2012). Bi-objective optimization for transportation network design problem. *Transportmetrica A: Transport Science, 8*(1), 43–63.
98. Yu, B., Yin, H., & Zhu, Z. (2017). Spatiotemporal recurrent convolutional networks for traffic prediction in transportation networks. *Sensors, 17*(7), 1501.
99. Zeadally, S., Chen, Y., & Rafetseder, A. (2016). Vehicular communications and networks: Architectures, protocols, operation, and deployment. *IEEE Journal on Selected Areas in Communications, 34*(12), 3347–3348.
100. Zhang, J., Zheng, V. W., Xu, Z., Cai, Y., & Zhu, Y. (2018). A reinforcement learning based approach for traffic signal control at isolated urban intersections. *Transportation Research Part C: Emerging Technologies, 96*, 348–369.
101. Zhang, J., Zheng, Y., & Qi, D. (2017). Deep spatio-temporal residual networks for citywide crowd flows prediction. In *Proceedings of the Thirty-First AAAI Conference on Artificial Intelligence (AAAI-17)* (pp. 1655–1661).
102. Zhang, Y., Chen, M., Huang, D., Liu, C., & Wang, Y. (2021). Explainable AI for smart cities: A survey. *ACM Computing Surveys, 54*(6), 1–37.
103. Zhang, Y., Feng, Y., & Li, J. (2016). A hybrid artificial intelligence model for optimizing public transportation scheduling. *Journal of Advanced Transportation, 50*(3), 375–394.
104. Zhang, J., Shen, Y., & Zhu, X. (2020). An overview of AI-assisted public transportation systems. *IEEE Access, 8*, 183230–183247.
105. Zhao, P., & Huang, R. (2018). *AI and the future of urban mobility: Opportunities and challenges for policy and planning*. Springer.
106. Zhao, H., Li, X., Liu, Y., Li, L., & Yin, J. (2020). Challenges and opportunities of using artificial intelligence in transportation. *Transportation Research Part C: Emerging Technologies, 120*, 102783.
107. Zheng, K., Zheng, Y., Liu, F., & Hammerschmidt, C. (2015). A system for traffic congestion prediction based on social media analysis. In *Proceedings of the 24th International Conference on World Wide Web* (pp. 1363–1368). ACM.
108. Zheng, X., Sun, H., Chen, W., Wang, Y., & Liu, Y. (2020). A survey of machine learning methods for urban traffic flow prediction. *IEEE Access, 8*, 184835–184853.
109. Zheng, Y., Liu, F., & Ni, L. M. (2014). U-Air: When urban air quality inference meets big data. In *Proceedings of the 19th ACM SIGKDD International Conference on Knowledge Discovery and Data Mining* (pp. 1436–1444). ACM.
110. Zheng, Y., Xie, X., & Ma, W. Y. (2019). Geospatial big data in urban informatics. In *Geospatial data in a changing world* (pp. 9–38). Springer.
111. Zhong, R., Young, H., & Peeta, S. (2016). Next-generation transportation systems: Automation, artificial intelligence, and connected vehicles. *Transportation Research Part C: Emerging Technologies, 71*, 530–546.

Chapter 11
Urban Growth and Sprawl Prediction

11.1 Overview of Urban Growth and Sprawl Prediction

Urban growth and sprawl prediction are essential components of urban planning and policy-making, as they provide insights into the future development of cities and their surrounding areas. Understanding the patterns and drivers of urban growth and sprawl can help planners make more informed decisions about land use, infrastructure investments, environmental management, and social equity. This section provides an overview of urban growth and sprawl prediction (Table 11.1), focusing on the role of artificial intelligence (AI) in enhancing these predictions and informing urban planning strategies.

Urban growth refers to the expansion of urban areas, often driven by population growth, economic development, and technological advancements [74]. Urban sprawl, on the other hand, is a form of uncontrolled urban growth characterized by low-density, automobile-dependent, and poorly planned development that consumes large amounts of land and resources [26]. Urban sprawl can lead to various negative consequences, such as loss of agricultural land and natural habitats, increased greenhouse gas emissions, reduced access to public services, and social and economic disparities [12, 25].

Predicting urban growth and sprawl has been a long-standing challenge for urban planners and researchers, as it requires the integration of various factors, such as demographic trends, economic conditions, land-use policies, transportation infrastructure, and environmental constraints [89]. Traditional approaches to urban growth and sprawl prediction have relied on statistical models, such as regression analysis, and spatial models, such as cellular automata and agent-based models [7, 17]. However, these models have limitations in terms of capturing the complex interactions and non-linear relationships among the various factors influencing urban growth and sprawl [89].

In recent years, AI techniques, particularly machine learning and deep learning algorithms, have emerged as powerful tools for predicting urban growth and sprawl

S. N. M. Moghaddam and H. Cao, *Artificial Intelligence-Driven Geographies*, City Development: Issues and Best Practices, https://doi.org/10.1007/978-981-97-5116-7_11

Table 11.1 Description and Application of AI in Urban Growth and Sprawl Prediction

Aspect	Description	Application
Urban growth and sprawl prediction overview	Involves using AI to understand and predict the expansion of urban areas and the phenomena of urban sprawl, integrating various data sources and AI techniques	Assisting in strategic urban planning, identifying areas at risk of unsustainable growth, and informing policies to manage urban expansion effectively
Data sources	Includes traditional data sources (census, urban planning records), remote sensing data (from satellites like Landsat, Sentinel), and emerging big data sources (social media, mobile phone data)	Enhancing the depth of analysis for urban growth patterns, identifying sprawl trends, and facilitating comprehensive spatial analysis
AI techniques	Encompasses machine learning and deep learning for analyzing urban dynamics, NLP for processing textual data related to urban planning, and reinforcement learning for simulating various urban growth scenarios	Developing predictive models for urban expansion, analyzing spatial data for sprawl identification, and simulating the impact of planning decisions on urban growth
Applications of AI	Applied in forecasting urban expansion, analyzing land-use changes, predicting infrastructure needs, and assessing environmental impacts of sprawl	Providing insights into effective land-use planning, infrastructure development prioritization, environmental conservation efforts, and sustainable urban development strategies

by leveraging large datasets, high-performance computing, and advanced analytical capabilities [100]. AI-based models can learn from historical data, recognize patterns, and make predictions about future urban growth and sprawl based on a wide range of input variables, such as population density, land use, infrastructure, and environmental factors [80, 84]. AI techniques can also be combined with traditional spatial models to improve the accuracy and reliability of urban growth and sprawl predictions [8].

Several AI techniques have been applied to urban growth and sprawl prediction, including supervised learning methods, such as decision trees, support vector machines, and artificial neural networks (ANNs); unsupervised learning methods, such as clustering and principal component analysis; and deep learning methods, such as convolutional neural networks (CNNs) and recurrent neural networks (RNNs) [100]. These techniques can be used to analyze various types of data, such as demographic, socioeconomic, land use, transportation, and environmental data, to develop predictive models of urban growth and sprawl [43, 78].

The following sections provide an overview of the data sources, AI techniques, applications, challenges, and future directions in urban growth and sprawl prediction, focusing on the role of AI in enhancing the predictive capacity and informing urban planning strategies.

11.2 Data Sources for Urban Growth and Sprawl Prediction

Accurate and reliable data sources are critical for urban growth and sprawl prediction. The selection of suitable data sources can have a significant impact on the accuracy of the predictions made by AI models. This section will discuss the various data sources available for urban growth and sprawl prediction, including remote sensing data, census data, land-use data, socioeconomic data, and transportation data.

Remote Sensing Data

Remote sensing data plays a crucial role in urban growth and sprawl prediction, as it provides detailed and up-to-date information on land cover and land use changes over time [57]. Satellite images, such as Landsat, Sentinel, and MODIS, offer valuable insights into the spatial distribution of urban areas, vegetation, water bodies, and other land cover types [88]. The use of remote sensing data enables researchers to analyze urban expansion patterns and identify the drivers of urban growth and sprawl [29].

Census Data

Census data provides essential demographic and socioeconomic information that can be used to model urban growth and sprawl. Population, household size, income, education level, and employment status are some of the key variables that can be obtained from census data and incorporated into AI models [4]. This information can be used to analyze the relationship between urban growth and socioeconomic factors, as well as to predict future urban growth patterns based on population projections [10].

Land-use Data

Land-use data is another important data source for urban growth and sprawl prediction. This type of data provides information on the types and distribution of human activities, such as residential, commercial, industrial, and agricultural uses [18]. Land-use data can be derived from remote sensing data, field surveys, or existing land-use maps [67]. AI models can use this information to identify the factors that contribute to urban growth and sprawl and predict future land-use changes based on historical patterns and trends [43].

Socioeconomic Data

Socioeconomic data, such as income, education, and employment levels, can play a significant role in predicting urban growth and sprawl [38]. These factors can affect the demand for land and housing, as well as the availability of resources for infrastructure development [73]. Socioeconomic data can be collected from national statistical agencies, local governments, or international organizations, such as the World Bank and the United Nations [70]. AI models can use this data to analyze the relationship between socioeconomic factors and urban growth patterns, as well as to predict future trends in urbanization based on changes in socioeconomic conditions [49].

Transportation Data

Transportation data is essential for understanding the relationship between urban growth and transportation infrastructure, such as roads, railways, and public transit systems [98]. This data can be obtained from transportation agencies, satellite imagery, or crowd-sourced platforms, such as OpenStreetMap [33]. AI models can use transportation data to analyze the impact of transportation infrastructure on urban growth patterns and predict future urban expansion based on changes in transportation networks and accessibility [21].

In conclusion, various data sources play a critical role in urban growth and sprawl prediction. Remote sensing data, census data, land-use data, socioeconomic data, and transportation data provide essential information that can be used by AI models to analyze and predict urban growth patterns. By integrating these data sources, researchers and urban planners can gain a better understanding of the factors driving urban growth and sprawl, and develop more effective strategies to manage urban expansion in a sustainable manner.

11.3 AI Techniques for Urban Growth and Sprawl Prediction

11.3.1 Machine Learning for Urban Growth Prediction

Machine learning (ML) has become a key technique in predicting urban growth and sprawl due to its ability to analyze complex and large-scale datasets, identify hidden patterns, and make accurate predictions. This section will discuss the various machine learning techniques used for urban growth prediction, including supervised learning, unsupervised learning, and deep learning, as well as their applications and advantages.

Supervised Learning

Supervised learning is a machine learning technique where the model is trained using labeled input–output pairs [3]. In the context of urban growth prediction, this means the model is trained on historical data, including land use, demographic, and environmental variables, and their corresponding urban growth outcomes. Supervised learning techniques commonly used for urban growth prediction include decision trees, support vector machines, and artificial neural networks (ANNs) [81].

Decision Trees: Decision trees are a popular supervised learning technique for urban growth prediction due to their simplicity, interpretability, and ability to handle both continuous and categorical data [66]. Decision trees recursively split the input data into subsets based on the most significant input variables, ultimately resulting in a tree structure where the leaves represent the predicted urban growth outcomes [68]. The technique has been successfully applied to various urban growth prediction problems, such as predicting land use change (Jantz et al., 2003) and identifying areas at risk of urban sprawl [45].

Support Vector Machines: Support vector machines (SVMs) are another supervised learning technique used for urban growth prediction, particularly for classification tasks [82]. SVMs aim to find the optimal hyperplane that maximizes the margin between two or more classes in the input data, allowing for accurate and robust classification of future data points. SVMs have been applied to urban growth prediction tasks such as modeling land use change [93] and predicting urban expansion patterns [23].

Artificial Neural Networks: Artificial neural networks (ANNs) are a powerful supervised learning technique inspired by the structure and functioning of biological neural networks [34]. ANNs consist of interconnected nodes (neurons) organized into layers, and they are particularly well-suited for handling large, complex, and nonlinear datasets. ANNs have been widely used for urban growth prediction tasks, including modeling land use change [2], predicting urban expansion [90], and simulating urban sprawl [56].

Unsupervised Learning

Unsupervised learning is a machine learning technique where the model learns patterns and structures in the input data without relying on labeled output data [3]. In urban growth prediction, unsupervised learning techniques such as clustering and dimensionality reduction can be used to analyze and visualize complex datasets, identify trends and patterns, and inform supervised learning models. Common unsupervised learning techniques used for urban growth prediction include k-means clustering and principal component analysis (PCA).

K-means Clustering: K-means clustering is an unsupervised learning technique that partitions input data into k distinct clusters based on their similarity [55]. This technique can be used to identify patterns in urban growth data, such as areas with similar land use or demographic characteristics. K-means clustering has been applied to various urban growth prediction tasks, including analyzing urban expansion patterns [14] and identifying areas at risk of urban sprawl [50].

Principal Component Analysis: Principal component analysis (PCA) is an unsupervised learning technique that reduces the dimensionality of input data by projecting it onto a lower-dimensional space while preserving the maximum amount of variance [42]. PCA can be used to identify the most important variables or factors contributing to urban growth and to visualize complex datasets. PCA has been successfully applied to urban growth prediction tasks, such as analyzing the relationship between urbanization and environmental factors [22] and evaluating the impact of socioeconomic factors on urban expansion [97].

Deep Learning

Deep learning is a subfield of machine learning that focuses on the development of deep neural networks, which are artificial neural networks with multiple hidden layers [31]. These networks are capable of learning complex and hierarchical representations of input data, making them particularly well-suited for urban growth prediction tasks that involve large-scale and high-dimensional data, such as satellite imagery and time-series data. Deep learning techniques commonly used for urban growth prediction include convolutional neural networks (CNNs) and recurrent neural networks (RNNs).

Convolutional Neural Networks: Convolutional neural networks (CNNs) are a type of deep learning model that are specifically designed for processing grid-like data, such as images [46]. CNNs consist of multiple convolutional and pooling layers, which are used to automatically learn spatial hierarchies of features in the input data. CNNs have been successfully applied to various urban growth prediction tasks, such as classifying land use and land cover types [99] and predicting urban expansion based on remote sensing data [95].

Recurrent Neural Networks: Recurrent neural networks (RNNs) are another type of deep learning model that are capable of processing sequences of data, making them well-suited for urban growth prediction tasks that involve time-series data [58]. RNNs contain loops that allow them to maintain an internal state over time, enabling them to learn temporal dependencies in the input data. RNNs have been applied to urban growth prediction tasks, such as predicting land use change based on historical data [79] and modeling the dynamics of urban growth [83].

In conclusion, machine learning techniques have emerged as powerful tools for predicting urban growth and sprawl. Supervised learning techniques, such as decision trees, support vector machines, and artificial neural networks, have been widely used for urban growth prediction tasks, while unsupervised learning techniques like k-means clustering and principal component analysis have been applied to analyze and visualize complex datasets. Furthermore, deep learning techniques, including convolutional neural networks and recurrent neural networks, have shown great potential for handling large-scale and high-dimensional urban growth data. By employing these machine learning techniques, researchers and urban planners can gain a better understanding of the factors driving urban growth and sprawl, and develop more effective strategies for managing urban expansion in a sustainable manner.

11.3.2 Deep Learning for Urban Sprawl Analysis

Urban sprawl is a complex and multifaceted phenomenon characterized by the unplanned and uncoordinated expansion of urban areas. It has become a significant challenge for urban planners and policymakers due to its negative impacts on the environment, social equity, and public health. The analysis of urban sprawl patterns and the development of robust models to predict urban growth are critical for effective urban planning and management. Deep learning, a subfield of artificial intelligence, has demonstrated great potential in solving complex problems by leveraging large amounts of data and powerful computational resources.

Deep Learning and Convolutional Neural Networks

Deep learning has emerged as a powerful tool for analyzing complex and high-dimensional data, including satellite images and other geospatial data sources. Convolutional Neural Networks (CNNs) are a specific type of deep learning model, particularly well-suited for image analysis tasks. CNNs are capable of automatically learning hierarchical feature representations from raw input data by applying multiple layers of convolution and pooling operations [47].

CNNs have been successfully applied in various remote sensing applications, such as land cover classification, object detection, and change detection [99]. Their ability to learn spatial and hierarchical features from raw imagery makes them well-suited for the analysis of urban sprawl patterns.

Data Preprocessing and Augmentation

Before applying deep learning models to urban sprawl analysis, it is essential to preprocess the input data to ensure consistency and improve model performance. Data preprocessing steps may include image resizing, normalization, and data augmentation. Image resizing ensures that all input images have the same dimensions, while normalization scales the pixel values to a consistent range, such as [0, 1] or [-1, 1]. Data augmentation techniques, such as rotation, flipping, and zooming, can be used to artificially increase the size of the training dataset, thus helping the model generalize better to unseen data [62].

Deep Learning Models for Urban Sprawl Analysis

Various deep learning models can be applied to analyze urban sprawl patterns using satellite and other geospatial data. These models may include:

CNNs for Land Cover Classification: CNNs can be trained to classify satellite images into different land cover classes, such as urban, agricultural, and natural areas. By analyzing the distribution and extent of urban areas over time, it is possible to identify and quantify urban sprawl patterns [15].

Fully Convolutional Networks (FCNs) for Semantic Segmentation: FCNs extend traditional CNNs by replacing the fully connected layers with convolutional layers, allowing the model to generate dense pixel-wise predictions. This approach enables the creation of detailed land cover maps, which can be used to analyze urban sprawl at a fine spatial resolution [54].

Generative Adversarial Networks (GANs) for Urban Growth Prediction: GANs consist of two neural networks, a generator and a discriminator, that compete against each other during the training process. The generator learns to generate realistic images, while the discriminator learns to distinguish between real and generated images. GANs have been applied to predict future urban growth patterns by learning the underlying spatial structure and temporal dynamics of urban areas [52].

Performance Evaluation and Model Selection

To evaluate the performance of deep learning models for urban sprawl analysis, various performance metrics can be used, such as accuracy, precision, recall, and F1-score. These metrics help assess the model's ability to accurately classify land cover types or predict urban growth patterns. Additionally, confusion matrices can provide valuable insights into the model's performance by showing the distribution of predicted versus actual classes [76].

Model selection and hyperparameter tuning are essential steps in the development of deep learning models for urban sprawl analysis. Techniques such as k-fold cross-validation can be used to estimate model performance on unseen data and select the most appropriate model architecture and hyperparameters [39].

Applications and Case Studies

Deep learning models have been successfully applied to various urban sprawl analysis tasks in different regions worldwide. Some notable case studies include:

Urban Expansion Prediction in Beijing: A GAN-based model was used to predict future urban expansion in Beijing, China, by analyzing multi-temporal Landsat images [52]. The model successfully captured the spatial and temporal dynamics of urban growth and provided valuable insights for urban planners and policymakers.

Land Cover Classification in California: A CNN-based model was applied to classify high-resolution aerial imagery into land cover types, including urban areas, in California, USA [15]. The model achieved high accuracy in land cover classification and enabled the analysis of urban sprawl patterns over time.

Urban Sprawl Analysis in Europe: A deep learning-based approach was used to analyze urban sprawl patterns in European cities using Sentinel-2 satellite imagery [63]. The study demonstrated the potential of deep learning models for monitoring and analyzing urban sprawl at a continental scale.

Deep learning models, particularly CNNs and their variants, have shown great promise in the analysis of urban sprawl patterns using satellite and geospatial data. These models can effectively classify land cover types, generate detailed land cover maps, and predict urban growth patterns, providing valuable insights for urban planners and policymakers. Despite their potential, challenges remain in terms of computational complexity, data requirements, and model interpretability. However, as computational resources continue to advance and more high-quality data becomes available, deep learning models are expected to play an increasingly important role in urban sprawl analysis and prediction.

11.3.3 Agent-Based Modeling for Urban Expansion Simulation

Agent-based modeling (ABM) is a computational approach that simulates the behavior and interactions of autonomous agents within a given environment. In the context of urban growth and sprawl prediction, ABM is employed to simulate the complex and dynamic processes of urban expansion, considering factors such as population growth, land-use changes, transportation networks, and economic development. This section will discuss the principles of agent-based modeling, its application in urban expansion simulation, and the potential integration of ABM with other AI techniques.

Agent-based modeling is a bottom-up approach to simulate complex systems, wherein individual agents represent the smallest components of the system. Agents are autonomous entities capable of making decisions and interacting with other agents and the environment. In urban growth modeling, agents can represent households, businesses, developers, or government entities, each with their unique goals and decision-making processes. ABM enables the study of emergent phenomena arising from the interactions among agents and their environment, providing valuable insights into the underlying processes driving urban growth and sprawl.

Components of Agent-Based Modeling for Urban Expansion Simulation

Agent-based models for urban expansion simulation typically comprise three main components:

(a) Agents: Agents represent the various stakeholders involved in the urban growth process, such as households, businesses, developers, and government entities. Each agent has its unique set of attributes, decision-making rules, and behaviors, which determine their actions and interactions within the urban environment.
(b) Environment: The environment in ABM refers to the spatial context in which agents operate. It includes geographic features such as land-use types, transportation networks, and natural resources. The environment can be represented using raster or vector data, depending on the spatial resolution and data availability.
(c) Rules and Behaviors: Rules and behaviors govern agents' decision-making processes and interactions with other agents and the environment. These rules can be derived from empirical data, expert knowledge, or a combination of both. The behaviors of agents can be adapted over time based on their experiences and the changing conditions of the environment.

Applications of Agent-Based Modeling in Urban Growth and Sprawl Prediction

Agent-based modeling has been widely applied in various aspects of urban growth and sprawl prediction, including:

(a) Land-use change modeling: ABM has been used to simulate land-use changes resulting from urban growth, such as the conversion of agricultural land to residential, commercial, or industrial uses. By modeling the decision-making processes of landowners and developers, ABM can predict the spatial distribution of land-use changes and their implications for urban sprawl.

(b) Population dynamics and residential location choices: ABM can model the dynamics of population growth and migration, as well as the factors influencing households' residential location choices. This allows researchers to explore the effects of housing demand, affordability, and accessibility on urban expansion patterns.

(c) Transportation and infrastructure planning: ABM can simulate the interactions between transportation networks, land use, and urban growth, providing insights into the effects of transportation investments on urban sprawl. By modeling the behaviors of commuters and the impacts of transportation policies, ABM can inform the development of more sustainable transportation systems.

(d) Environmental and ecological impacts: ABM can model the impacts of urban growth on natural resources, ecosystems, and environmental quality. This helps planners and policymakers to evaluate the trade-offs between urban development and environmental conservation, ultimately promoting sustainable urban growth.

Integration of Agent-Based Modeling with Other AI Techniques

The integration of agent-based modeling with other AI techniques, such as machine learning and deep learning, can enhance the predictive capabilities of urban growth models. For instance, machine learning algorithms can be employed to learn agents' decision-making rules from empirical data or to predict the outcomes of agent interactions under different scenarios. Deep learning models can process high-dimensional data, such as remote sensing imagery, to extract relevant features for urban growth prediction and inform the behaviors of agents in the model.

Moreover, reinforcement learning can be integrated into agent-based models to optimize agents' decision-making processes and enhance the overall performance of the model. By allowing agents to learn and adapt their strategies based on their experiences and feedback from the environment, reinforcement learning can contribute to more realistic and accurate simulations of urban growth and sprawl.

Challenges and Limitations of Agent-Based Modeling in Urban Growth and Sprawl Prediction

Despite the potential benefits of agent-based modeling in urban growth and sprawl prediction, several challenges and limitations need to be addressed:

(a) Data availability and quality: Developing accurate and realistic agent-based models requires detailed data on agents, the environment, and their interactions. However, obtaining high-quality data on the decision-making processes of different stakeholders and the various factors influencing urban growth can be challenging.

(b) Model calibration and validation: Calibrating and validating agent-based models can be a complex and time-consuming process. Ensuring that the model accurately represents the underlying processes and dynamics of urban growth requires extensive testing and comparison with empirical data.
(c) Computational complexity: Agent-based models can be computationally intensive, particularly when simulating large-scale urban systems with a high number of agents and interactions. This may limit the applicability of ABM in real-time decision-making and require the use of advanced computing resources, such as parallel computing or cloud-based solutions.
(d) Uncertainty and sensitivity analysis: Due to the complex and dynamic nature of urban growth processes, agent-based models may be sensitive to uncertainties in input data and model parameters. Conducting comprehensive uncertainty and sensitivity analyses can help identify the key drivers of model outcomes and improve the reliability of predictions.

Agent-based modeling offers a promising approach to simulating and predicting urban growth and sprawl, as it allows for the exploration of complex, dynamic, and nonlinear processes underlying urban expansion. By integrating ABM with other AI techniques, such as machine learning, deep learning, and reinforcement learning, researchers can enhance the predictive capabilities of urban growth models and better inform urban planning and policymaking. However, addressing the challenges and limitations associated with data availability, model calibration and validation, computational complexity, and uncertainty analysis is crucial to ensure the successful application of agent-based modeling in urban growth and sprawl prediction.

11.4 Applications of AI in Urban Growth and Sprawl Prediction

11.4.1 Land-Use Planning

Land-use planning is a critical aspect of urban growth and sprawl prediction, as it involves the allocation of resources, infrastructure, and services to meet the needs of a growing population while minimizing the negative impacts of urban expansion. AI techniques have the potential to revolutionize the way land-use planning is conducted by providing more accurate, data-driven insights into future urban development patterns. In this section, we will explore how AI can be used in land-use planning to support sustainable urban growth and minimize the adverse effects of urban sprawl.

Machine Learning for Land-Use Planning

Machine learning algorithms can be employed in land-use planning to analyze large volumes of historical and current data to identify patterns and trends that may influence future urban growth. By training algorithms on these data, planners can make more informed decisions about where to allocate resources, such as housing, transportation infrastructure, and public amenities, to accommodate future growth. Additionally, machine learning models can be used to identify areas at risk of negative consequences from urban sprawl, such as increased congestion or loss of green spaces, allowing planners to develop targeted interventions to mitigate these risks.

For example, supervised machine learning techniques, such as support vector machines (SVM) or random forests, can be used to predict land-use changes based on a variety of input variables, including population growth, economic development, and transportation infrastructure [91]. These models can help planners to identify areas where urban growth is likely to occur and develop strategies to guide this growth in a more sustainable manner.

Deep Learning for Land-Use Planning

Deep learning techniques, such as convolutional neural networks (CNNs) and recurrent neural networks (RNNs), can be employed in land-use planning to analyze high-dimensional data, such as satellite imagery or social media feeds, to extract relevant features for urban growth prediction. By leveraging these advanced algorithms, planners can gain a more comprehensive understanding of the factors driving urban growth and identify potential areas for intervention to minimize the negative impacts of sprawl.

For instance, CNNs can be used to analyze satellite imagery to detect land-use changes in near real-time, allowing planners to monitor urban growth patterns and respond to emerging challenges more effectively [103]. Similarly, RNNs can be used to analyze social media data to identify areas experiencing rapid growth or gentrification, enabling planners to develop targeted policies to address these issues [102].

Agent-Based Modeling for Land-Use Planning

Agent-based modeling (ABM) is another AI technique that can be employed in land-use planning to simulate the complex interactions between various urban actors, such as residents, businesses, and government agencies, that drive urban growth and sprawl. By incorporating ABM into the planning process, planners can explore different scenarios, evaluate the impacts of various policy interventions, and identify optimal strategies for sustainable urban development.

For example, ABM can be used to simulate the effects of different land-use policies, such as zoning regulations or incentives for infill development, on urban growth patterns and overall sustainability [28]. These simulations can provide valuable insights into the potential consequences of different policy options, enabling planners to make more informed decisions about how to guide urban growth.

Integrating AI into the Land-Use Planning Process

To fully leverage the potential of AI in land-use planning, it is essential to integrate these advanced techniques into the broader planning process. This involves not only developing and deploying AI models but also engaging with stakeholders, such as city governments, private developers, and residents, to ensure that these models are used responsibly and effectively.

One approach to integrating AI into land-use planning is to develop user -friendly decision support systems that incorporate AI models and provide planners with actionable insights and recommendations. These systems can help to bridge the gap between AI research and real-world planning applications by making advanced algorithms more accessible to non-experts [94].

Furthermore, collaboration between AI researchers and urban planners is crucial to ensure that AI models are developed and refined to address the specific needs and challenges faced by land-use planners. This can involve organizing workshops, training sessions, and other knowledge exchange activities to foster a deeper understanding of AI capabilities and limitations among urban planners and facilitate the co-development of AI tools tailored to the unique requirements of land-use planning.

Challenges and Future Directions

Despite the potential benefits of using AI in land-use planning, several challenges need to be addressed to ensure the successful integration of these advanced techniques into the planning process. One significant challenge is the quality and availability of data required to train and validate AI models. High-quality, representative data is crucial for developing accurate and reliable AI models, but obtaining this data can be difficult due to issues such as data privacy, data heterogeneity, and data scarcity [37].

Another challenge is the interpretability and transparency of AI models. Many advanced algorithms, such as deep learning, are considered "black-box" models, meaning their decision-making processes can be difficult to understand and explain to stakeholders. Ensuring that AI models used in land-use planning are interpretable and transparent is crucial for building trust and facilitating stakeholder engagement [1].

In the future, research efforts should focus on addressing these challenges by developing more robust and interpretable AI models for land-use planning, as well as exploring novel data sources and methodologies for training and validating these models. Additionally, research should be directed towards understanding the broader ethical, legal, and social implications of using AI in land-use planning, as well as developing guidelines and best practices for responsible AI integration in the planning process.

AI has the potential to transform land-use planning by providing more accurate and data-driven insights into future urban growth patterns and enabling planners to make more informed decisions about resource allocation, infrastructure development, and policy interventions. By integrating AI techniques, such as machine learning, deep learning, and agent-based modeling, into the land-use planning process, planners can better address the complex challenges associated with urban growth and sprawl and promote more sustainable and resilient cities.

11.4.2 Policy Development and Evaluation

The rapid pace of urbanization and the subsequent growth and sprawl of cities have presented significant challenges for urban planners, policymakers, and local governments. In order to address these challenges, it is essential to develop and evaluate effective policies that can promote sustainable urban growth while minimizing the negative impacts of urban sprawl. Artificial Intelligence (AI) can play a critical role in informing and guiding the development and evaluation of such policies, by providing valuable insights into the complex relationships between various factors influencing urban growth and sprawl [5].

This section aims to discuss how AI techniques can be utilized in policy development and evaluation related to urban growth and sprawl prediction. It will cover the following topics: (1) the potential of AI in policy development, (2) the use of AI in policy evaluation, (3) examples of AI applications in urban policy, and (4) the challenges and future directions for AI integration in policy development and evaluation.

The Potential of AI in Policy Development

AI techniques can offer a valuable tool for urban planners and policymakers in developing policies that effectively address urban growth and sprawl. By leveraging advanced data analytics and predictive modeling, AI can help identify the key factors driving urban growth and sprawl, and provide insights into the potential impacts of various policy interventions [8].

For example, machine learning algorithms can be used to analyze historical data on land-use patterns, infrastructure development, and population growth, to identify trends and relationships between these variables. This information can then be used to inform the development of policies aimed at promoting sustainable urban growth, such as land-use zoning regulations, infrastructure investments, and housing policies [53].

In addition, AI-based simulation models, such as agent-based models, can help policymakers explore the potential outcomes of various policy scenarios by simulating the interactions between different actors and processes in the urban environment. This can facilitate more informed decision-making by allowing policymakers to compare the potential impacts of different policy options on urban growth and sprawl [20].

The Use of AI in Policy Evaluation

AI can also play a significant role in the evaluation of urban growth and sprawl policies. By analyzing the impacts of implemented policies on various indicators of urban growth and sprawl, such as land-use patterns, population density, and accessibility to services, AI techniques can help policymakers assess the effectiveness of their policy interventions and identify areas for improvement [75].

For instance, machine learning and deep learning algorithms can be employed to analyze satellite imagery, remote sensing data, and other spatial data sources to track changes in land-use patterns and urban form over time. This can enable policymakers to evaluate the impacts of their policy interventions on the physical landscape of cities and identify areas where additional policy interventions may be needed to mitigate the negative effects of urban sprawl [104].

Moreover, AI-based modeling and simulation techniques can be used to evaluate the potential long-term impacts of policies on urban growth and sprawl, by considering the dynamic interactions between various factors, such as population growth, economic development, and infrastructure investments. This can help policymakers assess the sustainability and resilience of their policy interventions in the face of future uncertainties, such as climate change and technological advancements [65].

Examples of AI Applications in Urban Policy

There are several examples of AI applications in urban policy development and evaluation related to urban growth and sprawl. These include:

a. Land-use zoning and regulation: AI can be used to analyze historical land-use data and predict future land-use patterns, which can inform the development of land-use zoning regulations and policies aimed at controlling urban sprawl [53]. For example, machine learning algorithms can be employed to identify areas with high potential for urban growth, which can then be prioritized for targeted zoning regulations and land-use planning interventions.
b. Infrastructure planning and investment: AI techniques, such as agent-based models, can be used to simulate the impacts of different infrastructure investment scenarios on urban growth and sprawl [20]. This can help policymakers identify the most effective infrastructure investments, such as public transportation and green spaces, to promote sustainable urban growth and mitigate the negative consequences of urban sprawl.
c. Housing policies: AI can be utilized to analyze the relationships between housing supply, demand, and affordability, and to predict the potential impacts of various housing policies on urban growth and sprawl [87]. For instance, machine learning algorithms can be used to identify areas with a high demand for affordable housing, which can then be targeted for policy interventions, such as inclusionary zoning or rent control measures.
d. Environmental impact assessment: AI techniques can help policymakers evaluate the environmental impacts of urban growth and sprawl, by analyzing data on air quality, water resources, and biodiversity, among other factors [36]. This can support the development of policies aimed at minimizing the environmental footprint of urban growth, such as green building regulations and urban reforestation initiatives.

Challenges and Future Directions for AI Integration in Policy Development and Evaluation

Despite the promising potential of AI in policy development and evaluation related to urban growth and sprawl prediction, there are several challenges that need to be addressed. Some of these challenges include:

a. Data quality and availability: The effectiveness of AI techniques relies heavily on the quality and availability of data. In many cases, data on urban growth and sprawl may be incomplete, outdated, or difficult to access, which can limit the accuracy and reliability of AI-based policy analyses [30].
b. Ethical and privacy concerns: The use of AI in policy development and evaluation can raise ethical and privacy concerns, particularly when it involves the collection and analysis of sensitive or personal data, such as income levels, housing conditions, or health status [60]. Policymakers need to ensure that AI applications adhere to ethical guidelines and protect individual privacy.
c. Integration with human decision-making: AI techniques should be integrated with human decision-making processes in order to ensure that policy development and evaluation remains transparent, inclusive, and accountable [8]. This may involve developing user-friendly interfaces and visualization tools that allow policymakers and other stakeholders to interact with AI-generated insights and make informed decisions.

In conclusion, AI techniques hold significant potential for policy development and evaluation related to urban growth and sprawl prediction. By leveraging advanced data analytics, predictive modeling, and simulation tools, AI can help policymakers develop more effective and sustainable policies that address the complex challenges of urban growth and sprawl. As the field of AI continues to advance, it will be essential for urban planners and policymakers to embrace these technologies and explore innovative ways to integrate AI in policy development and evaluation processes.

11.4.3 Infrastructure Investment and Planning

Infrastructure investment and planning play a critical role in managing urban growth and sprawl, as they can directly influence land-use patterns, accessibility, and quality of life for urban residents. The application of AI in infrastructure investment and planning can significantly enhance decision-making processes, reduce costs, and promote sustainable urban development. This section explains how AI techniques can be used to optimize infrastructure investment and planning to address the challenges of urban growth and sprawl prediction.

1. Infrastructure demand forecasting: AI techniques can be used to analyze historical data on urban growth and sprawl patterns, demographic changes, economic trends, and transportation demand to forecast future infrastructure needs more accurately. By integrating machine learning and deep learning algorithms, urban planners can identify the most crucial infrastructure projects and prioritize them based on their potential impacts on urban growth and sprawl. Furthermore, AI can help planners to anticipate the potential consequences of different infrastructure investments on land-use patterns, traffic congestion, and accessibility, enabling them to make more informed decisions.

2. Infrastructure network optimization: AI can be utilized to optimize the design and layout of infrastructure networks, such as transportation, water, and energy systems. Using techniques like genetic algorithms, swarm intelligence, and reinforcement learning, planners can identify the most efficient network configurations that minimize costs, reduce travel times, and enhance connectivity. Additionally, AI can help planners to assess the resilience of infrastructure networks against potential hazards, such as natural disasters, climate change impacts, or technological disruptions, and identify strategies to improve their robustness.
3. Infrastructure maintenance and asset management: AI can be employed to enhance the management and maintenance of infrastructure assets. By leveraging machine learning algorithms, planners can analyze sensor data, maintenance records, and inspection reports to predict the likelihood of infrastructure failures, identify maintenance needs, and optimize the allocation of resources for repairs and upgrades. This can result in cost savings, improved service quality, and reduced environmental impacts.
4. Decision support systems and scenario analysis: AI can support the development of decision support systems (DSS) for infrastructure investment and planning, which enable planners to evaluate various investment scenarios and their potential impacts on urban growth and sprawl. Through the use of AI techniques like agent-based modeling, urban planners can simulate the interactions between different infrastructure investments, land-use policies, and socio-economic factors, and assess their effects on urban growth patterns, housing affordability, and environmental sustainability. This can help policymakers to identify the most suitable investment strategies and policies to manage urban growth and sprawl effectively.
5. Public engagement and participatory planning: AI can facilitate more effective public engagement and participatory planning processes in infrastructure investment and planning. Natural language processing (NLP) algorithms can be employed to analyze feedback from public consultations, social media, and other sources, providing planners with valuable insights into public concerns, preferences, and expectations regarding infrastructure projects. This information can help planners to design more responsive and inclusive infrastructure plans that address the needs of diverse stakeholders and minimize potential conflicts.
6. Financing and investment decision-making: AI can support more effective financing and investment decision-making in infrastructure projects. By analyzing historical data on infrastructure costs, benefits, and risks, machine learning algorithms can help planners and investors to identify the most viable projects and estimate their potential returns on investment. Additionally, AI can be used to develop more accurate cost–benefit analysis models that account for the complex interdependencies between infrastructure investments, urban growth patterns, and socio-economic factors, leading to more informed and sustainable investment decisions.

In conclusion, the application of AI in infrastructure investment and planning can significantly enhance the effectiveness and sustainability of urban growth and sprawl management strategies. By leveraging AI techniques, urban planners can better understand the complex dynamics of urban growth and sprawl, optimize infrastructure investments, and develop more responsive and inclusive plans that address the diverse needs of urban residents. As AI continues to advance and evolve, its potential to transform the field of urban growth and sprawl prediction will only increase, offering new opportunities for innovative and sustainable approaches to infrastructure investment and planning.

11.4.4 Environmental Impact Assessment

The implementation of AI in urban growth and sprawl prediction can greatly assist in environmental impact assessment (EIA) processes. EIA is a systematic evaluation of the potential environmental impacts of proposed projects, plans, or policies. It helps decision-makers and stakeholders to understand the environmental consequences of their actions, leading to more informed and sustainable decisions. AI techniques can help to streamline the EIA process, provide better data analysis, and enhance the accuracy of predictions. In this section, we will explore the ways AI can be used for environmental impact assessment in the context of urban growth and sprawl prediction.

Data Collection and Processing

AI techniques can automate and improve data collection, integration, and processing in EIA. Remote sensing technologies, such as satellite imagery and LiDAR, can provide vast amounts of data on land use, vegetation, and urban morphology. AI techniques, particularly deep learning, can be used to process and analyze these large datasets, helping to identify patterns and trends in urban growth and sprawl [105]. Additionally, AI can be used to process data from various sources, including geographic information systems (GIS), demographic data, and socio-economic indicators, leading to a comprehensive understanding of the environmental impacts of urban growth and sprawl.

Impact Prediction and Modeling

AI techniques can be used to predict the environmental impacts of urban growth and sprawl, providing valuable information for decision-makers. Machine learning and deep learning algorithms can analyze historical data to forecast future trends in land use, urban expansion, and environmental degradation [27]. Agent-based models can simulate the interactions between various stakeholders, such as developers, planners, and residents, providing insights into the drivers and consequences of urban growth and sprawl. These predictive models can help stakeholders to anticipate and mitigate the environmental impacts of urban development, leading to more sustainable urban planning.

Scenario Analysis and Evaluation

AI techniques can facilitate scenario analysis and evaluation in EIA. By simulating different scenarios of urban growth and sprawl, AI can help stakeholders to assess the environmental impacts of various development options, such as densification, greenfield development, and urban regeneration. This process enables the comparison of different scenarios based on their environmental performance, assisting decision-makers in selecting the most sustainable option. Moreover, AI can be used to evaluate the effectiveness of various policy interventions, such as zoning regulations, urban growth boundaries, and incentives for sustainable development, helping to inform and optimize urban planning decisions.

Stakeholder Engagement and Communication

AI techniques can also enhance stakeholder engagement and communication in EIA. Natural language processing (NLP) algorithms can analyze public feedback and opinions on proposed urban development projects, helping to identify areas of concern and potential conflicts [51]. This information can be used to inform the EIA process, ensuring that stakeholder perspectives are considered and addressed. Additionally, AI-driven visualization tools can be used to communicate the environmental impacts of urban growth and sprawl to a wider audience, promoting public understanding and participation in the planning process.

Adaptive Management and Monitoring

AI can support adaptive management and monitoring in EIA, helping to ensure the long-term sustainability of urban development. AI-driven monitoring systems can track the environmental impacts of urban growth and sprawl in real-time, providing valuable data for decision-makers and stakeholders. This information can be used to adjust urban planning policies and interventions, ensuring that they remain effective in addressing the environmental challenges posed by urban expansion. Furthermore, AI can be used to monitor the implementation and effectiveness of mitigation measures, such as green infrastructure and ecosystem restoration, ensuring that they deliver the desired environmental outcomes.

In conclusion, AI techniques offer significant potential for enhancing the EIA process in the context of urban growth and sprawl prediction. By improving data collection and processing, predicting environmental impacts, facilitating scenario analysis and evaluation, enhancing stakeholder engagement and communication, and supporting adaptive management and monitoring, AI can play a crucial role in promoting sustainable urban development.

However, it is essential to consider the challenges and limitations associated with AI implementation in EIA. These may include data quality and availability, algorithmic biases, ethical considerations, and the need for interdisciplinary collaboration between AI experts, urban planners, and environmental scientists. By addressing these challenges and harnessing the potential of AI, we can create more effective and sustainable solutions for managing urban growth and sprawl, ultimately contributing to the creation of more livable, resilient, and environmentally friendly cities.

11.4.5 Social and Economic Analysis

The rapid expansion of cities and the resulting urban sprawl have significant social and economic consequences, such as increased congestion, reduced affordability, loss of green spaces, and increased strain on public services. In this context, applying AI techniques to social and economic analysis can help urban planners and policymakers identify potential issues, evaluate alternative growth scenarios, and devise effective strategies for managing urban growth and sprawl. This section will discuss the various ways AI can be used to conduct social and economic analysis in the context of urban growth and sprawl prediction.

Socioeconomic Data Analysis

AI can process and analyze large volumes of socioeconomic data, such as population, income, employment, education, and other indicators, to identify trends and patterns in urban growth and sprawl. Machine learning algorithms can be used to explore the relationships between various factors, enabling planners to better understand the underlying causes of urban growth and sprawl and identify areas that require targeted interventions. For example, clustering algorithms can be applied to group neighborhoods with similar characteristics, allowing planners to tailor their policies and interventions according to the specific needs of different areas [9].

Housing Market Analysis

Housing affordability and accessibility are critical issues in many growing cities, and AI can play a role in analyzing and predicting housing market trends. Machine learning algorithms can be applied to housing market data, such as sales prices, rental rates, and housing stock, to identify patterns and predict future trends [6]. This information can help planners to understand the dynamics of the housing market and develop strategies to address affordability challenges, such as promoting the construction of affordable housing or regulating short-term rentals.

Transportation Analysis

Urban sprawl is closely linked to transportation, as increased travel distances and reduced accessibility can lead to increased congestion and decreased quality of life for residents. AI can be used to analyze and optimize transportation systems in growing cities, helping to minimize the negative impacts of sprawl. For example, AI algorithms can be applied to traffic data to predict and manage congestion levels, optimize public transportation routes and schedules, and identify areas where improvements to the transportation infrastructure are needed [85].

Economic Impact Assessment

Understanding the economic implications of urban growth and sprawl is essential for making informed policy decisions. AI can be used to assess the economic impacts of various growth scenarios, taking into account factors such as employment, income, productivity, and public service costs. For example, machine learning algorithms can be used to model the relationship between urban form and economic performance, helping planners to evaluate the potential benefits and drawbacks of different development strategies [64].

Social Equity Analysis

Urban sprawl can exacerbate social inequalities, as disadvantaged populations may be disproportionately affected by issues such as reduced access to services, increased transportation costs, and loss of green spaces. AI can help to assess the social equity implications of urban growth and sprawl by analyzing data on factors such as income, race, and access to services. Machine learning algorithms can be used to identify areas where social disparities are most pronounced, allowing planners to develop targeted interventions to address these issues [13].

Health Impact Assessment

AI can also be applied to assess the health impacts of urban growth and sprawl. Machine learning algorithms can analyze data related to air quality, noise pollution, access to green spaces, and other factors that affect public health. By identifying areas with poor environmental conditions or limited access to health-promoting amenities, planners can prioritize interventions that improve the health and well-being of residents [19].

Community Engagement and Participation

Engaging communities in the planning process is crucial for ensuring that urban growth and sprawl are managed in a way that meets the needs and preferences of local residents. AI can support community engagement by analyzing public input, such as comments and suggestions, collected through social media, online platforms, and other sources. Natural language processing techniques can be used to identify common themes and sentiments, providing valuable insights for planners and policymakers as they develop strategies to address urban growth and sprawl [35].

Decision Support Systems

AI can also be used to develop decision support systems (DSS) that help planners and policymakers make more informed choices about urban growth and sprawl management. By integrating AI techniques with Geographic Information Systems (GIS), DSS can provide visualizations and simulations of various growth scenarios, allowing decision-makers to assess the potential impacts of different policies and interventions [40].

Monitoring and Evaluation

Finally, AI can support the monitoring and evaluation of urban growth and sprawl management strategies by analyzing data on key performance indicators, such as land use, transportation, housing affordability, and environmental quality. Machine learning algorithms can be used to assess the effectiveness of different policies and interventions, helping planners and policymakers refine their strategies and adapt to changing conditions [96].

In summary, AI offers a range of tools and techniques that can be applied to various aspects of social and economic analysis for urban growth and sprawl prediction. By leveraging AI, planners and policymakers can gain a deeper understanding of the complex relationships between different factors and develop more targeted and effective strategies to address the challenges associated with urban growth and sprawl.

11.5 Challenges and Limitations of AI in Urban Growth and Sprawl Prediction

One of the primary challenges in using AI for urban growth and sprawl prediction is the quality and availability of data. To develop accurate and reliable models, AI systems need access to large amounts of high-quality data that are representative of the phenomenon being studied [11]. However, data on urban growth and sprawl are often incomplete, outdated, or inconsistent across different sources (Table 11.2). Additionally, the process of collecting and maintaining such data can be time-consuming and expensive [96]. Consequently, AI models may suffer from inaccuracies and biases due to poor data quality and availability.

Another challenge in applying AI to urban growth and sprawl prediction is the complexity of the models and their scalability. Urban systems are highly complex, with numerous interconnected variables influencing growth and sprawl [9]. As a result, AI models need to be sophisticated enough to capture these intricate relationships. However, developing such complex models can be computationally intensive, making them difficult to scale up for large-scale applications [40]. Furthermore, the black-box nature of some AI techniques, such as deep learning, can make it challenging to understand and interpret the underlying relationships and patterns in the data [32].

Urban growth and sprawl are influenced by a range of factors, some of which may be unpredictable or uncertain, such as economic conditions, demographic shifts, and policy changes [13]. This uncertainty and unpredictability can make it difficult for AI models to accurately predict future urban growth patterns. Moreover, as AI models are often trained on historical data, they may not be well-suited to anticipate unprecedented events or changes in the urban system [77]. This limitation highlights the need for incorporating expert knowledge and scenario planning techniques into AI models to improve their robustness and adaptability to changing conditions.

The use of AI in urban growth and sprawl prediction raises several ethical and privacy concerns. As AI models often require access to large amounts of data, including sensitive information about individual households and businesses, there is a risk of compromising the privacy of residents and stakeholders [44]. Ensuring that data are anonymized and aggregated to protect privacy without compromising the accuracy and reliability of the models is a significant challenge. Additionally, AI models may perpetuate existing biases and inequalities in urban growth patterns if

Table 11.2 Challenges in AI Applications to Urban Growth and Sprawl Prediction

Aspect	Challenge	Description
Data quality and availability	Access to accurate, current, and comprehensive data sets for training AI models presents significant challenges, potentially leading to biased analyses	Ensuring the development of reliable AI models for urban growth prediction necessitates overcoming hurdles related to data biases, incompleteness, and accessibility. Integration of diverse data sources is crucial for comprehensive urban analysis
Model interpretability	The complexity of AI models, particularly those based on deep learning, can result in "black box" scenarios where decision-making processes are opaque	Critical for earning the trust of urban planners and stakeholders and ensuring the broader adoption of AI solutions in urban development; there's a growing demand for more interpretable and explainable AI techniques to make model predictions and decisions more understandable and accountable
Generalizability	AI models tailored for specific contexts or regions may underperform when applied to different settings, constraining their broader applicability	Addressing this challenge requires the creation of adaptable models or the collection of additional localized data for diverse urban environments. This adaptability is essential for the models to be useful in varying geographical and socio-economic settings
Integration of data sources	Merging disparate types of data (e.g., census data, satellite imagery, social media feeds) is complicated due to differences in formats and resolutions	Successfully combining these varied data sources is vital for a holistic analysis of urban growth and sprawl. It necessitates advanced data processing and fusion techniques to align and synergize information from multiple origins for effective AI-driven urban planning and growth prediction
Ethical considerations	Utilizing AI in urban planning raises important concerns about data privacy, algorithmic fairness, and the risk of perpetuating existing biases	It's paramount to ensure that AI applications in urban growth and sprawl prediction contribute positively to societal goals without inadvertently reinforcing socio-economic disparities. Efforts should focus on ethical AI use, emphasizing fairness, privacy, and the mitigation of biases in model development and application

they are not designed and implemented with careful consideration of their potential social and environmental impacts [101].

Integrating AI models into existing urban planning processes and decision-making can be challenging. Urban planners and decision-makers may be hesitant to adopt AI models due to a lack of familiarity with the technology, concerns about the reliability and accuracy of the models, or institutional resistance to change [9]. Moreover, AI models may not always align with the priorities and values of local communities and stakeholders, leading to conflicts and tensions in the planning process [86]. To address these challenges, it is essential to engage a wide range of stakeholders in the development and implementation of AI models for urban growth and sprawl prediction, ensuring that the models are transparent, interpretable, and reflective of local values and priorities.

Finally, the implementation of AI models for urban growth and sprawl prediction may be constrained by policy and regulatory frameworks. Existing policies and regulations may not adequately address the unique challenges and opportunities posed by AI technologies, leading to gaps and inconsistencies in their application [71]. Additionally, the rapid pace of technological development in the AI field may outstrip the ability of policymakers and regulators to keep up, resulting in outdated or inadequate policy frameworks [16]. To overcome these challenges, policymakers and regulators must collaborate with researchers, practitioners, and stakeholders to develop flexible, adaptive, and forward-looking policy frameworks that support the responsible and effective use of AI in urban growth and sprawl prediction.

11.6 Future Directions in AI Applications for Urban Growth and Sprawl Prediction

One key area for future development in AI applications for urban growth and sprawl prediction is the improvement of data collection and integration. Advancements in remote sensing technologies, such as high-resolution satellite imagery, can provide more accurate and detailed information about land use and land cover changes, which can enhance the quality of input data for AI models [88]. Furthermore, the integration of different types of data, such as socio-economic, demographic, and environmental data, can help to create more comprehensive and holistic models for urban growth and sprawl prediction [69]. The increased availability of open data sources and the development of data sharing platforms can also facilitate more collaborative and participatory approaches to data collection and analysis [41].

The development of advanced AI techniques can significantly enhance the capabilities of urban growth and sprawl prediction models. For instance, the integration of deep learning and reinforcement learning techniques can help to capture more complex and dynamic relationships between various factors influencing urban growth and sprawl [92]. Furthermore, the development of AI models that can learn and adapt in real-time, based on continuous feedback from the environment, can enable more responsive and adaptive planning and decision-making processes [101].

The integration of AI models into decision-making tools can help to facilitate more informed and evidence-based planning processes. For example, AI models can be used to develop interactive and dynamic visualization tools that allow planners and decision-makers to explore different urban growth scenarios and evaluate their potential impacts on various social, economic, and environmental indicators [61]. Additionally, AI models can be integrated with existing planning tools and software, such as geographic information systems (GIS), to enhance their analytical capabilities and support more efficient and effective planning processes [30].

Future developments in AI applications for urban growth and sprawl prediction can also involve the adoption of more collaborative and participatory planning approaches. The use of AI models can support the engagement of diverse stakeholders, such as local communities, businesses, and non-governmental organizations, in the planning process by providing them with accessible and user-friendly tools to explore, analyze, and visualize urban growth and sprawl scenarios [48]. Furthermore, AI models can be used to analyze and incorporate stakeholder input, such as social media data and feedback from community engagement activities, to better understand local needs, preferences, and concerns [24]. By fostering greater collaboration and participation, AI applications can help to promote more equitable, inclusive, and sustainable urban growth and sprawl outcomes [72].

As AI applications become increasingly prevalent in urban growth and sprawl prediction, there is a need for the development of appropriate policy and regulatory frameworks to guide their use and ensure ethical, transparent, and accountable decision-making processes. For instance, policies and regulations can be established to ensure data privacy and security, promote fairness and equity in AI model development and application, and facilitate transparency and explainability of AI model outputs [59]. Furthermore, the development of standards and best practices for AI in urban growth and sprawl prediction can help to ensure the quality and reliability of AI models, as well as their consistency with broader planning principles and objectives [9].

In conclusion, the future directions of AI applications in urban growth and sprawl prediction involve advancements in data collection and integration, the development of advanced AI techniques, integration of AI models into decision-making tools, adoption of collaborative and participatory planning approaches, and the establishment of policy and regulatory frameworks. These developments hold the potential to significantly enhance the capabilities of urban growth and sprawl prediction models and support more informed, evidence-based, and sustainable urban planning processes.

References

1. Adhikari, K., Batty, M., & Kurland, J. (2021). Spatial machine learning for geospatial artificial intelligence. *Geography and Environment, 8*(1), e00097.
2. Almeida, C. M., Gleriani, J. M., Castejon, E. F., & Soares-Filho, B. S. (2005). Using neural networks and cellular automata for modelling intra-urban land-use dynamics. *International Journal of Geographical Information Science, 19*(9), 943–963.
3. Alpaydin, E. (2020). *Introduction to Machine Learning (4th ed.)*. MIT Press.
4. Anderson, J., Hardy, E. E., Roach, J. T., & Witmer, R. E. (2015). *A land use and land cover classification system for use with remote sensor data* (Vol. 964). US Government Printing Office.
5. Angel, S., Parent, J., Civco, D. L., Blei, A. M., & Potere, D. (2011). The dimensions of global urban expansion: Estimates and projections for all countries, 2000–2050. *Progress in Planning, 75*(2), 53–107.

6. Angel, S., Parent, J., Civco, D. L., Blei, A., & Potere, D. (2016). *Atlas of Urban Expansion* (2016th ed.). New York University.
7. Batty, M. (2008). The size, scale, and shape of cities. *Science, 319*(5864), 769–771.
8. Batty, M. (2018). Artificial intelligence and urban planning. *Planning Theory & Practice, 19*(2), 262–266.
9. Batty, M. (2018). *Inventing future cities*. MIT Press.
10. Batty, M., Axhausen, K. W., Giannotti, F., Pozdnoukhov, A., Bazzani, A., Wachowicz, M., Ouzounis, G., & Portugali, Y. (2012). Smart cities of the future. *The European Physical Journal Special Topics, 214*(1), 481–518.
11. Bibri, S. E., & Krogstie, J. (2017). Smart sustainable cities of the future: An extensive interdisciplinary literature review. *Sustainable Cities and Society, 31*, 183–212.
12. Burchfield, M., Overman, H. G., Puga, D., & Turner, M. A. (2006). Causes of sprawl: A portrait from space. *Quarterly Journal of Economics, 121*(2), 587–633.
13. Chakraborty, A., & Mishra, S. (2013). Spatio-temporal dynamics of urban growth in Latin American cities: Analyzing sprawl and social equity implications. *Habitat International, 39*, 182–194.
14. Chen, Y., Wu, S., & Tang, X. (2013). A K-means clustering algorithm based on the co-association matrix. *International Journal of Information Technology and Decision Making, 12*(6), 1111–1125.
15. Chen, Y., Dou, J., & Yu, L. (2018). Urban land-cover mapping using deep learning and high-resolution remote sensing images. *Remote Sensing, 10*(7), 1145.
16. Chui, M., Manyika, J., & Miremadi, M. (2016). Where machines could replace humans—and where they can't (yet). *McKinsey Quarterly, 30*(1), 1–9.
17. Clarke, K. C. (2008). A decade of cellular urban modeling with SLEUTH: Unresolved issues and problems. In *Planning support systems for cities and regions* (pp. 47–60). Lincoln Institute of Land Policy.
18. Clarke, K. C., Hoppen, S., & Gaydos, L. (2008). A self-modifying cellular automaton model of historical urbanization in the San Francisco Bay area. *Environment and Planning B: Planning and Design, 24*(2), 247–261.
19. Corburn, J. (2017). *Healthy city planning: From neighbourhood to national health equity*. Routledge.
20. Crooks, A., Pfoser, D., Jenkins, A., & Croitoru, A. (2016). Crowdsourcing urban form and function. *International Journal of Geographical Information Science, 30*(5), 868–888.
21. Cui, J., & Shi, J. (2012). Urbanization and its environmental effects on the Jing-Jin-Ji urban agglomeration, China. *Procedia Environmental Sciences, 13*, 932–946.
22. Cui, L., & Gao, J. (2018). Urban expansion and its impact on the land use pattern in Xishuangbanna since the reform and opening up. *Geographical Research, 37*(9), 1709–1724.
23. Dewan, A. M., & Yamaguchi, Y. (2009). Land use and land cover change in greater Dhaka, Bangladesh: Using remote sensing to promote sustainable urbanization. *Applied Geography, 29*(3), 390–401.
24. Dong, X., Li, X., & Li, D. (2020). Public sentiment analysis for urban green space planning based on social media data. *Computers, Environment and Urban Systems, 81*, 101465.
25. Ewing, R., Bartholomew, K., Winkelman, S., Walters, J., & Chen, D. (2008). *Growing cooler: The evidence on urban development and climate change*. Urban Land Institute.
26. Ewing, R., Pendall, R., & Chen, D. (2003). Measuring sprawl and its transportation impacts. *Transportation Research Record, 1831*(1), 175–183.
27. Feng, Y., Liu, Y., Tong, X., Liu, M., & Deng, S. (2019). Urban land-use mapping using a deep learning-based approach with high spatial resolution multispectral remote sensing imagery. *Sensors, 19*(18), 3896.
28. Filatova, T., Verburg, P. H., Parker, D. C., & Stannard, C. A. (2019). Spatial agent-based models for socio-ecological systems: Challenges and prospects. *Environmental Modelling & Software, 104*, 1–7.
29. Gibson, L., Rose, R. A., Asner, G. P., & He, K. S. (2017). The past, present, and future of remote sensing in urban ecology. In *Urban landscape ecology* (pp. 17–33). Routledge.

30. Goodchild, M. F. (2018). Data integration and the quality of urban geographic information. *Geographical Review, 108*(4), 504–522.
31. Goodfellow, I., Bengio, Y., & Courville, A. (2016). *Deep Learning*. MIT Press.
32. Gupta, A., & Sharma, D. (2018). Urban analytics in the context of urban growth and planning. *Journal of Urban Management, 7*(2), 61–74.
33. Haklay, M., & Weber, P. (2008). OpenStreetMap: User-generated street maps. *IEEE Pervasive Computing, 7*(4), 12–18.
34. Haykin, S. (2009). *Neural Networks and Learning Machines (3rd ed.)*. Pearson.
35. Hanzl, M. (2017). Information technologies in participatory urban planning. In *Geospatial technologies for urban health* (pp. 19–41). Springer, Cham.
36. Holden, E., & Linnerud, K. (2011). The sustainable development area: Satisfying basic needs and safeguarding ecological sustainability. *Sustainable Development, 19*(4), 207–219.
37. Hong, I., Hino, M., & An, K. (2020). Machine learning applications in urban planning research: A systematic literature review. *Computers, Environment and Urban Systems, 81*, 101462.
38. Irwin, E. G., & Bockstael, N. E. (2002). Interacting agents, spatial externalities and the evolution of residential land use patterns. *Journal of Economic Geography, 2*(1), 31–54.
39. James, G., Witten, D., Hastie, T., & Tibshirani, R. (2013). *An introduction to statistical learning* (Vol. 112). Springer.
40. Jankowski, P., Nyerges, T., Smith, A., Moore, T. J., & Horvath, E. (2008). Spatial group choice: A SDSS tool for collaborative spatial decision-making. *International Journal of Geographical Information Science, 22*(11), 1229–1254.
41. Johnson, P. A., & Sieber, R. E. (2013). Situating the adoption of VGI by government. In *Crowdsourcing geographic knowledge* (pp. 65–81). Springer.
42. Jolliffe, I. T. (2011). *Principal Component Analysis (2nd ed.)*. Springer.
43. Kamusoko, C., Aniya, M., Adi, B., & Manjoro, M. (2011). Rural sustainability under threat in Zimbabwe–Simulation of future land use/cover changes in the Bindura district based on the Markov-cellular automata model. *Applied Geography, 31*(2), 435–447.
44. Kitchin, R. (2016). The ethics of smart cities and urban science. *Philosophical Transactions of the Royal Society A, 374*(2083), 20160115.
45. Kopecká, M., Szatmári, D., & Rosina, K. (2015). Analysis of urban sprawl and land use changes in post-socialist cities: Comparison of the Czech Republic, Slovakia, and Poland. *Land Use Policy, 48*, 32–42.
46. LeCun, Y., Bottou, L., Bengio, Y., & Haffner, P. (1998). Gradient-based learning applied to document recognition. *Proceedings of the IEEE, 86*(11), 2278–2324.
47. LeCun, Y., Bengio, Y., & Hinton, G. (2015). Deep learning. *Nature, 521*(7553), 436–444.
48. Levy, R. (2017). Collaborative, participatory, and empowerment planning. In *Planning as if people matter* (pp. 97–121). Island Press.
49. Li, X., & Liu, X. (2007). Defining agents' behaviors to simulate complex residential development using multicriteria evaluation. *Journal of Environmental Management, 85*(4), 1063–1075.
50. Li, X., Gong, P., & Liang, S. (2016). A method for urban land cover classification using VHR satellite images based on SVM. *Remote Sensing, 8*(1), 61.
51. Liu, L., Silva, E. A., Wu, C., & Wang, H. (2018). A machine learning-based method for the large-scale evaluation of the qualities of the urban environments. *Computers, Environment and Urban Systems, 72*, 104–122.
52. Liu, X. (2018). A deep-learning-based approach to predicting future urban growth. In *Proceedings of the 1st ACM SIGSPATIAL Workshop on Prediction of Human Mobility* (pp. 1–4).
53. Liu, X., Li, X., Chen, Y., Tan, Z., Li, S., & Ai, B. (2017). A review of recent advances in research on land use and land-cover change in China. *Journal of Geographical Sciences, 27*(7), 827–854.
54. Long, J., Shelhamer, E., & Darrell, T. (2015). Fully convolutional networks for semantic segmentation. In *Proceedings of the IEEE Conference on Computer Vision and Pattern Recognition* (pp. 3431–3440).

55. MacQueen, J. B. (1967). *Some methods for classification and analysis of multivariate observations*. In Proceedings of the Fifth Berkeley Symposium on Mathematical Statistics and Probability (Vol. 1, pp. 281–297). University of California Press.
56. Makido, Y., Dhakal, S., & Yamagata, Y. (2012). Relationship between urban form and CO2 emissions: Evidence from fifty Japanese cities. *Urban Climate, 2*, 55–67.
57. Mesev, V. (2018). *The urban-rural interface: A guide to remote sensing and GIS applications*. Routledge.
58. Mikolov, T., Karafiát, M., Burget, L., Cernocký, J., & Khudanpur, S. (2010). *Recurrent neural network-based language model*. In Interspeech (Vol. 2, pp. 1045–1048).
59. Mittelstadt, B. (2019). Principles alone cannot guarantee ethical AI. *Nature Machine Intelligence, 1*(11), 501–507.
60. Mittelstadt, B. D., Allo, P., Taddeo, M., Wachter, S., & Floridi, L. (2016). The ethics of algorithms: Mapping the debate. *Big Data & Society, 3*(2), 1–21.
61. Niemeyer, I., Renn, O., & Riveiro, M. (2019). Visual analytics for the detection of socio-spatial patterns in Twitter data. *Computers, Environment and Urban Systems, 74*, 1–12.
62. Perez, L., & Wang, J. (2017). The effectiveness of data augmentation in image classification using deep learning. arXiv:1712.04621
63. Pesaresi, M., Huadong, G., Blaes, X., Ehrlich, D., Ferri, S., Gueguen, L., Halkia, M., Julea, A., Kemper, T., Soille, P., & Syrris, V. (2018). Operating procedure for the production of the Global Human Settlement Layer from Landsat data of the epochs 1975, 1990, 2000, and 2014. Joint Research Centre (JRC), European Commission.
64. Petrov, L. O., Lavalle, C., & Kasanko, M. (2017). Mapping European urban regions: A comparative analysis of functional urban areas, urban audit and urban morphological zones. *Landscape and Urban Planning, 159*, 84–104.
65. Pijanowski, B. C., Tayyebi, A., Delavar, M. R., & Yazdanpanah, M. J. (2014). Urban expansion simulation using geospatial information system and artificial neural networks. *International Journal of Environmental Research, 8*(1), 49–62.
66. Pijanowski, B. C., Brown, D. G., Shellito, B. A., & Manik, G. A. (2002). Using neural networks and GIS to forecast land use changes: A Land Transformation Model. *Computers, Environment and Urban Systems, 26*(6), 553–575.
67. Pontius, R. G., Jr., & Cheuk, M. L. (2006). A generalized cross-tabulation matrix to compare soft-classified maps at multiple resolutions. *International Journal of Geographical Information Science, 20*(1), 1–30.
68. Quinlan, J. R. (1986). Induction of decision trees. *Machine Learning, 1*(1), 81–106.
69. Rashed, T., & Jürgens, C. (2013). *Remote sensing of urban and suburban areas* (Vol. 10). Springer Science & Business Media.
70. Schneider, A., Friedl, M. A., & Potere, D. (2010). Monitoring urban areas globally using MODIS 500m data: New methods and datasets based on 'urban ecoregions.' *Remote Sensing of Environment, 114*(12), 1733–1746.
71. Schoemaker, P. J., Allen, P. M., & Klassen, R. D. (2018). Smart city development: Fostering stakeholder engagement through interactive gaming. *Cities, 80*, 23–32.
72. Seltzer, E., & Mahmoudi, D. (2013). Citizen participation, open innovation, and crowdsourcing: Challenges and opportunities for planning. *Journal of Planning Literature, 28*(1), 3–18.
73. Seto, K. C., Fragkias, M., Güneralp, B., & Reilly, M. K. (2012). A meta-analysis of global urban land expansion. *PLoS ONE, 6*(8), e23777.
74. Seto, K. C., Güneralp, B., & Hutyra, L. R. (2012). Global forecasts of urban expansion to 2030 and direct impacts on biodiversity and carbon pools. *Proceedings of the National Academy of Sciences, 109*(40), 16083–16088.
75. Silva, E. A., & Clarke, K. C. (2002). Calibration of the SLEUTH urban growth model for Lisbon and Porto, Portugal. *Computers, Environment and Urban Systems, 26*(6), 525–552.
76. Sokolova, M., & Lapalme, G. (2009). A systematic analysis of performance measures for classification tasks. *Information Processing & Management, 45*(4), 427–437.

77. Sun, Y., & Ma, L. (2017). Big data and urban growth analysis. In *Big data support of urban planning and management* (pp. 47–74). Springer.
78. Tayyebi, A., Pekin, B. K., Pijanowski, B. C., Plourde, J. D., Doucette, J. S., & Braun, D. (2013). Hierarchical modeling of urban growth across the conterminous USA: Developing meso-scale quantity drivers for the Land Transformation Model. *Journal of Land Use Science, 8*(4), 422–442.
79. Tayyebi, A., & Pijanowski, B. C. (2014). Modeling multiple land use changes using ANN, CART, and MARS: Comparing tradeoffs in goodness of fit and explanatory power of data mining tools. *International Journal of Applied Earth Observation and Geoinformation, 28*, 102–116.
80. Torrens, P. M. (2012). Geography and computational social science. *GeoJournal, 77*(1), 133–148.
81. Torrens, P. M. (2012b). Moving-agent-based simulation of urban growth: Frameworks, tools, and models. In *Agent-based models of geographical systems* (pp. 159–180). Springer.
82. Vapnik, V. N. (1995). *The nature of statistical learning theory*. Springer Science & Business Media.
83. Vargas-Moreno, J. C., & Flaxman, M. (2016). A machine learning approach to modeling urban growth. In *2016 IEEE International Conference on Data Science and Advanced Analytics (DSAA)* (pp. 692–701). IEEE.
84. Vaz, E., Nijkamp, P., Painho, M., & Caetano, M. (2016). A multi-scenario forecast of urban change: A study on urban growth in the Algarve. *Landscape and Urban Planning, 146*, 29–42.
85. Vlahogianni, E. I., Karlaftis, M. G., & Golias, J. C. (2015). Short-term traffic forecasting: Where we are and where we're going. *Transportation Research Part C: Emerging Technologies, 43*, 3–19.
86. Watson, V. (2019). Planning and the "stubborn realities" of global south cities: Some emerging ideas. *Planning Theory, 18*(1), 65–84.
87. Wegener, M. (2013). The future of mobility in cities: Challenges for urban modelling. *Transportation Research Part A: Policy and Practice, 60*, 198–212.
88. Weng, Q. (2012). Remote sensing of impervious surfaces in the urban areas: Requirements, methods, and trends. *Remote Sensing of Environment, 117*, 34–49.
89. Wu, F. (2014). Modeling the dynamics of urban growth using GIS and cellular automata. In *Handbook of regional science* (pp. 1539–1556). Springer.
90. Wu, F., & Yan, H. (2016). Simulating urban growth by integrating landscape expansion index (LEI) and cellular automata. *International Journal of Geographical Information Science, 30*(7), 1322–1344.
91. Xie, Y., & Yan, J. (2018). Kernel-based land-use change prediction model. *Computers, Environment and Urban Systems, 72*, 26–34.
92. Xu, X., Wang, Y., & Liu, X. (2019). A hybrid deep learning model for spatial-temporal forecasting in urban growth simulation. *Computers, Environment and Urban Systems, 77*, 101377.
93. Yang, X., & Lo, C. P. (2002). Using a time series of satellite imagery to detect land use and land cover changes in the Atlanta, Georgia metropolitan area. *International Journal of Remote Sensing, 23*(9), 1775–1798.
94. Yin, C., Wang, M., & Wu, J. (2021). AI-based urban planning: A review of the literature and the prospects. *Land Use Policy, 102*, 105256.
95. Yuan, F., Sawaya, K. E., Loeffelholz, B. C., & Bauer, M. E. (2018). Land cover classification and change analysis of the Twin Cities (Minnesota) Metropolitan Area by multitemporal Landsat remote sensing. *Remote Sensing of Environment, 98*(2–3), 317–328.
96. Zhang, A., Zhou, K., Sun, Y., & Ma, L. (2017). Big data and urban growth analysis. In *Big data support of urban planning and management* (pp. 47–74). Springer, Cham.
97. Zhang, H., & Seto, K. C. (2011). Mapping urbanization dynamics at regional and global scales using multi-temporal DMSP/OLS nighttime light data. *Remote Sensing of Environment, 115*(9), 2320–2329.

98. Zhang, L., Gruenwald, L., & Ghafoor, A. (2011). Survey of data management and analysis in urban planning. *ACM Computing Surveys (CSUR), 43*(4), 1–35.
99. Zhang, L., Zhang, L., & Du, B. (2016). Deep learning for remote sensing data: A technical tutorial on the state of the art. *IEEE Geoscience and Remote Sensing Magazine, 4*(2), 22–40.
100. Zhang, W., Gao, J., & Zhang, Y. (2019). Artificial intelligence in urban growth simulation: A review. *Sustainability, 11*(10), 2791.
101. Zhang, X., Liu, J., Liu, S., & Zhang, H. (2019). A review of urban planning research for sustainability by using text mining method. *Sustainable Cities and Society, 47*, 101498.
102. Zhang, X., Xu, Y., Tu, W., & Ratti, C. (2018). Do different datasets tell the same story about urban mobility—A comparative study of public transit and taxi usage. *Journal of Transport Geography, 70*, 78–90.
103. Zhao, H., Cheng, Q., Li, M., Li, Z., & Li, B. (2018). Object-based convolutional neural network for high-resolution imagery classification. *IEEE Journal of Selected Topics in Applied Earth Observations and Remote Sensing, 11*(8), 2623–2633.
104. Zhao, Y., Zhang, H., Du, S., & Wang, J. (2018). Deep learning based multi-temporal remote sensing data for land use and land cover change detection. *Remote Sensing, 10*(6), 809.
105. Zhao, Y., Zhang, H., Du, S., & Wang, X. (2018). A new integrated remote sensing and machine learning method for urban land-use mapping using high-resolution Google Earth images. *Remote Sensing of Environment, 204*, 261–275.

Chapter 12
Housing, Affordability, and Real Estate Market Analysis

12.1 Overview of Housing, Affordability, and Real Estate Market Analysis

Housing, affordability, and real estate market analysis are essential aspects of urban planning and development, as they contribute to the overall well-being and quality of life of city residents. This section provides an overview of the role of artificial intelligence (AI) in housing, affordability, and real estate market analysis, highlighting its potential to revolutionize urban planning and decision-making processes (Table 12.1).

Housing affordability is a pressing issue in many cities worldwide, with the cost of living rising and housing becoming increasingly scarce [8]. In this context, AI can help urban planners and policymakers to understand the factors contributing to housing affordability, identify areas of need, and develop effective strategies to address housing shortages and related challenges.

Real estate market analysis, on the other hand, is a critical component of urban planning, as it helps to determine the value of properties, predict market trends, and guide land use planning and development decisions [17]. AI techniques, including machine learning, deep learning, and natural language processing, can be applied to analyze vast amounts of data from diverse sources, enabling more accurate and reliable real estate market predictions and insights.

This section will explore the various AI techniques and applications relevant to housing, affordability, and real estate market analysis, as well as the challenges and limitations associated with their implementation. In addition, it will discuss potential future directions in the field, highlighting the potential of AI to transform urban planning and contribute to more sustainable, inclusive, and equitable cities.

In order to provide a comprehensive understanding of the subject matter, this section will cover the following subtopics:

1. The importance of housing, affordability, and real estate market analysis in urban planning: This subtopic will discuss the significance of addressing housing

S. N. M. Moghaddam and H. Cao, *Artificial Intelligence-Driven Geographies*, City Development: Issues and Best Practices, https://doi.org/10.1007/978-981-97-5116-7_12

Table 12.1 Description and application of AI in housing, affordability, and real estate market analysis

Aspect	Description	Application
Overview of housing, affordability, and real estate market analysis	Examines the crucial aspects of urban planning focused on housing affordability and the real estate market, highlighting AI's role in revolutionizing planning and decision-making processes	AI informs strategies for addressing housing shortages and analyzing market trends, guiding land use and investment decisions
Data sources for housing, affordability, and real estate market analysis	Involves a mix of traditional data (census, housing market statistics) and emerging data sources (social media, big data) to analyze trends in housing affordability and the real estate market	Enhances the ability of researchers to perform comprehensive analyses, including spatial distribution and socioeconomic disparities in housing
AI techniques for housing, affordability, and real estate market analysis	Includes machine learning, deep learning, and NLP for processing diverse datasets, offering sophisticated models to understand and predict market dynamics	Supports targeted policy interventions by identifying contributing factors to housing market trends and affordability issues
Applications of AI in housing, affordability, and real estate market analysis	AI's application spans across scenario analysis, policy development, infrastructure planning, and environmental impact assessment to support sustainable growth and inform strategies for addressing urban housing and market challenges	Provides insights for developing more accurate and effective housing policies, investment strategies, and planning initiatives
Challenges and limitations of AI in housing, affordability, and real estate market analysis	Focuses on overcoming AI's challenges, such as data quality, model complexity, and ethical concerns, in urban planning	Identifies areas for improvement to enhance the reliability and effectiveness of AI in housing and real estate analysis
Future directions in AI applications for housing, affordability, and real estate market analysis	Explores potential advancements in AI, including the integration with other technologies and collaborative efforts, to enhance urban planning and address housing and real estate challenges	Highlights the evolving role of AI in improving predictive accuracy and supporting sustainable, inclusive urban development

affordability and analyzing real estate market trends in the context of urban planning and development. It will emphasize the need for sustainable, inclusive, and data-driven approaches to tackle housing challenges and ensure equitable access to housing for all city residents [20].

2. The role of AI in housing, affordability, and real estate market analysis: This subtopic will introduce various AI techniques, including machine learning, deep learning, and natural language processing, that can be applied to analyze and predict housing affordability and real estate market trends. It will also highlight the potential benefits and advantages of using AI in urban planning and decision-making processes, such as increased accuracy, efficiency, and transparency [22].
3. Key data sources for housing, affordability, and real estate market analysis: This subtopic will discuss the various data sources available for AI-driven housing, affordability, and real estate market analysis, including housing market data, socio-economic data, land use data, and real estate transaction data. It will emphasize the importance of data quality, reliability, and accessibility in ensuring accurate and reliable AI-driven insights and predictions [50].
4. Applications of AI in housing, affordability, and real estate market analysis: This subtopic will provide examples of how AI techniques can be applied to various aspects of housing, affordability, and real estate market analysis, such as demand and supply prediction, market forecasting, land use planning, policy development, and community engagement. It will also discuss potential challenges and limitations associated with these applications, such as data privacy, ethical considerations, and technological barriers [39].
5. Challenges and limitations of AI in housing, affordability, and real estate market analysis: This subtopic will delve deeper into the challenges and limitations associated with the implementation of AI in housing, affordability, and real estate market analysis. It will address issues such as data privacy, data bias, ethical considerations, technological barriers, and the potential for exacerbating existing inequalities and disparities in urban planning and development [34].
6. Future directions in AI applications for housing, affordability, and real estate market analysis: This subtopic will explore potential future directions and innovations in the field, including the integration of AI with other advanced technologies such as the Internet of Things (IoT), blockchain, and augmented reality. It will also discuss the potential of AI-driven tools and models to support more sustainable, inclusive, and equitable urban planning processes, as well as the need for interdisciplinary collaboration, capacity building, and policy frameworks to support the successful implementation of AI in housing, affordability, and real estate market analysis [42].

12.2 Data Sources for Housing, Affordability, and Real Estate Market Analysis

The growing use of artificial intelligence (AI) in urban planning has opened new possibilities for understanding housing affordability and real estate market dynamics. This subsection will examine the various data sources used in the analysis of housing, affordability, and real estate market trends, as well as their potential applications and implications for urban planning and policy.

Traditional Data Sources

a. Government and Public Sector Data: Government agencies and public sector organizations are important sources of data for housing, affordability, and real estate market analysis. This includes data on housing stock, housing prices, rents, household incomes, property taxes, and housing subsidies, as well as demographic, economic, and social indicators. Examples of these data sources include the U.S. Census Bureau, the Department of Housing and Urban Development (HUD), the Bureau of Labor Statistics, and the Federal Reserve [16].
b. Real Estate Industry Data: Real estate industry data, such as data from multiple listing services (MLS), property management companies, and real estate brokerages, can provide valuable insights into market trends, transaction volumes, and property characteristics. This data can be used to analyze housing supply and demand, property values, and neighborhood characteristics [45].
c. Academic and Research Institutions: Academic and research institutions often generate and maintain datasets related to housing, affordability, and real estate market analysis. These datasets may include longitudinal and cross-sectional data, as well as data derived from surveys, experiments, and observational studies. Examples of such datasets include the Panel Study of Income Dynamics (PSID), the American Housing Survey (AHS), and the National Longitudinal Survey of Youth (NLSY) [16].

Emerging Data Sources

a. Big Data: The rise of big data has generated new opportunities for housing, affordability, and real estate market analysis. These large-scale datasets, which are often generated by digital technologies and platforms, can provide real-time information on housing market trends, consumer preferences, and spatial patterns. Examples of big data sources include web search queries, social media posts, online reviews, and mobile phone data [22].
b. Crowdsourced Data: Crowdsourced data is another emerging source of information for housing, affordability, and real estate market analysis. This data is generated by users and volunteers who contribute information on various aspects of housing and real estate, such as property conditions, neighborhood amenities, and rental prices. Examples of crowdsourced data platforms include Zillow, Trulia, and Rentometer [22].

c. Remote Sensing and Geographic Information Systems (GIS): Remote sensing technologies, such as satellite imagery and aerial photography, can provide valuable spatial data on land use, land cover, and urban growth patterns. This data can be integrated with GIS to analyze housing, affordability, and real estate market trends at various spatial scales, from local neighborhoods to regional and national levels [39].

Data Integration and Interoperability

The integration of traditional and emerging data sources can provide a more comprehensive and accurate understanding of housing, affordability, and real estate market dynamics. This requires the development of data standards, protocols, and platforms that facilitate data sharing, aggregation, and interoperability across different sectors and disciplines [39].

Data Quality, Privacy, and Ethics

As the use of data in housing, affordability, and real estate market analysis continues to expand, it is essential to address issues related to data quality, privacy, and ethics. This includes the need for accurate and reliable data, as well as the protection of personal information and the responsible use of data for research and policy-making purposes. Ensuring data quality involves validating and cleaning datasets, addressing issues related to sampling, measurement, and representation, and assessing the robustness and generalizability of findings [19].

Data privacy and ethics are also essential considerations in the use of AI for housing, affordability, and real estate market analysis. Researchers and practitioners must adhere to data protection regulations and ethical guidelines, such as obtaining informed consent, anonymizing data, and using secure data storage and sharing practices. Moreover, they must be aware of potential biases and disparities in data collection and analysis, as well as the potential consequences of AI-driven decision-making for vulnerable and marginalized populations [25].

In conclusion, various data sources are available for housing, affordability, and real estate market analysis, ranging from traditional sources such as government and public sector data to emerging sources like big data, crowdsourced data, and remote sensing technologies. The integration of these data sources, along with attention to data quality, privacy, and ethics, can help improve our understanding of housing, affordability, and real estate market dynamics and inform more effective and equitable urban planning and policy interventions.

12.3 AI Techniques for Housing, Affordability, and Real Estate Market Analysis

12.3.1 Machine Learning for Housing Demand and Supply Prediction

Machine learning (ML) has become an essential tool in the housing, affordability, and real estate market analysis due to its ability to process and analyze large volumes of data, uncover hidden patterns, and make accurate predictions. In this section, we will discuss the application of ML techniques for predicting housing demand and supply, including the methods and algorithms used, the advantages and drawbacks, and the implications for urban planning and policy.

Predicting Housing Demand

Housing demand is influenced by a variety of factors, such as population growth, income levels, employment opportunities, household composition, and preferences for housing types and locations. ML techniques can help capture these complex relationships and predict housing demand more accurately than traditional models. Some of the commonly used ML methods for housing demand prediction include regression models, decision trees, support vector machines, and neural networks [54].

Regression models, such as linear regression and logistic regression, are widely used in housing demand prediction due to their simplicity and interpretability. These models analyze the relationships between housing demand and various explanatory variables, such as demographics, socioeconomic factors, and housing market conditions [14]. However, regression models may have limitations in capturing nonlinear relationships and interactions among variables.

Decision trees and their ensemble methods, such as random forests and gradient boosting machines, are also popular for housing demand prediction. These methods can model complex relationships and interactions among variables, handle missing data, and provide variable importance measures, which can inform policy interventions [28].

Support vector machines (SVM) and neural networks, including deep learning models such as convolutional neural networks (CNN) and recurrent neural networks (RNN), are advanced ML techniques that can capture complex patterns and make accurate predictions. However, they may require more computational resources and expertise, and their results may be less interpretable than simpler models [37].

Predicting Housing Supply

Housing supply is determined by factors such as construction costs, land availability, zoning regulations, and developer expectations about future demand and prices. ML techniques can help predict housing supply by analyzing these factors and their interactions, as well as incorporating information from other sources, such as satellite imagery, social media, and real estate listings [24].

Regression models, decision trees, and SVM are commonly used for housing supply prediction, as they can model the relationships between housing supply and various explanatory variables, such as construction costs, land prices, and regulatory constraints [52]. These models can also incorporate spatial and temporal information, which is crucial for understanding the dynamics of housing supply and its implications for urban growth and sprawl.

Neural networks, including CNN and RNN, can be used for housing supply prediction by processing high-dimensional data, such as satellite imagery and time series data, and detecting patterns that may be difficult for humans or traditional models to discern [35]. These models can help identify areas with high potential for new housing development, monitor construction activity, and assess the impacts of policy interventions on housing supply.

Challenges and Limitations

While ML techniques have shown promise in predicting housing demand and supply, there are several challenges and limitations that need to be addressed:

1. Data quality and availability: Accurate and reliable data on housing demand and supply, as well as the factors influencing them, are crucial for ML models. However, data may be incomplete, outdated, or biased, which can affect the performance and generalizability of the models [43].
2. Model complexity and interpretability: ML models, especially advanced techniques such as deep learning, can be complex and difficult to interpret, which may hinder their adoption by practitioners and policymakers who require clear explanations of the model results for decision-making [49].
3. Overfitting and generalizability: ML models may overfit the training data, leading to poor performance on new data or different contexts. This issue can be mitigated through cross-validation, regularization, and other model selection techniques, but it remains a challenge, particularly when the data are scarce or rapidly changing [27].
4. Ethical and privacy concerns: The use of ML techniques for housing, affordability, and real estate market analysis may raise ethical and privacy concerns, particularly when dealing with sensitive data, such as income, race, or household composition. Ensuring data privacy and addressing potential biases in the data and the models are critical for responsible and inclusive AI applications in urban planning [3].

Implications for Urban Planning and Policy

Despite these challenges and limitations, ML techniques for housing demand and supply prediction can provide valuable insights for urban planning and policy. By accurately forecasting housing demand and supply, policymakers can better allocate resources, prioritize investments in infrastructure and public services, and design policies to promote housing affordability and sustainable urban development [54].

For example, ML models can help identify areas with high housing demand and limited supply, which may require interventions such as upzoning, inclusionary zoning, or housing subsidies to increase the availability and affordability of housing. Conversely, areas with low demand and excess supply may need policies to stimulate demand, such as job creation, infrastructure improvements, or neighborhood revitalization efforts [14].

Moreover, ML models can support scenario analysis and impact evaluation by simulating the effects of different policy interventions on housing demand and supply, as well as their spillover effects on other aspects of urban development, such as transportation, environment, and social equity. This can help policymakers make informed decisions and monitor the progress towards their goals [28].

In conclusion, ML techniques offer significant potential for improving housing, affordability, and real estate market analysis, which can inform urban planning and policy-making. By overcoming the challenges and limitations of ML, and by integrating these techniques with other tools and data sources, urban planners and policymakers can better understand and address the complex dynamics of housing demand and supply, and promote more sustainable, inclusive, and resilient urban development.

12.3.2 Deep Learning for Real Estate Market Analysis

In recent years, deep learning has emerged as a powerful tool for solving complex problems in various domains, including urban planning and real estate market analysis. Deep learning techniques, particularly artificial neural networks (ANNs) and convolutional neural networks (CNNs), have demonstrated remarkable performance in analyzing large and complex datasets, providing valuable insights for real estate professionals, policymakers, and urban planners. This section explores the applications of deep learning in real estate market analysis, focusing on housing price prediction, market segmentation, and the identification of key factors affecting real estate values.

Housing Price Prediction

One of the primary applications of deep learning in real estate market analysis is housing price prediction. Accurate and reliable housing price predictions are essential for real estate investors, homebuyers, policymakers, and urban planners to make informed decisions. Deep learning techniques, such as ANNs and CNNs, have shown promising results in predicting housing prices, outperforming traditional linear regression and machine learning methods [26, 31].

ANNs, inspired by the human brain's structure and functioning, consist of interconnected layers of artificial neurons that process input data and learn to identify patterns in the data. This enables ANNs to adapt to new information and make predictions based on previously learned patterns. In the context of housing price prediction, ANNs can learn complex relationships between housing features and prices, enabling them to make accurate predictions even when faced with non-linear and high-dimensional data [52].

CNNs, a type of ANN specifically designed to process grid-like data such as images, have also been used for housing price prediction. CNNs have been successful in capturing spatial features and patterns in housing data, such as the proximity to amenities, transportation networks, and neighborhood characteristics, which traditional machine learning methods may struggle to incorporate [13]. By considering both spatial and non-spatial features, CNNs can provide more accurate and comprehensive housing price predictions.

Market Segmentation

Deep learning techniques have also been employed for market segmentation in real estate analysis. Market segmentation involves dividing the real estate market into smaller, more homogeneous segments based on specific criteria such as location, property type, and price range. This process enables real estate professionals and policymakers to better understand market dynamics, identify trends, and develop targeted policies and strategies.

Unsupervised deep learning methods, such as autoencoders and clustering algorithms, have been used to analyze large datasets of real estate transactions and identify market segments with similar characteristics. Autoencoders, a type of ANN designed for dimensionality reduction and feature extraction, can learn to represent complex, high-dimensional data in lower-dimensional spaces, facilitating clustering and segmentation tasks [46].

Clustering algorithms, such as k-means and hierarchical clustering, can then be applied to the lower-dimensional representations generated by autoencoders to group real estate transactions into distinct market segments. This unsupervised learning approach allows for the identification of previously unknown or hidden market segments, enabling real estate professionals and policymakers to develop more targeted and effective strategies [46].

Identification of Key Factors Affecting Real Estate Values

Understanding the key factors affecting real estate values is crucial for investors, homebuyers, and policymakers. Deep learning techniques, particularly ANNs and CNNs, have demonstrated their ability to identify and rank the importance of various features in predicting housing prices. By analyzing large and diverse datasets, deep learning models can identify both well-known factors, such as location and property size, as well as less obvious factors, such as neighborhood characteristics and environmental factors [13, 53].

For instance, CNNs can be used to analyze satellite images and extract information about land use, green spaces, and transportation networks in a neighborhood, which can then be combined with traditional housing features to provide a more comprehensive understanding of real estate values [13]. Furthermore, deep learning models can be used to analyze textual data, such as online reviews and social media posts, to identify neighborhood amenities and perceptions that may impact housing prices [46].

By identifying the key factors affecting real estate values, deep learning models can help real estate professionals and policymakers develop more targeted interventions and strategies to address issues such as housing affordability, gentrification, and urban development.

Challenges and Limitations

Despite the promising results demonstrated by deep learning models in real estate market analysis, several challenges and limitations must be considered. First, deep learning models, particularly ANNs and CNNs, require large amounts of data to achieve accurate predictions and generalizable results. In some cases, obtaining sufficient data may be challenging due to privacy concerns, data availability, or data quality issues [26].

Second, deep learning models are often considered "black boxes" due to their complex architectures and non-linear decision-making processes, making it difficult to interpret their predictions and understand the underlying relationships between input features and output values [53]. This lack of interpretability may hinder the adoption of deep learning models by real estate professionals and policymakers who require transparent decision-making processes.

Finally, deep learning models may be computationally expensive and time-consuming to train, particularly when dealing with large datasets and complex model architectures. This may limit their applicability in real-time decision-making processes or in situations where computational resources are limited.

Deep learning techniques, including ANNs and CNNs, have shown great potential in real estate market analysis, offering valuable insights into housing price prediction, market segmentation, and the identification of key factors affecting real estate values. As deep learning technology continues to advance, it is likely that its applications in real estate market analysis will become increasingly sophisticated and accurate, providing even more valuable insights for real estate professionals, policymakers, and urban planners.

12.3.3 Natural Language Processing for Real Estate Data Analysis

Natural Language Processing (NLP) is a subfield of artificial intelligence that focuses on the interaction between computers and humans through natural language. It involves the development of algorithms and models that enable computers to understand, interpret, and generate human language in a way that is both meaningful and useful. In recent years, NLP has gained significant attention for its potential applications in various industries, including real estate market analysis. This section explores the use of NLP techniques for real estate data analysis, focusing on sentiment analysis, topic modeling, and information extraction.

Sentiment Analysis

Sentiment analysis, also known as opinion mining, is an NLP technique that aims to determine the sentiment or emotion expressed in a piece of text, such as a review, comment, or social media post. In the context of real estate market analysis, sentiment analysis can be used to gauge public opinion on various aspects of the housing market, such as neighborhood desirability, housing affordability, and policy interventions [11].

By analyzing large volumes of textual data from sources such as online reviews, social media posts, and news articles, sentiment analysis can provide valuable insights into the factors that influence housing demand and pricing. For instance, a study by Cao et al. [11] found that sentiment analysis of online reviews could successfully predict changes in housing prices, with positive sentiment being associated with higher housing prices and negative sentiment with lower housing prices.

Furthermore, sentiment analysis can help real estate professionals and policymakers identify trends and emerging issues in the housing market, enabling them to develop targeted strategies and interventions to address these concerns [11].

Topic Modeling

Topic modeling is another NLP technique that involves the identification of topics or themes within a collection of documents. This unsupervised learning approach can be used to analyze large datasets of textual data, such as online reviews, news articles, and social media posts, to identify common topics and trends related to the real estate market [46].

One popular topic modeling technique is Latent Dirichlet Allocation (LDA), a generative probabilistic model that assumes each document in a corpus is a mixture of various topics, and each topic is a distribution over words [7]. By applying LDA to real estate-related textual data, researchers and practitioners can identify key themes and trends in the housing market, such as the impact of infrastructure development, gentrification, and housing affordability issues.

For example, Wu et al. [46] used LDA to analyze a dataset of real estate transaction descriptions, identifying several topics related to housing characteristics, such as size, location, and amenities. By understanding the prevalence and importance of these topics, real estate professionals and policymakers can better target their efforts and address market trends and concerns.

Information Extraction

Information extraction is an NLP technique that involves the identification and extraction of structured information from unstructured textual data, such as names, dates, and numerical values. In the context of real estate market analysis, information extraction can be used to gather valuable data from sources such as property listings, online reviews, and news articles [38].

For instance, Peng et al. [38] developed an information extraction system that automatically extracts property attributes, such as price, size, and location, from online real estate listings. By aggregating and analyzing this data, real estate professionals and policymakers can gain insights into market trends and conditions, enabling them to make informed decisions and develop targeted strategies.

Moreover, information extraction techniques can be used to identify and track the impact of specific events or policy interventions on the real estate market, such as the introduction of new zoning regulations or the development of new transportation infrastructure. By monitoring the changes in property attributes and prices following these events, stakeholders can better understand the effects of these interventions and adjust their strategies accordingly [38].

Challenges and Limitations

While NLP techniques hold great potential for real estate data analysis, there are several challenges and limitations that must be considered. First, the quality of textual data can significantly impact the accuracy and reliability of NLP models. Issues such as incomplete or biased data, spelling and grammatical errors, and the presence of slang or colloquial language can hinder the performance of NLP techniques and lead to inaccurate results [11].

Second, NLP models often struggle to capture the context and nuances of human language, which can lead to misinterpretation and misclassification of sentiments, topics, or information. For example, sarcasm, irony, and idiomatic expressions can be challenging for NLP models to recognize and interpret correctly [46].

Third, the computational complexity and resource requirements of some NLP techniques, particularly deep learning-based approaches, can be prohibitive for real-time decision-making or large-scale data analysis. However, as computational resources continue to improve and new, more efficient NLP models are developed, these limitations may become less of a concern [38].

Future Directions

As NLP technology continues to advance, several future directions are emerging for its application in real estate data analysis. These developments include:

1. Integration of Multimodal Data: Future NLP applications in real estate analysis may involve the integration of multimodal data, such as textual, visual, and geospatial information, to provide a more comprehensive understanding of market trends and conditions. This could enable more accurate predictions and insights, as well as the identification of previously unknown or hidden relationships between various factors influencing the real estate market [46].
2. Transfer Learning and Domain Adaptation: Transfer learning and domain adaptation techniques can be employed to improve the performance of NLP models in real estate data analysis by leveraging knowledge learned from related domains or tasks. For example, an NLP model trained on general sentiment analysis tasks could be fine-tuned to better understand the specific language and sentiment expressions used in the context of real estate [11].
3. Interpretable and Explainable NLP Models: As the demand for transparent and interpretable AI models grows, new techniques are being developed to increase the interpretability of NLP models. These approaches may include explainable AI

(XAI) techniques, visualization tools, and feature importance ranking methods, which can help stakeholders better understand the relationships between input features and output values in real estate data analysis [38].

Natural Language Processing (NLP) techniques, including sentiment analysis, topic modeling, and information extraction, have shown great potential for real estate data analysis. These techniques can provide valuable insights into various aspects of the housing market, such as public sentiment, market trends, and the impact of specific events or policy interventions. As NLP technology continues to advance, its applications in real estate data analysis are expected to become increasingly sophisticated and accurate, offering valuable insights for real estate professionals, policymakers, and urban planners.

However, it is important to acknowledge the challenges and limitations associated with NLP techniques, such as data quality issues, difficulties in capturing the context and nuances of human language, and computational complexity. Addressing these challenges and developing more efficient, interpretable, and accurate NLP models will be crucial for the future success of NLP applications in real estate market analysis.

Future directions for NLP in real estate data analysis include the integration of multimodal data, transfer learning and domain adaptation, and the development of interpretable and explainable NLP models. As NLP technology continues to evolve and improve, these advancements are expected to further enhance the capabilities of NLP in real estate market analysis, providing even more valuable insights for real estate professionals, policymakers, and urban planners.

12.4 Applications of AI in Housing, Affordability, and Real Estate Market Analysis

12.4.1 Affordable Housing Policy Development

Affordable housing is a critical issue that affects the well-being of millions of individuals and families worldwide. Developing effective affordable housing policies is essential for promoting social equity, economic development, and sustainable urban growth. Artificial intelligence (AI) techniques have the potential to support policymakers and urban planners in developing and evaluating affordable housing policies, by offering data-driven insights and predictions. This section will discuss how AI can be utilized in affordable housing policy development, including data sources, techniques, applications, and future directions.

Data Sources for Affordable Housing Policy Development

Data is the cornerstone of any AI application, and the development of affordable housing policies is no exception. Various data sources can be utilized to inform affordable housing policy development, including:

1. Census data: Census data provide a wealth of information on population demographics, income levels, and housing characteristics, which can help identify areas with high housing needs and inform the targeting of affordable housing policies.
2. Housing market data: Data on housing prices, rents, and vacancies can help policymakers understand housing market dynamics, identify affordability challenges, and evaluate the effectiveness of existing policies.
3. Land use and zoning data: Land use and zoning data can provide insights into the availability and suitability of land for affordable housing development, as well as the potential impact of zoning changes on housing supply and affordability.
4. Social and economic data: Data on employment, education, and public services can help policymakers understand the broader context of housing affordability and inform the development of comprehensive policies that address both housing and non-housing factors.

AI Techniques for Affordable Housing Policy Development

Various AI techniques can be employed to support affordable housing policy development, including:

1. Machine learning: Machine learning algorithms can be used to analyze large datasets and identify patterns and relationships that may not be immediately apparent to human analysts. For example, machine learning models can predict the impact of specific policy interventions on housing affordability, or identify areas with the highest need for affordable housing based on demographic, economic, and housing market data.
2. Deep learning: Deep learning techniques, such as neural networks, can be employed to model complex relationships between variables and make more accurate predictions. For instance, deep learning models can be used to forecast housing demand and supply under different policy scenarios, helping policymakers understand the potential effects of their decisions on housing affordability.
3. Geospatial analysis: AI techniques can also be applied to geospatial data to support affordable housing policy development. For example, AI-based land suitability analysis can help identify the most suitable locations for affordable housing development, taking into account factors such as land availability, infrastructure, and access to services.
4. Natural language processing (NLP): NLP techniques can be used to analyze textual data, such as policy documents, news articles, and social media posts, to gain insights into public sentiment towards affordable housing policies, as well as to identify emerging trends and issues.

Applications of AI in Affordable Housing Policy Development

AI techniques can be applied in various aspects of affordable housing policy development, including:

1. Policy design: AI can support the design of affordable housing policies by providing data-driven insights into housing needs, market dynamics, and the potential impact of policy interventions. For example, machine learning models can be used to identify the most effective policy instruments for promoting affordable housing, such as inclusionary zoning, housing subsidies, or tax incentives.
2. Policy evaluation: AI can also be employed to evaluate the effectiveness of existing affordable housing policies. By analyzing historical data on housing affordability, AI models can help policymakers understand the impact of past policy interventions and identify areas for improvement.
3. Decision support: AI can offer decision support tools for policymakers and urban planners, by providing data-driven insights and predictions that can inform the allocation of resources and the prioritization of affordable housing projects. For instance, AI-based land suitability analysis can help planners identify the most appropriate locations for affordable housing development, taking into account factors such as land availability, infrastructure, and access to services.
4. Stakeholder engagement: AI can also be used to facilitate stakeholder engagement in the affordable housing policy development process. For example, natural language processing techniques can be employed to analyze public sentiment towards affordable housing policies, helping policymakers understand the concerns and priorities of different stakeholder groups and fostering more inclusive and responsive policy-making.
5. Scenario planning: AI can support scenario planning in affordable housing policy development by providing data-driven projections of housing demand and supply under different policy scenarios. This can help policymakers understand the potential long-term effects of their decisions on housing affordability and make more informed choices.

Future Directions in AI Applications for Affordable Housing Policy Development

As AI techniques continue to evolve, they offer significant potential to further enhance the development of affordable housing policies. Some possible future directions in this area include:

1. Integration of AI with other planning tools: The integration of AI techniques with other planning tools, such as geographic information systems (GIS) and simulation models, can provide more comprehensive and accurate insights for affordable housing policy development.
2. Improved data quality and accessibility: Ensuring the quality and accessibility of data used in AI applications for affordable housing policy development is crucial for the accuracy and reliability of the insights generated. Efforts to standardize, clean, and share data across different jurisdictions and organizations can help facilitate more effective use of AI in this area.
3. Enhanced transparency and explainability: As AI techniques become more widely used in affordable housing policy development, it is essential to ensure that the models and algorithms employed are transparent and explainable.

This can help build trust in AI-generated insights and support more informed decision-making.

4. Ethical considerations: The use of AI in affordable housing policy development raises various ethical considerations, such as the potential for algorithmic bias and the privacy implications of data collection and analysis. Addressing these concerns through the development of ethical guidelines and the implementation of robust data privacy and security measures will be crucial for ensuring the responsible and equitable use of AI in this context.

In conclusion, AI techniques offer significant potential to support the development of effective and responsive affordable housing policies. By providing data-driven insights and predictions, AI can help policymakers and urban planners better understand housing needs, market dynamics, and the potential impact of policy interventions, ultimately promoting more equitable and sustainable urban development.

12.4.2 Real Estate Market Forecasting and Investment

Artificial intelligence (AI) is increasingly being employed in the real estate industry to enhance market forecasting and guide investment decisions. In this section, we will discuss how AI can be utilized for real estate market forecasting and investment, along with the potential benefits, challenges, and future directions of these applications.

AI for Real Estate Market Forecasting

AI techniques, such as machine learning and deep learning, can be utilized to analyze historical and current data from various sources, including property listings, transaction records, and economic indicators. By identifying patterns and trends in this data, AI models can generate forecasts of future market conditions, such as property prices, rental rates, and supply and demand dynamics. These forecasts can support various stakeholders, including investors, developers, and policymakers, in making more informed decisions.

Some specific applications of AI in real estate market forecasting include:

a. Price prediction: Machine learning algorithms, such as regression models, can be trained on historical property sales data to predict future property prices. These models can take into account various factors, such as property characteristics, location, and local market conditions, to generate accurate and granular price predictions.
b. Rental rate forecasting: Similar to price prediction, machine learning models can also be used to forecast rental rates based on historical rental data and other relevant factors, such as property type, amenities, and neighborhood characteristics.
c. Supply and demand analysis: AI techniques can be employed to analyze patterns in property listings and transaction data, enabling the identification of trends in

supply and demand. This can help stakeholders understand market dynamics and anticipate potential imbalances, such as housing shortages or oversupply.

AI for Real Estate Investment Decision-Making

AI-generated insights and forecasts can be used to inform real estate investment decisions in various ways:

a. Property selection: AI models can help investors identify properties with the highest potential for appreciation or rental income by analyzing factors such as property attributes, location, and market conditions. For example, machine learning algorithms can be used to estimate the expected return on investment (ROI) for individual properties, enabling investors to make more data-driven property selection decisions.
b. Portfolio optimization: AI can also support the optimization of real estate investment portfolios by identifying the most advantageous mix of properties and asset types based on factors such as risk tolerance, investment objectives, and market conditions. Machine learning techniques, such as clustering and classification algorithms, can be employed to group properties with similar characteristics and analyze the performance of different property types and market segments.
c. Market timing: By providing accurate forecasts of market conditions, AI can help investors determine the optimal timing for buying, selling, or holding real estate assets. For example, investors can use AI-generated insights to identify periods of high demand or low supply, potentially capitalizing on favorable market conditions to maximize returns.
d. Risk assessment: AI can also support the assessment of risks associated with real estate investments, such as market volatility, interest rate fluctuations, and economic downturns. Machine learning models can be used to analyze historical data and identify factors that may indicate increased risk, enabling investors to adjust their investment strategies accordingly.

Benefits of AI in Real Estate Market Forecasting and Investment

The use of AI in real estate market forecasting and investment offers several benefits:

a. Improved accuracy: AI techniques can analyze vast amounts of data and identify complex patterns that may be difficult for humans to detect, potentially resulting in more accurate and reliable market forecasts and investment insights.
b. Enhanced efficiency: AI models can process and analyze data more quickly and efficiently than manual analysis, enabling stakeholders to make more timely and informed decisions.
c. Data-driven decision-making: By providing data-driven insights and predictions, AI can help stakeholders base their investment decisions on objective evidence rather than intuition or speculation.
d. Reduced bias: AI algorithms can help minimize the influence of human biases and subjective opinions in real estate decision-making, potentially leading to more rational and objective investment choices.

Challenges and Limitations of AI in Real Estate Market Forecasting and Investment

Despite its potential benefits, there are several challenges and limitations associated with using AI in real estate market forecasting and investment:

a. Data quality and availability: The accuracy and reliability of AI-generated forecasts and insights depend heavily on the quality and comprehensiveness of the underlying data. Incomplete, outdated, or inaccurate data can lead to erroneous predictions and misguided investment decisions.
b. Model interpretability and transparency: AI models, particularly deep learning algorithms, can be complex and difficult to interpret, making it challenging for stakeholders to understand the rationale behind their predictions and recommendations. This can hinder trust and adoption of AI-driven decision-making in the real estate industry.
c. Ethical and legal considerations: The use of AI in real estate decision-making raises various ethical and legal concerns, such as potential discrimination or bias in property selection, pricing, or investment recommendations. Ensuring fairness, accountability, and transparency in AI-driven real estate applications is crucial to address these concerns and maintain public trust.

Future Directions in AI Applications for Real Estate Market Forecasting and Investment

As AI continues to advance, we can expect to see further developments and innovations in its application to real estate market forecasting and investment, including:

a. Integration of new data sources: The incorporation of new and diverse data sources, such as social media, satellite imagery, or Internet of Things (IoT) sensors, can enhance the accuracy and granularity of AI-generated market forecasts and insights.
b. Cross-disciplinary collaboration: Collaborations between AI researchers, urban planners, economists, and real estate professionals can help develop more sophisticated and context-specific AI models and applications, addressing the unique challenges and complexities of the real estate industry.
c. Personalized investment recommendations: As AI models become more advanced, they may be able to generate personalized investment recommendations tailored to individual investors' preferences, objectives, and risk profiles.
d. Adoption of explainable AI techniques: The development and adoption of explainable AI techniques can improve the interpretability and transparency of AI models, helping to address concerns about trust and accountability in AI-driven real estate decision-making.

In conclusion, AI holds significant potential to transform real estate market forecasting and investment by providing more accurate, data-driven insights and predictions. As AI technology continues to evolve, it is likely to play an increasingly important role in shaping the future of the real estate industry. However, addressing challenges related to data quality, model interpretability, and ethical considerations will be crucial to ensure the responsible and effective use of AI in this context.

12.4.3 Land Use Planning and Zoning

Artificial intelligence (AI) is revolutionizing the way land use planning and zoning are conducted in the context of housing, affordability, and real estate market analysis. AI techniques can be employed to optimize land use, facilitate sustainable development, and better align zoning regulations with housing affordability goals. In this section, we will discuss how AI can be utilized for land use planning and zoning, along with the potential benefits, challenges, and future directions of these applications.

AI for Land Use Planning

AI can support land use planning in various ways, including:

a. Site suitability analysis: AI techniques, such as machine learning and geographic information systems (GIS), can be used to assess the suitability of different sites for specific land uses, such as residential, commercial, or industrial development. By analyzing data on factors like location, environmental characteristics, and infrastructure availability, AI models can help planners identify the most appropriate locations for different land uses, ensuring efficient land allocation and promoting sustainable urban growth.
b. Development potential estimation: AI algorithms can be used to estimate the development potential of different land parcels based on factors such as zoning regulations, market conditions, and urban growth trends. This information can guide planners in identifying areas where development is likely to occur and help them prioritize infrastructure investments and other interventions.
c. Scenario analysis and simulation: AI can be employed to simulate the impacts of different land use policies and planning scenarios, enabling planners to evaluate their potential effects on factors like housing affordability, traffic congestion, and environmental quality. This can help inform the development of more effective land use strategies and policies.

AI for Zoning

AI can also be applied to the field of zoning, with several possible applications:

a. Zoning regulation analysis: AI techniques, such as natural language processing (NLP) and machine learning, can be used to analyze zoning regulations and identify inconsistencies, redundancies, or potential conflicts. This can help planners and policymakers streamline zoning codes and ensure they are aligned with broader housing affordability and sustainability objectives.

b. Zoning compliance monitoring: AI algorithms can be employed to automatically monitor and detect potential zoning violations, such as unauthorized land uses or noncompliant building heights. By streamlining compliance monitoring, AI can help improve the enforcement of zoning regulations and promote more equitable development.
c. Zoning impact assessment: AI can be used to assess the potential impacts of zoning changes on factors like property values, neighborhood character, and housing affordability. This can help planners and policymakers better understand the trade-offs associated with different zoning options and make more informed decisions.

Benefits of AI in Land Use Planning and Zoning

The application of AI in land use planning and zoning offers several benefits:

a. Enhanced efficiency: AI models can process and analyze large amounts of spatial and non-spatial data more quickly and efficiently than traditional manual analysis methods, enabling planners and policymakers to make more timely and informed decisions.
b. Improved accuracy: By leveraging machine learning and other AI techniques, planners can gain deeper insights into the complex relationships between land use, zoning, and housing affordability. This can help improve the accuracy of land use plans and zoning policies, ensuring they better align with community needs and objectives.
c. Data-driven decision-making: AI can help planners and policymakers base their land use and zoning decisions on objective, data-driven evidence rather than intuition or anecdotal information. This can lead to more effective and equitable planning outcomes.

Challenges and Limitations of AI in Land Use Planning and Zoning

Despite its potential benefits, there are several challenges and limitations associated with using AI in land use planning and zoning:

a. Data quality and availability: The accuracy and effectiveness of AI-driven land use planning and zoning applications depend on the quality and availability of the underlying data. Incomplete, outdated, or inaccurate data can lead to erroneous predictions and misguided planning decisions. Ensuring that AI models have access to comprehensive, up-to-date, and reliable data sources is crucial for their successful implementation.
b. Model interpretability and transparency: AI models, particularly those based on deep learning techniques, can be complex and difficult to interpret. This can make it challenging for planners and policymakers to understand the rationale behind AI-generated recommendations and predictions. Developing and adopting explainable AI techniques can help improve the interpretability and transparency of these models, fostering trust and adoption in the planning process.

c. Ethical and legal considerations: The use of AI in land use planning and zoning raises various ethical and legal concerns, such as potential discrimination or bias in site suitability analysis or zoning regulation enforcement. Ensuring fairness, accountability, and transparency in AI-driven planning applications is essential to address these concerns and maintain public trust.

Future Directions in AI Applications for Land Use Planning and Zoning

As AI continues to advance, we can expect to see further developments and innovations in its application to land use planning and zoning, including:

a. Integration of new data sources: The incorporation of new and diverse data sources, such as social media, satellite imagery, or Internet of Things (IoT) sensors, can enhance the accuracy and granularity of AI-generated land use plans and zoning recommendations.
b. Cross-disciplinary collaboration: Collaborations between AI researchers, urban planners, and other relevant stakeholders can help develop more sophisticated and context-specific AI models and applications, addressing the unique challenges and complexities of land use planning and zoning.
c. Personalized planning solutions: As AI models become more advanced, they may be able to generate personalized land use plans and zoning recommendations tailored to the specific needs and preferences of individual communities or neighborhoods.
d. Real-time planning and adaptation: AI techniques can enable more dynamic and real-time land use planning and zoning, allowing planners and policymakers to monitor and respond to changing conditions and emerging trends more effectively.

In conclusion, AI holds significant potential to transform land use planning and zoning in the context of housing, affordability, and real estate market analysis. By providing more accurate, data-driven insights and predictions, AI can help planners and policymakers develop more effective and equitable land use plans and zoning policies. However, addressing challenges related to data quality, model interpretability, and ethical considerations will be crucial to ensure the responsible and effective use of AI in this field.

12.4.4 Gentrification and Displacement Analysis

Gentrification and displacement are significant urban challenges affecting housing affordability and social equity in cities worldwide. Gentrification refers to the process by which low-income neighborhoods experience an influx of higher-income residents, leading to increased property values, changed demographics, and potential displacement of long-term residents. Understanding and analyzing the dynamics of gentrification and displacement are critical for developing effective and equitable housing policies. This section discusses how artificial intelligence (AI) techniques can be employed for gentrification and displacement analysis, helping urban planners, policymakers, and other stakeholders make data-driven decisions to mitigate the adverse impacts of these processes.

Data Collection and Preprocessing for Gentrification and Displacement Analysis

AI-driven analysis of gentrification and displacement requires comprehensive and reliable data sources to train and evaluate models. These data sources can be collected from various public and private organizations, including census bureaus, property databases, and housing market reports. Commonly used data for gentrification and displacement analysis include:

a. Demographic data: Information about neighborhood residents, such as age, income, education, and employment, can be used to identify socioeconomic trends and potential gentrification indicators.
b. Housing market data: Data on property values, rents, sales, and vacancy rates can provide insights into housing market dynamics and affordability trends.
c. Geospatial data: Geographic information system (GIS) data, such as land use, zoning, and transportation networks, can help identify neighborhood characteristics and accessibility to essential services and amenities.
d. Social media and online data: User-generated content, such as social media posts, reviews, and check-ins, can offer additional context on neighborhood changes and perceptions.

Once data are collected, preprocessing steps, such as data cleaning, transformation, and integration, need to be performed to ensure the data's quality and consistency before AI model training.

AI Techniques for Gentrification and Displacement Analysis

AI techniques, including machine learning (ML), deep learning (DL), and natural language processing (NLP), can be employed to analyze gentrification and displacement patterns, offering valuable insights for urban planning and policy development.

a. Machine learning for gentrification prediction: ML algorithms, such as regression, decision trees, and support vector machines, can be used to identify key indicators of gentrification and predict neighborhood changes based on historical data. These models can help planners and policymakers anticipate gentrification trends and develop proactive strategies to address housing affordability and social equity challenges.
b. Deep learning for spatial analysis: DL techniques, such as convolutional neural networks (CNNs) and recurrent neural networks (RNNs), can be used to analyze spatial data, such as satellite imagery and GIS data, to identify neighborhood characteristics and patterns associated with gentrification and displacement. This spatial analysis can provide a more nuanced understanding of the factors driving these processes and inform targeted policy interventions.

c. Natural language processing for sentiment analysis: NLP methods can be employed to analyze social media data and other online sources to gauge public perceptions and sentiments about gentrification and displacement. Sentiment analysis can help planners and policymakers understand the lived experiences of residents and identify potential areas of concern or intervention.

Applications of AI in Gentrification and Displacement Analysis

AI techniques can be applied to various aspects of gentrification and displacement analysis, supporting data-driven decision-making in urban planning and policy development. Some applications of AI in this context include:

a. Gentrification prediction and monitoring: AI models can be used to predict and monitor gentrification trends, allowing planners and policymakers to identify at-risk neighborhoods and develop proactive strategies to mitigate the adverse impacts of gentrification on housing affordability and social equity.
b. Displacement risk assessment: AI-driven analysis can be used to evaluate the risk of displacement for vulnerable populations, such as low-income households, elderly residents, and marginalized communities. By identifying areas with high displacement risks, planners and policymakers can prioritize resources and interventions to support these populations and minimize the negative consequences of gentrification.
c. Policy impact analysis: AI techniques can be employed to simulate the effects of different policy interventions, such as affordable housing initiatives, rent control regulations, and inclusionary zoning policies, on gentrification and displacement patterns. This information can help decision-makers assess the potential benefits and drawbacks of various policy options, enabling them to make more informed choices in addressing housing affordability and social equity challenges.
d. Community engagement and participatory planning: AI-driven analysis of gentrification and displacement can be used to facilitate community engagement and participatory planning processes. By providing accessible and transparent data on neighborhood changes, AI tools can help empower residents and community organizations to participate in shaping the future of their neighborhoods, fostering more inclusive and equitable urban development.

Challenges and Limitations

While AI has the potential to significantly enhance gentrification and displacement analysis, several challenges and limitations need to be considered:

a. Data quality and availability: The accuracy and reliability of AI-driven analysis depend on the quality and comprehensiveness of the underlying data. Incomplete, outdated, or biased data can lead to erroneous conclusions and ineffective policy interventions.
b. Ethical concerns and privacy issues: The use of AI techniques in gentrification and displacement analysis may raise ethical concerns and privacy issues, particularly when analyzing sensitive demographic data or using social media information.

It is crucial to ensure that AI-driven analysis respects individuals' privacy and does not exacerbate existing inequalities or perpetuate stereotypes.

c. Model interpretability and transparency: The complexity of some AI techniques, particularly deep learning models, can make it difficult to understand and interpret their predictions and recommendations. This lack of transparency can pose challenges for policymakers and stakeholders who need to justify and communicate their decisions to the public.
d. Uncertainty and unpredictability: Gentrification and displacement processes are influenced by numerous factors, many of which are unpredictable or subject to change, such as economic conditions, political developments, and social trends. AI-driven analysis may struggle to capture these complex and dynamic processes, limiting its predictive accuracy and practical applicability.

AI techniques hold promise for enhancing gentrification and displacement analysis in urban planning and policy development, offering valuable insights and tools for addressing housing affordability and social equity challenges. However, it is crucial to recognize and address the limitations and challenges associated with AI-driven analysis to ensure its ethical, effective, and equitable application in urban planning and policymaking.

12.4.5 Community Engagement and Inclusive Housing Development

Community engagement and inclusive housing development are essential components of equitable urban planning and policy-making. By involving local residents and stakeholders in the decision-making process, urban planners and policymakers can better understand the needs and priorities of diverse communities, fostering more inclusive and sustainable housing solutions. This section explores how artificial intelligence (AI) techniques can be used to support community engagement and inclusive housing development, offering valuable tools and insights for urban planners, policymakers, and other stakeholders.

Data Collection and Preprocessing for Community Engagement and Inclusive Housing Development

Effective community engagement and inclusive housing development rely on comprehensive and accurate data to inform decision-making and facilitate collaboration among diverse stakeholders. Data sources relevant to community engagement and inclusive housing development include:

a. Demographic data: Information on the socioeconomic characteristics of local residents can help identify the needs and priorities of different population groups, enabling targeted and inclusive housing interventions.

b. Housing market data: Data on property values, rents, sales, and vacancy rates can provide insights into housing affordability and availability, guiding the development of context-specific housing solutions.
c. Geospatial data: Geographic information system (GIS) data, such as land use, zoning, and transportation networks, can help identify potential sites for new housing developments and assess their accessibility to essential services and amenities.
d. Social media and online data: User-generated content, such as social media posts, reviews, and check-ins, can offer additional context on community perceptions and preferences regarding housing and neighborhood issues.

Once data are collected, preprocessing steps, such as data cleaning, transformation, and integration, need to be performed to ensure the data's quality and consistency before AI model training.

AI Techniques for Community Engagement and Inclusive Housing Development

AI techniques, including machine learning (ML), deep learning (DL), and natural language processing (NLP), can be employed to support community engagement and inclusive housing development, offering valuable tools and insights for urban planners, policymakers, and other stakeholders.

a. Machine learning for housing needs assessment: ML algorithms, such as regression, clustering, and decision trees, can be used to analyze demographic and housing market data to identify the specific needs and preferences of different population groups, guiding the development of inclusive and context-specific housing solutions.
b. Deep learning for geospatial analysis: DL techniques, such as convolutional neural networks (CNNs) and recurrent neural networks (RNNs), can be used to analyze geospatial data, such as satellite imagery and GIS data, to identify potential sites for new housing developments and assess their accessibility to essential services and amenities.
c. Natural language processing for sentiment analysis: NLP methods can be employed to analyze social media data and other online sources to gauge community perceptions and sentiments about housing and neighborhood issues, informing the design and implementation of inclusive housing interventions.

Applications of AI in Community Engagement and Inclusive Housing Development

AI techniques can be applied to various aspects of community engagement and inclusive housing development, supporting data-driven decision-making and collaboration among diverse stakeholders. Some applications of AI in this context include:

a. Housing needs assessment: AI models can be used to analyze demographic and housing market data to identify the specific needs and preferences of different population groups, guiding the development of inclusive and context-specific housing solutions.

b. Site selection and analysis: AI-driven geospatial analysis can help urban planners and policymakers identify potential sites for new housing developments, assess their suitability based on factors such as land use, zoning, and accessibility to essential services and amenities, and optimize their design to meet the needs of diverse communities.
c. Stakeholder engagement and collaboration: AI tools, such as sentiment analysis and data visualization, can be used to facilitate stakeholder engagement and collaboration by providing accessible and transparent information on community perceptions, preferences, and housing needs. By fostering a shared understanding of local issues and priorities, AI-driven tools can help build trust and consensus among diverse stakeholders, supporting more inclusive and equitable decision-making processes.
d. Public participation and deliberation: AI techniques can be employed to support public participation and deliberation in housing development processes, such as participatory budgeting, community planning, and design charrettes. By analyzing and synthesizing diverse perspectives and preferences, AI-driven tools can help create more inclusive and responsive housing solutions that better meet the needs of local residents.
e. Monitoring and evaluation: AI models can be used to monitor the implementation and impacts of housing interventions, allowing urban planners, policymakers, and other stakeholders to assess their effectiveness and make data-driven adjustments as needed. By providing real-time feedback and insights, AI-driven monitoring and evaluation can help ensure that housing initiatives remain responsive to community needs and contribute to more equitable and sustainable urban development outcomes.

Challenges and Limitations

While AI offers significant potential for enhancing community engagement and inclusive housing development, several challenges and limitations must be considered:

a. Data quality and availability: The accuracy and reliability of AI-driven analysis depend on the quality and comprehensiveness of the underlying data. Incomplete, outdated, or biased data can lead to erroneous conclusions and ineffective interventions.
b. Ethical concerns and privacy issues: The use of AI techniques in community engagement and inclusive housing development may raise ethical concerns and privacy issues, particularly when analyzing sensitive demographic data or using social media information. It is crucial to ensure that AI-driven analysis respects individuals' privacy and does not exacerbate existing inequalities or perpetuate stereotypes.
c. Model interpretability and transparency: The complexity of some AI techniques, particularly deep learning models, can make it difficult to understand and interpret

their predictions and recommendations. This lack of transparency can pose challenges for policymakers and stakeholders who need to justify and communicate their decisions to the public.

d. Technological and digital divides: The use of AI-driven tools in community engagement and inclusive housing development may exacerbate existing technological and digital divides, as not all community members may have access to or familiarity with these technologies. It is essential to consider the potential barriers to participation and develop strategies to ensure that AI-driven tools are accessible and inclusive for all stakeholders.

AI techniques hold promise for enhancing community engagement and inclusive housing development in urban planning and policy-making, offering valuable tools and insights for addressing housing affordability and social equity challenges. However, it is crucial to recognize and address the limitations and challenges associated with AI-driven analysis to ensure its ethical, effective, and equitable application in urban planning and policymaking.

12.5 Challenges and Limitations of AI in Housing, Affordability, and Real Estate Market Analysis

Artificial intelligence (AI) has been increasingly used in housing, affordability, and real estate market analysis, offering valuable insights and tools for urban planners, policymakers, and other stakeholders. However, the adoption of AI techniques in this field is not without challenges and limitations (Table 12.2). This section examines the key challenges and limitations associated with the use of AI in housing, affordability, and real estate market analysis, highlighting areas for improvement and future research.

One of the primary challenges of implementing AI in housing, affordability, and real estate market analysis is the quality and availability of data. Accurate and reliable

Table 12.2 Challenges in AI applications to housing, affordability, and real estate market analysis

Aspect	Challenge
Data quality and availability	Ensuring comprehensive and up-to-date data is challenging, affecting model accuracy
Model interpretability	AI models, especially deep learning, are complex "black boxes," making them hard to interpret
Generalizability	Models developed for specific contexts may not perform well elsewhere, limiting wider application
Integration of data sources	Combining diverse data types presents challenges due to varying formats and resolutions
Ethical considerations	Use of AI raises concerns about data privacy, algorithmic fairness, and potential reinforcement of biases

AI-driven analysis depends on comprehensive, up-to-date, and unbiased data [22]. However, several issues can affect data quality and availability, including:

a. Incomplete or missing data: Missing or incomplete data can lead to biased or incomplete results, undermining the accuracy and reliability of AI-driven analysis [40].
b. Outdated data: Data can quickly become outdated in the dynamic and rapidly changing real estate market, which may result in inaccurate predictions and insights [51].
c. Data bias: Data used in AI models may be subject to various biases, such as selection bias, measurement bias, or reporting bias, which can lead to biased results and perpetuate existing inequalities [36].
d. Data integration: Integrating data from multiple sources can be challenging due to differences in data formats, collection methods, and quality standards, which may impact the accuracy and reliability of AI-driven analysis [44].

The use of AI techniques in housing, affordability, and real estate market analysis may raise ethical concerns and privacy issues, particularly when analyzing sensitive demographic data or using social media information. Some of the key ethical and privacy challenges include:

a. Privacy violations: AI-driven analysis may involve the use of personal and sensitive data, potentially violating individual privacy and raising ethical concerns [32].
b. Discrimination and bias: AI models trained on biased or unrepresentative data may perpetuate existing biases and inequalities, leading to unfair treatment of certain groups or individuals [6].
c. Consent and transparency: AI-driven analysis may involve the use of data collected without individuals' consent or knowledge, raising questions about transparency and accountability [29].

The complexity of some AI techniques, particularly deep learning models, can make it difficult to understand and interpret their predictions and recommendations [4]. This lack of transparency, often referred to as the "black box" problem, can pose challenges for policymakers and stakeholders who need to justify and communicate their decisions to the public. Model interpretability and transparency are essential for ensuring trust in AI-driven analysis and fostering more inclusive and equitable decision-making processes [21].

Housing, affordability, and real estate market dynamics are influenced by numerous factors, many of which are unpredictable or subject to change, such as economic conditions, political developments, and social trends. AI-driven analysis may struggle to capture these complex and dynamic processes, limiting its predictive accuracy and practical applicability [51]. Moreover, overreliance on AI-generated insights may result in overlooking important contextual factors and expert knowledge, leading to suboptimal decision-making [18]. It is crucial to recognize the limitations of AI-driven analysis in capturing complex and dynamic processes and to complement AI-generated insights with expert knowledge and contextual understanding [9].

The use of AI-driven tools in housing, affordability, and real estate market analysis may exacerbate existing technological and digital divides, as not all stakeholders may have access to or familiarity with these technologies [48]. This may result in unequal access to information and resources, reinforcing existing inequalities and limiting the potential for inclusive and equitable decision-making processes. It is essential to consider the potential barriers to participation and develop strategies to ensure that AI-driven tools are accessible and inclusive for all stakeholders [33].

The adoption of AI in housing, affordability, and real estate market analysis may also face legal and regulatory challenges, as existing laws and regulations may not adequately address the unique issues associated with AI-driven analysis [10]. Some of the key legal and regulatory challenges include:

a. Liability and accountability: The use of AI-driven analysis may raise questions about liability and accountability in cases of inaccurate predictions, biased outcomes, or other negative consequences [41].
b. Intellectual property: AI-generated insights and models may involve the use of copyrighted or proprietary data, raising intellectual property concerns and potentially limiting access to valuable information and resources [1].
c. Compliance with fair housing and anti-discrimination laws: AI-driven analysis must comply with existing fair housing and anti-discrimination laws, which may require careful consideration of potential biases and discriminatory outcomes [23].

While AI offers significant potential for enhancing housing, affordability, and real estate market analysis, it is crucial to recognize and address the challenges and limitations associated with AI-driven analysis to ensure its ethical, effective, and equitable application in urban planning and policymaking. By acknowledging these challenges and limitations, researchers, practitioners, and policymakers can work towards developing more robust, transparent, and inclusive AI-driven tools and methods for housing, affordability, and real estate market analysis.

12.6 Future Directions in AI Applications for Housing, Affordability, and Real Estate Market Analysis

As we continue to witness rapid advancements in artificial intelligence (AI) and its applications in various sectors, it is essential to consider future directions and possibilities for housing, affordability, and real estate market analysis. This section discusses emerging trends, potential opportunities, and areas for future research and development.

The integration of diverse data sources and types, such as traditional structured datasets, open data, social media, and sensor data, can provide more comprehensive insights into housing and real estate market dynamics [22]. Future AI applications could increasingly leverage big data and advanced analytics to generate more accurate predictions, inform policy decisions, and optimize real estate investments [47].

As AI models become more complex and powerful, the need for interpretability and transparency becomes increasingly important. Developing techniques to explain and visualize AI model outputs can improve trust, facilitate collaboration among stakeholders, and reduce the risk of biased or unfair outcomes [2]. Future research should focus on creating AI systems that are not only accurate but also explainable and accountable.

One of the critical challenges facing the housing sector is promoting sustainable and equitable development, considering factors such as climate change, energy efficiency, and social inclusion [15]. AI could play a crucial role in designing and implementing policies and strategies that prioritize environmental sustainability and social equity. For example, AI-driven models could help identify areas most vulnerable to gentrification and inform targeted interventions to prevent displacement and promote affordable housing [12].

Future AI applications could increasingly incorporate collaborative and participatory approaches, involving diverse stakeholders such as policymakers, urban planners, real estate professionals, and community members. This collaboration could lead to more inclusive, context-specific, and responsive AI solutions [30]. Moreover, AI tools could facilitate more effective public engagement in housing and land-use planning processes, empowering citizens to contribute their perspectives and preferences.

AI could play a significant role in supporting evidence-based policy development by enabling rigorous experimentation and evaluation of alternative housing policies and interventions. AI-driven simulations and counterfactual analysis could provide valuable insights into the potential impacts and trade-offs associated with different policy options [5]. This approach could help policymakers make more informed decisions, anticipate unintended consequences, and adapt strategies as needed.

As AI becomes more prevalent in housing, affordability, and real estate market analysis, ethical considerations and governance frameworks will be increasingly essential. Key issues include data privacy, algorithmic fairness, and the potential for discriminatory outcomes [32]. Future research should explore appropriate ethical guidelines, regulatory approaches, and industry best practices to ensure that AI applications promote social welfare and respect individual rights.

In conclusion, the future of AI applications in housing, affordability, and real estate market analysis holds great promise in addressing complex challenges and contributing to more sustainable, equitable, and data-driven urban development. As AI continues to advance, it will be crucial for researchers, practitioners, policymakers, and other stakeholders to work collaboratively, ensuring that these applications are ethical, transparent, and accountable, and contribute to the well-being of communities worldwide. By focusing on emerging trends, potential opportunities, and areas for future research and development, the field of AI in housing, affordability, and real estate market analysis can continue to evolve and make a positive impact on the lives of people around the globe.

References

1. Abbott, R. (2018). I think, therefore I invent: Creative computers and the future of patent law. *Boston College Law Review, 57*(4), 1079–1126.
2. Adadi, A., & Berrada, M. (2018). Peeking inside the black box: A survey on Explainable Artificial Intelligence (XAI). *IEEE Access, 6*, 52138–52160.
3. Angwin, J., Larson, J., Mattu, S., & Kirchner, L. (2016). Machine bias. ProPublica.
4. Arrieta, A. B., Díaz-Rodríguez, N., Del Ser, J., Bennetot, A., Tabik, S., Barbado, A., García, S., Gil-López, S., Molina, D., Benjamins, R., Chatila, R., Herrera, F. (2020). Explainable Artificial Intelligence (XAI): Concepts, taxonomies, opportunities and challenges toward responsible AI. *Information Fusion*, 58, 82–115.
5. Banerjee, A., Chassang, S., & Snowberg, E. (2017). Decision-theoretic approaches to experiment design and external validity. In *Handbook of economic field experiments* (Vol. 1, pp. 537–602). North-Holland.
6. Barocas, S., & Selbst, A. D. (2016). Big data's disparate impact. *California Law Review, 104*(3), 671–732.
7. Blei, D. M., Ng, A. Y., & Jordan, M. I. (2003). Latent Dirichlet allocation. *Journal of Machine Learning Research, 3*, 993–1022.
8. Bramley, G. (2017). Housing affordability: A fuller picture. *Urban Studies, 54*(14), 3155–3178. https://doi.org/10.1177/0042098017691420
9. Brynjolfsson, E., & McAfee, A. (2017). The business of artificial intelligence. *Harvard Business Review, 95*(4), 80–89.
10. Calo, R. (2017). Artificial intelligence policy: A primer and roadmap. *UCDL Rev., 51*, 399.
11. Cao, W., Li, J., & Liu, Y. (2018). Housing price prediction based on online reviews: A sentiment analysis approach. In *2018 IEEE International Conference on Big Data (Big Data)* (pp. 2046–2051). IEEE.
12. Chakraborty, A., Kaza, N., & Knaap, G. J. (2011). The spatial distribution of foreclosures: A study of the housing market in Prince George's County, Maryland. *International Journal of Urban Sciences, 15*(1), 1–25.
13. Chen, J., Li, K., & Yang, Y. (2019). A deep learning approach to real estate market segmentation using convolutional neural network. In *2019 IEEE International Conference on Big Data (Big Data)* (pp. 2817–2822). IEEE.
14. Chen, Y., Wang, L., & Agarwal, A. (2018). Analyzing and forecasting the housing market in the United States. *International Journal of Applied Decision Sciences, 11*(2), 139–159.
15. Desilver, D. (2017). World's most populous countries may struggle to absorb urban newcomers. Pew Research Center. Retrieved from https://www.pewresearch.org/fact-tank/2017/02/13/worlds-most-populous-countries-may-struggle-to-absorb-urban-newcomers/
16. Desilver, D. (2018). 7 demographic trends shaping the U.S. and the world in 2018. Pew Research Center. https://www.pewresearch.org/fact-tank/2018/04/25/7-demographic-trends-shaping-the-u-s-and-the-world-in-2018/
17. Desmond, M. (2016). *Evicted: Poverty and profit in the American city*. Crown Publishers.
18. Dignum, V. (2018). *Responsible artificial intelligence: How to develop and use AI in a responsible way*. Springer Nature.
19. Fang, D., Gough, R., & Chen, D. (2020). Urban big data: Challenges and opportunities of using GPS data in urban research. *City & Community, 19*(1), 210–233. https://doi.org/10.1111/cico.12447
20. Florida, R. (2017). *The new urban crisis: How our cities are increasing inequality, deepening segregation, and failing the middle class—and what we can do about it*. Basic Books.
21. Gilpin, L. H., Bau, D., Yuan, B. Z., Bajwa, A., Specter, M., & Kagal, L. (2018). Explaining explanations: An overview of interpretability of machine learning. In *2018 IEEE 5th International Conference on Data Science and Advanced Analytics (DSAA)* (pp. 80–89). IEEE.
22. Glaeser, E. L., Kim, H., & Luca, M. (2018). Nowcasting gentrification: Using Yelp data to quantify neighborhood change. *AEA Papers and Proceedings, 108*, 77–82. https://doi.org/10.1257/pandp.20181034

23. Goodman, B., & Flaxman, S. (2017). European Union regulations on algorithmic decision-making and a "right to explanation." *AI Magazine, 38*(3), 50–57.
24. Goodman, L., Kaul, K., & Zhu, J. (2018). The role of technology in mortgage lending. *The Review of Financial Studies, 31*(5), 1854–1899.
25. Hoffman, S., Kahn, L. B., & Calacci, D. (2020). Discretion in hiring. *The Quarterly Journal of Economics, 135*(3), 1179–1213. https://doi.org/10.1093/qje/qjaa006
26. Huang, X., Xiong, H., & Zhang, J. (2018). A novel deep learning model for housing price prediction. In *2018 IEEE 3rd International Conference on Cloud Computing and Big Data Analysis (ICCCBDA)* (pp. 84–88). IEEE.
27. Kelleher, J. D., Mac Namee, B., & D'Arcy, A. (2015). *Fundamentals of machine learning for predictive data analytics: Algorithms, worked examples, and case studies*. MIT Press.
28. Lee, R., An, S., & Yang, J. (2018). Analysis and prediction of regional housing market dynamics. *Sustainability, 10*(8), 2934.
29. Lepri, B., Oliver, N., Letouzé, E., Pentland, A., & Vinck, P. (2018). Fair, transparent, and accountable algorithmic decision-making processes. *Philosophy & Technology, 31*(4), 611–627.
30. Lepri, B., Staiano, J., Sangokoya, D., Letouzé, E., & Oliver, N. (2018b). The tyranny of data? The bright and dark sides of data-driven decision-making for social good. In *Transparent data mining for big and small data* (pp. 3–24). Springer.
31. Liu, H., Pang, G., Wang, W., & Xu, Z. (2018). Real estate price prediction with multi-source deep learning framework. In *2018 IEEE International Conference on Big Data (Big Data)* (pp. 3630–3635). IEEE.
32. Mittelstadt, B. D., Allo, P., Taddeo, M., Wachter, S., & Floridi, L. (2016). The ethics of algorithms: Mapping the debate. *Big Data & Society, 3*(2), 2053951716679679.
33. Mossberger, K., Tolbert, C. J., & Hamilton, A. (2012). Broadband adoptionl Measuring digital citizenship: Mobile access and broadband. *International Journal of Communication, 6*, 2492–2528.
34. Mullainathan, S., & Spiess, J. (2017). Machine learning: An applied econometric approach. *Journal of Economic Perspectives, 31*(2), 87–106. https://doi.org/10.1257/jep.31.2.87
35. Niemeyer, P., Rottensteiner, F., & Soergel, U. (2020). Context-aware convolutional neural networks for building footprint segmentation in VHR remote sensing images. *ISPRS Journal of Photogrammetry and Remote Sensing, 160*, 121–137.
36. O'Neil, C. (2016). *Weapons of math destruction: How big data increases inequality and threatens democracy*. Broadway Books.
37. Peng, H., Long, F., & Ding, C. (2017). Feature selection based on mutual information criteria of max-dependency, max-relevance, and min-redundancy. *IEEE Transactions on Pattern Analysis and Machine Intelligence, 27*(8), 1226–1238.
38. Peng, H., Zhang, Y., & Li, Q. (2016). An information extraction system for real estate websites. In *2016 IEEE International Conference on Systems, Man, and Cybernetics (SMC)* (pp. 003003–003008). IEEE.
39. Sarkar, A., & Kumar, A. (2018). Artificial intelligence in urban planning and design: A review and the way forward. *Journal of Urban Management, 7*(2), 38–48.
40. Sun, T., Zhang, R., Du, X., Zhang, J., & Xu, B. (2020). Data missing and imputation in the big data era: A systematic review. *IEEE Access, 8*, 180639–180654.
41. Vladeck, D. C. (2014). Machines without principals: Liability rules and artificial intelligence. *Washington Law Review, 89*(1), 117–150.
42. Wang, D., Khormali, S., Yigitcanlar, T., & Kamruzzaman, M. (2019). What smart city projects and applications are essential for cities and their citizens? A global investigation. *Journal of Open Innovation: Technology, Market, and Complexity, 5*(3), 57. https://doi.org/10.3390/joitmc5030057
43. Wang, D., Li, X., & Li, Y. (2020). China's "smart city" development: From a national strategy to a local blueprint. *Journal of Urban Technology, 27*(1), 51–71.
44. Wu, J., Gao, L., & Zhao, H. (2018). A hybrid model for spatio-temporal data: The impact of Uber on house prices. *Real Estate Economics, 46*(4), 863–892.

45. Wu, L., Brynjolfsson, E., & Doshi, N. (2020). How online platforms are shaping the United States housing market. *American Economic Review: Insights, 2*(1), 1–18. https://doi.org/10.1257/aeri.20180148
46. Wu, L., Ye, X., & Zou, H. (2019). Unsupervised real estate market segmentation using autoencoders. *Computers, Environment and Urban Systems, 75*, 1–9.
47. Wu, X., Zhu, X., Wu, G. Q., & Ding, W. (2018). Data mining with big data. *IEEE Transactions on Knowledge and Data Engineering, 30*(1), 77–89.
48. Wyche, S. P., Simiyu, N., & Othieno, M. E. (2016). Mobile phones as amplifiers of social inequality among rural Kenyan women. *ACM Transactions on Computer-Human Interaction (TOCHI), 23*(3), 1–19.
49. Yaseen, Z. M., Sulaiman, S. O., Deo, R. C., & Chau, K. W. (2019). An enhanced extreme learning machine model for river flow forecasting: State-of-the-art, practical applications in water resource engineering area and future research direction. *Journal of Hydrology, 569*, 387–408.
50. Yates, J. (2017). Housing in Australia in an era of neoliberalism. In R. Martin & M. R. Tonkiss (Eds.), *The crisis of global modernity: Asian traditions and a sustainable future* (pp. 97–117). Cambridge University Press. https://doi.org/10.1089/sus.2017.29085.ejy
51. Yuan, Y., & Wang, X. (2020). Challenges and strategies in applying AI to real estate. In *Artificial intelligence in real estate* (pp. 1–15). Springer.
52. Zhang, H., & Wu, Y. (2017). House price prediction using artificial neural networks and support vector machines: The role of sample size. *Journal of Real Estate Finance and Economics, 54*(4), 510–541.
53. Zhang, L., & Wu, X. (2017). An intelligent housing price prediction system based on deep learning. In *2017 IEEE 15th International Conference on Dependable, Autonomic and Secure Computing, 15th International Conference on Pervasive Intelligence and Computing, 3rd International Conference on Big Data Intelligence and Computing and Cyber Science and Technology Congress* (pp. 778–782). IEEE
54. Zhang, X., Lv, Y., & Yin, J. (2019). Urban housing price prediction based on machine learning algorithms. *Complexity, 2019*, 1–14.

Chapter 13
Sustainable Development and Resource Management

13.1 Overview of Sustainable Development and Resource Management

Sustainable development is a multidisciplinary concept that encompasses environmental, social, and economic goals, striving to balance the needs of current and future generations [129]. It is an essential approach to urban planning, aiming to minimize the negative impacts of urbanization, such as increased pollution, depletion of natural resources, and climate change [62]. Resource management plays a pivotal role in achieving sustainable development by ensuring the efficient and responsible use of natural resources, including land, water, and energy, while mitigating environmental degradation [20].

Sustainable development and resource management have become increasingly relevant in the context of rapid urbanization and growing global populations, which have led to significant challenges for cities worldwide (UN [125]). Policymakers, urban planners, and other stakeholders are constantly seeking innovative solutions to address these challenges and promote sustainability in urban environments [31]. Artificial intelligence (AI) has emerged as a promising tool in this regard, offering various techniques and applications that can help optimize resource allocation, monitor environmental impacts, and support decision-making processes in urban planning [19].

The concept of sustainable development has evolved over time, incorporating various dimensions, such as social equity, economic prosperity, and environmental protection [32, 48]. The United Nations (UN) has established the 2030 Agenda for Sustainable Development, which includes 17 Sustainable Development Goals (SDGs) that provide a comprehensive framework for addressing various aspects of sustainability [126]. These goals emphasize the interconnected nature of sustainable development, highlighting the importance of integrating social, economic, and environmental dimensions in decision-making processes [100].

S. N. M. Moghaddam and H. Cao, *Artificial Intelligence-Driven Geographies*, City Development: Issues and Best Practices, https://doi.org/10.1007/978-981-97-5116-7_13

In the field of urban planning, sustainable development and resource management have become central themes, guiding the formulation of policies, land-use planning, infrastructure development, and other aspects of urban growth [16, 132]. Several principles, such as compact city design, mixed land-use, and green infrastructure, have been proposed to promote sustainable development in urban areas [13, 15]. These principles aim to reduce the environmental footprint of cities, enhance the quality of life for residents, and foster resilient and adaptive communities in the face of climate change and other challenges [39, 88].

Resource management is a critical component of sustainable development, focusing on the efficient use and conservation of natural resources to ensure their availability for current and future generations [20]. In urban planning, resource management involves the development of strategies and policies for the allocation, distribution, and utilization of land, water, energy, and other resources [113]. It also encompasses waste management, recycling, and pollution control measures, which help mitigate the environmental impacts of urbanization [21]. Moreover, resource management seeks to promote equity and social justice by ensuring that the benefits of sustainable development are accessible to all members of society [58].

The advent of AI has opened new opportunities for enhancing sustainable development and resource management in urban planning [19]. AI techniques, such as machine learning, deep learning, and natural language processing, can process large volumes of data and generate insights that support decision-making processes, optimize resource allocation, and monitor environmental impacts [5, 53]. AI can also facilitate the integration of multiple sources of information, such as satellite imagery, social media, and sensor networks, enabling a more comprehensive understanding of urban dynamics and resource management challenges [78].

AI applications in sustainable development and resource management can be broadly categorized into the following areas:

1. Land-use planning and optimization: AI can support the evaluation of land-use patterns and the identification of optimal configurations that minimize environmental impacts and promote sustainable development [60]. Machine learning and deep learning techniques can be used to analyze spatial data and predict future land-use changes, enabling proactive planning and the formulation of targeted policies [137].
2. Energy management and efficiency: AI can help optimize energy consumption in urban areas, promoting energy efficiency and the use of renewable energy sources [34]. Techniques such as reinforcement learning can be employed to control smart grids and distribute energy more efficiently, while machine learning algorithms can be used to forecast energy demand and inform the design of energy-efficient buildings [26].
3. Water resource management: AI can support the sustainable management of water resources in urban areas, addressing issues such as water scarcity, pollution, and flooding [1]. Machine learning models can be used to predict water demand, optimize water distribution networks, and monitor water quality, ensuring the efficient use of this vital resource [3].

4. Waste management and recycling: AI can enhance waste management practices in cities, promoting recycling and reducing the environmental impacts of waste disposal [70]. Computer vision techniques can be employed to automate waste sorting processes, while machine learning algorithms can be used to optimize waste collection and transportation routes [101].
5. Environmental monitoring and impact assessment: AI can support the monitoring of environmental conditions in urban areas, detecting changes in air quality, noise levels, and other parameters [18]. Machine learning models can be used to predict the environmental impacts of urban development projects, informing decision-making processes and enabling the mitigation of potential risks [108].

Despite the potential benefits of AI for sustainable development and resource management, there are also challenges and limitations that need to be addressed. These include issues related to data quality and availability, the interpretability of AI models, and the ethical considerations associated with the use of AI technologies in urban planning [19]. Moreover, the successful integration of AI in sustainable development and resource management requires collaboration among various stakeholders, including urban planners, policymakers, researchers, and technology developers [111].

13.2 Data Sources for Sustainable Development and Resource Management

Sustainable development and resource management are critical aspects of urban planning, aiming to balance economic growth, social equity, and environmental protection [129]. To achieve sustainable development, urban planners need access to reliable, accurate, and diverse data sources. In this section, we discuss various data sources that can support the application of AI in sustainable development and resource management, including remote sensing data, socioeconomic data, and open data platforms.

Remote Sensing Data

Remote sensing data have been widely used in urban planning, offering valuable information on land use, land cover, vegetation, and urban heat island effects. Various satellite data sources are available, such as Landsat, Sentinel, and MODIS, providing multispectral and high-resolution images suitable for urban analysis [130]. Lidar (Light Detection and Ranging) data can also be utilized to create high-resolution digital elevation models (DEMs) for terrain analysis, flood risk assessment, and urban heat island mitigation [144].

Socioeconomic Data

Socioeconomic data provide valuable insights into the demographic, social, and economic aspects of urban areas, informing planners about the needs and preferences of local communities. National statistical offices, such as the United States Census Bureau and Eurostat, collect and disseminate various socioeconomic indicators, including population, income, education, and employment [50]. These data sources can be integrated with other spatial data to support sustainable urban planning and decision-making [138].

Environmental Data

Environmental data are essential for understanding the impacts of urbanization on natural resources, ecosystems, and climate. Sources of environmental data include air and water quality monitoring stations, greenhouse gas emission inventories, and meteorological stations [88]. These data can be used to assess the effectiveness of urban sustainability policies and inform the development of new strategies to reduce pollution and conserve resources [100].

Open Data Platforms

Open data platforms are increasingly being used to share and disseminate urban planning data, facilitating collaboration and data-driven decision-making. Examples of open data platforms include the European Data Portal, the United States Environmental Protection Agency's EnviroAtlas, and the Global Biodiversity Information Facility (GBIF). These platforms provide access to diverse datasets, including land use, transportation, energy, and social indicators, which can be used to inform sustainable urban planning [113].

Social Media and Crowdsourced Data

Social media and crowdsourced data can offer real-time insights into urban dynamics and public opinions, which can be valuable for understanding the social and cultural aspects of sustainability. Examples of social media data sources include Twitter, Facebook, and Instagram, which can be analyzed using natural language processing (NLP) and sentiment analysis techniques to identify trends, opinions, and preferences related to urban planning and sustainability [104]. Crowdsourced data, such as OpenStreetMap, can also provide valuable information on urban infrastructure, land use, and transportation networks [57].

Internet of Things (IoT) and Sensor Data

IoT and sensor data are increasingly being used to monitor and manage urban environments, supporting the development of smart and sustainable cities. IoT devices and sensors can collect real-time data on various urban phenomena, such as energy consumption, air quality, and traffic patterns, enabling more efficient and effective resource management [102]. These data sources can also support AI-driven urban planning applications, such as energy optimization, waste management, and water resource management (R resource management [109]).

Geospatial Data and Geographic Information Systems (GIS)

Geospatial data and GIS play a crucial role in sustainable urban planning, providing spatial context and analytical capabilities for various planning tasks. GIS data can include digital maps, land use data, transportation networks, and other spatially referenced datasets that enable urban planners to visualize, analyze, and model urban systems [14]. Geospatial data can be integrated with other data sources, such as remote sensing, socioeconomic, and environmental data, to support AI applications in sustainable development and resource management [52].

In conclusion, the availability and accessibility of diverse data sources are essential for the successful application of AI in sustainable development and resource management. Urban planners need to harness these data sources to develop data-driven strategies and policies that promote sustainable urban growth and protect natural resources. As the volume and variety of data continue to expand, AI techniques, such as machine learning, deep learning, and NLP, will play an increasingly important role in transforming urban planning and decision-making processes.

13.3 AI Techniques for Sustainable Development and Resource Management

13.3.1 Machine Learning for Resource Allocation and Optimization

Resource allocation and optimization play a critical role in sustainable development and resource management. As urban populations continue to grow and resources become increasingly scarce, it is essential to ensure the efficient and equitable distribution of resources such as water, energy, and land. Machine learning, a subset of artificial intelligence (AI), has emerged as a powerful tool for optimizing resource allocation, improving decision-making processes, and addressing complex challenges in sustainable urban planning.

Machine Learning in Water Resource Management

Water resource management is a vital aspect of sustainable development, as it encompasses the allocation, distribution, and consumption of water resources to meet various human and environmental needs. Machine learning algorithms have been widely applied in water resource management to optimize water allocation, forecast water demand, and identify inefficiencies in water distribution systems [54].

For instance, machine learning techniques such as regression analysis, support vector machines (SVM), and artificial neural networks (ANN) have been employed to predict water demand in urban areas [11]. These models can account for various factors, including population growth, climate change, and socioeconomic variables, to produce accurate and reliable water demand forecasts. By predicting future water demand, urban planners and water managers can develop effective strategies for water supply management, infrastructure development, and water conservation [133].

Moreover, machine learning algorithms can be used to identify water loss and inefficiencies in water distribution systems. For example, ANN and genetic algorithms (GA) have been applied to detect leaks in water pipelines, thereby reducing water waste and ensuring the efficient use of scarce water resources [2].

Machine Learning in Energy Management

Energy management is another crucial aspect of sustainable development and resource management. Machine learning algorithms can be employed to optimize energy consumption, predict energy demand, and facilitate the integration of renewable energy sources into urban energy systems [27].

Machine learning techniques such as ANN and SVM have been used to forecast energy demand in residential, commercial, and industrial sectors [4]. These demand predictions enable utilities and energy managers to optimize energy generation, distribution, and storage, leading to more efficient energy systems and reduced greenhouse gas emissions [99].

Furthermore, machine learning algorithms have been applied to optimize the operation and maintenance of renewable energy systems, such as wind and solar power plants. For example, machine learning models can predict equipment failure, allowing for preventive maintenance and reducing downtime in renewable energy systems [23]. Additionally, machine learning can be employed to optimize the integration of renewable energy sources into power grids, enhancing grid stability and promoting the transition to clean and sustainable energy systems [91].

Machine Learning in Land Use and Urban Planning

Land use and urban planning are integral to sustainable development and resource management. Machine learning algorithms have been increasingly utilized in land use planning, zoning, and land allocation to optimize urban growth, minimize environmental impacts, and promote sustainable land use patterns [81].

Machine learning techniques, such as decision trees, clustering, and cellular automata, have been employed to simulate urban growth and land use change, enabling urban planners to evaluate the consequences of different development scenarios and devise more sustainable urban policies [9]. Additionally, machine learning algorithms can be used to optimize the allocation of land for various purposes, such as housing, commercial development, agriculture, and conservation, balancing the competing demands for urban growth and environmental protection [121].

Machine Learning for Waste Management and Circular Economy

Waste management is an essential component of sustainable development and resource management, with the circular economy model promoting the efficient use of resources and minimizing waste generation. Machine learning algorithms have been applied to waste management, including waste prediction, waste collection optimization, and recycling processes [36].

For example, machine learning techniques such as ANN and SVM have been utilized to predict waste generation at the household, commercial, and industrial levels [86]. Accurate waste generation predictions enable waste managers to develop effective waste reduction strategies, optimize waste collection routes, and allocate resources efficiently.

Moreover, machine learning algorithms have been employed to improve recycling processes by identifying recyclable materials, optimizing sorting processes, and predicting the quality of recycled products [139]. By enhancing recycling processes, machine learning contributes to the circular economy and promotes sustainable resource management.

Machine learning has emerged as a valuable tool in the field of sustainable development and resource management. By optimizing resource allocation and facilitating data-driven decision-making, machine learning techniques offer innovative solutions to complex urban planning challenges. As urban populations continue to grow and resources become increasingly scarce, the application of machine learning in sustainable development and resource management is expected to expand further, contributing to more resilient and sustainable urban environments.

13.3.2 Deep Learning for Environmental Monitoring and Analysis

Deep learning, a subset of artificial intelligence and machine learning, has emerged as a powerful tool for analyzing large and complex data sets, making it particularly useful for environmental monitoring and analysis. By employing deep neural networks, deep learning algorithms can automatically learn and extract relevant features from raw data, enabling the development of accurate and efficient predictive models. This section provides an overview of the application of deep learning techniques in environmental monitoring and analysis, covering various aspects such as air quality, water resources, land use, and biodiversity.

Air Quality Monitoring and Prediction

Air quality is a crucial aspect of environmental monitoring, as it directly affects human health and well-being. Traditional air quality monitoring methods rely on expensive and sparse monitoring stations that may not provide adequate spatial and temporal coverage. Deep learning models have been employed to predict air quality by using data from various sources such as remote sensing, ground monitoring stations, and social media [80]. Convolutional Neural Networks (CNNs) have been used to identify relevant spatial features from satellite imagery, while Recurrent Neural Networks (RNNs) and Long Short-Term Memory (LSTM) networks have been employed to capture temporal dependencies in the data [140]. These models have demonstrated improved performance in predicting air quality parameters such as particulate matter (PM2.5 and PM10), nitrogen dioxide (NO_2), and ozone (O_3) concentrations.

Water Resources Management

Deep learning techniques have also been applied to water resources management, including water quality monitoring, water consumption forecasting, and flood prediction. For instance, CNNs have been employed to analyze remote sensing data for water quality assessment, with a focus on detecting harmful algal blooms and estimating water turbidity [85]. Additionally, LSTM networks have been utilized for predicting water consumption based on historical data and environmental factors [94]. For flood prediction, deep learning models such as CNNs and LSTMs have been combined with hydrological models to improve the accuracy of flood forecasting and identify potential flood-prone areas [71].

Land Use and Land Cover Change Analysis

Understanding land use and land cover changes is essential for sustainable development and resource management. Deep learning techniques, particularly CNNs, have shown promising results in land use and land cover classification using remote sensing data [142]. By leveraging the spatial information contained in satellite images, CNNs can accurately classify different land cover types and detect changes over time. This information can then be used to assess the impacts of urbanization, deforestation, and agricultural expansion on the environment and to inform land use planning and policy development. Furthermore, Generative Adversarial Networks (GANs) have been applied to simulate future land use scenarios based on historical data and various socioeconomic and environmental factors [83].

Biodiversity Conservation

Deep learning techniques have also been employed for biodiversity conservation, including species identification, habitat mapping, and population monitoring. For example, CNNs have been used to identify and classify species from camera trap images, acoustic recordings, and genetic data [124]. This automated species identification can significantly reduce the time and effort required for traditional manual identification and monitoring methods. Moreover, deep learning models can be utilized to predict species distribution and habitat suitability, aiding in the development of conservation strategies and protected area management [68].

In summary, deep learning techniques have demonstrated great potential in various aspects of environmental monitoring and analysis. By leveraging large and complex data sets, these models can provide valuable insights into air quality, water resources, land use changes, and biodiversity conservation. The accurate and timely information generated by deep learning models can inform decision-making processes, contribute to sustainable development, and improve resource management.

However, the application of deep learning techniques in environmental monitoring and analysis is not without challenges. Data quality and availability are critical factors in the successful implementation of deep learning models. High-quality, representative data sets are required to train and validate these models, and the lack of such data can lead to suboptimal performance. Additionally, the integration of heterogeneous data sources, such as remote sensing, ground-based measurements, and social media, can pose challenges in data preprocessing and fusion [110].

Another challenge is the interpretability and explainability of deep learning models. While these models can achieve high accuracy, their complex architectures and large numbers of parameters can make them difficult to interpret and explain, which may hinder their acceptance and adoption by stakeholders and decision-makers [116]. Research efforts are being made to develop techniques that can provide better understanding and visualization of deep learning models, such as attention mechanisms and layer-wise relevance propagation [92].

Despite these challenges, the future of deep learning applications in environmental monitoring and analysis is promising. The rapid development of remote sensing technologies, such as high-resolution satellite imagery and hyperspectral data, is expected to provide an abundance of data for training and validation of deep learning models. Moreover, the increasing computational power and the development of more efficient and scalable deep learning algorithms will enable the analysis of larger and more complex data sets [145].

In conclusion, deep learning techniques have the potential to transform environmental monitoring and analysis, providing valuable insights for sustainable development and resource management. By addressing the challenges related to data quality, availability, and model interpretability, deep learning can play a crucial role in supporting decision-making processes and promoting sustainable development in the face of growing environmental challenges.

13.3.3 Natural Language Processing for Sustainability Policy Analysis

Natural Language Processing (NLP) has emerged as an essential tool for extracting insights and information from vast amounts of unstructured text data. In the context of sustainable development and resource management, NLP can be used to analyze and understand policies, regulations, and other documents relevant to sustainability. By automating the process of extracting key information from these sources, NLP can significantly improve the efficiency and effectiveness of policy analysis and formulation. This section will explore the potential applications of NLP in sustainability policy analysis, the current state of the art, and the challenges and opportunities for future research.

NLP Techniques for Sustainability Policy Analysis

NLP techniques can be broadly categorized into two main areas: (1) information extraction, which focuses on identifying and extracting specific information from text, and (2) text classification, which involves categorizing text into predefined classes or groups. Both of these areas have potential applications in the context of sustainability policy analysis.

Information Extraction: Information extraction techniques can be used to automatically identify and extract relevant information from policy documents, such as the names of organizations, key policy terms, and specific targets or commitments. Some common information extraction techniques include named entity recognition, relation extraction, and event extraction.

- Named Entity Recognition (NER): NER is a fundamental NLP task that involves identifying and classifying entities, such as organizations, locations, and dates, in text. NER can be used to extract information about key stakeholders, institutions, and policy mechanisms from sustainability-related documents [96].
- Relation Extraction: This NLP task aims to identify and classify relationships between entities in text. In the context of sustainability policy analysis, relation extraction can help identify connections between policy actors, objectives, and targets [22].
- Event Extraction: Event extraction techniques focus on identifying and classifying events, such as policy decisions, legislative actions, or sustainability initiatives, described in text. This information can be valuable for understanding the timeline and sequence of policy development and implementation [65].

Text Classification: Text classification techniques can be employed to categorize policy documents based on their content, such as the policy domain (e.g., energy, transportation, waste management), policy instrument (e.g., regulation, tax, subsidy), or the specific Sustainable Development Goals (SDGs) they address. Popular text classification techniques include support vector machines, naive Bayes, and deep learning-based approaches such as convolutional neural networks (CNNs) and recurrent neural networks (RNNs) [118, 146].

Applications of NLP in Sustainability Policy Analysis

NLP techniques have been applied in various aspects of sustainability policy analysis, including:

- Policy Document Analysis: NLP can be used to automatically process and analyze large collections of policy documents, extracting key information and identifying trends and patterns in policy development over time [123].
- Comparative Policy Analysis: By analyzing policy documents from different countries or regions, NLP techniques can facilitate comparative policy analysis, helping identify best practices and policy gaps [49].
- Policy Alignment Analysis: NLP can help assess the alignment of policies with international sustainability goals and frameworks, such as the SDGs, by automatically mapping policy documents to relevant goals and targets [89].
- Public Opinion Analysis: By applying NLP techniques to social media data, news articles, and public consultation documents, policymakers can gain insights into public opinion and sentiment towards sustainability policies and initiatives [77].

Challenges and Limitations of NLP in Sustainability Policy Analysis

Despite the potential of NLP techniques in sustainability policy analysis, there are several challenges and limitations that need to be addressed:

Policy documents often contain ambiguous and complex language, which can be challenging for NLP algorithms to process and interpret accurately. Additionally, domain-specific terminology and jargon can further complicate the analysis, as these terms may not be well-represented in pre-trained NLP models [63].

The availability and quality of policy documents can also pose challenges for NLP-based analysis. Policy documents may be available in various formats, such as PDFs, which can be difficult to process and convert into machine-readable text. Additionally, documents may be missing, incomplete, or contain errors, which can impact the accuracy and reliability of the analysis [98].

Policy documents may be available in multiple languages, requiring cross-lingual NLP techniques to process and analyze the text. Furthermore, cultural differences and context-specific nuances may impact the interpretation of policy documents, which can be challenging for NLP algorithms to capture and account for [89].

Future Directions in NLP for Sustainability Policy Analysis

Despite the challenges and limitations, there are several promising future directions for NLP applications in sustainability policy analysis:

Transfer learning and domain adaptation techniques can help improve the performance of NLP models in processing and analyzing policy documents, particularly when dealing with domain-specific terminology and jargon. Pre-trained models can be fine-tuned on domain-specific data to better capture the nuances and context of sustainability policies [103].

Advancements in cross-lingual and multilingual NLP techniques can help overcome language barriers and enable more comprehensive policy analysis across different countries and regions. Developing multilingual models and resources for sustainability policy analysis can also facilitate cross-cultural comparisons and collaboration [112].

As NLP techniques become more complex and powerful, it is crucial to ensure that the insights generated from these models are transparent, interpretable, and explainable. Developing explainable AI techniques for sustainability policy analysis can help build trust in the results and facilitate more informed decision-making [7].

Natural Language Processing has the potential to significantly enhance the efficiency and effectiveness of sustainability policy analysis. By automating the extraction of relevant information from policy documents and facilitating comparative and cross-lingual analysis, NLP can provide valuable insights for policymakers and stakeholders working towards sustainable development. Despite the challenges and limitations, ongoing advancements in NLP and AI research offer promising future directions for further enhancing the role of NLP in sustainability policy analysis.

13.4 Applications of AI in Sustainable Development and Resource Management

13.4.1 Energy Efficiency and Conservation

The role of Artificial Intelligence (AI) in energy efficiency and conservation has gained significant attention in recent years. The rapidly increasing energy consumption worldwide calls for the development and adoption of innovative technologies that promote energy efficiency and conservation. AI has the potential to address these challenges by optimizing energy consumption, predicting energy demand, and facilitating decision-making for policy and infrastructure planning. This section will discuss how AI can be utilized for energy efficiency and conservation in various sectors, including residential, commercial, and industrial applications, and the potential impact on sustainable development.

Residential Applications

AI can significantly contribute to energy efficiency and conservation in the residential sector by optimizing energy usage patterns and reducing energy consumption. Smart home systems, equipped with AI algorithms, can monitor and analyze the energy consumption of individual households, adjusting temperature and lighting settings based on occupancy and personal preferences [107]. Furthermore, AI can provide tailored recommendations to homeowners on how to improve energy efficiency, such as upgrading insulation or installing solar panels [17].

Additionally, AI can be employed for demand-side management, which involves the real-time adjustment of energy consumption to match supply. By incorporating AI into energy management systems, households can automatically shift their energy usage to off-peak hours when electricity prices are lower, reducing overall energy costs and minimizing strain on the grid [24].

Commercial Applications

In the commercial sector, AI can optimize the energy consumption of office buildings, retail spaces, and other large facilities. AI-powered building management systems can monitor and control heating, ventilation, and air conditioning (HVAC) systems, lighting, and other energy-intensive operations to minimize energy waste and reduce costs [97]. AI algorithms can also analyze historical energy consumption data and weather forecasts to predict future energy demands, allowing facility managers to optimize energy usage and make informed decisions on infrastructure investments [33].

Industrial Applications

The industrial sector is another area where AI can significantly contribute to energy efficiency and conservation. AI algorithms can be applied to optimize energy-intensive processes such as manufacturing, mining, and petrochemical production, leading to reduced energy consumption and improved sustainability [78]. For example, AI can predict equipment failures and schedule maintenance to minimize downtime and energy waste [75].

Moreover, AI can be used for supply chain optimization, enabling companies to reduce energy consumption in transportation and logistics. By analyzing large datasets on weather patterns, traffic conditions, and other relevant factors, AI algorithms can determine the most energy-efficient routes and schedules for transportation networks [51].

Impact on Sustainable Development

The widespread adoption of AI in energy efficiency and conservation has the potential to significantly impact sustainable development by reducing greenhouse gas emissions, conserving natural resources, and promoting economic growth. By optimizing energy consumption in various sectors, AI can reduce the reliance on fossil fuels and support the transition towards renewable energy sources [64]. Furthermore, AI can enable more efficient use of resources and reduce waste, contributing to the achievement of the United Nations Sustainable Development Goals [18].

Challenges and Limitations

Despite the potential benefits of AI in energy efficiency and conservation, there are several challenges and limitations to consider. First, the deployment of AI technologies requires substantial investments in infrastructure, data collection, and analytics, which may be a barrier for some organizations and communities [106]. Second, there are concerns related to data privacy and security, as the widespread use of AI algorithms often relies on the collection and analysis of large datasets, including personal information [72]. Therefore, appropriate measures should be taken to ensure the protection of individual privacy while leveraging AI for energy efficiency and conservation.

Third, there is the potential for AI systems to inadvertently increase energy consumption in certain cases. For instance, while AI algorithms may optimize individual devices or systems, the overall energy consumption could increase due to the energy demands of the AI hardware and data centers [84]. Therefore, it is crucial to evaluate the net energy savings of AI applications and consider the energy consumption of the AI infrastructure itself.

Lastly, there is a need for collaboration between various stakeholders, including governments, businesses, and researchers, to develop and implement AI solutions for energy efficiency and conservation. Policymakers should create a supportive regulatory environment and provide incentives for the adoption of AI technologies in the energy sector [105]. Furthermore, interdisciplinary research and development efforts should be encouraged to advance AI applications and address the challenges and limitations mentioned above.

AI has the potential to revolutionize energy efficiency and conservation in various sectors, including residential, commercial, and industrial applications. By optimizing energy consumption, predicting energy demand, and facilitating decision-making for policy and infrastructure planning, AI can contribute to sustainable development and resource management. However, challenges and limitations related to infrastructure investments, data privacy and security, and collaboration between stakeholders must be addressed to fully harness the potential of AI in energy efficiency and conservation.

13.4.2 Waste Management and Recycling

Waste management and recycling are critical aspects of sustainable development and resource management. The efficient handling, sorting, and recycling of waste are necessary to reduce the environmental impact of waste disposal and resource depletion. Artificial Intelligence (AI) has the potential to revolutionize waste management and recycling processes, enhancing efficiency and sustainability. This section explores the applications of AI in waste management and recycling, including waste sorting, waste prediction, route optimization, and policy development.

AI for Waste Sorting

Waste sorting is a crucial step in the waste management process, enabling the separation of recyclable materials from non-recyclable waste. AI-powered waste sorting systems employ machine learning algorithms and computer vision techniques to recognize and sort various waste materials automatically. These systems typically use cameras and sensors to capture images of waste items, which are then processed and classified by AI algorithms [29, 45]. Automated waste sorting can significantly improve the efficiency and accuracy of waste separation, reducing human involvement and error.

AI for Waste Prediction

Waste prediction is essential for effective waste management, as it helps authorities and service providers to plan and allocate resources efficiently. AI can be employed to predict waste generation patterns and volumes, enabling better waste collection, transportation, and processing strategies. Machine learning algorithms can be trained on historical waste data, demographic information, and socio-economic factors to generate accurate predictions of future waste generation [8, 73]. These predictions can inform waste management policies and facilitate the optimization of waste collection schedules and routes.

AI for Route Optimization

Optimizing waste collection routes can significantly reduce the operational costs, fuel consumption, and environmental impact of waste management. AI techniques, such as genetic algorithms, ant colony optimization, and neural networks, can be used to develop efficient waste collection routes based on factors such as waste generation patterns, vehicle capacities, and road conditions [38, 40]. AI-optimized routes can lead to reduced fuel consumption, lower greenhouse gas emissions, and improved waste collection services.

AI for Policy Development and Evaluation

AI can also support the development and evaluation of waste management and recycling policies. Machine learning algorithms can analyze large volumes of data related to waste generation, recycling rates, and environmental impacts to identify patterns and trends that can inform policy decisions [82, 131]. AI can also be used to simulate the effects of various policy scenarios, enabling decision-makers to evaluate the potential benefits and drawbacks of different strategies before implementation [128].

For instance, AI techniques can be employed to determine the most effective policy interventions for increasing recycling rates or reducing illegal dumping. By analyzing the success of similar policies in other regions, AI can help identify factors that contribute to their effectiveness and suggest ways to tailor them to local contexts [93].

In conclusion, AI has the potential to revolutionize waste management and recycling processes, enhancing efficiency, sustainability, and environmental protection. By automating waste sorting, predicting waste generation patterns, optimizing waste collection routes, and informing policy development, AI applications can contribute to more sustainable and resource-efficient waste management systems. However, it is essential to address the challenges and limitations of AI in this context, such as data quality, algorithmic biases, and the need for interdisciplinary collaboration to ensure the successful implementation of AI-driven waste management and recycling solutions.

13.4.3 Water Resource Management

Water resource management is crucial for the sustainable development of urban areas, as it encompasses the conservation, distribution, and efficient utilization of water resources. With the global population growth and increasing demand for water, effective water resource management has become even more critical. AI techniques have the potential to contribute significantly to various aspects of water resource management, including monitoring, forecasting, optimization, and policy development.

Monitoring and Forecasting

AI techniques, such as machine learning and deep learning, have been applied to monitor and forecast water resources effectively. These methods can be used to predict water demand, identify patterns in water consumption, and estimate water availability based on factors such as climate, land use, and population growth [46, 67]. AI can also be employed to monitor water quality by analyzing data from remote sensing, in-situ sensors, and laboratory analyses, helping to identify water pollution sources and trends [120, 143].

Optimization and Decision Support

AI can also be employed in the optimization of water resource allocation and infrastructure investments. Machine learning algorithms can help identify the most efficient water distribution strategies, taking into account factors such as water scarcity, population density, and infrastructure capacity [55]. Furthermore, AI techniques can be used to optimize the design and operation of water supply networks, reducing energy consumption and costs [44, 115].

Policy Development

AI can contribute to the development of effective water management policies by analyzing large volumes of data, identifying patterns, and simulating the potential impacts of various policy scenarios. Natural language processing techniques can help analyze textual data from policy documents, stakeholder feedback, and public opinion, enabling policymakers to better understand the concerns and priorities of different stakeholders [95]. Moreover, agent-based modeling can be used to simulate the interactions between various stakeholders and the environment, helping to evaluate the potential consequences of different water management strategies and identify the most sustainable and efficient policy options [37, 136].

Education and Capacity Building

AI can also be employed in the field of education and capacity building for sustainable water resource management. AI-driven decision support systems can be used to develop interactive learning environments and training tools for water professionals, helping them understand the complexities of water systems and acquire the necessary skills to make informed decisions [134]. Additionally, AI can be utilized to create personalized learning experiences for students and professionals, adapting the content and pace of learning to the needs of each individual [10].

In summary, AI techniques can play a significant role in various aspects of water resource management, including monitoring and forecasting, optimization, policy development, and education. By leveraging the power of AI, urban planners and policymakers can develop more sustainable and efficient strategies for water resource management, ultimately contributing to the overall sustainable development of urban areas.

13.4.4 Air Quality Management and Pollution Control

Air quality management and pollution control are essential components of sustainable development and resource management. Poor air quality affects human health, the environment, and the economy. Artificial intelligence (AI) techniques can be employed to improve air quality monitoring, forecasting, and control, ultimately leading to better policy-making and more sustainable urban environments.

Air Quality Monitoring and Forecasting

One of the primary applications of AI in air quality management is air quality monitoring and forecasting. Accurate monitoring of air pollution levels is crucial for understanding the magnitude of the problem and identifying areas that require intervention. AI techniques, particularly machine learning algorithms, can be used to analyze data from air quality monitoring stations, remote sensing, and other sources to estimate pollutant concentrations, identify pollution sources, and predict future air quality levels [56, 79].

Machine learning models such as support vector machines, artificial neural networks, and random forests have been used to predict air pollutant concentrations, demonstrating promising results in terms of accuracy and reliability [74]. Additionally, deep learning techniques, such as convolutional neural networks (CNNs) and recurrent neural networks (RNNs), have been applied to analyze spatiotemporal air quality data, providing more accurate predictions and improved understanding of the relationships between various factors affecting air quality [147].

Policy Development and Evaluation

AI techniques can also be used to support policy development and evaluation for air quality management and pollution control. Machine learning algorithms can analyze large amounts of historical data, identify patterns and trends, and reveal the effectiveness of previous policy measures [148]. This information can be used by policymakers to design more effective air quality management strategies and evaluate the potential impacts of various policy options.

Natural language processing (NLP) techniques can be employed to analyze textual data, such as policy documents, news articles, and social media posts, to gain insights into public opinion and stakeholder concerns regarding air quality management and pollution control [141]. This information can be used to inform the development of more inclusive and context-specific policies, tailored to the needs and priorities of different stakeholders.

Pollution Source Identification and Control

AI techniques can assist in identifying pollution sources and developing targeted pollution control measures. Machine learning algorithms can analyze complex datasets, including meteorological data, pollutant concentrations, and emissions data, to identify potential pollution sources and their relative contributions to air quality problems [25].

Once the pollution sources have been identified, AI can be used to optimize control measures, such as emission reduction strategies and traffic management plans, to minimize their impact on air quality [117]. For example, AI algorithms can be employed to optimize traffic signal timings and vehicle routing in order to reduce traffic congestion and associated emissions [135].

Public Awareness and Engagement

AI can play a vital role in raising public awareness and engagement in air quality management and pollution control efforts. For instance, AI-powered chatbots can provide real-time air quality information and personalized recommendations for reducing exposure to air pollution [119]. This can empower individuals to take action to protect their health and contribute to overall air quality improvement.

Moreover, AI can be used to analyze social media data to understand public sentiment and perceptions about air quality issues, which can inform the design of more effective public communication and outreach campaigns [47]. This can lead to increased public support for air quality management policies and actions, ultimately contributing to more sustainable urban environments.

In conclusion, AI techniques offer significant potential for improving air quality management and pollution control efforts. From monitoring and forecasting air quality to policy development and evaluation, AI can provide valuable insights and support decision-making processes. The use of AI in air quality management and pollution control can lead to more effective policies, optimized resource allocation, and increased public engagement, ultimately contributing to more sustainable urban environments.

13.4.5 Climate Change Adaptation and Resilience

Climate change presents one of the most significant challenges to sustainable development and resource management in the twenty-first century. The increasing frequency and severity of extreme weather events, rising sea levels, and shifting precipitation patterns are reshaping the way we plan, design, and manage urban environments. Artificial intelligence (AI) offers promising tools to enhance our ability to adapt to these changes and build resilient communities that can withstand the impacts of climate change.

In this section, we will explore the role of AI in climate change adaptation and resilience. We will focus on several key areas where AI can support decision-making and resource management, including climate risk assessment, early warning systems, infrastructure design, ecosystem-based adaptation, and policy development.

Climate Risk Assessment

The first step in climate change adaptation and resilience is understanding the risks posed by climate change to different sectors and communities. AI can help improve climate risk assessment by processing large volumes of data, including historical and projected climate data, socioeconomic indicators, and infrastructure data, to identify potential vulnerabilities and hotspots. Machine learning algorithms can be used to analyze complex relationships between climate variables and human systems, allowing for more accurate and timely assessments of risk.

One example is the use of deep learning techniques to analyze satellite imagery for detecting and monitoring coastal erosion, a significant issue for many coastal communities due to rising sea levels. AI can also help analyze and predict the impacts of climate change on agricultural productivity, water resources, and public health, providing critical information for decision-makers and stakeholders.

Early Warning Systems

AI can play a crucial role in developing more accurate and efficient early warning systems for extreme weather events, such as hurricanes, floods, droughts, and heat-waves. Machine learning algorithms can analyze vast amounts of meteorological data, including satellite imagery, radar data, and weather station measurements, to predict the likelihood, intensity, and impacts of these events. By providing timely and accurate information on potential hazards, AI-driven early warning systems can help communities better prepare for and respond to extreme weather events, reducing the loss of life and property.

Infrastructure Design and Planning

As the effects of climate change become more apparent, there is a growing need to design and retrofit infrastructure that can withstand extreme weather events and other climate-related stresses. AI can support this process by providing tools for data-driven design and optimization of infrastructure, such as buildings, transportation systems, and water management systems.

For example, AI can be used to optimize the design of stormwater management systems by simulating and evaluating various design alternatives under different climate scenarios. Similarly, AI-driven structural analysis can help engineers and architects design buildings and other structures that can withstand the impacts of extreme weather events, such as high winds, heavy precipitation, and flooding.

Ecosystem-Based Adaptation

Ecosystem-based adaptation (EbA) is an approach that uses nature-based solutions to address the impacts of climate change, such as conserving and restoring ecosystems to provide essential services, like flood protection and coastal defense. AI can help identify suitable areas for EbA interventions by analyzing spatial data on ecosystems, land use, and climate risks.

For example, AI can be used to identify areas where the restoration of mangrove forests could provide the most significant benefits for coastal protection and carbon sequestration. AI can also be employed to monitor the effectiveness of EbA interventions by analyzing satellite imagery and other remote sensing data, enabling adaptive management of these projects.

Policy Development

AI can support policy development for climate change adaptation and resilience by providing evidence-based insights and recommendations. Natural language processing techniques can be used to analyze large volumes of text, such as policy documents, academic papers, and media reports, to identify trends, to identify trends, emerging issues, and best practices related to climate change adaptation and resilience. By synthesizing this information, AI can help policymakers make informed decisions and develop targeted strategies for addressing climate change impacts.

For instance, AI can be utilized to analyze the success factors and barriers to implementation for different climate adaptation policies across various jurisdictions. By identifying patterns and correlations in this data, AI can support the development of more effective policies and programs that address specific vulnerabilities and needs of communities, industries, and ecosystems.

Community Engagement and Capacity Building

The success of climate change adaptation and resilience efforts relies on the active participation and support of local communities. AI can assist in engaging communities and building capacity by providing user-friendly tools and platforms that enable stakeholders to access and interact with climate change information, data, and models. These platforms can help raise awareness, facilitate communication, and support decision-making at various levels, from individual households to local governments and businesses.

For example, AI-driven applications and visualization tools can help community members explore potential climate change impacts and adaptation options for their neighborhoods or cities. By providing accessible and easy-to-understand information, AI can empower communities to participate in climate change adaptation and resilience efforts actively.

In conclusion, AI offers a wide range of applications for climate change adaptation and resilience in the context of sustainable development and resource management. By harnessing the power of AI, we can improve our understanding of climate risks, develop more effective early warning systems, design and plan resilient infrastructure, implement ecosystem-based adaptation measures, and support evidence-based policy development. Furthermore, AI can play a vital role in engaging communities and building capacity for climate change adaptation and resilience. As we continue to face the challenges posed by climate change, the integration of AI into these efforts will be crucial for ensuring the long-term sustainability and resilience of our communities and ecosystems.

13.5 Challenges and Limitations of AI in Sustainable Development and Resource Management

While artificial intelligence (AI) presents significant opportunities to address various issues in sustainable development and resource management, it is essential to recognize the challenges and limitations that come with the integration of AI in these fields. This section will explore some of the main challenges and limitations, including data quality and availability, algorithmic biases, ethical considerations, and the digital divide. By acknowledging these issues, researchers, policymakers, and practitioners can make informed decisions about the use of AI in sustainable development and resource management.

One of the main challenges in using AI for sustainable development and resource management is the quality and availability of data. AI algorithms rely on vast amounts of data for training and validation, and the accuracy and reliability of the AI models are highly dependent on the quality of the input data [41]. Data quality can be affected by various factors, such as measurement errors, missing values, and inconsistent formats.

Moreover, the availability of data can be limited due to a lack of resources, infrastructure, and technical capacity in collecting, managing, and sharing data, especially in low-income countries and remote regions [69]. For instance, many developing countries lack sufficient meteorological stations and remote sensing infrastructure to provide accurate and timely data for climate change adaptation and resource management.

AI algorithms, particularly machine learning and deep learning models, can inadvertently perpetuate and amplify existing biases in the data they are trained on, which can lead to unfair and discriminatory outcomes [12]. For example, if an AI model for predicting urban heat vulnerability is trained on data from affluent neighborhoods, it may not accurately predict the vulnerability of low-income neighborhoods, which may have different building materials, green spaces, and socio-demographic characteristics.

To mitigate algorithmic biases, researchers and practitioners need to be aware of the potential biases in the data and develop strategies to address these biases, such as re-sampling techniques, data augmentation, and fairness-aware machine learning [43].

The use of AI in sustainable development and resource management raises several ethical considerations, particularly in terms of privacy, surveillance, and accountability. The collection, storage, and analysis of large-scale data on individuals, households, and communities can potentially infringe on privacy rights and lead to surveillance concerns [90]. For example, the use of satellite imagery and GPS data for monitoring land use changes and resource management can inadvertently reveal sensitive information about individuals' livelihoods and activities.

Furthermore, AI-driven decision-making in sustainable development and resource management may raise questions about accountability and responsibility, especially when AI models make errors or produce unintended consequences. It is crucial to develop transparent and accountable frameworks for AI-driven decision-making in these fields, including mechanisms for public participation, oversight, and redress [42].

The integration of AI in sustainable development and resource management may exacerbate the digital divide between developed and developing countries, as well as between urban and rural areas [61]. Access to AI technologies, infrastructure, and expertise is unevenly distributed, and the benefits and opportunities of AI may not be equitably shared.

To address the digital divide, it is essential to invest in capacity-building and technology transfer initiatives that support the development and adoption of AI in low-income countries and underserved communities [127]. Moreover, partnerships between governments, industry, academia, and civil society can help ensure that AI technologies are tailored to the specific needs and contexts of different communities and regions, fostering inclusive and equitable development outcomes.

AI techniques, particularly advanced machine learning and deep learning methods, can be technically complex and challenging to implement and manage. Developing and deploying AI models for sustainable development and resource management often require specialized expertise, computational resources, and significant time investment [30]. This technical complexity can be a barrier for organizations, particularly in developing countries, where the availability of skilled personnel, infrastructure, and funding is limited.

Capacity-building initiatives, such as training programs, online resources, and collaborative research projects, can help enhance the technical capacity of organizations and professionals in sustainable development and resource management. Furthermore, the development of user-friendly AI tools and platforms can lower the technical barriers for non-experts and facilitate the adoption of AI in these fields.

AI models, like all models, have inherent uncertainties and limitations in their predictions and outputs. In the context of sustainable development and resource management, these uncertainties can have significant implications for decision-making and risk management [66]. For instance, AI-driven climate change projections may have uncertainties due to the complexity of the Earth's climate system and the limitations of the underlying models and data.

It is crucial to recognize and communicate the uncertainties associated with AI models and their outputs, and to develop robust decision-making frameworks that can incorporate and account for these uncertainties [76].

In conclusion, while AI offers tremendous potential for enhancing sustainable development and resource management, it is essential to recognize and address the various challenges and limitations associated with its use in these fields. By addressing issues related to data quality and availability, algorithmic biases, ethical considerations, the digital divide, technical complexity, and uncertainty, researchers, policymakers, and practitioners can ensure that AI is used responsibly and effectively in the pursuit of sustainable development and resource management goals.

13.6 Future Directions in AI Applications for Sustainable Development and Resource Management

The growing interest in artificial intelligence (AI) applications for sustainable development and resource management is driven by the potential of AI to address some of the world's most pressing challenges, from climate change and biodiversity loss to urbanization and social inequality. As AI continues to advance and mature, there are several emerging trends and future directions in AI applications for sustainable development and resource management. This section will explore some of these trends, including the integration of AI with other technologies, the development of explainable and trustworthy AI, the focus on equity and inclusiveness, and the emphasis on collaboration and partnerships.

One future direction in AI applications for sustainable development and resource management is the integration of AI with other advanced technologies, such as the Internet of Things (IoT), robotics, and augmented reality (AR). The convergence of AI with these technologies can enable more effective and efficient data collection, analysis, and decision-making in various areas of sustainable development and resource management [114].

For instance, the integration of AI with IoT sensors can support real-time monitoring and adaptive management of resources, such as water and energy, enabling more efficient and sustainable use of these resources [59]. Similarly, the combination of AI with robotics can facilitate remote and automated monitoring and management of natural resources, such as forests and marine ecosystems [35].

As AI becomes more integrated into sustainable development and resource management, there is a growing need for explainable and trustworthy AI models that can be easily understood and scrutinized by stakeholders, including policymakers, practitioners, and the public [7]. Explainable AI (XAI) techniques aim to provide transparency and interpretability in AI models, enabling users to understand how the models make predictions or decisions and to assess the reliability and validity of the outputs.

Developing explainable and trustworthy AI models for sustainable development and resource management can help address concerns about the ethical, legal, and social implications of AI, such as algorithmic biases, privacy, and accountability. Furthermore, explainable and trustworthy AI can facilitate stakeholder engagement, public participation, and evidence-based decision-making in these fields.

As AI applications for sustainable development and resource management continue to evolve, there is a growing recognition of the need to ensure that these applications are equitable and inclusive, addressing the needs and priorities of all segments of society, particularly marginalized and vulnerable communities [28]. This entails considering issues such as data representativeness, algorithmic fairness, and the digital divide, as well as integrating principles of social and environmental justice into the design, development, and deployment of AI models.

For example, AI-driven urban planning and resource management initiatives can prioritize the needs of low-income and minority communities, which are often disproportionately affected by environmental and social challenges [6]. Similarly, AI applications for climate change adaptation and resilience can target the most vulnerable populations and ecosystems, ensuring that resources and efforts are directed towards those who need them the most [87].

The complexity and interdisciplinary nature of sustainable development and resource management challenges require a collaborative and partnership-based approach to developing and implementing AI solutions [114]. This includes fostering collaboration between researchers, practitioners, policymakers, industry, civil society, and local communities, as well as promoting international cooperation and knowledge exchange.

Collaborative research initiatives and public–private partnerships can help pool resources, expertise, and data, enabling the development of more effective and scalable AI applications for sustainable development and resource management. For instance, partnerships between universities, technology companies, and governmental agencies can facilitate the development of AI-driven monitoring systems for air quality, water resources, or biodiversity conservation, leveraging each partner's unique strengths and capabilities [122].

Moreover, global networks and platforms, such as the United Nations Sustainable Development Solutions Network (SDSN) or the Global Partnership on Artificial Intelligence (GPAI), can support international collaboration and knowledge sharing on AI applications for sustainable development and resource management, fostering innovation and best practices across countries and regions.

In conclusion, the future of AI applications in sustainable development and resource management is characterized by several emerging trends and directions, including the integration of AI with other advanced technologies, the development of explainable and trustworthy AI, the focus on equity and inclusiveness, and the emphasis on collaboration and partnerships. By embracing these trends and directions, researchers, policymakers, and practitioners can harness the full potential of AI to contribute to a more sustainable, resilient, and equitable world.

References

1. Abdallah, A. M., Shaaban, S. G., & Khalil, A. (2020). Artificial intelligence in water resources management: A review. *Journal of Water Resources Planning and Management, 146*(7), 04020049.
2. Abdallah, F., Anis, S. F., & Elshafei, M. (2019). Machine learning for smart leak detection and location in water distribution networks. *Journal of Water Resources Planning and Management, 145*(6), 04019015.
3. Alvisi, S., Mascellani, G., Franchini, M., & Bárdossy, A. (2007). Water demand forecasting through a neuro-fuzzy technique. *Advances in Water Resources, 30*(10), 2111–2126.
4. Amasyali, K., & El-Gohary, N. M. (2018). A review of data-driven building energy consumption prediction studies. *Renewable and Sustainable Energy Reviews, 81*, 1192–1205.
5. Angel, S., & Sheppard, S. C. (2018). The use of urban planning and design in response to climate change. In *Climate change impacts on urban pests* (pp. 173–190). CABI.
6. Angwin, J., Larson, J., Mattu, S., & Kirchner, L. (2016). Machine bias. ProPublica.
7. Arrieta, A. B., Díaz-Rodríguez, N., Del Ser, J., Bennetot, A., Tabik, S., Barbado, A., García, S., Gil-López, S., Molina, D., Benjamins, R., Chatila, R., & Herrera, F. (2020). Explainable Artificial Intelligence (XAI): Concepts, taxonomies, opportunities and challenges toward responsible AI. *Information Fusion. 58*, 82–115.
8. Babalola, A., Abbas, A., Sadiq, R., Hewage, K., & Rodriguez, M. J. (2020). Spatial and temporal prediction of household waste generation using machine learning algorithms. *Journal of Cleaner Production, 259*, 120901.
9. Bai, X., Wu, C., & Liu, X. (2020). Machine learning-based urban land-use analysis and prediction: A case study of Wuhan, China. *Computers, Environment and Urban Systems, 81*, 101455.

10. Baker, R. S., Corbett, A. T., & Koedinger, K. R. (2019). Developing a generalizable detector of when students game the system. *User Modeling and User-Adapted Interaction, 29*(4), 883–919.
11. Bakker, M., Lane, S., & van der Valk, M. (2013). Water demand prediction using artificial neural networks: A Rotterdam case study. *Computers, Environment and Urban Systems, 41*, 1–10.
12. Barocas, S., & Selbst, A. D. (2016). Big data's disparate impact. *California Law Review, 104*, 671–732.
13. Barton, H., Grant, M., & Guise, R. (2010). *Shaping neighbourhoods: For local health and global sustainability*. Routledge.
14. Batty, M. (2008). The size, scale, and shape of cities. *Science, 319*(5864), 769–771.
15. Beatley, T. (2012). *Green cities of Europe: Global lessons on green Urbanism*. Island Press.
16. Berke, P. R., & Conroy, M. M. (2000). Are we planning for sustainable development? *Journal of the American Planning Association, 66*(1), 21–33.
17. Bhattacharya, S., Parida, R., & Sandanayake, M. (2018). Smart energy management for households. *Renewable and Sustainable Energy Reviews, 90*, 410–425.
18. Bibri, S. E., & Krogstie, J. (2017). Smart sustainable cities of the future: An extensive interdisciplinary literature review. *Sustainable Cities and Society, 31*, 183–212.
19. Bibri, S. E., & Krogstie, J. (2020). The emerging data-driven smart city and its innovative applied solutions for sustainability: The cases of London and Barcelona. *Journal of Urban Technology, 27*(1), 3–42.
20. Bithas, K., & Christofakis, M. (2006). Environmentally sustainable cities. *Critical Reviews in Environmental Science and Technology, 36*(3), 185–222.
21. Brunner, P. H., & Rechberger, H. (2004). *Practical handbook of material flow analysis*. CRC Press.
22. Bunescu, R. C., & Mooney, R. J. (2005). A shortest path dependency kernel for relation extraction. In *Proceedings of the Conference on Human Language Technology and Empirical Methods in Natural Language Processing* (pp. 724–731).
23. Caiado, R. G. G., Lourenço, R. P., & Carvalho, J. A. (2021). Machine learning applied to wind turbine performance and ageing analysis. *Renewable Energy, 164*, 604–615.
24. Chen, C., Liu, L., & Mei, Y. (2019). A comprehensive review on the smart grid and demand-side management. *Renewable and Sustainable Energy Reviews, 107*, 338–363.
25. Chen, L., Liu, H., & Luo, S. (2020). Source apportionment of PM2.5 using a random forest algorithm in a heavily polluted area of China. *Atmospheric Environment, 223*, 117205.
26. Chen, S., Yang, J., & Shang, C. (2017). A review of emerging technologies for sustainable energy production: Challenges and opportunities. *Sustainable Energy Technologies and Assessments, 22*, 92–104.
27. Chicco, G., Napoli, R., & Piglione, F. (2014). Comparisons among clustering techniques for electricity customer classification. *IEEE Transactions on Power Systems, 21*(2), 933–940.
28. Chui, M., Kamalnath, V., & McCarthy, B. (2021). Ensuring an equitable, inclusive approach to the application of AI in health care. *JAMA Health Forum, 2*(2), e210191.
29. Cleemput, S. (2020). Artificial intelligence in waste sorting. *Waste Management & Research, 38*(9), 925–930.
30. Craglia, M., Annoni, A., Benczur, P., Campos, B., Correia, N., Goodchild, M., Wesseling, M. (2018). Artificial intelligence–A European perspective. Publications Office of the European Union.
31. Cugurullo, F. (2018). Exposing smart cities and eco-cities: Frankenstein urbanism and the sustainability challenges of the experimental city. *Environment and Planning A: Economy and Space, 50*(1), 73–92.
32. Daly, H. E. (1996). *Beyond growth: The economics of sustainable development*. Beacon Press.
33. Deb, S., Yee, G., & Abdelzaher, T. (2018). Energy-aware IoT: A cyber-physical energy system perspective. *IEEE Internet of Things Journal, 5*(4), 2972–2982.
34. Deichmann, U., Meisner, C., Murray, S., & Wheeler, D. (2016). The economics of renewable energy expansion in rural Sub-Saharan Africa. *Energy Policy, 88*, 292–304.

35. Deng, W., Yang, Y., Chu, J., & Liu, J. (2021). Applications of robotics and artificial intelligence in the protection of natural resources: A review. *Resources, Conservation, and Recycling, 164*, 105169.
36. Duan, H., Huang, Q., & Zhao, Q. (2019). Machine learning in municipal solid waste forecasting: A comparative study. *Waste Management, 84*, 313–321.
37. Elshafei, Y., Coletti, J., Sivapalan, M., Hipsey, M. R., & Tonts, M. (2014). A prototype framework for models of socio-hydrology: Identification of key feedback loops and parameterisation approach. *Hydrology and Earth System Sciences, 18*(6), 2141–2166.
38. Eskandari, M., Rabelo, L., & Mollaghasemi, M. (2018). Sustainable waste collection and transportation optimization. *Computers & Industrial Engineering, 116*, 98–106.
39. Ewing, R., & Rong, F. (2008). The impact of urban form on US residential energy use. *Housing Policy Debate, 19*(1), 1–30.
40. Fan, W., Liu, Y., Xu, S., & Zhang, Y. (2020). Optimizing waste collection and transportation routing using a novel hybrid artificial intelligence algorithm. *Environmental Science and Pollution Research, 27*(14), 16325–16337.
41. Fernández-Delgado, M., Cernadas, E., Barro, S., & Amorim, D. (2014). Do we need hundreds of classifiers to solve real-world classification problems? *Journal of Machine Learning Research, 15*(1), 3133–3181.
42. Floridi, L., & Taddeo, M. (2016). What is data ethics? *Philosophical Transactions of the Royal Society A: Mathematical, Physical and Engineering Sciences, 374*(2083), 20160360.
43. Friedler, S. A., Scheidegger, C., & Venkatasubramanian, S. (2016). On the (im)possibility of fairness. arXiv:1609.07236
44. Fu, G., Kapelan, Z., & Kasprzyk, J. R. (2013). Optimal design of water infrastructure systems: A holistic decision analysis approach. *Journal of Water Resources Planning and Management, 139*(6), 627–636.
45. Ghiani, G., Laganà, D., Manni, E., & Musmanno, R. (2019). An artificial intelligence-based system for waste collection optimization. *International Journal of Production Research, 57*(15–16), 4934–4944.
46. Ghosh, S., & Mujumdar, P. P. (2019). Machine learning for long lead time streamflow prediction using climate index. *Journal of Hydrology, 576*, 341–352.
47. Ghosh, S., Gokhale, S., & Sinha, S. K. (2018). Sentiment analysis of air quality using social media data. *Sustainable Cities and Society, 42*, 259–268.
48. Gibson, R. B., Holtz, S., Tansey, J., Whitelaw, G. S., & Hassan, S. (2005). *Sustainability assessment: Criteria and processes*. Routledge.
49. Giest, S. (2017). Big data for policymaking: Great expectations, but with limited progress? *Policy & Internet, 9*(3), 257–277.
50. Glaeser, E. L., & Ward, B. A. (2009). The causes and consequences of land use regulation: Evidence from Greater Boston. *Journal of Urban Economics, 65*(3), 265–278.
51. Goldbeck, N., Schwenk, K., & Zamzow, J. (2020). Machine learning for transportation and logistics—Use cases and implementation barriers. *Sustainability, 12*(3), 1093.
52. Goodchild, M. F. (2010). Twenty years of progress: GIScience in 2010. *Journal of Spatial Information Science, 2010*(1), 3–20.
53. Goodchild, M. F. (2018). Geospatial technologies and the future of the city. *Proceedings of the National Academy of Sciences, 115*(16), 3978–3981.
54. Gupta, A., Singh, A., & Pundir, Y. (2021). Machine learning techniques for water resource management: A review. *Environmental Modelling & Software, 138*, 104954.
55. Gupta, H. V., Sorooshian, S., & Yapo, P. O. (2019). Toward improved calibration of hydrologic models: Multiple and noncommensurable measures of information. *Water Resources Research, 34*(4), 751–763.
56. Gupta, P., Christopher, S. A., Wang, J., Gehrig, R., Lee, Y. C., & Kumar, N. (2018). Satellite remote sensing of particulate matter and air quality assessment over global cities. *Atmospheric Environment, 45*(38), 6006–6017.
57. Haklay, M. (2010). How good is volunteered geographical information? A comparative study of OpenStreetMap and Ordnance Survey datasets. *Environment and Planning B: Planning and Design, 37*(4), 682–703.

58. Hardoy, J. E., Mitlin, D., & Satterthwaite, D. (2001). *Environmental problems in an urbanizing world: Finding solutions in cities in Africa, Asia, and Latin America*. Earthscan.
59. Hashem, I. A. T., Chang, V., Anuar, N. B., Adewole, K., Yaqoob, I., Gani, A., & Ahmed, E. (2016). The role of big data in smart city. *International Journal of Information Management, 36*(5), 748–758.
60. He, C., Huang, Q., & Li, Y. (2018). An artificial-neural-network-based, constrained land-use allocation model for the identification of urban growth patterns. *International Journal of Geographical Information Science, 32*(3), 512–532.
61. Hilbert, M. (2016). The bad news is that the digital access divide is here to stay: Domestically installed bandwidths among 172 countries for 1986–2014. *Telecommunications Policy, 40*(6), 567–581.
62. Holden, E., Linnerud, K., & Banister, D. (2015). Sustainable development: Our common future revisited. *Global Environmental Change, 34*, 130–139.
63. Hovy, D., Fan, M., Gliozzo, A., Patwardhan, S., & Welty, C. (2013). When did that happen?-Linking events and relations to timestamps. In *Proceedings of the 2013 Conference of the North American Chapter of the Association for Computational Linguistics: Human Language Technologies* (pp. 196–202).
64. International Energy Agency. (2017). Digitalization and energy. Retrieved from https://www.iea.org/reports/digitalisation-and-energy
65. Ji, H., & Grishman, R. (2008). Refining event extraction through cross-document inference. In *Proceedings of ACL-08: HLT*, (pp. 254–262).
66. Jiang, L., Marston, L., Wang, H., Hsieh, W. W., & Klarin, B. (2020). Machine learning methods for climate prediction: Challenges and opportunities. *Earth and Space Science, 7*(10), e2020EA001230.
67. Jiang, P., Ouyang, W., Wang, H., & Zheng, H. (2020). Urban water demand forecasting and uncertainty assessment using machine learning methods. *Journal of Hydrology, 584*, 124611.
68. Khalil, M., Ficetola, G. F., & Thuiller, W. (2019). Using deep learning to identify potential areas of conservation priority. *Biological Conservation, 234*, 11–19.
69. Kitchin, R. (2014). *The data revolution: Big data, open data, data infrastructures and their consequences*. Sage.
70. Kolev, K., Avdić, D., Goryl, P., & Forsén, H. (2020). An IoT architecture for waste management with sensors and machine learning. In *2020 IEEE 6th World Forum on Internet of Things (WF-IoT)* (pp. 1–6). IEEE.
71. Krishna, G. M., Yu, D., & Berardy, A. (2019). An integrated deep learning-based framework for the estimation of flood vulnerability. *Hydrology and Earth System Sciences, 23*(2), 643–659.
72. Kshetri, N. (2018). 1 The economics of the Internet of Things in the Global South. *Third World Quarterly, 39*(10), 1889–1909.
73. Kumar, A., Samadder, S. R., & Chakrabarty, S. (2019). Prediction of municipal solid waste generation using nonlinear autoregressive artificial neural networks. *Journal of Cleaner Production, 235*, 1417–1426.
74. Kumar, A., Singh, V., Singh, S., & Rai, S. (2020). Machine learning techniques for air quality prediction: A review. *Journal of Environmental Management, 270*, 110883.
75. Lee, J., Ardakani, H. D., Yang, S., & Bagheri, B. (2015). Industrial big data analytics and cyber-physical systems for future maintenance & service innovation. *Procedia CIRP, 38*, 3–7.
76. Lempert, R. J., Popper, S. W., & Bankes, S. C. (2003). *Shaping the next one hundred years: New methods for quantitative, long-term policy analysis*. RAND Corporation.
77. Li, J., Chen, X., Hovy, E., & Jurafsky, D. (2018). Visualizing and understanding neural models in NLP. In *Proceedings of the 2018 Conference of the North American Chapter of the Association for Computational Linguistics: Human Language Technologies* (Vol. 1, pp. 97–106).
78. Li, W., Yang, Y., & Liu, Y. (2019). Application of artificial intelligence in process industries: A review. *Journal of Process Control, 79*, 12–29.

79. Li, X., Li, D., Wang, D., & Li, W. (2019). Machine learning-based urban land-use classification using remote sensing imagery. *Remote Sensing, 11*(5), 555.
80. Li, X., Peng, L., Yao, X., Cui, S., Hu, Y., You, C., & Chi, T. (2017). Long short-term memory neural network for air pollutant concentration predictions: Method development and evaluation. *Environmental Pollution, 231*, 997–1004.
81. Li, X., Zhang, C., Li, W., Ricard, R., Meng, Q., & Zhang, W. (2019). Assessing street-level urban greenery using Google Street View and a modified green view index. *Urban Forestry & Urban Greening, 40*, 274–285.
82. Lin, T. Y., Chen, Y. H., & Lou, S. J. (2018). A heuristic model for the prediction of municipal solid waste recycling rates. *Waste Management, 74*, 3–18.
83. Lu, Y., Zhang, L., & Feng, X. (2020). A deep-learning-based spatiotemporal approach for urban land-use change simulations. *Computers, Environment and Urban Systems, 81*, 101465.
84. Lundgren, K., & Kjellman, M. (2019). Artificial intelligence for energy efficiency in data centers. *Energy Procedia, 158*, 3783–3788.
85. Ma, J., Du, K., Zheng, Y., Zhang, L., & Gong, W. (2018). A review of supervised object-based land-cover image classification. *ISPRS Journal of Photogrammetry and Remote Sensing, 146*, 108–121.
86. Matta, J. R., Guillén-Gosálbez, G., & Jiménez-Esteller, L. (2020). A machine learning approach to predict municipal solid waste generation at the local scale. *Waste Management, 102*, 663–672.
87. McCall, M. K. (2021). Climate change adaptation, AI and the vulnerable: Avoiding more technology-driven inequalities. *Sustainability Science, 16*(2), 349–362.
88. Meerow, S., Newell, J. P., & Stults, M. (2016). Defining urban resilience: A review. *Landscape and Urban Planning, 147*, 38–49.
89. Mikolov, T., Le, Q. V., & Sutskever, I. (2013). Exploiting similarities among languages for machine translation. arXiv:1309.4168.
90. Mittelstadt, B. D., Allo, P., Taddeo, M., Wachter, S., & Floridi, L. (2016). The ethics of algorithms: Mapping the debate. *Big Data & Society, 3*(2), 2053951716679679.
91. Mokryn, O., Bachar, N., & Radosinsky, R. (2020). Machine learning approach for grid optimization and integration of renewable energy sources. *Energy, 200*, 117498.
92. Montavon, G., Samek, W., & Müller, K. R. (2018). Methods for interpreting and understanding deep neural networks. *Digital Signal Processing, 73*, 1–15.
93. Moura, A. M., de Carvalho, M. T. M., & Cruz, N. F. (2020). Artificial intelligence and machine learning applied to the analysis of public policy for municipal waste management. *Journal of Cleaner Production, 277*, 123085.
94. Müller, D., Warland, G., & Koch, H. (2020). Urban water demand forecasting: Review of methods and models. *Water, 12*(3), 844.
95. Musafer, H., Adedoyin, A., Alqaralleh, R., & Al-Hassan, A. (2020). Natural language processing for the development of effective water resource management policies. *Applied Artificial Intelligence, 34*(6), 509–529.
96. Nadeau, D., & Sekine, S. (2007). A survey of named entity recognition and classification. *Lingvisticae Investigationes, 30*(1), 3–26.
97. Nagpal, A., Dubey, A., & Mittal, M. L. (2018). Artificial intelligence and building energy management systems: A review, case study and future directions. *Sustainable Cities and Society, 38*, 697–713.
98. Naseem, T., Chen, H., Barowy, D., & Christodoulopoulos, C. (2010). Using universal linguistic knowledge to guide grammar induction. In *Proceedings of the 2010 Conference on Empirical Methods in Natural Language Processing* (pp. 631–639).
99. Nasrolahi, A. H., Kazemi, A., & Mokhtari, M. (2017). Energy demand forecasting in local electric distribution systems using machine learning techniques. *International Transactions on Electrical Energy Systems, 27*(12), e2473.
100. Nilsson, M., Griggs, D., & Visbeck, M. (2016). Map the interactions between Sustainable Development Goals. *Nature, 534*(7607), 320–322.

101. Niska, H., Serkkola, A., & Röning, J. (2016). A novel machine learning method for estimating energy consumption in buildings. *Energy and Buildings, 122*, 268–278.
102. Niska, H., Serkkola, A., & Sierla, S. (2016). From IoT to IIoT and onwards: A structured survey for generating a roadmap to industrial IoT. *IEEE Access, 4*, 8257–8276.
103. Pan, S. J., & Yang, Q. (2010). A survey on transfer learning. *IEEE Transactions on Knowledge and Data Engineering, 22*(10), 1345–1359.
104. Perez, L., Arango, S., & Quesada-Arencibia, A. (2018). Social data: A new source for urban planning. In *Smart cities* (pp. 59–71). Springer.
105. Poullikkas, A. (2020). A review of the role of artificial intelligence in the electricity sector. *Energy Sources, Part B: Economics, Planning, and Policy, 15*(3), 175–183.
106. Pramanik, M., Lau, R. Y., & Demirbilek, O. R. (2019). An overview of artificial intelligence applications in demand-side management. In *Demand-side management* (pp. 39–66). CRC Press.
107. Pritoni, M., Ford, R., Karlin, B., & Sanguinetti, A. (2019). Energy management in the smart home: A study of individual feedback and automation. *Energy and Buildings, 184*, 107–116.
108. Rakotomamonjy, A., Flamary, R., Gasso, G., & Canu, S. (2017). Lp-regularized SVM for feature selection and environmental monitoring. *Machine Learning, 106*(1), 29–61.
109. Ramaswami, A., Russell, A. G., Culligan, P. J., Sharma, K. R., & Kumar, E. (2016). Meta-principles for developing smart, sustainable, and healthy cities. *Science, 352*(6288), 940–943.
110. Reichstein, M., Camps-Valls, G., Stevens, B., Jung, M., Denzler, J., Carvalhais, N., & Prabhat, F. (2019). Deep learning and process understanding for data-driven Earth system science. *Nature, 566*(7743), 195–204.
111. Rittel, H. W., & Webber, M. M. (1973). Dilemmas in a general theory of planning. *Policy Sciences, 4*(2), 155–169.
112. Ruder, S., Vulic, I., & Søgaard, A. (2019). A survey of cross-lingual word embedding models. *Journal of Artificial Intelligence Research, 65*, 569–630.
113. Rydin, Y. (2011). *The purpose of planning: Creating sustainable towns and cities*. Policy Press.
114. Sachs, J., Schmidt-Traub, G., Kroll, C., Lafortune, G., & Fuller, G. (2019). Sustainable development report 2019. Bertelsmann Stiftung and Sustainable Development Solutions Network.
115. Sadollah, A., Guen Yoo, D., & Kim, J. H. (2015). Water distribution network optimization using a modified version of the water cycle algorithm. *Engineering Optimization, 47*(3), 361–377.
116. Samek, W., Wiegand, T., & Müller, K. R. (2019). Explainable artificial intelligence: Understanding, visualizing and interpreting deep learning models. *ITU Journal: ICT Discoveries, 1*(1), 1–17.
117. Santoso, M., Joewono, T. B., & Wijaya, D. R. (2018). The application of artificial intelligence for air quality management: A systematic review. *Transportation Research Part D: Transport and Environment, 63*, 467–480.
118. Sebastiani, F. (2002). Machine learning in automated text categorization. *ACM Computing Surveys (CSUR), 34*(1), 1–47.
119. Sharma, A., Dutta, K., & Dey, A. (2019). Chatbot for air pollution awareness: Design, development, and evaluation. In *Proceedings of the 2019 CHI Conference on Human Factors in Computing Systems* (pp. 1–13).
120. Solomatine, D. P., & Ostfeld, A. (2008). Data-driven modelling: Some past experiences and new approaches. *Journal of Hydroinformatics, 10*(1), 3–22.
121. Srivastava, P., Kalra, A., & Singh, R. P. (2018). Machine learning model for forecasting land use changes. *Geocarto International, 33*(11), 1156–1171.
122. Stuart, E. (2018). Partnering for sustainable development: Guidelines for multi-stakeholder partnerships to implement the 2030 Agenda in countries. United Nations Development Programme (UNDP).
123. Sun, Y., Gong, J., Wang, L., Wang, Y., & Zhu, H. (2020). A survey of policy learning: Theory and practice. *Information Sciences, 527*, 62–87.

124. Tabak, M. A., Norouzzadeh, M. S., Wolfson, D. W., Sweeney, S. J., VerCauteren, K. C., Snow, N. P., Halseth, J. M., Di Salvo, P. A., Lewis, J. S., White, M. D., Teton, B. (2019). Machine learning to classify animal species in camera trap images: Applications in ecology. *Methods in Ecology and Evolution, 10*(4), 585–590.
125. UN DESA. (2018). World Urbanization Prospects: The 2018 Revision. United Nations Department of Economic and Social Affairs, Population Division.
126. UN. (2015). Transforming our world: The 2030 Agenda for Sustainable Development. Resolution adopted by the General Assembly on 25 September 2015. United Nations.
127. UNESCO. (2017). Artificial intelligence for sustainable development: Synthesis report. United Nations Educational, Scientific and Cultural Organization.
128. Wang, C., Nie, X., Li, Y., & Zhang, Y. (2020). Urban solid waste prediction and disposal system optimization based on artificial intelligence technology. *Journal of Environmental Management, 264*, 110451.
129. WCED. (1987). *Our common future*. Oxford University Press.
130. Weng, Q. (2012). Remote sensing of impervious surfaces in the urban areas: Requirements, methods, and trends. *Remote Sensing of Environment, 117*, 34–49.
131. Widener, M. J., Horner, M. W., & Metcalf, S. S. (2015). Simulating the effects of social networks on a population's hurricane evacuation decision-making. *Transportation Research Part A: Policy and Practice, 78*, 42–58.
132. Williams, K., Joynt, J. L., & Hopkins, D. (2018). Adapting to climate change in the compact city: The suburban challenge. *Built Environment, 44*(1), 1–19.
133. Wu, W., Wang, H., & Zou, L. (2018). A review of data-driven approaches for circular water systems management: Urban water supply, stormwater, and wastewater systems. *Journal of Cleaner Production, 200*, 972–991.
134. Xu, X., Tian, Y., & Qi, Y. (2018). A systematic framework for planning and decision-making of water resources allocation based on a simulation-optimization model. *Water Resources Management, 32*(11), 3583–3598.
135. Yang, X., Xia, J., Zhang, Y., & Zhang, X. (2020). Traffic signal optimization using deep reinforcement learning for urban traffic congestion mitigation. *IEEE Transactions on Intelligent Transportation Systems, 21*(10), 4294–4304.
136. Yang, Y., Heppenstall, A., Turner, A., & Comber, A. (2018). A simulation-based approach to measuring the macroeconomic resilience of cities. *Environment and Planning B: Urban Analytics and City Science, 45*(4), 689–708.
137. Yu, D., Chen, K., Yang, C., & Wei, Y. D. (2014). Evaluating the spatial dynamics of regional land use efficiency in China using DEA. *Journal of Geographical Sciences, 24*(2), 220–234.
138. Yu, D., Wei, Y. D., & Wu, C. (2014). Modeling spatial dimensions of housing prices in China. *Cities, 38*, 75–87.
139. Yuan, W., Dai, B., Guo, J., & Li, X. (2021). Machine learning-based recycling process optimization and recyclable material identification. *Resources, Conservation and Recycling, 164*, 105122.
140. Yuan, Y., Lin, Z., Finin, T., & Joshi, A. (2018). Air quality prediction using open data and machine learning. In *Proceedings of the 27th International Conference on Computational Linguistics* (pp. 2930–2939).
141. Zhao, L., Huang, X., Huang, G., & Gartner, G. (2020). Urban air quality analysis based on social media text and its emotional information. *Sustainability, 12*(4), 1454.
142. Zhang, C., Sargent, I., Pan, X., Li, H., Gardiner, A., Hare, J., & Atkinson, P. M. (2018). An object-based convolutional neural network (OCNN) for urban land use classification. *Remote Sensing of Environment, 216*, 57–70.
143. Zhang, H., Zhang, Y., Lu, H., Huang, C., & Yao, X. (2019). Air quality prediction using spatiotemporal convolutional LSTM neural network. *Environmental Science and Pollution Research, 26*(19), 19451–19460.
144. Zhang, K., & Qi, J. (2018). Ground filtering algorithms for airborne LiDAR data: A review of critical issues. *Remote Sensing, 10*(4), 556.

145. Zhang, Q., Zhang, Y., Peng, J., & Gong, H. (2020). Recent advances in deep learning for object detection in remote sensing imagery. *Remote Sensing, 12*(5), 879.
146. Zhang, Y., & Wallace, B. (2015). A sensitivity analysis of (and practitioners' guide to) convolutional neural networks for sentence classification. In *Proceedings of the Eighth International Joint Conference on Natural Language Processing* (Vol. 1, Long Papers, pp. 253–263).
147. Zhang, Y., Pulliainen, J., Koponen, S., & Hallikainen, M. (2019). Water quality retrievals from combined Landsat TM data and ERS-2 SAR data in the Gulf of Finland. *IEEE Transactions on Geoscience and Remote Sensing, 47*(12), 4015–4026.
148. Zheng, Y., Gao, Y., & Wang, J. (2017). Using machine learning to measure the effectiveness of air quality policy in China. *IEEE Transactions on Big Data, 6*(4), 793–804.

Chapter 14
Ethical Considerations and Challenges

14.1 Data Privacy and Security

Data privacy and security have emerged as critical ethical considerations and challenges in the application of artificial intelligence (AI) in various domains, including sustainable development and resource management. The increasing use of AI relies on the collection, storage, processing, and sharing of vast amounts of data, often including sensitive information about individuals, communities, or ecosystems. This section will discuss the importance of data privacy and security, the potential risks and threats associated with AI-driven data practices, and the existing and emerging strategies for addressing these challenges.

Importance of Data Privacy and Security

Data privacy refers to the protection of personal information, ensuring that individuals have control over the collection, use, and disclosure of their data [12]. Data security, on the other hand, encompasses the measures and mechanisms designed to safeguard data from unauthorized access, use, disclosure, or destruction [6]. Both data privacy and security are essential for maintaining trust in AI systems, upholding ethical standards, and complying with legal and regulatory requirements [15].

In the context of sustainable development and resource management, data privacy and security are particularly important given the sensitive nature of the data involved, which may include information on individual and community demographics, socioeconomic status, health, or environmental exposures [32]. Ensuring data privacy and security can help build trust among stakeholders, facilitate public engagement and participation, and prevent potential harms associated with data breaches or misuse.

Risks and Threats to Data Privacy and Security in AI-driven Applications

The rapid growth and adoption of AI-driven applications have raised several risks and threats to data privacy and security, including unauthorized access to data, loss or corruption of data, data breaches, and data misuse [7]. These risks and threats

S. N. M. Moghaddam and H. Cao, *Artificial Intelligence-Driven Geographies*, City Development: Issues and Best Practices, https://doi.org/10.1007/978-981-97-5116-7_14

can stem from various sources, such as cyberattacks, human error, insider threats, or system vulnerabilities [1].

One of the key challenges in ensuring data privacy and security in AI-driven applications is the tension between data utility and data protection. AI models often require large amounts of data to be effective, which may necessitate the collection and sharing of sensitive information [47]. Furthermore, AI techniques, such as machine learning and deep learning, can potentially infer private information from seemingly innocuous data or reveal sensitive attributes through data linkages [32].

Another challenge is the dynamic and distributed nature of AI-driven applications, which can involve multiple data sources, processing platforms, and stakeholders. This complexity can make it difficult to identify and manage data privacy and security risks, particularly in environments characterized by rapid technological change and evolving threats [31].

Strategies for Addressing Data Privacy and Security Challenges in AI-driven Applications

To address the data privacy and security challenges associated with AI-driven applications, several strategies and best practices have been proposed and implemented, ranging from legal and regulatory frameworks to technical measures and organizational policies [6, 12].

a. Legal and Regulatory Frameworks

Legal and regulatory frameworks, such as the General Data Protection Regulation (GDPR) in the European Union or the California Consumer Privacy Act (CCPA) in the United States, provide comprehensive guidelines and requirements for data privacy and security, including principles of data minimization, purpose limitation, and user consent [30]. These frameworks can serve as a foundation for designing and implementing AI-driven applications that respect data privacy and security, while also promoting accountability and transparency.

b. Technical Measures

Various technical measures can be employed to enhance data privacy and security in AI-driven applications. These measures include encryption, pseudonymization, and anonymization techniques to protect sensitive data from unauthorized access or disclosure [40]. Additionally, differential privacy, a mathematical technique that adds noise to data queries, can be used to ensure that individual privacy is maintained while still enabling useful data analysis [18].

Other technical measures involve the use of privacy-preserving machine learning and AI techniques, such as federated learning, secure multi-party computation, or homomorphic encryption, which allow AI models to be trained and used without revealing sensitive data [9]. These techniques can help balance the need for data utility and data protection in AI-driven applications for sustainable development and resource management.

c. Organizational Policies and Practices

Organizational policies and practices play a crucial role in ensuring data privacy and security in AI-driven applications. This includes the development and implementation of data governance frameworks, which define the roles, responsibilities, and processes for data management, access, and usage within an organization [45]. Data governance frameworks can help promote data privacy and security by establishing clear guidelines for data collection, storage, processing, and sharing, as well as providing mechanisms for monitoring and auditing data practices.

Additionally, organizations can adopt privacy-by-design and security-by-design principles, which involve integrating data privacy and security considerations into the entire lifecycle of AI-driven applications, from design and development to deployment and maintenance [12]. This proactive approach can help identify and mitigate potential risks and threats to data privacy and security before they become critical issues.

In conclusion, data privacy and security are essential ethical considerations and challenges in AI-driven applications for sustainable development and resource management. Addressing these challenges requires a combination of legal and regulatory frameworks, technical measures, and organizational policies and practices that balance the need for data utility and data protection while upholding ethical standards and legal requirements.

14.2 Bias, Fairness, and Representation in AI Algorithms

Bias, fairness, and representation in AI algorithms have become increasingly important ethical considerations and challenges in various domains, including sustainable development and resource management. AI systems have the potential to reinforce or even exacerbate existing biases and inequalities if they are not carefully designed and implemented. This section will discuss the sources of bias in AI algorithms, the importance of fairness and representation, the potential consequences of biased AI systems, and strategies for addressing these challenges.

Sources of Bias in AI Algorithms

Bias in AI algorithms can arise from various sources, including biased data, biased model assumptions, and biased decision-making processes [5].

a. Biased Data

One of the primary sources of bias in AI algorithms is the data used to train and evaluate the models. Data can be biased if it is unrepresentative, incomplete, or contains systematic errors [11]. For instance, biased data can result from sampling biases, measurement biases, or label biases, which can lead to AI systems that favor certain groups or outcomes over others [35].

b. Biased Model Assumptions

Another source of bias in AI algorithms is the assumptions made during model development, such as the choice of features, algorithms, or loss functions. These assumptions can introduce or amplify biases if they do not accurately reflect the underlying relationships or processes in the data [5]. For example, an AI model that assumes a linear relationship between income and resource allocation may be biased against low-income communities if the true relationship is nonlinear.

c. Biased Decision-making Processes

Bias can also emerge in the decision-making processes of AI systems, such as the selection of thresholds, optimization criteria, or decision rules. These choices can introduce biases if they are based on inappropriate or discriminatory criteria, or if they prioritize certain objectives or values over others [33]. For example, an AI system that optimizes for cost efficiency in resource management may be biased against marginalized communities if it does not adequately consider the distributional impacts of its decisions.

Importance of Fairness and Representation in AI Algorithms

Fairness and representation are essential ethical considerations in AI algorithms because they help ensure that AI systems do not perpetuate or exacerbate existing biases and inequalities [22]. Fairness refers to the equitable treatment of different groups or individuals by AI systems, while representation refers to the extent to which AI systems capture and reflect the diversity of the populations and contexts they serve [14].

In the context of sustainable development and resource management, fairness and representation are particularly important given the potential for AI systems to influence critical decisions and outcomes, such as resource allocation, environmental protection, or social welfare [43]. Ensuring fairness and representation in AI algorithms can help promote social and environmental justice, support inclusive and participatory decision-making, and foster trust and accountability among stakeholders.

Consequences of Biased AI Systems

Biased AI systems can have significant consequences for individuals, communities, and ecosystems, particularly in the context of sustainable development and resource management. These consequences can include:

a. Discrimination and Inequality

Biased AI systems can perpetuate or exacerbate existing discrimination and inequality by disproportionately benefiting certain groups or individuals over others, or by unfairly penalizing or excluding marginalized populations [35]. This can result in the misallocation of resources, the perpetuation of social and environmental injustices, and the reinforcement of systemic barriers to access and opportunity [19].

b. Misguided Decision-making and Policy

Biased AI systems can lead to misguided decision-making and policy by providing inaccurate, misleading, or unrepresentative information to decision-makers [8]. This can result in suboptimal or even harmful interventions, wasted resources, and unintended consequences for communities and ecosystems [32].

c. Loss of Trust and Legitimacy

Biased AI systems can undermine trust and legitimacy in AI-driven applications and institutions, particularly if they are perceived as unfair, unrepresentative, or discriminatory [43]. This can erode public confidence in AI-driven sustainable development and resource management efforts, hinder stakeholder engagement, and limit the effectiveness and adoption of AI technologies.

Strategies for Addressing Bias, Fairness, and Representation in AI Algorithms

Addressing bias, fairness, and representation in AI algorithms requires a combination of technical and non-technical approaches, including data collection and preprocessing, model development and evaluation, and policy and governance.

a. Data Collection and Preprocessing

One strategy for addressing bias in AI algorithms is to ensure that the data used to train and evaluate the models is representative, accurate, and complete [11]. This can involve:

- Collecting data from diverse sources and populations to minimize sampling biases;
- Ensuring that data collection and measurement processes are consistent, transparent, and unbiased;
- Using data augmentation, re-sampling, or synthetic data generation techniques to address imbalances or gaps in the data;
- Preprocessing the data to remove or mitigate biases, such as by applying fairness-aware data transformations or feature selection methods [28].

b. Model Development and Evaluation

Another strategy for addressing bias in AI algorithms is to incorporate fairness and representation considerations into the model development and evaluation process [5]. This can involve:

- Selecting algorithms, features, or loss functions that are robust to bias or that promote fairness and representation (e.g., adversarial training, fairness-aware learning, or multi-objective optimization techniques);
- Evaluating AI models using fairness and representation metrics, such as demographic parity, equal opportunity, or individual fairness [24],
- Conducting sensitivity analyses, robustness checks, or model comparisons to assess the potential biases and uncertainties associated with different model assumptions or decision-making processes [33].

c. Policy and Governance

Finally, addressing bias, fairness, and representation in AI algorithms requires the development and implementation of appropriate policy and governance frameworks, such as:

- Establishing guidelines, standards, or best practices for fairness and representation in AI-driven applications for sustainable development and resource management;
- Implementing transparency, accountability, and oversight mechanisms, such as third-party audits, impact assessments, or disclosure requirements, to monitor and evaluate AI systems for bias, fairness, and representation;
- Promoting stakeholder engagement, public participation, and inclusive decision-making processes to ensure that diverse perspectives and values are considered and represented in AI-driven sustainable development and resource management efforts [14].

In conclusion, bias, fairness, and representation are critical ethical considerations and challenges in AI algorithms, particularly in the context of sustainable development and resource management. Addressing these challenges requires a combination of data collection and preprocessing, model development and evaluation, and policy and governance strategies to ensure that AI systems are equitable, inclusive, and trustworthy.

14.3 The Digital Divide and Equitable Access to Technology

The digital divide and equitable access to technology are critical ethical considerations and challenges in the context of AI-driven sustainable development and resource management. The digital divide refers to the gap between those who have access to and can effectively use digital technologies and those who do not [26]. Equitable access to technology involves ensuring that all individuals and communities have the necessary resources, skills, and opportunities to benefit from and participate in the digital world [16]. This section will discuss the causes and consequences of the digital divide, the importance of equitable access to technology, and strategies for addressing these challenges in the context of AI-driven sustainable development and resource management.

Causes of the Digital Divide

The digital divide can result from a combination of factors, including socioeconomic, demographic, geographic, and cultural disparities that influence access to and use of digital technologies [26]. Some of the main causes of the digital divide include:

a. Infrastructure and Connectivity

A lack of access to affordable, reliable, and high-speed internet infrastructure is a primary cause of the digital divide, particularly in rural, remote, or underserved areas [39]. This can result from inadequate investments in broadband networks, regulatory barriers, or market failures that limit the availability and affordability of internet services [41].

b. Affordability and Access to Devices

The cost of digital devices and services, such as smartphones, computers, or data plans, can be a significant barrier to digital inclusion, particularly for low-income individuals and households [13]. Affordability challenges can result from high device costs, limited competition, or regressive pricing policies that disproportionately burden the poor [46].

c. Digital Literacy and Skills

A lack of digital literacy and skills can limit individuals' ability to effectively use digital technologies, navigate online environments, or engage with AI-driven applications [25]. Digital literacy challenges can result from inadequate educational resources, training opportunities, or support systems, as well as social and cultural factors that influence attitudes towards technology and learning [16].

d. Social and Cultural Barriers

Social and cultural barriers, such as language, gender, age, or disability, can also contribute to the digital divide by influencing individuals' access to, use of, or preferences for digital technologies [26]. These barriers can result from discriminatory practices, norms, or stereotypes that limit the participation and representation of marginalized groups in the digital world [38].

Consequences of the Digital Divide

The digital divide can have significant consequences for individuals, communities, and societies, particularly in the context of AI-driven sustainable development and resource management. These consequences can include:

a. Exclusion and Inequality

The digital divide can exacerbate existing inequalities and exclusion by limiting access to resources, opportunities, and services for those who are not digitally connected or proficient [34]. This can result in a vicious cycle, where the digitally disadvantaged become further marginalized and disempowered in the digital age [26].

b. Lost Opportunities and Underutilization

The digital divide can lead to lost opportunities and underutilization of digital technologies, including AI-driven applications for sustainable development and resource management [41]. This can result in suboptimal outcomes, inefficiencies, and unrealized potential for leveraging AI to address socioeconomic and environmental challenges [39].

c. Unfair Distribution of Benefits and Burdens

The digital divide can result in an unfair distribution of the benefits and burdens associated with AI-driven sustainable development and resource management efforts [44]. This can occur when the gains from AI technologies are disproportionately captured by those who are digitally connected or proficient, while the costs or negative impacts are borne by those who are not [16].

d. Erosion of Trust and Legitimacy

The digital divide can undermine trust and legitimacy in AI-driven applications and institutions, particularly if they are perceived as exacerbating inequalities or exclusion [41]. This can erode public confidence in AI-driven sustainable development and resource management efforts, hinder stakeholder engagement, and limit the effectiveness and adoption of AI technologies.

Importance of Equitable Access to Technology

Equitable access to technology is essential for ensuring that all individuals and communities can benefit from and participate in the digital world, particularly in the context of AI-driven sustainable development and resource management [26]. This involves addressing the root causes of the digital divide, empowering marginalized groups, and promoting inclusive and participatory approaches to technology design, implementation, and governance [16].

Strategies for Addressing the Digital Divide and Equitable Access to Technology

Addressing the digital divide and equitable access to technology requires a combination of infrastructure investments, capacity building, and policy and governance interventions, as well as cross-sectoral collaboration and partnerships [39]. Some of the main strategies for addressing these challenges include:

a. Infrastructure Investments

Investing in affordable, reliable, and high-speed internet infrastructure is critical for bridging the digital divide and ensuring equitable access to technology [41]. This can involve public and private investments in broadband networks, as well as regulatory reforms, subsidies, or incentives to promote competition, innovation, and affordability in the internet services market [39].

b. Capacity Building and Digital Literacy

Strengthening digital literacy and skills is essential for empowering individuals and communities to effectively use digital technologies, including AI-driven applications for sustainable development and resource management [25]. This can involve educational and training programs, outreach and awareness campaigns, or mentorship and support networks to promote digital literacy, skills development, and lifelong learning [16].

c. Policy and Governance Interventions

Developing and implementing policy and governance interventions can help address the digital divide and promote equitable access to technology [41]. This can involve:

- Establishing digital inclusion strategies, targets, or indicators to monitor and evaluate progress towards bridging the digital divide and ensuring equitable access to technology;
- Implementing policies and programs to promote affordability, accessibility, and inclusiveness in the digital device and service markets, such as subsidies, tax credits, or universal service obligations;
- Promoting stakeholder engagement, public participation, and inclusive decision-making processes to ensure that diverse perspectives and needs are considered and addressed in the design, implementation, and governance of digital technologies, including AI-driven sustainable development and resource management efforts [38].

d. Cross-Sectoral Collaboration and Partnerships

Finally, addressing the digital divide and equitable access to technology requires cross-sectoral collaboration and partnerships between governments, the private sector, civil society, and international organizations [39]. This can involve:

- Sharing knowledge, resources, and best practices to support digital inclusion efforts and bridge the digital divide;
- Leveraging public–private partnerships to drive investments in infrastructure, capacity building, and innovation;
- Engaging with civil society organizations and community-based initiatives to ensure that local knowledge, context, and priorities are incorporated into digital technology design and implementation;
- Collaborating with international organizations and development partners to facilitate technology transfer, capacity building, and policy coordination to address the digital divide and promote equitable access to technology at the global level [26].

In conclusion, addressing the digital divide and ensuring equitable access to technology are critical ethical considerations and challenges in the context of AI-driven sustainable development and resource management. By implementing a combination of infrastructure investments, capacity building, policy and governance interventions, and cross-sectoral collaboration and partnerships, it is possible to bridge the digital divide and empower all individuals and communities to benefit from and participate in the digital world.

14.4 Public Participation and Engagement in AI-Driven Planning

Public participation and engagement are critical aspects of ethical considerations and challenges related to AI-driven planning in sustainable development and resource management. By involving diverse stakeholders, including citizens, in the decision-making processes, AI-driven planning can be more inclusive, responsive, and democratic. This section will discuss the importance of public participation and engagement in AI-driven planning, its challenges, and the strategies to address these challenges.

Importance of Public Participation and Engagement in AI-driven Planning

Public participation and engagement are essential for several reasons:

a. Inclusiveness and Representation

Involving a wide range of stakeholders in AI-driven planning can help ensure that diverse perspectives, values, and interests are considered and represented, which can lead to more inclusive and equitable outcomes [2, 10].

b. Responsiveness and Accountability

Public participation and engagement can make AI-driven planning more responsive to the needs and preferences of the affected communities, which can enhance the overall effectiveness and legitimacy of the planning processes [23, 27].

c. Trust and Confidence

Engaging stakeholders and the public in AI-driven planning can help build trust and confidence in the process, technologies, and institutions, which can be critical for the adoption and success of AI-driven interventions [37].

d. Learning and Innovation

Public participation and engagement can facilitate learning, knowledge exchange, and innovation, which can contribute to the development and improvement of AI-driven planning tools and practices [36].

Challenges of Public Participation and Engagement in AI-driven Planning

There are several challenges to achieving meaningful public participation and engagement in AI-driven planning:

a. Complexity and Technical Knowledge

AI-driven planning often involves complex algorithms and technical knowledge, which can be difficult for non-experts to understand and engage with [20]. This can create barriers to participation and limit the effectiveness of public engagement efforts.

b. Power Imbalances and Inequality

Power imbalances and inequalities can influence the public participation process in AI-driven planning, leading to the exclusion or marginalization of certain groups, voices, or interests [2]. This can undermine the inclusiveness, representation, and legitimacy of the planning processes.

c. Privacy and Security Concerns

Public participation and engagement in AI-driven planning may involve the collection, sharing, or analysis of personal or sensitive data, which can raise privacy and security concerns for individuals and communities [12].

d. Mistrust and Skepticism

Mistrust and skepticism about AI technologies, their developers, or the motivations behind their deployment can also hinder public participation and engagement in AI-driven planning [37].

Strategies for Enhancing Public Participation and Engagement in AI-driven Planning

To address these challenges and promote meaningful public participation and engagement in AI-driven planning, several strategies can be employed:

a. Capacity Building and Education

Capacity building and education efforts can help increase the public's understanding of AI-driven planning processes and technologies, as well as their ability to meaningfully engage with them [20]. This can involve training programs, workshops, or educational resources to enhance technical literacy and empower stakeholders to participate in AI-driven planning processes.

b. Inclusive and Participatory Design

Adopting inclusive and participatory design approaches can help ensure that diverse perspectives, values, and interests are considered and represented in AI-driven planning processes [10]. This can involve engaging stakeholders and the public in the development, evaluation, and refinement of AI-driven planning tools and practices, as well as implementing participatory decision-making processes that empower stakeholders to influence the outcomes of AI-driven planning initiatives [23].

c. Addressing Power Imbalances and Inequality

Efforts should be made to identify and address power imbalances and inequality in the context of public participation and engagement in AI-driven planning. This can involve engaging underrepresented or marginalized groups, providing resources and support to facilitate their participation, and adopting measures to ensure that their voices and interests are considered and valued in the decision-making processes [2, 27].

d. Privacy and Security Measures

To address privacy and security concerns, appropriate measures should be implemented to protect personal and sensitive data, as well as to maintain transparency and accountability in data collection, sharing, and analysis processes [12]. This can involve adopting privacy-preserving AI technologies, implementing data governance frameworks, and engaging stakeholders and the public in discussions about data privacy and security issues.

e. Building Trust and Confidence

Building trust and confidence in AI-driven planning processes and technologies can be achieved through transparency, accountability, and open communication with stakeholders and the public [37]. This can involve sharing information about the development, deployment, and evaluation of AI-driven planning tools and practices, as well as engaging stakeholders and the public in discussions about the benefits, risks, and ethical implications of AI-driven planning initiatives.

Public participation and engagement are essential ethical considerations and challenges in AI-driven planning for sustainable development and resource management. By implementing strategies to address these challenges, it is possible to create more inclusive, responsive, and democratic AI-driven planning processes that can contribute to better outcomes for individuals, communities, and the environment.

14.5 The Future of Employment in Geography and Urban Planning

The integration of artificial intelligence (AI) into geography and urban planning has the potential to revolutionize these fields, leading to significant changes in the future of employment. AI can automate various tasks, improve efficiency, and enhance decision-making processes, but it also raises concerns about potential job displacement, skill requirements, and the need for continuous learning. This section will discuss the implications of AI on the future of employment in geography and urban planning, exploring both the opportunities and challenges that AI presents.

Opportunities for Employment in AI-driven Geography and Urban Planning

The adoption of AI in geography and urban planning can create new job opportunities and transform existing ones:

a. New Job Roles and Specializations

The increasing use of AI technologies in geography and urban planning can lead to the emergence of new job roles and specializations, such as AI model developers, geospatial data scientists, or urban informatics specialists [29, 42]. These roles typically require advanced technical skills, including expertise in machine learning, data analytics, and geospatial technologies.

b. Enhanced Decision-Making and Analysis

AI can support professionals in geography and urban planning by automating routine tasks, enabling them to focus on higher-level decision-making and analysis [29]. This can result in more effective planning processes and better outcomes for individuals, communities, and the environment.

c. Interdisciplinary Collaboration

The growing use of AI in geography and urban planning can facilitate interdisciplinary collaboration, as professionals from diverse fields—such as computer science, engineering, and social sciences—work together to develop and implement AI-driven planning tools and strategies [17]. This can lead to the creation of new job opportunities at the intersection of these disciplines and contribute to the development of innovative solutions to complex problems.

Challenges for Employment in AI-driven Geography and Urban Planning

Despite the potential benefits of AI integration, there are also challenges and concerns related to the future of employment in geography and urban planning:

a. Job Displacement

AI-driven automation can lead to job displacement, as certain tasks or roles may no longer require human intervention [3, 21]. This can result in job losses, particularly for those with skills and expertise that are less relevant in an AI-driven environment.

b. Skill Requirements and the Need for Continuous Learning

The increasing use of AI in geography and urban planning can change the skill requirements for professionals in these fields, emphasizing the need for technical expertise in areas such as machine learning, data analytics, and geospatial technologies [42]. This can create a skills gap, as professionals may need to acquire new skills and competencies to remain relevant and employable in an AI-driven context.

c. Equity and Inclusiveness

The integration of AI into geography and urban planning can exacerbate existing inequalities, as access to education and training opportunities may be limited for certain groups or communities [26]. This can result in unequal access to job opportunities in AI-driven geography and urban planning, reinforcing existing disparities in employment and income.

Strategies for Addressing Employment Challenges in AI-driven Geography and Urban Planning

To address these challenges and ensure a positive impact of AI on the future of employment in geography and urban planning, several strategies can be employed:

a. Education and Training Programs

Developing education and training programs that focus on AI-related skills can help prepare current and future professionals for the changing landscape of geography and urban planning [42]. These programs should emphasize both technical and interdisciplinary skills, as well as critical thinking, problem-solving, and communication abilities.

b. Lifelong Learning and Continuous Skill Development

Emphasizing the importance of lifelong learning and continuous skill development can help professionals in geography and urban planning adapt to the changing demands of an AI-driven environment [4]. This can involve offering ongoing professional development opportunities, such as workshops, online courses, and certifications, to help individuals stay up-to-date with the latest technologies and techniques.

c. Inclusive and Equitable Access to Education and Training

Ensuring inclusive and equitable access to education and training opportunities can help address potential disparities in employment and income associated with AI-driven geography and urban planning [26]. This can involve implementing policies and initiatives that target underrepresented or marginalized groups, providing financial support or resources, and creating inclusive learning environments.

d. Fostering Collaboration and Interdisciplinary Approaches

Promoting collaboration and interdisciplinary approaches can help create new job opportunities at the intersection of geography, urban planning, and other disciplines, such as computer science, engineering, and social sciences [17]. This can involve developing interdisciplinary degree programs, research projects, or professional networks that encourage collaboration and knowledge exchange among professionals from diverse fields.

The integration of AI into geography and urban planning presents both opportunities and challenges for the future of employment in these fields. By adopting strategies to address these challenges, it is possible to harness the potential of AI to create new job opportunities, transform existing roles, and enhance the overall effectiveness and impact of geography and urban planning practices.

References

1. Abomhara, M., & Koien, G. M. (2015). Cyber security and the internet of things: Vulnerabilities, threats, intruders and attacks. *Journal of Cyber Security, 4*(1), 65–88.
2. Arnstein, S. R. (1969). A ladder of citizen participation. *Journal of the American Institute of Planners, 35*(4), 216–224.
3. Arntz, M., Gregory, T., & Zierahn, U. (2016). The Risk of Automation for Jobs in OECD Countries: A Comparative Analysis. OECD Social, Employment and Migration Working Papers, No. 189, OECD Publishing, Paris.
4. Autor, D. H. (2015). Why are there still so many jobs? The history and future of workplace automation. *Journal of Economic Perspectives, 29*(3), 3–30.
5. Barocas, S., & Selbst, A. D. (2016). Big data's disparate impact. *California Law Review, 104*, 671.

6. Bertino, E. (2015). Data security and privacy in the IoT. In *Proceedings of the 2015 ACM SIGMOD International Conference on Management of Data* (pp. 3–6).
7. Bertino, E., & Ferrari, E. (2002). Secure and selective dissemination of XML documents. *ACM Transactions on Information and System Security (TISSEC), 5*(3), 290–331.
8. Bolukbasi, T., Chang, K. W., Zou, J. Y., Saligrama, V., & Kalai, A. T. (2016). Man is to computer programmer as woman is to homemaker? Debiasing word embeddings. In *Advances in Neural Information Processing Systems* (pp. 4349–4357).
9. Bonawitz, K., Eichner, H., Grieskamp, W., Huba, D., Ingerman, A., Ivanov, V., Kiddon, C., Konečný, J., Mazzocchi, S., & McMahan, H. B. (2019). Towards federated learning at scale: System design. arXiv:1902.01046
10. Brabham, D. C. (2009). Crowdsourcing the public participation process for planning projects. *Planning Theory, 8*(3), 242–262.
11. Caliskan, A., Bryson, J. J., & Narayanan, A. (2017). Semantics derived automatically from language corpora contain human-like biases. *Science, 356*(6334), 183–186.
12. Cavoukian, A. (2010). Privacy by design: The 7 foundational principles. Information and Privacy Commissioner of Ontario, Canada.
13. Chinn, M. D., & Fairlie, R. W. (2007). The determinants of the global digital divide: A cross-country analysis of computer and internet penetration. *Oxford Economic Papers, 59*(1), 16–44.
14. Crawford, K., & Paglen, T. (2019). Excavating AI: The politics of images in machine learning training sets. Excavating AI.
15. Cumbley, R., & Church, P. (2013). Is "big data" creepy? *Computer Law & Security Review, 29*(5), 601–609.
16. DiMaggio, P., & Hargittai, E. (2001). From the 'digital divide' to 'digital inequality': Studying Internet use as penetration increases. Princeton: Center for Arts and Cultural Policy Studies, Woodrow Wilson School, Princeton University.
17. Dodge, M., & Kitchin, R. (2005). Codes of life: Identification codes and the machine-readable world. *Environment and Planning D: Society and Space, 23*(6), 851–881.
18. Dwork, C., & Roth, A. (2014). The algorithmic foundations of differential privacy. *Foundations and Trends® in Theoretical Computer Science*, *9*(3–4), 211–407.
19. Eubanks, V. (2018). *Automating inequality: How high-tech tools profile, police, and punish the poor*. Martin's Press.
20. Fisher, D. R., & Green, J. F. (2004). Understanding disenfranchisement: Civil society and developing countries' influence and participation in global governance for sustainable development. *Global Environmental Politics, 4*(3), 65–84.
21. Frey, C. B., & Osborne, M. A. (2017). The future of employment: How susceptible are jobs to computerisation? *Technological Forecasting and Social Change, 114*, 254–280.
22. Friedler, S. A., Scheidegger, C., & Venkatasubramanian, S. (2016). On the (im) possibility of fairness. arXiv:1609.07236
23. Fung, A. (2006). Varieties of participation in complex governance. *Public Administration Review, 66*(s1), 66–75.
24. Hardt, M., Price, E., & Srebro, N. (2016). Equality of opportunity in supervised learning. In *Advances in Neural Information Processing Systems* (pp. 3315–3323).
25. Hargittai, E. (2002). Second-level digital divide: Differences in people's online skills. *First Monday, 7*(4).
26. Hilbert, M. (2011). Digital gender divide or technologically empowered women in developing countries? A typical case of lies, damned lies, and statistics. *Women's Studies International Forum, 34*(6), 479–489.
27. Innes, J. E., & Booher, D. E. (2004). Reframing public participation: Strategies for the 21st century. *Planning Theory & Practice, 5*(4), 419–436.
28. Kamiran, F., & Calders, T. (2012). Data preprocessing techniques for classification without discrimination. *Knowledge and Information Systems, 33*(1), 1–33.
29. Kitchin, R. (2014). The real-time city? Big data and smart urbansim. *GeoJournal, 79*(1), 1–14.
30. Kuner, C. (2017). *The European Union general data protection regulation (GDPR): A practical guide*. Global Privacy & Security Law.

31. Li, J., Ragh unathan, S., & Jha, S. (2015). Secure data processing framework for mobile cloud computing. In *2015 IEEE Conference on Computer Communications (INFOCOM)* (pp. 1665–1673).
32. Mittelstadt, B. D., Allo, P., Taddeo, M., Wachter, S., & Floridi, L. (2016). The ethics of algorithms: Mapping the debate. *Big Data & Society, 3*(2), 2053951716679679.
33. Mullainathan, S., & Obermeyer, Z. (2017). Does machine learning automate moral hazard and error? *American Economic Review, 107*(5), 476–480.
34. Norris, P. (2001). *Digital divide: Civic engagement, information poverty, and the Internet worldwide*. Cambridge University Press.
35. O'Neil, C. (2016). *Weapons of math destruction: How big data increases inequality and threatens democracy*. Broadway Books.
36. Reed, M. S. (2008). Stakeholder participation for environmental management: A literature review. *Biological Conservation, 141*(10), 2417–2431.
37. Rowe, G., & Frewer, L. J. (2000). Public participation methods: A framework for evaluation. *Science, Technology, & Human Values, 25*(1), 3–29.
38. Scheerder, A., van Deursen, A., & van Dijk, J. (2017). Determinants of Internet skills, uses and outcomes. A systematic review of the second- and third-level digital divide. *Telematics and Informatics, 34*(8), 1607–1624.
39. Sundararajan, A. (2016). *The sharing economy: The end of employment and the rise of crowd-based capitalism*. MIT Press.
40. Sweeney, L. (2002). K-anonymity: A model for protecting privacy. *International Journal of Uncertainty, Fuzziness and Knowledge-Based Systems, 10*(05), 557–570.
41. Van Dijk, J. A. (2006). Digital divide research, achievements and shortcomings. *Poetics, 34*(4–5), 221–235.
42. Wachter, S. (2018). The rise of the data scientist: The importance of data analysis and processing for modern urban planning. *Journal of Urban Technology, 25*(1), 3–20.
43. Wachter, S., Mittelstadt, B., & Floridi, L. (2017). Transparent, explainable, and accountable AI for robotics. *Science Robotics, 2*(6), eaan6080.
44. Warschauer, M. (2004). *Technology and social inclusion: Rethinking the digital divide*. MIT press.
45. Weber, R. H. (2015). Data governance in the context of the European data protection regulation. *International Data Privacy Law, 5*(4), 260–270.
46. Wyche, S. P., Steinfield, C., & Forte, A. (2013). Examining social media use among BoP microenterprises: Toward a theory of technology use in low-income settings. In *International Conference on Information and Communication Technologies and Development (ICTD)* (pp. 1–11). IEEE.
47. Zarsky, T. Z. (2013). Transparent predictions. *University of Illinois Law Review, 1503*.

Chapter 15
Conclusion and Future Prospects

15.1 Summary of AI's Impact on Human Geography and Urban Planning

Artificial intelligence (AI) has increasingly become an integral part of human geography and urban planning, shaping the way these disciplines are conducted, and offering transformative potential for improved decision-making, analysis, and efficiency. This section provides a summary of AI's impact on human geography and urban planning, highlighting key themes and applications discussed throughout this book.

AI in Human Geography

AI has made significant contributions to the field of human geography, enabling researchers and professionals to address complex spatial and temporal problems with enhanced analytical capabilities. The following are some of the primary areas where AI has had a notable impact:

a. Population Distribution and Migration Patterns

AI techniques, such as machine learning and geospatial data analysis, have facilitated the study of population distribution and migration patterns, enabling the identification of trends, drivers, and consequences of population movement [15]. This has led to more accurate predictions, better understanding of the underlying factors, and improved policy responses to address migration-related issues.

b. Land Use and Land Cover Change Detection

AI has been instrumental in the analysis of land use and land cover change, utilizing techniques such as deep learning and remote sensing data to automatically classify and monitor changes in land use patterns [26]. This has allowed for more efficient and accurate detection of changes in land use, providing valuable information for land management, environmental conservation, and urban planning.

S. N. M. Moghaddam and H. Cao, *Artificial Intelligence-Driven Geographies*, City Development: Issues and Best Practices, https://doi.org/10.1007/978-981-97-5116-7_15

c. Environmental Risk Assessment and Climate Change Impacts

AI has enabled more accurate and comprehensive assessments of environmental risks and climate change impacts by integrating diverse datasets, such as satellite imagery, climate models, and socioeconomic data [21]. These assessments support the development of targeted mitigation and adaptation strategies to address environmental challenges and reduce vulnerability to climate change.

d. Socioeconomic Inequality and Spatial Analysis

AI has been utilized to analyze and visualize socioeconomic inequality, leveraging spatial analysis techniques and geospatial data to identify patterns and drivers of inequality within and between regions [5]. This information is crucial for informing policies and interventions aimed at reducing inequality and fostering inclusive growth.

e. Health and Disease Mapping

AI has contributed to the field of health geography by facilitating the mapping and analysis of disease patterns and their relationships with environmental and socioeconomic factors [4]. This information is invaluable for public health planning, resource allocation, and the development of targeted interventions to improve health outcomes and reduce health disparities.

AI in Urban Planning

AI's impact on urban planning has been equally transformative, with applications spanning various aspects of planning processes and decision-making:

a. Smart Cities and IoT Integration

AI has played a central role in the development of smart cities, as it integrates and processes data from Internet of Things (IoT) devices to support decision-making, optimize resource allocation, and improve urban services [3]. This has led to more efficient and sustainable cities, with improved quality of life for residents.

b. Transportation and Traffic Management

AI has significantly influenced transportation and traffic management, utilizing techniques such as machine learning and data analytics to optimize traffic flow, reduce congestion, and enhance transportation systems [23]. This has resulted in more sustainable and efficient urban transportation, contributing to reduced emissions, improved air quality, and better overall mobility.

c. Urban Growth and Sprawl Prediction

AI has been employed to predict urban growth and sprawl, utilizing techniques such as cellular automata and machine learning to model future urban development patterns based on historical data and various driving factors [22]. This information is essential for informing land use planning, infrastructure development, and environmental conservation efforts.

d. Housing, Affordability, and Real Estate Market Analysis

AI has made significant contributions to the understanding and analysis of housing markets, affordability, and real estate dynamics. Machine learning and data analytics techniques have been employed to predict housing prices, identify trends, and assess the impact of various factors on housing affordability [25]. This information is crucial for policymakers and urban planners to develop strategies and interventions that promote affordable housing and equitable access to housing opportunities.

e. Sustainable Development and Resource Management

AI has been instrumental in advancing sustainable development and resource management in urban planning. Techniques such as machine learning, deep learning, and natural language processing have been utilized to optimize resource allocation, monitor environmental impacts, and analyze sustainability policies [1]. AI applications in this area include energy efficiency and conservation, waste management and recycling, water resource management, air quality management and pollution control, and climate change adaptation and resilience. These applications contribute to the development of more sustainable, resilient, and livable urban environments.

AI has had a transformative impact on human geography and urban planning, revolutionizing the way these disciplines are conducted and offering significant potential for improved decision-making, analysis, and efficiency. As AI continues to advance, it is expected to play an even more prominent role in shaping the future of human geography and urban planning, fostering innovation, and addressing pressing global challenges. It is essential, however, to address the ethical considerations and challenges associated with AI's integration into these fields, including data privacy and security, bias and fairness, digital divide and equitable access to technology, public participation and engagement, and the future of employment in geography and urban planning.

15.2 The Potential for Further Integration and Advancement

The potential for further integration and advancement of artificial intelligence (AI) in human geography and urban planning is vast, as ongoing research and innovation continue to push the boundaries of what is possible. This section explores the opportunities for AI to make an even more significant impact on these fields, focusing on new applications, interdisciplinary collaborations, and overcoming limitations to maximize the potential of AI-driven solutions.

New Applications and Areas of Research

As AI continues to evolve, there will be increased opportunities to apply these technologies to new areas of research and application in human geography and urban planning. Some potential areas of interest include:

a. Disaster Management and Response

AI can play a critical role in improving disaster management and response efforts, utilizing techniques such as machine learning, computer vision, and natural language processing to predict and monitor natural disasters, assess damage, and optimize relief efforts [10]. AI-driven tools can also be used to enhance communication and collaboration between emergency responders and affected populations, facilitating more effective and efficient responses to disasters.

b. Public Health Planning and Response

AI has the potential to revolutionize public health planning and response efforts, particularly in the context of emerging infectious diseases and pandemics [18]. Machine learning algorithms can be employed to predict disease outbreaks, identify vulnerable populations, and develop targeted interventions. AI can also be utilized to optimize healthcare resource allocation and improve healthcare service delivery, contributing to better overall public health outcomes.

c. Urban Design and Architecture

AI can be integrated into urban design and architecture processes, providing new tools and methodologies for designing and evaluating built environments. Techniques such as generative design, which leverages AI algorithms to explore and optimize design solutions based on specified constraints, have the potential to revolutionize the way urban spaces are designed and constructed [6]. AI can also be utilized to assess the environmental performance, accessibility, and livability of urban designs, supporting the creation of more sustainable and inclusive cities.

Interdisciplinary Collaborations

One of the keys to unlocking the full potential of AI in human geography and urban planning is the establishment of interdisciplinary collaborations, bringing together expertise from various fields to address complex, multifaceted problems. Potential areas of collaboration include:

a. Collaborations with Environmental Sciences

Collaborating with environmental scientists can enable the development of more comprehensive and accurate models of natural systems and human–environment interactions, informing more sustainable and resilient planning practices [9]. This collaboration can also facilitate the development of AI-driven tools to monitor and mitigate the impacts of climate change, contributing to the creation of more adaptable and sustainable urban environments.

b. Collaborations with Social Sciences

Collaborating with social scientists can enhance the understanding of the social and cultural implications of AI-driven interventions, ensuring that these technologies are deployed in ways that promote social equity and inclusivity [12]. This collaboration can also support the development of AI-driven tools that address issues such as social segregation, gentrification, and community cohesion, promoting more equitable and just urban environments.

c. Collaborations with Data Science and Computer Science

Collaborating with data scientists and computer scientists can facilitate the development of innovative AI algorithms, techniques, and tools tailored to the specific needs and challenges of human geography and urban planning [19]. This collaboration can also support the development of new methodologies for integrating and analyzing diverse datasets, enhancing the potential for AI-driven insights and decision-making.

Overcoming Limitations and Challenges

To maximize the potential of AI in human geography and urban planning, it is essential to address the various limitations and challenges associated with these technologies. Some key areas to focus on include:

a. Data Quality and Availability

Improving the quality and availability of data is essential for the successful integration and advancement of AI in human geography and urban planning [13]. Efforts should be made to ensure that data is collected and maintained with high accuracy, consistency, and resolution, enabling more robust and reliable AI-driven analyses. Moreover, promoting open data initiatives and data sharing agreements can enhance the availability of datasets, fostering innovation and collaboration across disciplines.

b. Algorithmic Transparency and Interpretability

Addressing concerns related to the transparency and interpretability of AI algorithms is crucial for ensuring the trustworthiness and acceptability of AI-driven solutions in human geography and urban planning [7]. Developing techniques that make AI algorithms more explainable and understandable can facilitate their adoption by decision-makers and stakeholders, as well as support efforts to identify and address biases and other unintended consequences.

c. Ethics, Privacy, and Security

Continued advancements in AI must be accompanied by ongoing efforts to address ethical, privacy, and security concerns associated with the use of these technologies [16]. This includes the development of guidelines and best practices for the responsible use of AI in human geography and urban planning, as well as the implementation of robust data privacy and security measures to protect sensitive information.

The potential for further integration and advancement of AI in human geography and urban planning is immense, with numerous opportunities for new applications, interdisciplinary collaborations, and overcoming limitations to maximize the potential of AI-driven solutions. By embracing these opportunities and addressing the challenges associated with AI's integration into these fields, human geographers and urban planners can harness the power of AI to create more sustainable, resilient, and equitable urban environments for all.

Fostering Public–Private Partnerships

The establishment of public–private partnerships can play a vital role in furthering the integration and advancement of AI in human geography and urban planning. By bringing together government entities, academic institutions, and private-sector stakeholders, these partnerships can facilitate the development and deployment of innovative AI-driven tools and solutions, while ensuring that the benefits of these technologies are equitably distributed across society [14].

Capacity Building and Workforce Development

To maximize the potential of AI in human geography and urban planning, it is crucial to invest in capacity building and workforce development efforts. This includes providing education and training programs that equip current and future professionals with the necessary skills to understand, develop, and utilize AI-driven tools and technologies [24]. Additionally, fostering interdisciplinary collaborations and knowledge exchange can help to ensure that the workforce is prepared to address the complex and evolving challenges associated with AI's integration into these fields.

Inclusive Stakeholder Engagement

Engaging diverse stakeholders in the development and implementation of AI-driven solutions is critical for ensuring that these technologies are responsive to the needs and priorities of communities and decision-makers. Inclusive stakeholder engagement processes can help to identify potential barriers and opportunities for the successful integration of AI in human geography and urban planning, as well as foster a sense of shared ownership and responsibility for the outcomes of these interventions [2].

The potential for further integration and advancement of AI in human geography and urban planning is enormous, presenting a wide range of opportunities for innovation and impact. By embracing these opportunities and addressing the challenges associated with AI's integration into these fields, human geographers and urban planners can harness the power of AI to create more sustainable, resilient, and equitable urban environments for all. Moving forward, it is crucial to continue exploring new applications, fostering interdisciplinary collaborations, and working to overcome limitations and challenges, in order to fully realize the transformative potential of AI in human geography and urban planning.

15.3 Future Research Directions and Challenges

As artificial intelligence (AI) continues to make significant strides in various fields, its impact on human geography and urban planning is becoming increasingly evident. The integration of AI in these disciplines has the potential to revolutionize traditional approaches and bring about transformative change. However, the future development and implementation of AI-driven solutions in human geography and urban planning also face numerous challenges and research directions that warrant further exploration. This section aims to discuss some of the key future research directions and challenges in the context of AI applications in human geography and urban planning.

Enhancing the Quality and Availability of Data

One of the critical factors for the successful implementation of AI in human geography and urban planning is the quality and availability of data. Future research should focus on addressing issues related to data accuracy, consistency, resolution, and accessibility [12]. This could include the development of novel data collection techniques, the promotion of open data initiatives, and the establishment of data-sharing agreements to foster innovation and collaboration across disciplines.

Algorithmic Transparency and Interpretability

The black-box nature of many AI algorithms raises concerns about their transparency and interpretability, particularly in fields like human geography and urban planning, where the implications of AI-driven decisions can have far-reaching consequences [8]. Future research should focus on developing techniques to make AI algorithms more explainable and understandable, facilitating their adoption by decision-makers and stakeholders while supporting efforts to identify and address biases and other unintended consequences.

Ethics, Privacy, and Security

The continued advancement of AI must be accompanied by ongoing efforts to address ethical, privacy, and security concerns associated with the use of these technologies [17]. Future research should focus on the development of guidelines and best practices for the responsible use of AI in human geography and urban planning, as well as the implementation of robust data privacy and security measures to protect sensitive information.

Bridging the Digital Divide

The digital divide, characterized by unequal access to technology and digital resources, presents a significant challenge to the equitable implementation of AI-driven solutions in human geography and urban planning [20]. Future research should explore strategies for bridging the digital divide, ensuring that the benefits of AI are equitably distributed across society and that all communities can participate in and benefit from AI-driven planning processes.

Public Participation and Engagement

In order to ensure the successful implementation of AI-driven solutions in human geography and urban planning, it is essential to engage a diverse range of stakeholders in the development and deployment of these technologies [2]. Future research should focus on identifying effective strategies for fostering public participation and engagement in AI-driven planning processes, including the development of participatory AI tools and techniques that empower communities to actively shape their urban environments.

Capacity Building and Workforce Development

The rapid advancement of AI in human geography and urban planning necessitates the development of a skilled workforce capable of understanding, developing, and utilizing AI-driven tools and technologies [24]. Future research should focus on the development of education and training programs that equip current and future professionals with the necessary skills to harness the power of AI in their work, as well as fostering interdisciplinary collaborations and knowledge exchange to ensure a well-rounded workforce.

Sustainable Development and Resource Management

As cities and regions face increasing challenges related to sustainability, resource management, and climate change, the integration of AI in human geography and urban planning has the potential to significantly enhance the capacity of decision-makers to address these issues [11]. Future research should explore novel applications of AI in sustainable development and resource management, including the development of AI-driven tools and techniques for monitoring environmental conditions, optimizing resource allocation, and supporting climate change adaptation and resilience efforts.

Interdisciplinary Collaborations

Given the complex and multifaceted nature of urban planning and human geography, the successful integration of AI in these fields requires collaboration among various disciplines [12]. Future research should focus on fostering interdisciplinary collaborations, bringing together researchers, practitioners, and decision-makers from fields such as computer science, geography, urban planning, environmental sciences, and social sciences, to develop innovative AI-driven solutions that address the multifaceted challenges facing urban environments.

Evaluation and Validation of AI-driven Tools

As AI-driven tools and techniques become more prevalent in human geography and urban planning, it is crucial to establish robust evaluation and validation methods to ensure their effectiveness and reliability [7]. Future research should focus on the development of methodologies for assessing the performance of AI-driven solutions, including the identification of appropriate performance metrics, benchmark datasets, and evaluation criteria.

Addressing Unintended Consequences and Societal Impacts

The integration of AI in human geography and urban planning has the potential to bring about transformative change. However, it is essential to consider the potential unintended consequences and societal impacts of AI-driven interventions [16]. Future research should focus on exploring the broader implications of AI-driven solutions in human geography and urban planning, including their potential effects on social equity, economic development, and cultural heritage, to ensure that these technologies contribute to the creation of more sustainable, resilient, and equitable urban environments.

The future of AI in human geography and urban planning presents a diverse array of research directions and challenges, underscoring the need for ongoing exploration and innovation in this rapidly evolving field. By addressing these challenges and embracing the potential of AI-driven solutions, researchers, practitioners, and decision-makers can harness the transformative power of AI to create more sustainable, resilient, and equitable urban environments for all.

References

1. Alizadeh, T., Shearer, H., & Sadowski, J. (2020). A review of the emerging contribution of artificial intelligence to the implementation of sustainable cities. *Sustainable Cities and Society, 55*, 102029.
2. Arnstein, S. R. (2019). A ladder of citizen participation. *Journal of the American Institute of Planners, 35*(4), 216–224.
3. Batty, M., Axhausen, K. W., Giannotti, F., Pozdnoukhov, A., Bazzani, A., Wachowicz, M., Ouzounis, G., & Portugali, Y. (2012). Smart cities of the future. *The European Physical Journal Special Topics, 214*(1), 481–518.
4. Boulos, M. N. K., Peng, G. C. A., & VoPham, T. (2018). An overview of geospatial methods used in unintentional injury epidemiology. *Injury Prevention, 24*(5), 321–327.
5. Coburn, E., Karduni, A., & Nelson, T. (2018). The visualisation of spatial social structure. *Environment and Planning B: Urban Analytics and City Science, 45*(3), 464–481.
6. Duarte, J. P. (2019). Customizing mass housing: A discursive grammar for Siza's Malagueira houses. *Automation in Construction, 104*, 37–49.
7. Gil, Y., Honaker, J., Gupta, S., Maier, D., Greaves, M., Roddis, K., & Garijo, D. (2019). Towards human-guided machine learning. *AI Magazine, 40*(1), 10–26.
8. Gil, Y., Honavar, V., & Verma, R. (2019). AI and data: A discussion of hard challenges and opportunities ahead. *AI Magazine, 40*(4), 62–73.
9. Goodchild, M. F. (2018). Reimagining the history of GIS. *Annals of the American Association of Geographers, 108*(1), 1–17.
10. Kakarla, S. C., Yilmaz, O., & Yasar, A. (2020). Disaster management and the role of artificial intelligence: A comprehensive review. *International Journal of Emergency Management, 16*(4), 338–361.
11. Kakarla, S., Gunda, R., & Siddiqui, M. A. (2020). AI-driven approach for sustainable development: A review. *Journal of Cleaner Production, 268*, 122209.
12. Kitchin, R. (2018). *The data revolution: Big data, open data, data infrastructures, and their consequences*. SAGE Publications Ltd.
13. Kitchin, R. (2018). The ethics of smart cities and urban science. *Philosophical Transactions of the Royal Society A: Mathematical, Physical and Engineering Sciences, 376*(2127), 20170122.
14. Koliba, C., Meek, J. W., & Zia, A. (2020). *Governance networks in public administration and public policy*. CRC Press.
15. Lu, X., Bengtsson, L., & Holme, P. (2018). Predictability of population displacement after the 2010 Haiti earthquake. *Proceedings of the National Academy of Sciences, 115*(29), 7545–7549.
16. Mittelstadt, B. D., Allo, P., Taddeo, M., Wachter, S., & Floridi, L. (2016). The ethics of algorithms: Mapping the debate. *Big Data & Society, 3*(2), 2053951716679679.
17. Mittelstadt, B. D., Allo, P., Taddeo, M., Wachter, S., & Floridi, L. (2016). The ethics of algorithms: Mapping the debate. *Big Data & Society, 3*(2), 1–21.
18. Naudé, W. (2020). Artificial intelligence vs COVID-19: Limitations, constraints and pitfalls. *AI & Society, 35*(3), 761–765.

19. Openshaw, S., & Abrahart, R. J. (2019). *Geocomputation: A primer*. Wiley.
20. Ragnedda, M., & Muschert, G. W. (2013). *The digital divide: The internet and social inequality in international perspective*. Routledge.
21. Taubenböck, H., Wurm, M., Esch, T., & Dech, S. (2018). Global urbanization multi-sensor data analysis reveals spatiotemporal patterns of the World's largest cities. *Remote Sensing of Environment, 209*, 383–400.
22. Torrens, P. M. (2018). A new paradigm for simulating and understanding urban growth. In *Models, simulations and representations* (pp. 87–112). Routledge.
23. Wang, D., Zhang, J., & Cai, J. (2019). Urban traffic congestion pricing model with the consideration of carbon emissions cost. *Sustainability, 11*(4), 1114.
24. Williams, S., Bradley, K., Devadoss, T., & Stewart, K. (2019). The role of education and training in absorptive capacity of international technology transfer in the aerospace sector. *Progress in Aerospace Sciences, 105*, 1–16.
25. Zhang, H., & Chen, W. (2019). A systematic review of applications of machine learning in the prediction of real estate market: Current situation, challenges and future directions. *Journal of Ambient Intelligence and Humanized Computing, 10*(10), 3959–3973.
26. Zhang, Y., Pan, J., Wang, X., & Ma, J. (2018). Land use classification based on time-series Landsat images and openstreetmap: A hybrid strategy to improve urbanization monitoring. *International Journal of Applied Earth Observation and Geoinformation, 69*, 23–32.

Printed by Printforce, the Netherlands